Macmillan Interdisciplinary Handbooks

Posthumanism

The Future of Homo Sapiens

Macmillan Interdisciplinary Handbooks

Posthumanism

The Future of Homo Sapiens

Michael Bess and Diana Walsh Pasulka
EDITORS

MACMILLAN REFERENCE USA
A part of Gale, a Cengage Company

GALE
A Cengage Company

Farmington Hills, Mich • San Francisco • New York • Waterville, Maine
Meriden, Conn • Mason, Ohio • Chicago

GALE
A Cengage Company

Posthumanism: The Future of Homo Sapiens

Michael Bess and Diana Walsh Pasulka, Editors

Project Editor: Jonathan Vereecke

Product Design: Kristine Julien

Associate Publisher, Macmillan Reference USA: Hélène Potter

For product information and technology assistance, contact us at
Gale Customer Support, 1-800-877-4253.
For permission to use material from this text or product,
submit all requests online at **www.cengage.com/permissions.**
Further permissions questions can be emailed to
permissionrequest@cengage.com.

Cover photograph: Colin Anderson/Getty Images.

LIBRARY OF CONGRESS CATALOGING-IN-PUBLICATION DATA

Names: Bess, Michael, editor. | Pasulka, Diana Walsh, editor.
Title: Posthumanism : the Future of Homo Sapiens /
Michael Bess and Diana Walsh Pasulka, editors.
Description: Farmington Hills, Mich. : Macmillan Reference USA, a part of Gale, a Cengage Company, 2018. | Series: Macmillan interdisciplinary handbooks | Includes bibliographical references and index.
Identifiers: LCCN 2017035520| ISBN 9780028664477 (hardcover) | ISBN 9780028664484 (ebook)
Subjects: LCSH: Philosophical anthropology. | Human beings. | Humanism. | Human body–Technological innovations.
Classification: LCC BD450 .H8653 2018 | DDC 128–dc23
LC record available at https://lccn.loc.gov/2017035520

Gale, a Cengage Company
27500 Drake Rd.
Farmington Hills, MI 48331-3535

ISBN 978-0-02-866447-7 (this volume)

This title is also available as an e-book.

ISBN: 978-0-02-866448-4

Contact your Gale sales representative for ordering information.

Printed in Mexico
2 3 4 5 6 7 22 21 20 19 18

Editorial Board

Contents

Introduction: Trans-, Post-, and Emerging Humans: What Do We Mean?

THE ANTHROPOCENE TURNS IN ON ITSELF

Scientists who study the great time spans in Earth's history have coined a new term for the most recent phase, following the Pleistocene (Ice Age) and Holocene (which began around 11,700 years ago): they call it the Anthropocene, and it encompasses the past three centuries since the Industrial Revolution of the eighteenth century. A mere wink of time when compared to the eons that our planet has traversed, it has nonetheless witnessed profound transformations of the biosphere, most of them directly caused by humans' use (and abuse) of technology. The term *Anthropocene* means "shaped by humankind," and it refers to the powerful impacts that our species has exerted on the physical environment that surrounds and sustains us.

Since the 1960s, however, a new meaning of Anthropocene has also emerged, for it is no longer just the natural world that bears the brunt of human technological ingenuity. The contemporary transformation is intimate, as it involves our own bodies and minds. "Shaped by humankind" is no longer a phrase that refers only to trees, oceans, animals, and sky: it is we ourselves who are becoming the objects of change, the targets for our own formidable creative powers. Our species has entered a new era, in which we are increasingly using our technologies to redesign our own most basic traits and capabilities, from physical features to cognition and from emotions to intelligence. The Anthropocene has turned inward.

In this book a distinguished group of scholars comes together to chart the broad contours of this epochal transformation, describing some of the paths it seems likely to take and assessing the impacts it may have on our lives as individuals, family members, and citizens. Over the coming century, every dimension of our human life-world will feel the effects of this process, as our relationships with nature, with technology, and with each other come to be incrementally reworked. Once this wave of change has passed through our civilization, many of the fundamental features that we take for granted in our present-day society will likely have fallen by the wayside—replaced by startling new challenges, capabilities, problems, and opportunities. Among the concrete possibilities that lie ahead are:

- human bodies that live longer, healthier lives extending decades (and perhaps more) beyond current limits;
- new physical capabilities, from novel ways of sensing and perceiving to greatly enhanced strength and dexterity;

- innovative avenues for controlling and modulating our moods and emotions in fine-grained ways, without unwanted side effects;
- new methods for connecting with advanced machines, using brain-machine interfaces to powerfully supplement and extend our physical and mental worlds; and
- cognitive enhancements that take our abilities to remember, analyze, and create information to unprecedented levels.

It is likely that these changes will come through innovations in pharmaceuticals, bioelectronic technologies, and genetic or epigenetic modifications that will be offered to people in the same way that new apps for smartphones are marketed today. What will such a society look like, and what will it feel like to live in such a world? Will these bioenhancements be equally accessible to all human beings, or will a new form of biologically based caste system gradually come into being? Will people come to regard themselves (and each other) in the manner of commodities or partial products, comparing each other's performance the way people compare cars or laptop computers today? What will happen to human dignity, and to the spiritual sense that each of us forms part of a community larger than our individual self? What are the hidden dangers lying latent in such a social order? What new and tantalizing possibilities await us, over the horizon of the coming decades? These are among the questions posed by the authors of the chapters that follow.

A NOTE ON TERMINOLOGY

The authors of the present volume make frequent use of the terms *human*, *transhuman*, and *posthuman*, so here are brief working definitions of these key terms.

Human: For the purposes of this volume, the term *human* refers to individuals who—like most people in present-day society—have not engaged in significant alterations that take them markedly beyond the species-typical performance range of *Homo sapiens.* To be sure, no human being today can claim to be wholly untouched by modern biotechnology: whether we drink coffee, wear glasses, or take medications, we are all engaged in various levels of deliberate self-alteration. Even the air we breathe has been altered by technology in subtle (or not-so-subtle) ways. But we are *relatively* unmodified, as compared with the categories of the transhuman and posthuman.

Transhuman: This term refers to humans who have begun using pharmaceutical, bioelectronic, genetic, or other technologies to make significant alterations to their physical, emotional, or cognitive capabilities. If they once played basketball at the level of an average high school varsity player, now they are approaching the performance level of LeBron James. If they used to score about average on IQ tests, now their various bioenhancements are granting them scores that are 50 percent higher. If they used to show the telltale signs of aging, now their rejuvenation treatments have delivered many more decades of full physical fitness and sharp mental acuity, well past the age of 100. In short, these "transhumans" have boosted their performance well beyond levels that have typified *Homo sapiens* over the course of history, taking them into an unprecedented range of novel capabilities. Nevertheless, these modified humans still undertake most of the same kinds of activities that have long characterized the human life-world: they still get enjoyment from the types of undertakings and sensations that have always pleased people; they still suffer the same kinds of pains, disappointments, and struggles that have always characterized the human condition. Their performance lies well beyond the traditional human range, but it is still recognizably commensurate with the list of "human universals" that anthropologists have identified in most societies throughout history.

The behaviors of these transhumans are described by many humans as remarkable, impressive, awe inspiring—but most people (modified or unmodified) still regard transhuman persons as falling clearly within the category of humankind.

Posthuman: This term refers to persons who have undertaken deep and systematic forms of biotechnological self-modification that take them into qualitatively new realms lying radically beyond the range of species-typical human performance. Their existence has heretofore been described only in religious literature and science fiction. Their capabilities are so advanced that many kinds of activities undertaken by humans and transhumans no longer appeal to these posthuman persons. They do not play basketball at all, because throwing a ball through a hoop from 50 feet away seems so trivially easy that it is like scratching an itch or turning one's head. Their cognitive performance—which in many cases relies on an entirely new brain architecture—operates at such high levels of dimensional dexterity, mathematical complexity, parallel processing, immense working memory, and analytical nuance that many of their ideas are no longer comprehensible by mere humans or transhumans. Their bodies now incorporate forms of technology that are not yet conceivable by present-day inventors and deliver forms of sensation, intuition, emotion, and interpersonal communication that are hard for present-day humans to even imagine. Whereas transhuman individuals have undergone a significant *transformation*, these posthuman individuals have undergone a radical *qualitative metamorphosis*—a full-scale transmogrification. Their behaviors are described by many humans and transhumans as strange, unsettling, or baffling, and it is no longer clear to many people whether these radically different beings still qualify as forming part of humankind.

Although all the authors in the present volume use these three terms in ways that reflect the above definitions, it is important to note that the broader literature on human bioenhancement does not always follow this three-level hierarchy of capabilities. Nevertheless, this tripartite hierarchy does reflect the implicit understanding under which many writers in the field are operating (whether they acknowledge it explicitly or not). Thus, for example, the transhumanist organization Humanity Plus posits the following on its website (2017):

> Many transhumanists wish to follow life paths which would, sooner or later, require growing into posthuman persons: they yearn to reach intellectual heights as far above any current human genius as humans are above other primates; . . . to have unlimited youth and vigor; to exercise control over their own desires, moods, and mental states; to be able to avoid feeling tired, hateful, or irritated about petty things; to have an increased capacity for pleasure, love, artistic appreciation, and serenity; to experience novel states of consciousness that current human brains cannot access. . . .
>
> Some posthumans may find it advantageous to jettison their bodies altogether and live as information patterns on vast super-fast computer networks. Their minds may be not only more powerful than ours but may also employ different cognitive architectures or include new sensory modalities that enable greater participation in their virtual reality settings.

SUMMARY OF CHAPTER CONTRIBUTIONS

Chapter 1: Between Progress and Armageddon: The Stakes of Our Times, by R. S. Deese. This chapter situates the rise of biotechnology and human enhancement technologies within the broader context of modern industrial and technological society, with a particular emphasis on developments since the Enlightenment of the eighteenth century. The chapter highlights the contrast between Enlightenment visions of social and technological progress and the increasingly uncertain and even pessimistic visions that have gained currency in the wake of World War I, the Holocaust, the atomic bomb, and the environmental crisis. It describes the fears and hopes reflected in utopian and dystopian literatures that explored transhumanist and posthumanist themes during the twentieth century.

Chapter 2: Essential Concepts of Science and Technology Studies (STS), by Robert G. W. Kirk. In this chapter, the author describes the history of science and technology studies (alternately known as science, technology, and society) as a field and the reasons for its growing influence in academia, including its displacement of older explanatory models of scientific and technological change. Then the principal concepts and methodological approaches of the field are laid out, each with its own subhead. Among these concepts one finds (for example) the scientific method, the concept of the paradigm shift, the strong program, the social construction of science and technology, actor-network theory, feminist science studies, and the governance of science and technology.

Chapter 3: Can We Predict the Middle-Term Future? By David Orrell. This chapter describes the challenges involved in identifying the key social, economic, and technological trends that are likely to characterize the coming decades, through the middle of the twenty-first century. It surveys the difficulties and failures of past efforts at prognostication, citing specific examples of highly inaccurate predictions. The chapter surveys the fields of future studies and forecasting, showing how researchers have developed methodologies for overcoming or mitigating some of the obstacles to successful prognostication. From weather forecasters to economists making long-term projections, much depends on the quality and sophistication of the assumptions that underpin models of historical process and causation.

Chapter 4: Is Mind Uploading a Real Possibility? By Patrick D. Hopkins. The central focus of this chapter is the complex assumption that philosophers commonly call the "ghost in the machine"—the notion that the human mind and body are in some sense separate and even severable. This idea pervades popular culture, and this chapter references two films, *All of Me* (1984) and *Transcendence* (2014), to illustrate the concept. In each film, a person's consciousness is moved into another medium that houses it—in *All of Me* a woman ends up sharing a man's body, and in *Transcendence* a man is "uploaded" into a computer system. This is called "mind uploading," and this chapter assesses common assumptions and misunderstandings about this potential process. By distinguishing between an "artificial mind" and "preserving an existing mind," the author explores how scientists and scholars approach the complex notion of mind uploading.

Chapter 5: The Prehistory of the Posthuman, by Diana Walsh Pasulka. This chapter surveys the historical and religious precedents for ideas of the posthuman. A common assumption is that posthumanism arises from the contemporary digital and technological infrastructure and that the image of humans rising to superhuman levels—becoming so spiritually and physically perfected that they no longer seem human—is unprecedented. This assumption, however, is incorrect. From Dante Alighieri's *Divine Comedy*, to William Shakespeare's play *The Tempest*, and finally to works of the early twentieth century, the posthuman has been presaged in literature in ways that are descriptive, beautiful, and frightening. The posthuman has always been with the human, implicit and explicit in religious history, art, and modern theology.

Chapter 6: Pharmaceuticals, by Nancy D. Campbell. This chapter describes the technologies for enhancing human traits and capabilities through pharmaceuticals. It surveys a broad range of physical, cognitive, and affective modifications, as well as modifications designed to improve life span and health span. It describes the history of such modifications over the past century, lays out current trends, and offers an extrapolation from current trends into the coming decades. It also explores how pharmaceutical bioenhancements and "pharmaceutical cyborgs" have been portrayed in utopian, dystopian, and feminist literature since the early industrial era.

Chapter 7: Bioelectronics, by Chris Hables Gray. This chapter describes the technologies for enhancing human traits and capabilities through bioelectronics (technologies that allow for increasingly seamless and direct interfaces between informatic machines and the human sensorium and nervous system). It surveys the history of such devices, touching on early prostheses and sensory augmentation systems before laying out the full range of such devices in use today (e.g., brain-machine interfaces, virtual-reality headsets). It offers an overview of the academic field of cybernetics, which has in turn generated the more popular term *cyborg*. Then it describes the future possibilities for these technologies over the coming decades, identifying the sorts of devices that are most likely to be developed given current research trends in fields such as photonics and optogenetics.

Chapter 8: Genetics and Epigenetics, by Michael Bess. This chapter describes the technologies for enhancing human traits and capabilities through genetic or epigenetic modifications. It starts with a description of genetic causation, detailing what scientists know today about how genes and environmental factors interact to produce the traits that characterize humans. Then it describes past efforts at genetic engineering, starting with trial-and-error methods used by farmers in the past, and ranging up to the more sophisticated methods that have come into widespread practice in the present day. It offers an extrapolation from current research trends, laying out some of the possibilities and obstacles surrounding the creation of designer babies and artificial chromosomes. It also describes the burgeoning new field of epigenetics and explores some of the ways in which epigenetic modifications could be used to achieve enhancement goals. The chapter concludes by surveying some of the moral questions raised by these genetic enhancement practices.

Chapter 9: Rejuvenation and Radically Increased Health Spans, by Michael G. Zey. This chapter describes the biotechnologies for boosting human longevity and (especially) the human health span. It starts with a description of the three historical stages of the "superlongevity revolution," surveying what scientists know today about how and why human bodies begin to break down, with increasing physical and mental deficits, after having reached a developmental peak in the mid-twenties. The chapter lays out the state of the art in contemporary medical approaches to mitigating the aging process and describes the various avenues of research being pursued in this effort. It offers a survey of experts' opinions about how far the human health span—as well as basic traits, such as physical or cognitive capabilities—may be increased over the coming decades and centuries. Finally, the chapter offers a glimpse of some of the possible societal and cultural consequences that might follow if a doubling or tripling of the human health span were to become a reality.

Chapter 10: Runaway AI, by Curry I. Guinn. The focus of this chapter is on a human future run by artificial intelligence (AI). The author considers why humans would cede control to

AI, and how AI would likely bring much-needed change to certain problems that humans currently face: economic inequality, scarce resources, and environmental crises. This idea is posed as the technological singularity hypothesis, which claims that humans will create machines that will have more cognitive abilities than do humans, and that those machines in turn will create even more sophisticated machines. The author explores the plausibility of this thesis and describes some of the potential consequences, such as the extinction of the human race. The author also explores less dire consequences, such as the possibility that humans will develop an ethics of AI that will ensure that humans and their machines will coexist amicably.

Chapter 11: A Skeptic's Perspective: Is This Actually Going to Happen? By Michael Bess. This chapter surveys the forecasts and predictions made by transhumanists and other writers about the future of bioenhancement, ranging from modest modifications to the full-blown Singularity. It then describes the skeptical literature that has emerged in response to such transhumanist predictions, laying out the principal arguments that skeptical writers have advanced. These include charges of technological determinism, succumbing to hype, claims of overly optimistic interpretations of past trend lines, claims of cherry-picking evidence regarding technological innovation, oversimplified historical causal models, and uncritical teleological assumptions.

Chapter 12: Buddhist Uploads, by Beverley F. McGuire. Remarks by the current Dalai Lama, who is arguably one of the most famous Buddhists in the eyes of Westerners, have suggested that he is amenable to "mind uploading," which is the idea that a person's consciousness might someday be downloaded or uploaded, or in a real sense transferred, into a computer or some other substrate. While he did make remarks on this process, the author of this chapter clarifies the Dalai Lama's position, and the position of Buddhism more generally, on this posthuman idea. By showing that Buddhist values sometimes contradict the values espoused by many posthumanists, the author suggests that many Buddhists would likely *not* embrace the process of mind uploading, should it become a real possibility. A core tenet of Buddhist philosophy lies in embracing the spiritual growth potentials opened up by human suffering, rather than seeking to escape suffering through technological enhancements.

Chapter 13: The Russian Cosmists: Evolving into Space, by George M. Young. This chapter considers the late nineteenth- and early twentieth-century Russian scientists and religious thinkers who believed that humans would overcome the historic constraints of their planetary home and evolve beyond Earth, colonizing outer space. These thinkers were not just entertaining abstract speculations, as many of them grounded their ideas in science and were among the founders of the Russian space program. The author explores the fascinating ideas of such thinkers as Nikolai F. Fedorov, Pavel Florensky, and Vladimir Solovyov, all of whom developed innovative concepts relating to genetic engineering, artificial organs, and human longevity. This chapter presents a history of posthuman ideas that predate the contemporary posthumanist movement.

Chapter 14: Virtual Religions and Real Lives, by Carole M. Cusack. While complicating the dualism inherent in the term *virtual religions*, the author of this chapter explores new religions that have emerged since the 1950s within digital environments such as the Internet. The principal connection between these religions, which the author terms *invented religions*, and posthuman ideas lies in the fact that they both reject the binary oppositions of Judeo-Christian religion and Enlightenment rationalism, while embracing fluid and nonconclusive theories and philosophies. The chapter corrects common misconceptions about the nature of religions and provides a survey of new religious movements, traditional religions, and invented religions. The author suggests that humans, and even posthumans, can be religious in a variety of nontraditional ways.

Chapter 15: The Spectrum of Human Techno-hybridity: The Total Recall Effect, by Diana Walsh Pasulka. This chapter explores the human as a cyborg. Scholars such as Donna Haraway and N. Katherine Hayles have argued that humans and technologies have always been inextricably connected and that these connections (and human blindness to them) are important facets of human identity. The chapter delves into the spectrum of these connections and addresses how modern media as ubiquitous as movies and video games change human bodies and minds. It also explores how biotechnologies can affect human bodies in more radical ways than contemporary media, as well as how some of the most innovative thinkers of the contemporary posthuman landscape predict an ever-more intimate enmeshment between humans and technology in the future.

Chapter 16: The Frontiers of Immortality, by Jamie L. Brummitt. This chapter argues that the notions of immortality espoused by contemporary posthumanists, such as young billionaire Dmitry Itskov and innovative futurist Ray Kurzweil, have a long history behind them, and it situates that history within American religion and its relationship to technology. It explores the concept of immortality posited by these thinkers, premised on the notion that the human mind can exist indefinitely if it is housed within a computer or another silicon or silicon-like substrate. In the history of American religions, similar ideas have been espoused by traditions such as Spiritualism, in which people thought they could contact the spirits of the deceased through technologies such as the spiritual telegraph. Additionally, the author surveys notions of secular immortality and Kurzweil's idea of spiritual machines.

Chapter 17: The Catholic Tradition and Posthumanism: A Matter of How to Be Human, by James F. Caccamo. This chapter explores a Catholic response to several of the ideas espoused or implied by posthumanism. Referencing the novel *Altered Carbon*, by Richard Morgan, the author argues that the idea that the human body can be separated from the human mind challenges the Catholic notion of incarnation. For Catholics (as well as people adhering to many other religious traditions), incarnation implies that the human body is essential to the essence of the human being. It cannot be discarded like an old skin. *Altered Carbon* seems to suggest that Catholics are the only ones who reject the option of discarding the human body; but the author of this chapter offers a more complex analysis, suggesting that the tenets of Catholicism are perfectly compatible with some posthuman technologies and incompatible with others.

Chapter 18: Buddhist Biohackers: The New Enlightenment, by Julian R. Keith. Through the examples of two rigorously challenging practices, Buddhist meditation and professional cycling, this chapter explores the benefits and costs of biohacking—the application of technologies and foreign agents to enhance the capabilities of one's body. The author is guided by his love of the sport of cycling to consider the ethics and the effects of biohacking. He considers the fate of one of the heroes of cycling, Lance Armstrong, and how Armstrong's use of biohacking prompted the media and sport's regulatory agencies to decry Armstrong and led to him being stripped of his sports victories, including those in the famed Tour de France. The author also considers biotechnologies such as CRISPR, which is essentially a biohacking tool that will enable humans to fight disease and disorders, and explores the hypothesis that humanity's definitions of ethics and biohacking are historically relative in nature. In the chapter's concluding section, the author describes his own use of biohacking technologies to enhance his practice of Buddhist meditation.

Chapter 19: What Is a Person? By Linda MacDonald Glenn. This chapter provides an overview of the shifting definitions of personhood within the Western philosophical tradition. It shows how emerging biotechnologies, as well as posthuman ideas, challenge conventional notions and even legal definitions of personhood. As persons are beings who have moral and legal rights, the creation of new entities that blur the boundaries between what is technological, animal, and human call into question conventional definitions of personhood. The author surveys definitions inherited from the Western philosophical tradition and finds that these are inadequate in light of emerging technologies and their effects on human and animal lives.

Chapter 20: The Debates over Enhancement, by Walter Glannon. This chapter surveys the debates that have emerged since the 1990s over the moral implications of bioenhancing human traits, focusing particularly on cognitive enhancements achieved via psychoactive drugs and neurostimulation devices. It describes the two main schools—the pro-enhancement camp and anti-enhancement camp—identifying the principal thinkers and authors associated with these rival positions. The chapter emphasizes the importance of the concept of human nature in these debates, as well as the concepts of authenticity and perfectibility. It concludes by evaluating some of the more radical and extreme forms of enhancement, achieved by integrating human biology with advanced informatic machines.

Chapter 21: Commodification of Human Traits: The Body as Industrial Product, by Daryl Wennemann. This chapter describes the ways in which bioenhancement technologies may be expected to produce a blurring of the boundary between "person" and "product," thereby posing a danger of commodifying human traits. The author emphasizes the importance of the distinction between the concepts of dignity and exchange value in how people categorize other human beings. Bioenhancement technologies will inherently tend to have many "product-like" qualities: they will be available for sale on the open market, they will become obsolete over time and will require continual upgrades, they will confer significant competitive advantages and social prestige on those who adopt them, they will have distinctive performance profiles, and they will be subject to constant comparisons by people who adopt them. Therefore, if these "product-like" qualities are extended implicitly or explicitly to the human beings who adopt such technologies, the result would be a profound dehumanization of the persons involved.

Chapter 22: Gender and Bioenhancement, by Colleen A. Reilly. In this chapter, the author surveys the theoretical assumptions implicit in some of the key concepts used by posthumanist writers, such as gender, disability, health, and normality. Leaving behind the idea that the worlds of humans, objects, and animals are separated by distinct and often binary boundaries, this essay suggests that objects and humans coexist as part of the same ecosystem, "inseparable and interconnected." The chapter explores how different communities arrive at their definitions of the key ideas underpinning bioenhancements, arguing that these fundamental concepts are unavoidably socially constructed and historically specific. In order to illustrate these points, the author pays particularly close attention to the "treatment versus enhancement" distinction, as well as to the technological possibility of creating artificial wombs.

Chapter 23: Animal Bioenhancement, by Amy Defibaugh. This chapter explores the ethical dimensions of the relationship between humans and nonhuman animals in the arena of biomedical enhancement. It explores xenotransplantation—the transplantation of animal organs or tissues into humans—and other procedures that use animals in the creation of enhancement technologies. It also examines procedures that produce transgenic animal/human bodies, as well as chimeras. Within this new technological frontier, the author proposes the necessity of an updated form of ethics that is capable of addressing the moral status of humans and nonhuman animals.

Chapter 24: Human Flourishing: A Moral Yardstick for Evaluating Specific Bioenhancements, by Michael Bess. This chapter describes the concept of human flourishing as laid out in two contemporary research fields: positive psychology and the "capabilities approach" in developmental economics. The chapter briefly describes the genesis of the field of positive psychology, the scholars most closely associated with it, and its principal tenets, paying particular attention to the work of Martin E. P. Seligman, Mihaly Csikszentmihalyi, and Jonathan Haidt. The chapter then describes the capabilities approach, focusing especially on the work of Amartya Sen and Martha C. Nussbaum. The chapter concludes by describing how the concepts laid out by scholars in these two fields provide fruitful frameworks for evaluating the ethical pros and cons of specific bioenhancement technologies.

Chapter 25: Enhancing Moral Behavior, by Scott M. James. This chapter surveys arguments opposed to and in favor of the controversial techniques and practices known as moral bioenhancement. The author explores the following core questions: If a person can improve herself morally, shouldn't she always do so? What arguments suggest that she should not, and what are their flaws? Do the means used for such improvement—whether behavioral or technological in nature—make a difference that matters morally? The author describes various behavioral programs of moral enhancement—for example, adopting traditional adages such as doing unto others as you would want to them to do unto you—showing that such methods are generally considered praiseworthy and uncontroversial. Why, then, the author asks, should pharmaceutical programs of moral enhancement be considered problematic? Is there something intrinsically wrong about them, as opposed to behavioral modifications? The author surveys the arguments, both pro and con, showing the many complexities and nuances that lie implicit in these debates.

Chapter 26: Synthetic Biology: The Digital Creation of Life, by Gaymon Bennett. This chapter explores how the nascent field of synthetic biology has created a lexicon and a social imaginary infused by the ideas of a digital frontier. The basic assumption of synthetic biologists is that biological entities function like digital entities and thus can be coded, and changed, at will. One of the questions that guides the author of this chapter is this: what are the potential societal consequences of this worldview? All things, according to this imaginary, can be reduced to a digital life: human beings, their thought processes, and life itself. The author traces the short history and ascendancy of this particular worldview, how it came to permeate popular culture, and how the language of this worldview determines the way in which many contemporary technological innovators describe and create reality.

Chapter 27: Trans/Posthuman Epistemology: Doubt and Back Again, by Steven Benko and Amelia Hruby. The focus of this chapter is on the epistemological underpinnings of posthumanism. Epistemology is the branch of philosophy that considers theories of knowledge, or how humans arrive at what they consider to be facts and certainty. Specifically, the authors of this chapter outline trans/posthuman epistemology by positioning the known and the knower in the context of nonhuman others, such as animals, computers, and potential trans- and posthumans. This unique moment in history, the authors suggest, allows philosophers and other thinkers to reconsider conventional notions of epistemology and knowledge, and even the agent, or the "one who knows," who has traditionally been conceived in a very narrow way in the past. New conceptions, as well as new "agents of knowledge," expose the limits of conventional theories of knowledge.

Chapter 28: What Relevance Does Eugenics Have Today? By Nicholas Agar. This chapter describes the history of eugenics as a concept and as a societal and political movement since its genesis in the late nineteenth century. It starts with the theories of Francis Galton, then surveys the eugenic practices and laws put into place in various European nations and the United States during the early twentieth century, culminating with Nazi eugenic ideas and policies. This history reveals eugenics to be a deeply morally problematic practice. To say eugenics is morally problematic is to say that it always raises moral issues. The author argues that it resembles experimentation on human subjects in this way but that this does not necessarily make it uniformly morally wrong. The author argues, for example, that many current uses of preimplantation genetic diagnosis (PGD) can be understood as both eugenic and morally justified. He proposes that it is important to continue to use the term *eugenics* in describing morally problematic technologies such as PGD, because the use of this fraught word underscores the potentially dangerous and immoral tendencies lying latent in all efforts to "make better people."

Chapter 29: Human-Robot Relationships, by Raya A. Jones. Human-robot relationships are a topic amply explored in popular movies and fiction, and the author of this chapter argues that robots are rapidly moving from the realm of sci-fi depictions into concrete societal reality. The advent of this new class of servants and coworkers will challenge conventional assumptions about personhood, identity, and the moral status of machines. The author references a science fiction short story by Ray Bradbury about a grandmother robot to illustrate the complex and unexpected relationships that can develop between humans and their robots. The chapter also explores the notion of the social robot and how a social model of selfhood—the idea that identity is constituted through relationships—can help humans adapt more fruitfully to their relationships, emotional and social, with robots. The chapter also surveys the growing academic research about the emotional and psychological bonds formed by humans with robots.

Chapter 30: Art and the Posthuman, by Kevin LaGrandeur. Art is not just a representation or manifestation of beauty; it also reflects and challenges viewers to think, feel, and see in new ways. In this chapter the author reviews how artists incorporate new conceptions of the posthuman into their work and reveal how humans are struggling with the challenges posed by new technologies, as well as rejoicing in them. Society is saturated with technology, and it is so ubiquitous as to be almost unnoticeable. Yet, artists rely on several key concepts to depict the human relationship with technology, some of which include the idea that the body is a substructure, and even a prosthesis of what is claimed to be the "true essence" of the human—namely, patterns of information. Therefore, posthuman artists experiment with changes in human bodies, which are conceived in many cases as being interchangeable and even disposable. Another key concept among artists of the posthuman involves the cyborg—the melding of the human and the machine.

Chapter 31: The Fragmenting of Humankind? By Maxwell J. Mehlman. This chapter describes a variety of future scenarios in which the advent of bioenhancement technologies could lead to undesired outcomes of socioeconomic and cultural fragmentation. The first such scenario is that of unequal access to enhancement technologies, leading to an ever-widening gap between haves and have-nots: if the wealthy are able to use advanced bioenhancements for themselves and their offspring, they would possess even more significant advantages than they do today in the competition for societal resources. Conversely, if poor people cannot gain access to such enhancements, their existing disadvantages would be exacerbated. A second scenario envisions increasingly severe fragmentation between those who adopt enhancements and those who (for a variety of ethical or religious reasons) refuse all forms of bioenhancement. A third scenario envisions growing levels of reproductive isolation as the enhanced become increasingly unable to interbreed successfully with the unenhanced. Ultimately this could lead to the creation of separate humanoid species competing for scarce planetary resources and dominance. The chapter concludes by surveying ideas that have been put forth in the bioenhancement literature for mitigating these sorts of centrifugal tendencies in tomorrow's social order.

Chapter 32: Regulating Bioenhancements: Law and Governance, by Michael Bess. This chapter surveys the space for human agency in influencing the development of bioenhancement technologies over time. It starts by describing the principal obstacles that would undermine efforts to enact an across-the-board ban on such technologies. Then it surveys various avenues available to citizens for shaping the course of technological innovation: state action (e.g., laws, regulations, tax policy, subsidies, patent law, executive action, international treaties); self-regulation by scientists and technology developers (e.g., the Asilomar conference of 1975) and direct consumer action along the lines of Ralph Nader's campaign for automobile safety or the campaign to discourage cigarette smoking. The chapter then lays out the array of obstacles that stand in the way of human agency in affecting the course of technological development: the difficulty of reaching political consensus; the challenge of educating the voting public about the technical and moral issues involved in bioethics regulation; and the short-term nature of political processes, as compared with the long-term nature of many forms of technological change. The chapter describes the history of the Office of Technology Assessment (a former US federal agency) as an example of constructive steps taken in the past to regulate new technologies.

Chapter 33: The Boundaries of Legal Personhood, by Fabrice Jotterand. In this chapter, the author describes the complex moral and legal challenges that are likely to emerge over the coming decades, as new forms of sophisticated entities (both biological and nonbiological) are generated through biotechnological advance. Such entities include animals with augmented physical and mental capabilities, pushing them up toward the human range of behaviors; modified humans possessing unprecedented new capabilities that distinguish them sharply from unmodified humans; and new machine beings possessing sophisticated capabilities that render it unclear whether they should be considered sentient entities possessing interests and rights of their own. The chapter starts by describing past efforts to define the legal concept of personhood, surveying gray areas in contemporary legal practice; then it shows how difficult it has been, in practice, to ground conceptions of personhood on the alleged uniqueness of human beings. Finally, the chapter assesses the possible challenges humanity will face in the coming decades, as increasingly humanlike robots become more common features of people's everyday lives.

Chapter 34: War and Terrorism, by Jesse L. Kirkpatrick and Sarah W. Denton. This chapter describes the implications of bioenhancement technologies for military policy and international security. It describes the principal forms of weaponry that are likely to characterize the mid-twenty-first century (e.g., military robots, swarms of miniaturized drones, bioenhanced microbes targeted to specific populations, super soldiers, cyber criminals who target bioelectronic implants), and the challenges that such new weapons are likely to pose. One such challenge is likely to revolve around the creation of lethal autonomous weapon systems, or military robots, and the question of who is legally responsible if such a system were to malfunction or otherwise engage in unintended destructive behavior.

Chapter 35: Trans-ecology and Post-sustainability, by Svein Anders Noer Lie and Fern Wickson. This chapter addresses the ecological implications of a social order in which biotechnology in general—and bioenhancement technologies in particular—has been adopted by billions of persons. The chapter starts out with an analysis of the continuities between humanistic thought and transhumanist ideas, emphasizing the central roles played by the concepts of emancipation and autonomy. Both these traditions are predicated on the existence of a rational, bounded human self that is fundamentally separate from the rest of the natural world and social milieu. These are problematic assumptions, the authors argue, especially when it comes to understanding the relationship of humankind with the ecological substrate that sustains all life on Earth. The authors then consider an alternative way of framing human identity—namely, as an embedded process coexisting in an intimate ongoing relationship with the myriad other processes unfolding across the planet. This "embedded posthumanism," they argue, offers a more positive grounding for the human place within nature over the coming decades.

Michael Bess
Chancellor's Professor of History
Vanderbilt University, Nashville, TN

Diana Walsh Pasulka
Professor and Chair, Department of Philosophy and Religion
University of North Carolina Wilmington

BIBLIOGRAPHY

Humanity Plus. "Transhumanist FAQ." Accessed October 13, 2017. http://humanityplus.org/philosophy/transhumanist-faq/.

How to Think about Science, Technology, and the Future?

CHAPTER 1

Between Progress and Armageddon: The Stakes of Our Time

R. S. Deese
Lecturer, Division of Social Sciences
Boston University, MA

For most of human history, the notion that we could transform ourselves for the better has belonged to the realm of religion and a belief in the supernatural. If there is a better version of ourselves that we might know, we will discover it only in a future life or in eternity. If there is a more glorious and just community in which we might live, it was a given that we would see it only when we had crossed the far horizons of time or space. In the here and now, there has long been an abundance of evidence that we are terribly flawed creatures and that the communities we create, whether they are as small as a village or as vast as an empire, are destined to reflect our tragic flaws. While some Hindu sects emphasized individual perfection through spiritual and physical self-discipline, and a few classical philosophers such as Plato and Plotinus sketched blueprints for ideal human communities, such visions of human perfectibility in this world were exceptionally rare. With the advent of the scientific revolution in the seventeenth century, however, a growing number of individuals dared to imagine that it might be possible to improve the human condition in fundamental ways, as well as to aim at the perfection of society and the individual. This notion of progress became more robust during the second half of the eighteenth century and was reflected in the lofty goals and rhetoric of new political and social movements in Europe and the Americas. The idea of progress gathered new force during the rapid industrial expansion of the nineteenth century and came to enjoy widespread acceptance in economically advanced countries by the dawn of the twentieth century.

In the first half of the twentieth century, however, the calamities of two world wars, the Holocaust, and the advent of nuclear weapons all dealt a serious blow to the gospel of progress. While these horrors dimmed the optimism of the early twentieth century, new developments in science presented a fundamental challenge to the vision of rational and linear progress that had fueled Western optimism since the late eighteenth century. The liberal vision of progress had been predicated on the assumption that human beings were fundamentally rational creatures and that the rational reorganization of human societies would lead to steady improvement of the human condition over time. As long-standing assumptions about our species, our psychological nature, and even time itself were upended in the twentieth century, the vision of steady historical progress that had come to prevail in the late nineteenth century seemed increasingly naive.

In the wake of the catastrophes of the first half of the twentieth century, many returned to the tragic view of history and the human condition that had prevailed in earlier centuries.

Among those who continued to hope that human beings might rationally create a better future, however, a subtle but very significant change emerged. Whereas previous apostles of progress had proposed the redesign of social institutions, a small but influential group of thinkers in the twentieth century shifted their attention to the reformation of human beings themselves. In the 1950s biologist Julian Huxley (1887–1975) coined the term *transhumanism* to describe this way of thinking, and by the end of the twentieth century it had become the name of a growing and influential social movement that advocated the technological enhancement of the human mind and body. By the dawn of the twenty-first century, the term *posthumanism* had also gained currency. The term was first used in cultural studies to challenge established conceptions of what it means to be human. By the early twenty-first century, the concept of posthumanism was embraced by advocates of technological enhancement to denote ambitious innovations, such as artificial intelligence, machine-brain interfaces, or the "uploading" of one's consciousness to a computer network, that would entail a complete and fundamental departure from what we currently think of as human.

A series of stunning technological innovations in the fields of genetics, computer science, and nanotechnology lent a growing credibility to the rise of posthumanist thought in the early twenty-first century, but these technological trends were not alone in fostering the growth of this movement. The increasing popularity of schemes to reinvent the human mind and body since the late twentieth century is directly related to the loss of faith in our ability to reinvent our social, political, and economic institutions. In the books, films, and video games that represent our collective dreams about our common future, visions of dystopia or postapocalyptic ruin are so ubiquitous as to be cliché, whereas visions of utopia are the rare vestiges of a forgotten genre. When the tragic view of history dominates our visions for the future of human societies, it is natural for human beings to seek some source of escape. In medieval times the soul's ascent to paradise offered such a vision. In the twenty-first century, the transformation of one's brain and body into a technological artifact offers a secular and ostensibly more plausible vision of escape from the sorrows and pitfalls of the human condition.

THE GOSPEL OF PROGRESS

Aristotle (c. 384–322 BCE) observed that "man is by nature a political animal." Because we are born helpless and remain so for far longer than many other species, we are dependent on the social unit of the family to survive, and families, throughout human history, have been dependent on such larger social units as tribes, city-states, nations, and empires. For this reason, ideal visions of the human condition have almost always been expressed in social terms, such as the harmonious society that Greeks and Romans imagined as the Golden Age of humankind or the bounteous community envisioned by the votaries of Pure Land Buddhism across large sections of East Asia. In his seminal work, *The City of God*, early Christian theologian Augustine of Hippo (354–430) imagined paradise not as a solitary state of bliss but as a bright and eternal city that would far outshine the earthly grandeur of Rome.

These religious visions of a blessed community were both dazzling and beyond our mortal reach. The Golden Age envisioned by Greek and Roman poets was lost in the distant past, much like the Garden of Eden envisioned by the Abrahamic religions. Whether visions of the ideal human society were set in a distant and irrecoverable past or in another world beyond this one, all shared the implicit assumption that human beings would not be capable of constructing such a society in the here and now. In the Christian tradition, the flawed nature of humanity came to be identified as "original sin," but other religious and philosophical traditions entailed a comparable sense of human limitations when we are not aided by supernatural gods or divine teachings.

The idea that human beings could improve their lives *without* depending on the assistance of a deity or supernatural forces is relatively new in the long history of our species. In a break from centuries of pessimism, human potential became a celebrated theme in some notable works of Renaissance literature, such as Pico della Mirandola's (1463–1494) *Oration on the Dignity of Man* (1486). Although the humanism of the Renaissance was still tempered by the Christian doctrine of original sin, scholars such as Thomas More (1478–1535) and Francis Bacon (1561–1626) opened new doors when they began to imagine that human society could be radically improved, not in the next world but in this one. More's *Utopia*, published in 1516, envisioned a distant island on which poverty and strife had been eliminated by the reforms of a wise philosopher king, creating a society as prosperous, stable, and just as any earthly kingdom could hope to be. A century later, Bacon's *New Atlantis* (1627) depicted another imaginary island kingdom that drew its opulence, wisdom, and power not from egalitarian social reforms but from the deliberate and systematic pursuit of scientific knowledge. The visions articulated in these books remained influential long after their authors had died. More's *Utopia* not only gave birth to the literary genre that bears its name but also became a foundational text for socialist thought in succeeding centuries. Bacon's *New Atlantis*, with its emphasis on state-sponsored scientific and technical research, helped to fuel the scientific revolution of the seventeenth century and inspired the creation of new institutions dedicated to the pursuit of scientific and technical research, such as the Royal Society, which was chartered by British king Charles II, in 1662.

In the eighteenth century, the idea that human beings could radically improve their lives by applying their capacity for reason and understanding animated a political and social movement that has come to be known as the Enlightenment. Inspired by Isaac Newton's (1642–1727) discovery of clear and comprehensible laws governing the cosmos, English philosopher John Locke (1632–1704) attempted to discover a similar set of principles about the nature of the human mind and human societies. Conceiving of the human mind at birth as a tabula rasa, or blank slate, Locke argued that environment and education are the most powerful forces that shape a person's character. In light of this principle, the structure of society has a profound impact on the character of its citizens. Building on his premise that the environment played a key role in shaping human character, Locke argued for a state that protected the fundamental rights of life, liberty, and property while regulating its own power through a carefully constructed system of checks and balances.

Locke's vision proved to be remarkably influential and furnished a key inspiration for both the Declaration of Independence in America and the revolutionary Declaration of the Rights of Man and of the Citizen in France. Following Locke's reasoning to its logical conclusion, the most ambitious revolutionaries and reformers on both sides of the Atlantic dared to imagine that fundamental reforms in society could create a new and enlightened citizenry freed from the follies and vices that earlier generations had simply accepted as human nature. In the early phases of the French Revolution, the Marquis de Condorcet (1743–1794) took this idea perhaps further than any of his contemporaries. In his book *Sketch for a Historical Picture of the Progress of the Human Mind* (1794), Condorcet argued that the perfection of human society would ultimately lead to the perfection of the individual human being. Although he fell victim to the Reign of Terror, Condorcet

Utopia. *By sketching an ideal society in his 1516 novel* Utopia, *Thomas More created a genre of speculative fiction that lasted for centuries. In the late twentieth century, however, the idea of redesigning human beings themselves began to eclipse the idea of redesigning human society.* ART COLLECTION 2/ALAMY STOCK PHOTO.

articulated a vision of the perfectibility of human nature through political and social reform that would influence generations of social reformers and revolutionaries in the early nineteenth century, including Henri de Saint-Simon (1760–1825) and Auguste Comte (1798–1857).

Comte, who attempted to make the study of human society a scientific discipline in its own right, embraced the principle that a fundamental change in society would lead to an equally fundamental change in human nature itself. German political philosopher Karl Marx (1818–1883), although he rejected the classical liberalism of the eighteenth century, retained a Lockean conception of human nature as essentially neutral and therefore subject to change with the emergence of new economic and social conditions. By the late nineteenth century, even popular utopian writers, such as American novelist and social reformer Edward Bellamy (1850–1898), had instilled in their readers a faith that a change in the structure of society could transform human nature for the better. Bellamy's 1888 novel *Looking Backward*, which envisioned an egalitarian economic system that would unlock "the unbounded possibilities of human nature," became a global best seller and led to the creation of myriad societies for social and economic reform that lasted into the early twentieth century.

The Darwinian revolution, however, would present new challenges to the idea that the reform of political, economic, and social institutions could yield improvements in human nature. Some social reformers, such as Lester Ward (1841–1913) and Charlotte Perkins Gilman (1860–1935), used the logic of English naturalist Charles Darwin's (1809–1882) theory of evolution to argue that improved social conditions would lead to healthier and better educated individuals, while also rapidly aiding the evolution of the human species. However, this optimistic view of biological evolution through social progress was challenged by the discovery, confirmed by German biologist August Weismann (1834–1914) in the early twentieth century, that acquired traits are not heritable. Even before it was proven untenable, this reformist view of evolution proved less widely accepted than the laissez-faire interpretation advocated by social Darwinists, such as Herbert Spencer (1820–1903) and William Graham Sumner (1840–1910). According to the paradigm of the social Darwinists, the progress of our species had always been advanced by what Spencer memorably called "the survival of the fittest." Because the fierce struggle for survival was the very engine of evolution, reasoned the social Darwinists, any new social reforms designed to help the weak survive and reproduce would actually hinder the progress of our species. When combined with the varieties of racial pseudoscience that proliferated in the late nineteenth century, this distorted interpretation of Darwinian thinking provided a handy justification for the dramatic expansion of colonialism in the late nineteenth century and laid a foundation for racist ideologies that would have a catastrophic impact in the twentieth century.

The biggest blow to the vision of progress through social reform came not from any single discovery in science but from the political and cultural earthquakes caused by two world wars. World War I, which raged across Europe and in parts of Asia and Africa from 1914 to 1918, led to the collapse of czarist Russia as well as the Ottoman and Austro-Hungarian Empires, redrawing the map of Europe. More significantly, however, it inaugurated an age of industrial warfare that claimed the lives of over ten million in the space of a few years and transformed a significant section of western Europe into a desolate and lifeless "no-man's-land." A painting by British artist and infantry soldier Paul Nash (1889–1946) sardonically illustrates the impact that World War I had on the vision of progress that had prevailed at the dawn of the twentieth century. Its title, *We Are Making a New World*, evokes the sunny rhetoric of progress that had become commonplace before the war, but its image is nothing but the complete and utter desolation of no-man's-land.

The ninety-nine years of relative peace that followed the close of the Napoleonic Wars in 1815 had seen enough social and technological progress to warrant the widespread optimism that had prevailed before the advent of World War I, especially in Europe and North America. In the realm of social progress, the world had seen the general abolition of slavery, a widespread rise in literacy, and the establishment of international standards for commerce and communication that would have seemed unimaginable in earlier eras. In terms of technological progress, the revolutions wrought by railroads, steamships, and telegraphy were epoch making, and the new revolutions to be wrought by radio and powered flight promised to be even more profound. Because World War I was a human-made catastrophe, it may have dealt its harshest blow to the idea that human beings were rational creatures who could guide themselves forward, given the chance, on the path of social progress.

Yet, because the fierce competition of the war had accelerated the development of a wide array of technical innovations, the widespread fascination with technological progress became only more powerful as a result of this conflict. In Italy, iconoclastic painter Filippo Tommaso Marinetti (1876–1944) came to celebrate the affinity between such wartime innovations as gas masks, tanks, and flamethrowers and the worldview of his Futurist movement in the arts, declaring, "War is beautiful because it initiates the dreamt-of metalization of the human body." While Marinetti would soon lend his support to Italian fascism, a number of prominent intellectuals on the left also drew inspiration from the burst of technological innovation that the war had instigated. At Cambridge University, noted scientist and Marxist J. B. S. Haldane (1892–1964) recalled his experiences on the modern battlefield as instrumental in shaping his conviction that there are few limits to the power of new technologies to change the world. In the hands of a progressive state, Haldane reasoned, it would be possible for new technologies to eliminate poverty, end hunger, and even rationalize human reproduction in the way that the Industrial Revolution had rationalized the production of manufactured goods.

Although Haldane and Marinetti came from opposite sides of the political spectrum, they both drew inspiration from the prospect that technology itself might transform human nature, and the visions they articulated would resonate in succeeding decades. Haldane's concept of artificial wombs, known as ectogenesis, was later advocated by prominent geneticists such as Hermann Joseph Muller (1890–1967) and became the defining feature of Aldous Huxley's (1894–1963) dystopian masterpiece *Brave New World* (1932). Decades later, some feminist writers, such as Elisabeth Mann Borgese (1918–2002) and Shulamith Firestone (1945–2012), even embraced the concept of ectogenesis as a way to liberate women from the burdens of pregnancy and childbirth. In a similar fashion, Marinetti's "dreamt-of metalization of the human body" anticipated the concept of the cybernetic organism, or cyborg, that would become a staple of popular science fiction in the second half of the twentieth century.

Dreams of a better future did not go out of style after the human-made calamities of the early twentieth century, but the focus of those dreams drifted from the reformation of social, economic, and political institutions to the transformation of the human species itself. In the late nineteenth century, German philosopher Friedrich Nietzsche (1844–1900) had anticipated this shift in his musings about the *Übermensch* (superman) and his prophecy that "man is something that shall be overcome." In the twentieth century it became the stated goal of totalitarian regimes of both the right and left to create a new kind of human. For the Nazis, who ignored many of Nietzsche's ideas but exalted those that served their purposes, the ideal of *Übermensch* justified not only eugenics but also mass murder and the demented experiments of pseudoscientists such as Joseph Mengele (1911–1979). For the Soviets, especially under Joseph Stalin (1879–1953), the vision of creating a new and ideal "Soviet Man" justified the brutality of forced collectivization and enshrined another form of pseudoscience by exalting the disproven Lamarckian claims of geneticist Trofim Lysenko (1898–1976). The Nazis stressed racial "purity" while the Soviets emphasized stringent environmental conditioning, but both shared a common vision of using totalitarian methods to create a new and radically improved version of *Homo sapiens.*

In the popular culture of the United States, images of a new and improved human species proliferated in the twentieth century. In his 1930 novel titled *Gladiator,* Philip Wylie depicted a young man born with superhuman strength because his mother had been unwittingly given a special serum when she was pregnant with him. The superhuman powers depicted in this popular novel provided the inspiration for the first Superman comics and helped establish the superhero genre that would proliferate across the entire spectrum of mass media in the decades to come. As the superhero genre grew, human enhancement through new technologies became a staple plot device. In his 1941 debut, Captain America gained his powers by imbibing a special serum developed by the US military to create a super soldier. During the nuclear arms race of the mid-twentieth century, a growing number of superheroes, from the Amazing Spider-Man to the Incredible Hulk, derived their powers from experiments involving radiation. Given the global reach of American popular culture, the story of an ordinary human being who attains superhuman powers through the accidental or intentional impact of a new technology would become a familiar narrative to children and adults across the world.

DYSTOPIA AND ARMAGEDDON

As narratives of a future utopia became less plausible to readers, depictions of dystopian states and apocalyptic destruction proliferated in literature, and eventually in popular culture, during the twentieth century. *The Sleeper Wakes* (1899) by H. G. Wells (1866–1946), *The Iron Heel* (1908) by Jack London (1896–1916), and *We* (1921) by Yevgeny Zamyatin (1884–1937) pioneered the dystopian genre, but they were soon eclipsed in their influence by the dystopian masterpieces of Aldous Huxley's *Brave New World* (1932) and George Orwell's (1903–1950) *Nineteen Eighty-Four* (1949). Huxley, who had actually been Orwell's French instructor at Eton late in the second decade of the twentieth century, imagined a world where human beings were conceived and gestated in bottles, birthed in factories inspired by the efficiency gospel of American automaker Henry Ford, and programmed to love their servitude through a steady stream of Pavlovian conditioning, ubiquitous propaganda, and a very pleasing drug called soma. *Brave New World* presents a moral thought experiment to the reader, by imagining a society in which war, crime, disease, and overpopulation have all been eliminated but only through the sacrifice of such cherished human qualities as love, wonder, and individuality. In contrast to the shallow and hedonistic dystopia imagined by his former teacher, Orwell imagined a dystopia in which constant warfare, atavistic propaganda, inescapable surveillance, and torture would keep the masses firmly under control. In many respects, Huxley's dystopia reflects the nascent culture of consumerism that had emerged in Western countries during the Jazz Age of the 1920s, whereas Orwell's dystopia was colored by the escalating violence and ubiquitous propaganda that had characterized the life of civilians in all the major industrial powers during World War II (1939–1945). Upon reading *Nineteen Eighty-Four,* Huxley quibbled with his former pupil about whose dystopian nightmare was more probable, but later visions of dystopia have borrowed liberally from both these works. In the decades after Orwell and Huxley exchanged their ideas about the more likely path to a terrible future, the dystopian genre that they helped create would become a staple of popular culture in the form of young adult fiction and movie franchises.

Visions of Armageddon also flourished during the twentieth century, inspired for the most part by the destructive power of modern weaponry. Whereas ancient depictions of the end of the world had relied on the wrath of the deity for their fireworks, twentieth-century visions of the apocalypse incorporated new technologies, such as guided missiles, nuclear bombs, and biological weapons. In the immediate aftermath of the atomic bombings Hiroshima and Nagasaki, Aldous Huxley published *Ape and Essence* (1948), which depicted Los Angeles a few generations after a total war involving nuclear and biological weapons. In Huxley's postapocalyptic landscape, the future citizens of Los Angeles are genetically damaged creatures who burn books for fuel, worship Satan as the obvious master of this world, and entertain themselves with rituals of infanticide followed by

Brave New World. *Aldous Huxley's 1932 novel* Brave New World *brought revolutionary and disturbing concepts such as ectogenesis and designer babies into the mainstream of public discourse.* THE ADVERTISING ARCHIVES/ALAMY STOCK PHOTO.

sexual orgies at the Los Angeles Coliseum. Although at the time of its publication many critics regarded *Ape and Essence* as excessive and revolting, it established a number of tropes that would become standard elements in depictions of the apocalypse during the Cold War era and beyond. Among the finest of these depictions, Cormac McCarthy's (1933–) novel *The Road* (2006) wastes no time detailing the catastrophe that brought on the apocalypse but sharpens its focus on the single most terrifying threat posed by the reckless use of technology—namely, the destruction of the ecological foundations of our existence. In the world of *The Road* there are no living plants and animals, so humans are left to eat each other. The concept of pervasive ecological destruction, or ecocide, evoked in McCarthy's prose has also become a central premise of science fiction films with a broad global appeal, such as *Avatar* and the Road Warrior movies.

Few would submit to a coin toss in which both heads and tails will result in a loss, but the most popular depictions of our collective future present us with these miserable odds. In one scenario, we destroy ourselves through warfare and the destruction of our environment, while, in the other scenario, we sacrifice our fundamental humanity to an all-powerful state for the sake of stability. Both these scenarios are frightening, but their success in the marketplace reflects what may be a tectonic shift in how we think about our collective future. In the late nineteenth century a sunny vision of social and technological progress such as *Looking Backward* could become a global best seller. More than a century later, a few optimistic visions of the future, such as the *Star Trek* television and film franchise, remain part of the cultural mix, but they are not nearly as ubiquitous as their more pessimistic rivals. When it comes to visions of our future as a species, visions of dystopia and Armageddon now sell the most books, video games, and movie tickets.

IMAGINING A POSTHUMAN WORLD

It may be that optimism, like nature, abhors a vacuum. Hope for a better future is an adaptive trait, and so it is quite likely an irreducible aspect of our genetic and cultural inheritance. If endemic cynicism about the modern state makes it untenable for us to place our hopes in political change, our hopes will fly to the transformation of ourselves. The concept of self-improvement has a long history in Western culture, although it has usually been seen in terms of religious redemption. In his *Divine Comedy*, Italian poet Dante Alighieri (1265–1321) coined the verb *trasumanar*, meaning "to transhumanize." For Dante this was the spiritual transformation that allowed him to visit paradise. In the mid-1950s, when biologist Julian Huxley coined the term *transhumanism,* he placed the concept in a secular frame. Two decades before he coined the term, Huxley had been part of a community of British scientists and writers, including Haldane and John Desmond Bernal (1901–1971), who advocated the improvement of the human species through the direct application of technology. In his 1931 book *What Dare I Think?* Huxley echoed the techno-optimism of Haldane and Bernal, even speculating that ectogenesis would free us from the cramping restrictions of the birth canal and allow our species to hatch babies with bigger brains. Haldane and Bernal were both radical Marxists who admired the purported Soviet program of transforming nature in the service of proletarian revolution. For his part, Huxley's vision of the future reflected the Fabian socialist leanings of his friend H. G. Wells. Huxley was an ardent admirer of ambitious government projects such as the Tennessee Valley Authority, and his ideal was far closer to the rational world state envisioned by Wells than to the libertarian ethos that would permeate the transhumanist movement that would emerge in subsequent decades.

Paradiso Canto 31. *The root of the word* transhuman *comes from the verb* transumunar *coined by Dante in his epic poem* The Divine Comedy. *For Dante, the human condition was to be transcended spiritually, rather than through the application of technology.* ART COLLECTION 3/ALAMY STOCK PHOTO.

Throughout the industrialized world, the second half of the twentieth century saw the proliferation of cosmetic surgery, new psychoactive drugs, and stunning advances in electronics, and these developments inspired the growing transhumanist movement. Mirroring the rise of neoliberal economics from the 1980s onward, the conception of *how* the enhancement of human beings might take place shifted from the modern state to the private sector. From a Mensa initiative collecting the sperm of high-IQ donors to the declaration of Silicon Valley icon Ray Kurzweil (1948–) that he could escape death by funding his own research program in nanotechnology and computer science, the most ambitious initiatives in transforming the human condition came from wealthy individuals, acting alone or in concert with private organizations or corporations such as Google. The future, like so much else in the late twentieth century, was aggressively privatized.

Some observers have speculated that the embrace of transhumanist goals by the very wealthy may one day lead to a bifurcation of the human race, allowing a small portion of the human race to thrive as a result of artificial enhancement while the vast majority of humankind struggles to survive in a society in which they are no longer fit to compete. Some of the more ambitious goals of the transhumanist movement would have impacts too profound for us to imagine fully. For example, if individuals were able to augment their own intelligence with artificial intelligence or to "upload" themselves to computers or robotic devices, the distinction between the enhanced and the unenhanced would most likely become an unbridgeable chasm. Surveying the vast distance between what we call human today and what some of our descendants could become, Oxford University philosopher Nick Bostrom (1973–) has argued that the term *posthuman* would be the best term for describing such radical advances.

Summary

In 1969 American inventor and futurist R. Buckminster Fuller (1895–1983) published *Utopia or Oblivion*, comprising the lectures he had given to college students across the country about the future of humanity. Fuller's popularity on the college lecture circuit reflected the heady mix of apocalyptic and utopian thought that characterized a decade that saw both the horrors of the Cuban missile crisis and the war in Vietnam and the promise of the civil rights movement and the Apollo moon landing. Fuller's message was idiosyncratic in many respects, but he presented a binary choice about technology that was commonplace during the Cold War, arguing that we could either use our technological know-how to end poverty and allow all human beings to live fulfilling lives, or we could use it to annihilate ourselves. With the end of the Cold War, however, this stark and binary conception of the future seemed less plausible than it had in the 1960s. In a multipolar world, fewer people worried about a full-scale nuclear exchange between two superpowers. With the rise of neoliberal economics, fewer people imagined that an enlightened technocratic state could somehow create a society without poverty across the world. Excitement about new technology shifted from collective achievements to the products of the marketplace. In the mid-twentieth century, the emblems of progress had been large state-sponsored operations, such as superhighways, dams, or the spectacles of Sputnik and Apollo. By the early twenty-first century the innovations that did the most to excite the public imagination, such as mobile phones, social media, and driverless cars, came largely from the private sector. Initially, the basic infrastructure of the Internet, which had been conceived by the US Defense Advanced Research Projects Agency at the height of the Cold War, was the product of a highly centralized, state-sponsored approach to technological progress. However, when public use of the Internet exploded in the 1990s, it fostered a technological culture that favored a radically privatized vision of the future, geared less toward our common aspirations as citizens and more toward our multifarious desires as consumers.

During the first half the twentieth century, the most popular venue for showcasing visions of progress was probably the world's fair. Following a template established by the spectacular expositions in Paris and Chicago during the last decades of the nineteenth century, the world's fairs of the early twentieth century were obsessed with the city of the future, perhaps best exemplified by the gleaming towers, broad highways, and verdant suburbs depicted in the Futurama exhibit at the 1939 New York World's Fair. During the first decades of the twenty-first century, by contrast, the most popular venue for showcasing visions of progress became the consumer electronics show, the most famous of which is held in Las Vegas, Nevada, every winter. Here, the vision of the future on display is not shared but solipsistic, not a brilliant city but a shiny cocoon.

In some key respects, the drive to redesign the human brain and body has more in common with the privatized vision of the future now displayed every year in Las Vegas than with the collective vision of the future that characterized the world's fair exhibits of the early twentieth century. Of course, the problems that we can solve only collectively, such as epidemics, wars, and ecological collapse, still remain. When Martin Luther King Jr. accepted the Nobel Peace Prize for his work on civil rights in 1964, he pointed to a dangerous gap between our technological achievements and our ability to cooperate that is no less problematic in the age of human enhancement than it was in the decade of the space race: "We have learned to fly the air like birds and swim the sea like fish, but we have not learned the simple art of living together as brothers." It would not be fair to judge the movement for human enhancement while it is still in its infancy, but these words point to an enduring standard for how it will be evaluated in the future.

Bibliography

Claeys, Gregory, ed. *The Cambridge Companion to Utopian Literature*. Cambridge: Cambridge University Press, 2010.

Deese, R. S. *We Are Amphibians: Julian and Aldous Huxley on the Future of Our Species*. Oakland: University of California Press, 2015.

Hayles, N. Katherine. *How We Became Posthuman: Virtual Bodies in Cybernetics, Literature, and Informatics*. Chicago: University of Chicago Press, 1999.

Huxley, Aldous. *Brave New World; & Brave New World Revisited*. New York: Harper & Row, 1995.

Istvan, Zoltan. *The Transhumanist Wager*. Reno, Nevada: Futurity Imagine Media, LLC.

Jacoby, Russell. *Picture Imperfect: Utopian Thought for an Anti-utopian Age*. New York: Columbia University Press, 2005.

Kurzweil, Ray. *The Singularity Is Near: When Humans Transcend Biology*. London: Duckworth, 2016. Originally published in 2005.

More, Max, and Natasha Vita-More, eds. *The Transhumanist Reader: Classical and Contemporary Essays on the Science, Technology, and Philosophy of the Human Future*. Chichester, UK: Wiley-Blackwell, 2013.

O'Connell, Mark. *To Be a Machine: Adventures Among Cyborgs, Utopians, Hackers, and the Futurists Solving the Modest Problem of Death*. New York: Doubleday, 2017.

Stapledon, Olaf. *Last and First Men*. New York: Dover, 1968.

CHAPTER 2

Essential Concepts of Science and Technology Studies (STS)

Robert G. W. Kirk
Lecturer in Medical History and Humanities
University of Manchester, UK

What is science? One definition, provided by *Merriam-Webster's Collegiate Dictionary*, is "knowledge or a system of knowledge covering general truths or the operation of general laws especially as obtained and tested through scientific method." This mirrors a commonsense view that, while acknowledging it contains widespread differences across fields of application, nevertheless assumes science to be a distinct form of knowledge because of its unparalleled access to "truth." Relative to other academic fields, science does employ a more systematic and formal method for aligning observed data with theory to reach agreement on what is true of the natural world. In science, facts correspond to theory, knowledge to reality. Within the philosophy of science, the scientific method has traditionally been explained through "logical positivism" and "falsification" (Okasha 2016). Logical positivism emerged in the 1930s in the work of a group of philosophers known as the Vienna Circle. It portrays science as a predominantly inductive process moving from the accumulation of observed factual data to the development and testing of theory to eventual consensus on truth. Falsificationism, a concept associated with Austrian-born British philosopher Karl Popper (1902–1994), accounts for the distinctiveness of scientific knowledge because of its capacity to be demonstrably false. The best scientific theories are those that, when tested, make accurate predictions. To be *scientific*, however, it must always be possible that the theory can fail to predict phenomena in the natural world. For Popper, scientific truth was always provisional because it held the potential to be falsified in the light of new data (which is how science progressed). In contrast, knowledge that was impossible to disprove was a matter of faith, not science. In sum, scientific knowledge is expected to be empirically derived, reliable, testable, and agreed on by at least a majority of scientists (Chalmers 2013). We expect science to be "true." But what if truth is not quite what we think it to be? What if truth is made rather than found?

In the late twentieth century, anthropologists, historians, sociologists, and other scholars from across the humanities and social sciences followed philosophers in making science the object of their study. The result was a series of loosely related fields of inquiry that collectively form a community of "science studies" known as STS. Rather than following philosophers in asking "what is science?" however, STS instead asks "how does science work?" (Biagioli 1999, xii). Answers to this question have revealed science to be very different from what the commonsense view of it might suggest.

DEFINING STS: SCIENCE AND TECHNOLOGY STUDIES OR SCIENCE, TECHNOLOGY, AND SOCIETY?

From the 1980s STS was gradually established as the collective name for the interdisciplinary field of science studies. Although all work within STS is characterized by having made science its object of study, the relationship between works within STS is fluid and uncertain. STS scholarship is eclectic, making for a multi- if not interdisciplinary field. Differences are most pronounced in the sets of tools and methods employed to study science, variously drawn from anthropology, ethnography, geography, history, and sociology, to name but a few of the more prominent influences. There is little consensus on approaches to the study of science or on what science is. An absence of agreement on what the initialism STS actually stands for is illustrative. For some, STS is science and technology studies, an interdisciplinary field of scholarly study. For others, STS refers to the same field but evokes the three objects of science studies: science, technology, and society. Emerging in the 1960s across Western universities, the latter is the older term and is instructive in the prominence it gives to society. STS coheres around the study of science not just in relation to society but as a social process in and of itself. Indeed, if science were not a social process, the humanities and social sciences would have little justification in making it an object of study. In many ways, STS is a diverse and divided field characterized by often spectacular disagreement on the nature of science and technology and their relationships to each other and society. Rather than seeing this as a weakness, however, STS valorizes the cultivation of difference, plurality, and uncertainty.

One definition of STS is the study of how social, political, economic, cultural, and other values shape and are shaped by science and technology. To claim that science is a social process has sometimes been taken as an attack on the credibility of the sciences. If one suggests that scientific knowledge is shaped by factors beyond observed facts, one challenges the direct correspondence of scientific theory to reality—which in turn appears to undermine scientific assertions of truth. For this reason, the relationship between practitioners of STS and the natural sciences is often tense and occasionally hostile. Distrust is magnified by misapprehension rooted in very different ways of thinking about knowledge. Generally speaking, the natural sciences build elegant and simple accounts of the world that explain phenomena by the simplest hypothesis possible. Simplicity is crucial to the natural sciences as it forms part of the falsification criteria. Simple statements can be easily tested and refuted, whereas the addition of additional clauses serves only to make it more difficult if not impossible to falsify a claim. In contrast, STS produces messy accounts of the world, celebrating complexity, contingency, and the multiplication of explanations and theories (Law and Mol 2002).

At worst, these differences in approach have led the natural sciences and STS into mutually damaging arguments. An example is the ferocious intellectual clashes of the 1990s known as the "science wars," in which "realist" scientists, feeling threatened by "constructivist" or "postmodern" scholarship, confronted STS in response to perceived attacks on the credibility of scientific knowledge (Hacking 1999). At the heart of these disagreements was a failure to comprehend what STS is all about. It is important to understand, for example, that to show how science is made is not to assert that science is made up (cf. Gane 2006). Although the relationship between STS and the natural sciences has markedly improved in the second decade of the twenty-first century, STS is nonetheless characterized by a spirit of serious concern about the societal consequences of its work.

Writing with climate change in mind, French philosopher Bruno Latour, one of the most influential voices within STS, reflected that

> dangerous extremists are using the very same argument of social construction to destroy hard-won evidence that could save our lives. Was I wrong to participate in the invention of this field known as science studies? Is it enough to say that we did not really mean what we said? Why does it burn my tongue to say that global warming is a fact whether you like it or not? (2004, 227)

In the early twenty-first century, as politics is widely acknowledged to have entered an era of post-truth, the questions raised by Latour are more urgent than ever.

ESSENTIAL CONCEPTS IN STS

The history of STS is a story of diversification and the broadening of the scope of inquiry as interest in the sociological study of scientific knowledge widened to address how gender and race (Harding 1998), institutions (Hull 1988), technologies (Bijker, Hughes, and Pinch 2012), material cultures (Latour 1987), experimental systems (Rheinberger 1997), and professions, expertise, regulation, and governance (Jasanoff 2005) shaped science, as well as, ultimately, addressing the nature of objectivity itself (Daston and Galison 2007). Although the diversification of interests within STS has led to ever-more sophisticated understandings of science, the consequent plurality has inevitably led to a decline of common reference points. From the outside, STS can appear to be an incoherent and messy collective. STS looks no different from the inside, but seen from within, the "mess" is not just an expectation but a positive finding (Law 2004). If society and the natural world are complex, vague, contingent, and frequently incoherent, should this not be reflected to some extent in accounts of them? In any case, a linear and comprehensive account of the development of STS would oversimplify an otherwise complex story (for lengthier attempts at summarizing STS, see Sismondo 2010; Yearley 2005). What follows is necessarily a concise and partial overview of some of the essential concepts that give shape to STS.

PARADIGMS AND SCIENTIFIC REVOLUTIONS

The concept of *paradigms* as an explanatory frame for revolutions in scientific thought is closely associated with the work of Thomas S. Kuhn (1922–1996), who questioned the presumption that science progressed through the accumulation of facts and the building, falsification, and improvement of theories. Rather than viewing science as developing in a linear positivist fashion, Kuhn saw in the history of science a process of periodic revolutionary changes that he named "paradigm shifts" (1962, 89). The concept of the paradigm shift fundamentally transformed how the history of science was understood. Previously, science had been thought to develop in a literally progressive way, moving forward by discovering ever more truthful knowledge of the world. For this reason, the history of science was teleological, assuming that the science of the past necessarily progressed toward the more truthful science of the present. However, the notion of revolutionary, paradigmatic shifts in scientific thought provided a radically new model for scientific change over time.

Kuhn described three phases of scientific activity: normal science, crisis, and the paradigm shift. Together they formed a cyclical process of scientific change. In periods of normal science, all scientists worked within a paradigm that provided a shared way of observing and understanding the natural world. Dominant scientific truths within a paradigm, such as the

Ptolemaic model of the universe, Newtonian mechanics, Darwinian evolution, or Mendelian genetics, shaped activities within a scientific community. The paradigm provided scientific work with meaning and purpose, ultimately shaping the scientific worldview. In periods of normal science, the dominant paradigm provided a rich foundation for creative theorizing. Observations that conflicted with expectations were considered anomalous "puzzles" of little consequence (Kuhn 1962). The metaphor of science as puzzle solving evoked the notion that a paradigm provided a frame within which to place scientific knowledge akin to how a picture operates in a jigsaw puzzle. Within normal science, pieces of knowledge fit together neatly to build an ever-clearer picture of the natural world. However, the accumulation of anomalies over time eventually can lead to a tipping point where the paradigm enters into crisis. Soon after, it is displaced by a new paradigm ushering in a new period of normal science. A paradigm shift was a revolutionary change in how the natural world was understood, leaving no common reference points across paradigms. Kuhn claimed that the scientific knowledge of the displaced paradigm would be incommensurable with the new normal science. Truth in one paradigm became meaningless in another, as truth is dependent on a common understanding particular to a paradigm. Accordingly, the perception of scientific progress over time is apparent only during a period of normal science.

By changing how science was understood to develop over time, Kuhn's work opened up new ways to think about what science is. Most importantly, in demonstrating that science makes sense only within the context of a paradigm, Kuhn recast science as a social process. This made possible new approaches to the question of how scientific knowledge was made and how consensus within a scientific community was achieved. These approaches did not presume correspondence to the natural world to be the delineator of truth.

In spite of its importance to the early formation of science studies, Kuhn's work was not accepted uncritically. The idea that paradigms were incommensurable has been strongly contested, as has the oversimplification of the unity of scientific culture. STS has shown how the production of scientific knowledge changes quite radically across specialized fields and disciplines. Indeed, STS has shown that the varied disciplines that make up the natural sciences are so different as to constitute distinct "epistemic cultures ... amalgams of arrangements and mechanisms—bonded through affinity, necessity and historical coincidence—which in a given field, make up *how we know what we know*" (Knorr Cetina 1999, 1). Consequently, the question of *how* scientists communicate, construct common understanding, and build consensus within and across fundamentally different areas of science is a fundamental question within STS literature (e.g., Galison 1997; Star and Griesemer 1989).

THE STRONG PROGRAM AND SYMMETRY

The *strong program* was a formative contribution to the sociology of scientific knowledge, a forerunner to STS that was established to show that science could and should be studied as a social phenomenon. Developed during the 1970s by an Edinburgh-based collective of historians, philosophers, and sociologists, the strong program is alternatively known as the Edinburgh school. It constituted a critical response to previous accounts of science that restricted sociological explanation to failed scientific theories (or pseudoscience). This was possible because it had been assumed that successful science was self-explanatory in that it revealed a true account of the natural world. In contrast, the strong program insisted that the truth or falsity of scientific knowledge was as much an outcome of social processes as the result of engagement with the natural world. Accordingly, all science, successful and unsuccessful, could be subjected to and explained by sociological study in the same way. This was not intended as a critique of scientific validity. Instead, it was an acceptance of the principle that all knowledge is inevitably constructed by and within human culture. As such, science was expected to reflect and be shaped by social factors. Understanding the sociological factors that enabled successful science was equally as important as understanding the reasons why science sometimes failed.

Methodological symmetry, or the assertion that all science, whether "successful" or not, could be sociologically studied in the same way, was one of four guiding principles set out for the strong program by David Bloor (1976). The others were a focus on *causality* (the social and cultural conditions that made scientific knowledge possible), a commitment to *impartiality* (whether a theory was thought true or false it would be approached in the same way), and the concept of *reflexivity* (explanatory frameworks would apply equally to sociological knowledge itself). As a foundational component to the sociology of scientific knowledge, the strong program made important and influential contributions to the development of STS—but not without attracting criticism. In particular, the strong program has been taken to task for having embraced a radically relativistic standpoint on truth (e.g., Sokal and Bricmont 2003), although its adherents would respond that relativism is a methodological stance and not a theoretical or explanatory claim.

SOCIAL CONSTRUCTION

While STS as a field is remarkably varied in its approaches almost to the point of appearing disorganized to the casual outsider, one common commitment is to the study of how scientific knowledge and objects are constructed (Hacking 1999). *Social construction* embodies a number of claims that are fundamental to STS. By emphasizing that knowledge is constructed in a social context, social construction asserts that science is a social process. Moreover, the language of construction invokes agency. Scientific knowledge is *made* within situated social contexts and not simply *found* by observing nature. Taken further, social construction suggests that the objects identified and studied by science are not to be thought of as natural (if "natural" means in some way located outside of the social). On the contrary, scientific objects are mediated to varying extents by the social context of their construction. As STS has grown in confidence as a field, social construction has shifted from an argument to an implicit assumption. One finds fewer arguments for the social construction of this or that in twenty-first century literature as the constructedness of science (and humanity's knowledge of nature) is taken to be the raison d'être of STS. This is not to suggest that the constructedness of scientific knowledge is uncontested. As a founding pillar of STS, social construction has been vehemently disputed by "realist" sciences. At the same time, although all STS supports some form of the constructedness of knowledge, there are many views and frequent disagreements as to what social construction actually means.

Social construction directly contests the notion of a direct correspondence between scientific knowledge and the reality of the natural world. Scientific facts are situated in social contexts and thus become contingent. Most differences within STS around the constructedness of knowledge revolve about the extent to which the truth of a fact is shaped by social factors, cultural values, and natural phenomena. A frequently cited example is H. M. Collins's (1985) analysis of Joseph Weber's claim to have detected gravity waves using a novel large antenna in the early 1970s. If confirmed, this would have been a major event in physics. Other scientists, however, struggled to replicate Weber's results. In explaining the subsequent controversy within physics as to whether Weber had or had not detected gravity waves, Collins provided an empirical demonstration that experimental results gained meaning only within a wider social context. Questions can be raised as to the competence of the scientists involved, the skill set of the technicians, the quality of the experimental system and whether or not it achieves what is claimed and/or faithfully reproduces the original experimental system that it seeks to test, and a whole manner of other factors. To assess whether Weber had detected gravity waves, physicists needed to know if they exist. To know whether gravity waves exist, they needed a means to detect them. In this way, scientific observations of new objects in the natural world can quickly become caught in a circular process. As all experimental reenactments will differ to some degree from the original work, there will always be room for disagreement. Following extended argument over a number of years, physicists eventually concluded that Weber had not observed gravity waves. A judgment was made, and the process of making it had to be social as there was evidence for and against Weber's assertion. In this way, experimental observations of the natural world fall back on social processes of negotiation and agreement in order to ascertain their validity. Collins called this example of social construction the "experimenters' regress" (1985, 84).

In a radical form, social construction becomes philosophical nominalism, a branch of philosophy that disputes that there is any correspondence between knowledge and reality. As such, it presents an understanding of science that appears to be the exact of opposite of how the natural sciences understand themselves. A scientist would appeal to nature to assert his or her theory as the only possible outcome of scientific inquiry, whereas a radical social constructivist would counter that multiple theories were possible but that social contingencies shaped this particular outcome. This is not a denial of the reality of scientific knowledge and objects. It does, however, propose a different understanding of what people take to be real; the real is only contingently real. A person's experience of reality is, to varying degrees, constructed within and through social processes. Construction is not, then, a denial of reality. Nor is it a claim that social factors shape scientific knowledge in a linear way. STS has shown that science and society shape each other over time through diverse processes of mutual construction or coproduction (e.g., Jasanoff 2004). This is yet another reason why STS embraces complexity and as a result can appear unfocused and incoherent. Unlike science, which builds anchoring points for knowledge through the supposition of an unchanging external referent (the natural world) as the cause and arbiter of truth, STS claims no constant referent. Everything is dependent on, contingent to, and shaped by everything else. Consequently, every element is constantly changing in relation to every other element. For this reason STS embraces the metaphor of "mess" and frequently talks of entanglement, relationships, and mutual constitution.

Periodically, strong arguments are made asserting radical forms of social constructivism (e.g., Hacking 1999). These are usually framed defensively as responses to critiques from outside, most often from the realist sciences. Such defenses are illustrative of the radically different ideas of science emerging from STS. Nevertheless, within STS, too, one can find disconcerted voices speaking against radical constructivism. The not infrequent attempts to include material culture within STS are in one way or another responses to extreme constructivist accounts of science. Robert E. Kohler (1994), for instance, has shown how the biology of the *Drosophila* fruit fly shaped not only the trajectory of the early science of genetics but also the values and social expectations of the community of scientists who used the *Drosophila* fly to understand genetics. The turn to technology and approaches such as actor-network theory, discussed later in this chapter, are similarly attempts to develop more balanced accounts giving due weight to both social and natural factors in the shaping of science.

SOCIAL CONSTRUCTION OF TECHNOLOGY (SCOT)

Whereas the philosophy of science possessed a long tradition mirroring the history of science itself, very little attention had been given to philosophical inquiry into technology until the late twentieth century. In part, the privileged status of science over technology reflected the social value and cultural prestige that science had established for itself in contrast to technology (technical labor was often portrayed passively as merely the application of science). Starting in the 1960s, however, historians of technology began to question the assumption that technology as the application of science was one of the major forces driving historical change. So-called technological determinism was critically undermined in the 1980s in what Steve Woolgar (1991) has termed "the turn to technology." The *social construction of technology* (SCOT) approach suggested that technology, like scientific knowledge, was socially constructed (Bijker, Hughes, and Pinch 2012; Mackenzie and Wajcman 1999). Early SCOT approaches applied the symmetry principle to show that the same kind of explanation could be sought regardless of whether a technology succeeded or failed. Technology was shown to hold different meanings in different contexts for different users. A further outcome of SCOT, as well as subsequent work within STS, has been the blurring of the boundary between science and technology. STS literature, inflected by an interest in material culture and the philosophy of technology, subsequently emphasizes the interconnectedness of science and technology and has adopted the term *technoscience* to express their interdependence (e.g., Latour 1987; Ihde and Selinger 2003; Haraway 2004).

Interest in how science, technology, and society mutually constitute each other has led to a more respectful understanding of how and why individuals may choose to "resist" a new technology. The linear and passive relationship between technological innovation and end user has been displaced in favor of a more relational understanding of technological production premised on coproduction (Oudshoorn and Pinch 2003). The ways in which users engage with new technology, and as a consequence transform and are transformed by it, are particularly prominent within health and medicine. Studies of new technologies of cancer treatment, such as genetics in diagnosis, therapy, and prevention, for instance, have been shown to have fundamentally reshaped the types of relationships people have with the disease and individuals' sense of identity or selfhood (e.g., Stemerding and Nelis 2005). The technology to test for the abnormal gene *BRCA2* (standing for BReast CAncer gene 2), linked to a genetic predisposition to breast cancer, has led actress Angelina Jolie and many other women to choose a double mastectomy, a procedure more conventionally associated with the treatment of breast cancer as opposed to a preventive measure. In this way, new technologies result in otherwise healthy individuals facing life-changing choices regarding how to respond to and manage diseases that they do not and may never have.

SCOT has not been without its critics. Langdon Winner (1993) accused SCOT of focusing on technological innovation to the detriment of examining the consequences of technology. Furthermore, Winner contends that SCOT fails to address the impact of technology on social groups who are excluded from its development but are nevertheless affected by it. As such, SCOT is incapable of engaging meaningfully with the moral implications of technology. All too often, SCOT approaches imply, however inadvertently, that technology is essentially value neutral and leads to morally consequential outcomes only once it is applied in practice. In contrast, Winner has argued that both design and technology intrinsically embody politics. Citing the unusually low bridges over parkways on Long Island, New York, Winner (1980) explained how they were designed to allow the passage of automobiles but were low enough to prevent the presence of buses. Consequently, the low parkway bridges embodied and acted to reinforce class and racial boundaries. They allowed free passage for the white automobile-owning middle class, while blocking the public transportation used by the lower classes. Although by no means representing a return to technological determinism, Winner does serve as an example of those STS scholars who emphasize the ways in which technology inherently embodies political and moral values and thereby acts to shape behavior, society, and culture (see also Verbeek 2011).

ACTOR-NETWORK THEORY (ANT)

Actor-network theory (ANT) is one of the most influential yet poorly named contributions to STS. It is less a theory than a method for approaching and thinking about how science is practiced (Latour 2005). In its original form, ANT was a response to the uncritical use of interest as an explanatory framework within the early sociology of knowledge (the idea that specific scientific claims and/or technologies in some way served a particular social group who were therefore advancing their own narrow social interests). The strong program in particular had established an analytic framework for illustrating science as a social process. First, a scientific claim or "discovery" was identified that had proved particularly divisive. Second, the differing positions taken on the scientific claim were mapped to particular social positions. Finally, the arguments presented by participants in the scientific debate were shown to match the corresponding social position of the advocate. In this way interest connected scientific and social worlds and allowed social explanations for scientific findings (e.g., Shapin 1975).

ANT asserted, however, that interests cannot serve external explanatory referents because interests are neither stable nor self-evident (Callon and Law 1982). Rather than existing outside the scientific process, interests are fashioned, refashioned, and eventually stabilized in the process of practicing science. Accordingly, ANT adopted a more dynamic approach to interests, emphasizing the use, manipulation, and "translation" of interests. ANT mapped how a scientist actively worked to assert how his or her work would serve the interest of a particular social group. For instance, in *The Pasteurization of France* (1988), Bruno Latour recast the story of French biologist Louis Pasteur (1822–1895), which was hitherto a story of a great advance in the understanding and prevention of disease. From an ANT perspective, Pasteur was successful because he effectively translated the interest of nineteenth-century farmers to reflect his scientific needs. Within ANT, interests were transformed from a passive to active form; science became the enrollment of others to form networks of support. Pasteur "enrolled" French farmers by promising a means to reduce animal disease if they collaborated with him and supported his work. In this way, the advancement of Pasteur's science was translated to be in the interest of livestock farmers. Interest was a consequence and not the cause of Pasteur's success.

Like SCOT, ANT also contributed a major corrective to the relative neglect of scientific objects and the material world in STS. Motivated by the desire to incorporate objects and technology into accounts of science, ANT took a strong materialist stance insisting on a radical methodological symmetry. The role of human and nonhuman actors was to be analyzed in the same way. A classic early example was Michel Callon's (1999) analysis of a group of scientists who sought to establish artificial scallop farming in St. Brieuc Bay in northwestern France. Callon described how, through the process of "translation," fishermen were enrolled into the scientist's project by establishing, for instance, that the future sustainability of their trade depended on a successful means to farm scallops. The fishermen did not change their aim, which was to harvest scallops. Their concerns, however, were translated so as to enroll them as allies of the scientists. Following the principle of symmetry, scallops, too, had to be enrolled, with their "interests" translated to align with those of the scientists. This was achieved by developing an artificial infrastructure in the form of fine netted bags within which scallops could feed and grow. Callon's study provides a clear example of how ANT recast the social study of science into a method of tracing a process in which "both social and natural entities are shaped and consolidated" (1999, 74).

A major point of confusion and contention within ANT is the way in which agency is distributed across a network of allies with no apparent concern as to whether actors are human or nonhuman (Sayes 2014). Although sociologists can, with some degree of credibility, claim expertise in the study of human social behavior, no such claim can be made regarding the behavior of nonhumans, such as scallops. Adherents of ANT respond that they do not need to know why nonhuman actors behave as they do but only how they behaved in a given situation. Yet, this can often result in ANT appearing to take scientific accounts of the natural world almost uncritically. There is no obvious way to independently know how Pasteur's microbes behaved any more than how the scallops acted in St. Brieuc Bay. In both instances, Latour (1988) and Callon (1999) rely on the scientists' own observations and accounts.

More problematic for many is the way ANT flattens the differences between human and nonhuman actors. Such an approach masks the vast differences in the range and degree of actions humans might take in comparison to, say, scallops. Nevertheless, when Latour (2005, 245) declared that nonhumans made up the "missing masses" of STS, he had a point. ANT was one mechanism for securing the reintroduction into STS of the material (and to some extent the "natural") world. Within ANT, the boundaries between human and nonhuman actors dissolve as agency is distributed across networks. Perhaps more than any other area of STS, this radical, albeit largely methodological, move has contributed to a rethinking of what it is to be (and not to be) human.

POLITICS, GOVERNANCE, AND POLICY

For some, STS's commitment to advocacy and social change, alongside its relevance to science policy and governance, had been a fault line dividing the field. Steve Fuller and James H. Collier (2003, xii) describe the "High Church" of STS, concerned with the question of how science operates and why it enjoys a privileged status in society, and the "Low Church," which is more focused on helping science work better. Frequently, one finds High Church STS accused of being incapable of saying anything meaningful about the fundamental nature of science because its findings are too narrow, contingent, and limited to the particular situation under study; often overly complex and disorganized; and all too frequently guilty of hiding behind a veil of acronyms and jargon. In contrast, Low Church STS can often appear too accepting of science's own accounts of itself and thus be unable to produce meaningful critical accounts of science as a social process.

Although Fuller and Collier's characterization has some merit, STS as a whole (one that is admittedly eclectic) shares a commitment to, and interest in, understanding and shaping the politics of science, as well as its governance and policy. The work of Sheila Jasanoff (e.g., 1990, 2004) provides a pertinent example. In a wide ranging study of biotechnology across the United States, Britain, and Germany, Jasanoff (2004) identified distinctive "civic epistemologies" that shape how science and technology are practiced in different national cultures. Jasanoff (2005) describes civic epistemologies as being made up of the following six dimensions:

1. participatory styles of public knowledge making,
2. methods to ensure accountability,
3. practices of public demonstration,
4. preferred registers of objectivity,
5. accepted basis of expertise, and
6. visibility of expert bodies.

Each dimension differs in different national contexts. In Germany, Jasanoff found consensus making to be the dominant civic epistemology, shaped by high levels of trust, whereas in the United States processes were more contentious, formalized, and legalistic. These different democratic approaches to science and technology explain why biotechnology developed and has been received in different ways across national cultures.

Many claims within STS have fundamental ramifications for politics and society. By challenging the assumption that nature and society are separate, for instance, STS invites a radical rethinking of the place of science in public discourse. If what humans think they know of the natural world has in fact been mediated by social concerns—as the constructedness of science suggests—then democratic politics must adapt to what is in effect a new reality. In his more recent work, Latour develops this point to argue for the critical *political* importance of recognizing what science really is and the urgent need to blur "the distinction between nature and society *durably*" (2004, 36).

FEMINIST SCIENCE STUDIES

The history of science, like the history of Western culture, presents a predominantly male account, operating to privilege the continuation of a patriarchal society. Critically responding to this, the study of the relationship between science and gender has been one of the most politically important and intellectually creative areas of STS. In 2017 the United Nations Educational, Scientific and Cultural Organization estimated that women accounted for only 30 percent of the world's researchers in the natural sciences and an even lower percentage at higher decision-making levels. *Feminist science studies* has produced significant accounts of the structural and social factors that contribute to the underrepresentation of women in science. Moreover, it has gone much further. If one assumes that science has privileged access to "truth" owing to a direct correspondence to the natural world, then a scientist's gender makes no difference to the content of scientific knowledge. Yet, STS asserts that scientific knowledge is constructed and thus is mediated and shaped by social factors. Developing this perspective has allowed feminist science studies to reveal the impact of gender on both the methods of scientific practice and the content of scientific knowledge (Wyer et al. 2014).

A formative contribution in this regard was Carolyn Merchant's reappraisal of the historical roots of Western science, *The Death of Nature: Women, Ecology, and the Scientific Revolution* (1980). Merchant described how the reframing of the natural world from an organismic to a mechanistic model served as a condition for the scientific revolution (as well as the wider emergent capitalist system). But it also had fundamental consequences for women, as nature was metaphorically represented and understood as female. In *The Death of Nature,* Merchant presented a historical account of how feminine metaphors for nature were transformed through the advent of science. Between 1500 and 1700 nature ceased to be imagined as a nurturing mother deserving of respect. Instead, nature was presented as disordered, wild, and in need of subduing through the imposition of rational (male) control. Science was the mechanism through which nature would be ordered and bent to the will of industry. In this way, science established a logic that entangled the changing status of women in society with the exploitation of the natural world. Although frequently critiqued for having overstretched her hand, Merchant makes a compelling albeit somewhat jumbled argument for the interconnectedness of science, society, the natural world, and the status of women. Her thesis continues to resonate in the light of contemporary concerns for the environment and climate change.

Feminist science studies, alternatively known as feminist technoscience, is, then, avowedly political. As such it challenges the division of STS into High Church (academic and intellectually oriented investigations) and Low Church (studies aiming to contribute to societal change). One of the major contributions of feminist science studies is to have charted a meaningful political consequence for the STS claim that scientific knowledge is socially situated. Feminist standpoint theory asserts that marginal groups are better placed to be aware of, and ask questions about, the impact of social relations than nonmarginalized groups (Harding 2004). By advancing an epistemological claim that identity, social positioning, and the construction of knowledge are interconnected, standpoint theory offers a different approach to studying the social construction of science—an approach that privileges the marginalized. Donna Haraway, as an example, has shown how gender, race, and imperial/colonial history shaped the twentieth-century science of primatology. In *Primate Visions* (1989), Haraway carefully analyzed the contributions of four white, female, North American primatologists working in the late twentieth century, revealing how each brought new and radical ways of thinking to the previously male-dominated field of primatology. In sum, feminist science studies suggests that the place a scientist occupies in society shapes not only what he or she can observe of the natural world but also what he or she does not observe. Such a claim, which can imply a form of relativism, has proven controversial. Nevertheless, the intent is not to undermine science by revealing it to be somehow "tainted" by social value. On the contrary, the hope is to develop better and more just understandings of the scientific production of knowledge of the natural world.

An interesting element of feminist science studies is that it is haunted by its own name, which inadvertently equates *women* with *gender*. The reasons for this are historical, and political. Within science and the modern world generally, gender limits women far more so than it does men. Yet feminist science studies insists on being acutely self-reflective and takes care to remain aware of the reasons for and consequences of this contradiction (e.g., Keller 1995). The intent is to challenge dichotomies such as female and male rather than entrench or assert essentialist difference. Feminist science studies has revealed science to be gendered. As such, gender provides a powerful lens through which to observe how social values and power relations shape the construction of scientific knowledge and result in meaningful and often negative consequences for society.

STS AND POSTHUMANISM

STS has contributed new, varied, and diverse ways of understanding what humanity knows of, and how humanity goes about knowing, the natural world. STS challenges the assumption that science has privileged access to the nature of reality. But it does not dispute that science remains a preeminent and important means of making knowledge of the natural world. By revealing the production of scientific knowledge to be a creative process, shaped by social and cultural values, STS equally challenges the fundamental categories that order, and are commonly thought to make up, the natural world. Categories such as nature/culture, science/society, male/female, and human/nonhuman no longer appear as separate and opposed and can no longer be thought of as rigid dichotomies. Instead, they are understood as complex, entangled, mutually constituting historical processes. Haraway famously drew on the image of the cyborg to make this point. Moreover, the cyborg serves equally as a rallying call, providing a reminder that science and technology hold the capacity not only to establish and police categorical boundaries but also to transgress and remake them. As such, for Haraway, engaging with and understanding science through STS serves as a resource for advancing a radical politics of social justice.

Within STS, the potential to rupture the categorical structure of how humans think is experienced when one tries to write about it; the cyborg challenges straightforward categorization as a he, she, or it. In doing so, the cyborg resists the language humans possess to represent the world and calls for change. Revealing and pushing the limits of language is one of the most interesting, provocative, and at times frustratingly challenging features of STS. Newcomers to STS may be puzzled by such conjunctions as *technoscience, natureculture,* and *FemaleMan,* each term deliberately resisting and thus challenging the embedded knowledge that is assumed to reflect the reality of the world. Working at the limit of language can also be exhilarating, as each challenge is an invitation to practice a new way of thinking. The cyborg is a posthuman figure, a hybrid embodying a radical transgression of the boundaries of nature/culture, male/female, and human/nonhuman (Hayles 1999). For some, the cyborg serves as a signpost, if not a clarion call, toward a posthuman future yet to be thought (cf. Grebowicz and Merrick 2013).

In an interview titled "When We Have Never Been Human, What Is to Be Done?" (Gane 2006), Haraway reflected on how her work had evolved since she wrote "A Cyborg Manifesto" (1984). The title is evocative in three ways. The suggestion that "we" have *never been human* is an intentional gesture to Latour's (1993) claim that those living in modernity have *never been modern.* Haraway and Latour have much in common, not least a shared desire to include nonhuman agency (Latour) or liveliness (Haraway) in their accounts of technoscience. Although their work diverges in certain significant ways, the two align around a fundamental characteristic of STS—namely, challenging dualism through the assertion of hybridity. In claiming that *we have never been modern,* Latour responded to what he saw at the time as the foremost contemporary challenge to Western science and democracy. Prominent political concerns, such as biotechnology, genetically modified food, HIV/AIDS, emergent disease, and climate change, each entangled science and society and nature and culture, so that it was increasingly difficult to maintain rigid and clear distinctions. Rather than seeing this as something new, Latour explained it to be no more than a public recognition of the fact that the categorical distinctions fundamental to modernity had always been artifactual. Humans have never been modern in an essentialist sense, because modernity had manufactured itself through the power of science to diligently separate and maintain dichotomous categories. Evoking a now familiar STS way of looking at the world, Latour called for thinking in terms of "hybrids," which embody intertwined aspects of natural and social phenomena. Ultimately, Latour arrived at a similar (though in many ways different) position to where Haraway had been a decade earlier in evoking the cyborg as a figure that embodied hybridity as a challenge to fixed and essentialist categories.

Returning to the 2006 interview, the question *what is to be done?* is a second equally important provocation. Haraway is not content to comment on technoscience as though located impartially, on the outside, and so divested of interests. Echoing Russian Communist leader Vladimir Lenin (1870–1924), she asks if this is so, then what is to be done? Haraway is acutely concerned with the fundamental challenge of how to develop understandings of science that are attentive to how knowledge is situated within specific social contexts while retaining its value as an account of the natural and thus material world (e.g., Haraway 1991). Consequently, Haraway is cautious in her use of the language of "human/posthuman," as it is too easily appropriated to support fanciful claims as to the next, teleologically imagined, evolutionary stage of transhumanist technoenhancement (Haraway 2004). Haraway is far too committed to retaining the material and the historical in her engagements with the world to follow Ray Kurzweil (2005) and other transhumanists in speculating that a posthuman future will involve transcending the body and thus the biological self.

In Haraway's hands, STS provides tools to challenge the centrality of the human-bounded subject in accounts of both society and the natural world. Haraway insists on a more inclusive understanding of society, in which social relationships include both nonhumans and humans as socially active partners (Haraway 1997). This is why the "we" of the title "When We Have Never Been Human, What Is to Be Done?" is critically important. Who and what makes up the "we" is often at the forefront of Haraway's thought as a means to remember who and what is excluded. Haraway presents a highly situated and contingent account of the human that, although allowing for the fluid reconfiguration of what the human is in itself and in its relations to nonhumans, evades the need to posit a teleological historical development from human to "posthuman." But it is no less radical for that.

Summary

STS provides a fundamentally different way of thinking about science, technology, and humanity's knowledge of the natural world. STS challenges the commonsense view of science as occupying a privileged epistemological space. It questions the claim that the truth of scientific knowledge rests in its direct correspondence to the "reality" of the natural world. Instead, STS presents science as a social process, producing highly situated knowledge that is no less a "true" representation of reality for having being assembled from a hybrid of natural and social elements. Importantly, this constitutes an effective erasure of "nature" (and/or the "human") as an objective stable referent on which moral and ethical argument can be grounded. Put another way, from an STS perspective, the claim that nothing is sacred becomes a positive affirmation. As such, STS invites a reframing of the public debate and ethical concerns about technoscience.

If one accepts the idea that what is natural and what is artifice is a manufactured contingency as opposed to a fixed essence, then how does one determine what is to be done? For some, STS paves the way for radical interventions in the material and conceptual makeup of the human, thereby opening up pathways toward new posthuman futures. For others, STS reveals only what the human has always been—a category without essence that makes up its form through its relations to all that is nonhuman. Neither reading is more nor less radical than the other. Both accept that the human has, is, and will always be historically situated and thus a contingent product of natural and social relations. Whether one finds in STS an opportunity to rework the content of the category "human" or legitimacy to move beyond the human into the posthuman, one thing is certain. What the human means today, and what one might want it to mean tomorrow, is an always open question.

Bibliography

Biagioli, Mario, ed. *The Science Studies Reader.* New York: Routledge, 1999.

Bijker, Wiebe E., Thomas P. Hughes, and Trevor Pinch, eds. *The Social Construction of Technological Systems.* Anniversary ed. Cambridge, MA: MIT Press, 2012.

Bloor, David. *Knowledge and Social Imagery.* London: Routledge and Kegan Paul, 1976. 2nd ed., Chicago: University of Chicago Press, 1991.

Callon, Michel. "Some Elements of a Sociology of Translation: Domestication of the Scallops and the Fishermen of

St. Brieuc Bay." In *The Science Studies Reader*, edited by Mario Biagioli, 67–83. New York: Routledge, 1999.

Callon, Michel, and John Law. "On Interests and Their Transformation: Enrolment and Counter-enrolment." *Social Studies of Science* 12, no. 4 (1982): 615–625.

Chalmers, A. F. *What Is This Thing Called Science?* St. Lucia, Australia: University of Queensland Press, 2013.

Collins, H. M. *Changing Order: Replication and Induction in Scientific Practice*. London: Sage, 1985.

Daston, Lorraine, and Peter Galison. *Objectivity*. New York: Zone Books, 2007.

Felt, Ulrike, Rayvon FouchÕ, Clark A. Miller, and Laurel Smith-Doerr, eds. *The Handbook of Science and Technology Studies*. 4th ed. Cambridge, MA: MIT Press, 2017.

Fuller, Steve, and James H. Collier. *Philosophy, Rhetoric, and the End of Knowledge: A New Beginning for Science and Technology Studies*. 2nd ed. Mahwah, NJ: Lawrence Erlbaum, 2003.

Galison, Peter. *Image and Logic: A Material Culture of Microphysics*. Chicago: University of Chicago Press, 1997.

Gane, Nicholas. "When We Have Never Been Human, What Is to Be Done? Interview with Donna Haraway." *Theory, Culture, and Society* 23, nos. 7–8 (2006): 135–158.

Grebowicz, Margret, and Helen Merrick. *Beyond the Cyborg: Adventures with Donna Haraway*. New York: Columbia University Press, 2013.

Hacking, Ian. *The Social Construction of What?* Cambridge, MA: Harvard University Press, 1999.

Haraway, Donna. *The Haraway Reader*. New York: Routledge, 2004.

Haraway, Donna. *Modest_Witness@Second_Millennium.FemaleMan_Meets_OncoMouse: Feminism and Technoscience*. New York: Routledge, 1997.

Haraway, Donna. *Primate Visions: Gender, Race, and Nature in the World of Modern Science*. New York: Routledge, 1989.

Haraway, Donna. "Situated Knowledge: The Science Question in Feminism and the Privilege of Partial Perspective." In *Simians, Cyborgs, and Women: The Reinvention of Nature*, 183–201. New York: Routledge, 1991.

Harding, Sandra, ed. *The Feminist Standpoint Theory Reader: Intellectual and Political Controversies*. New York: Routledge, 2004.

Harding, Sandra. *Is Science Multicultural? Postcolonialisms, Feminisms, and Epistemologies*. Bloomington: Indiana University Press, 1998.

Hayles, N. Katherine. *How We Became Posthuman: Virtual Bodies in Cybernetics, Literature, and Informatics*. Chicago: University of Chicago Press, 1999.

Hess, David J. *Science Studies: An Advanced Introduction*. New York: New York University Press, 1997.

Hull, David L. *Science as a Process: An Evolutionary Account of the Social and Conceptual Development of Science*. Chicago: University of Chicago Press, 1988.

Ihde, Don, and Evan Selinger. *Chasing Technoscience: Matrix for Materiality*. Bloomington: Indiana University Press, 2003.

Jasanoff, Sheila. *Designs on Nature: Science and Democracy in Europe and the United States*. Princeton, NJ: Princeton University Press, 2005.

Jasanoff, Sheila. *The Fifth Branch: Science Advisers as Policy-makers*. Cambridge, MA: Harvard University Press, 1990.

Jasanoff, Sheila, ed. *States of Knowledge: The Co-production of Science and Social Order*. London: Routledge, 2004.

Keller, Evelyn Fox. "The Origin, History, and Politics of the Subject Called 'Gender and Science': A First Person Account." In *Handbook of Science and Technology Studies*, rev. ed., edited by Sheila Jasanoff, Gerald E. Markle, James C. Peterson, and Trevor Pinch, 80–94. Thousand Oaks, CA: Sage, 1995.

Knorr Cetina, Karin. *Epistemic Cultures: How the Sciences Make Knowledge*. Cambridge, MA: Harvard University Press, 1999.

Kohler, Robert E. *Lords of the Fly:* Drosophila *Genetics and the Experimental Life*. Chicago: University of Chicago Press, 1994.

Kuhn, Thomas S. *The Structure of Scientific Revolutions*. Chicago: University of Chicago Press, 1962.

Kurzweil, Ray. *The Singularity Is Near: When Humans Transcend Biology*. New York: Viking, 2005.

Latour, Bruno. *The Pasteurization of France*. Translated by Alan Sheridan and John Law. Cambridge, MA: Harvard University Press, 1988.

Latour, Bruno. *Politics of Nature: How to Bring the Sciences into Democracy*. Translated by Catherine Porter. Cambridge, MA: Harvard University Press, 2004.

Latour, Bruno. *Reassembling the Social: An Introduction to Actor-Network-Theory*. Oxford: Oxford University Press, 2005.

Latour, Bruno. *Science in Action: How to Follow Scientists and Engineers through Society*. Cambridge, MA: Harvard University Press, 1987.

Latour, Bruno. *We Have Never Been Modern*. Translated by Catherine Porter. Cambridge, MA: Harvard University Press, 1993.

Law, John. *After Method: Mess in Social Science Research*. London: Routledge, 2004.

Law, John, and Annemarie Mol, eds. *Complexities: Social Studies of Knowledge Practices*. Durham, NC: Duke University Press, 2002.

Mackenzie, Donald, and Judy Wajcman, eds. *The Social Shaping of Technology* 2nd ed. Buckingham, UK: Open University Press, 1999.

Merchant, Carolyn. *The Death of Nature: Women, Ecology, and the Scientific Revolution*. San Francisco: Harper and Row, 1980.

Okasha, Samir. *Philosophy of Science: A Very Short Introduction*. 2nd ed. Oxford: Oxford University Press, 2016.

Oudshoorn, Nelly, and Trevor Pinch. *How Users Matter: The Co-construction of Users and Technology*. Cambridge, MA: MIT Press, 2003.

Rheinberger, Hans-Jörg. *Toward a History of Epistemic Things: Synthesizing Proteins in the Test Tube*. Stanford, CA: Stanford University Press, 1997.

Sayes, Edwin. "Actor-Network Theory and Methodology: Just What Does It Mean to Say That Nonhumans Have Agency?" *Social Studies of Science* 44, no. 1 (2014): 134–149.

Shapin, Steven. "Phrenological Knowledge and the Social Structure of Early Nineteenth-Century Edinburgh." *Annals of Science* 32, no. 3 (1975): 219–243.

Sismondo, Sergio. *An Introduction to Science and Technology Studies*. 2nd ed. Chichester, UK: Wiley-Blackwell, 2010.

Society for Social Studies of Science. http://4sonline.org.

Sokal, Alan, and Jean Bricmont. *Intellectual Impostures: Postmodern Philosophers' Abuse of Science*. London: Profile Books, 2003.

Star, Susan Leigh, and James R. Griesemer. "Institutional Ecology, 'Translations,' and Boundary Objects: Amateurs and Professionals in Berkeley's Museum of Vertebrate Zoology, 1907–39." *Social Studies of Science* 19, no. 3 (1989): 387–420.

Stemerding, Dirk, and Annemiek Nelis. "Choices and Choosing in Cancer Genetics." In *Inside the Politics of Technology: Agency and Normativity in the Co-production of Technology and Society*, edited by Hans Harbers, 109–124. Amsterdam: Amsterdam University Press, 2005.

United Nations Educational, Scientific and Cultural Organization. "Gender and Science." Accessed August 23, 2017. http:// www.unesco.org/new/en/natural-sciences/priority-areas/ gender-and-science.

Verbeek, Peter-Paul. *Moralizing Technology: Understanding and Designing the Morality of Things*. Chicago: University of Chicago Press, 2011.

Winner, Langdon. "Do Artifacts Have Politics?" *Daedalus* 109, no. 1 (1980): 121–136.

Winner, Langdon. "Upon Opening the Black Box and Finding It Empty: Social Constructivism and the Philosophy of Technology." *Science, Technology, and Human Values* 18, no. 3 (1993): 362–378.

Woolgar, Steve. "The Turn to Technology in Social Studies of Science." *Science, Technology, and Human Values* 16, no. 1 (1991): 20–50.

Wyer, Mary, Mary Barbercheck, Donna Cookmeyer, et al., eds. *Women, Science, and Technology: A Reader in Feminist Science Studies*. 3rd ed. New York: Routledge, 2014.

Yearley, Steven. *Making Sense of Science: Understanding the Social Study of Science*. London: Sage, 2005.

FILMS

Blade Runner. Dir. Ridley Scott. 1982. Quintessential science fiction film addressing a posthuman future where advances in science and biotechnology have removed all certainty as to what is natural, what is artifice, what is human, and what not. The story traces the forlorn attempts of a deeply uncertain humanity to police the boundary between human and nonhuman replicant before all sense of it is lost.

Containment. Dir. Peter Galison and Robb Moss. 2015. Examines the attempts by various governments to contain millions of gallons of toxic radioactive waste that will remain deadly for 10,000 years. Excellent example of a noted STS scholar (Galison) working in the medium of film.

Merchants of Doubt. Dir. Robert Kenner. 2014. Adaptation of the book of the same name by Naomi Oreskes and Erik M. Conway examining the strategic use of scientists and scientific knowledge by vested interests to cast doubt on scientific claims about the health impact of tobacco, the toxic effects of flame retardants, and climate change.

CHAPTER 3

Can We Predict the Middle-Term Future?

David Orrell
Principal
Systems Forecasting, Toronto, Canada

Humans have always wanted to look into the future. Today, we have even started to imagine a posthuman future—one in which our own role in the world, and our relationship with nature and our own bodies, will have fundamentally changed. In 2012, for example, the National Intelligence Council (NIC) released a report that anticipated a transhuman future by 2030. In this vision, implants, prosthetics, and powered exoskeletons will improve on human limbs, while human brains will be enhanced with psychostimulants and brain implants. But the NIC report also notes that "the future is not set in stone, but is malleable, the result of an interplay among megatrends, game-changers and, above all, human agency" (2012, under "Dear Reader").

To understand how such factors may influence the future, one must first understand the past, and this applies to the practice of prediction itself. This chapter looks at the past, present, and future of forecasting. It describes the difficulties and failures of past efforts at prognostication; examines the challenges involved in identifying the key social, economic, and technological trends that are likely to characterize the coming decades; and surveys the fields of future studies and forecasting, showing how researchers have developed methodologies for overcoming or mitigating some of the obstacles to successful prognostication.

EARLY PREDICTORS: A BRIEF HISTORY OF PREDICTION

The profession of forecasting the future has a long past. Its mythological roots in the Western tradition go back at least to Delphi in ancient Greece. According to legend, the first oracles were supplied by the Earth goddess Gaia at a site guarded by her daughter, the snake Python. At some point, though, the oracle was taken over by Apollo, who killed Python in a battle. From then on, the oracles were read by a woman known as the Pythia (named after Python and also known as the Oracle of Delphi).

This Apollonian version of the oracle, which dates to the eighth century BCE, went on to become the most successful forecasting operation in history, lasting for almost a thousand years (Wood 2003). The predictions were often rather vague or confusing. For example, King Croesus (d. c. 546 BCE) asked whether he should go ahead with a military operation against the Persians. The oracle said that if he did, a great empire would be destroyed. He neglected to ask which one (it was his). Deliberate vagueness has remained a constant in forecasts, whether those forecasts have come from astrologers, political pundits, or even the US Federal Reserve.

Of course the oracle did not have a monopoly on predictions. The practice of astrology, for example, is just as ancient and was enhanced by the development of mathematical models of the cosmos, which enabled mathematicians and astrologers to determine and predict the locations of planets and important events such as eclipses. These models eventually led to the concept of numerical forecasting (Orrell 2007, 2012a). The Greek models of the cosmos were based on two assumptions: that Earth was at the center of the universe and that everything else moved in circles, which the Greeks believed were the most beautiful shapes because of their symmetry. These assumptions seemed reasonable for the sun or moon or stars, which seemed to rotate around Earth, but did not work so well for planets such as Mercury, which followed a more complicated path (the word *planet* comes from a Greek word meaning "wanderer"). To address this problem, mathematicians introduced smaller circles known as epicycles, which in turn rotated around larger circles. These mathematical models took on a life of their own; Greek philosopher Aristotle (384–322 BCE), for example, argued that the celestial bodies were actually encased in crystalline spheres made of ether, which rotated around Earth.

This geocentric (Earth-centered) model was later adopted by the Catholic Church and remained more or less unquestioned until the Renaissance, even though improved observations were revealing problems. The first cracks in its facade appeared in the sixteenth century when Polish astronomer Nicolaus Copernicus (1473–1543) proposed that it might be simpler, in mathematical terms at least, if Earth rotated around the sun (a heliocentric model), rather than vice versa. Danish astronomer Tycho Brahe (1546–1601) later tracked a comet and showed that it should have crashed through Aristotle's crystalline spheres, had they existed. Then, English physicist and mathematician Isaac Newton (1642–1727) combined German astronomer Johannes Kepler's (1571–1630) theory of planetary motion with Italian astronomer Galileo Galilei's (1564–1642) study of the motion of falling objects to derive his three laws of motion and the law of gravity; the circle-based model was finally replaced by one based on mechanistic equations.

Newton believed that matter was made up of "solid, massy, hard, impenetrable, movable particles" (Newton 1952, 400) governed by physical laws—that is, atoms. To understand and predict a system, it was therefore necessary only to break it down into its component parts, figure out the equations that govern them, and solve. This mechanistic approach was adopted in many areas as the template for a mathematical model, with French mathematician Pierre-Simon Laplace (1749–1827) famously postulating the following:

> An intellect which at a certain moment would know all forces that set nature in motion, and all positions of all items of which nature is composed, if this intellect were also vast enough to submit these data to analysis, it would embrace in a single formula the movements of the greatest bodies of the universe and those of the tiniest atom; for such an intellect nothing would be uncertain and the future just like the past would be present before its eyes. (1951, 4)

Even today, this mechanistic style of prediction is dominant in everything from climate modeling to economics. To predict something, the standard approach is to look at the individual components, simulate the underlying processes, express them as mathematical equations, and solve. This model acts as a kind of predictive exoskeleton for humankind, allowing us, in principle, to see far into the future. Unfortunately, its vision remains blurry at best. The reasons for this hold interesting lessons for every type of prediction, even ones in which mathematical equations are not explicitly used.

PREDICTING STORMS

One of the earliest applications of numerical forecasting was to weather prediction. The first weather models were produced by English mathematician Lewis Fry Richardson (1881–1953) during World War I (1914–1918), but the complex computations required meant that the method became feasible only in the 1950s with the development of fast computers. Much of the initial work was funded by the US military, which saw in the weather a potential weapon of war. The weather, however, turned out to be even harder to predict, let alone control, than imagined.

Although there have since been huge advances in weather observations (e.g., via satellites), computers, and the models themselves, the improvements in forecasting accuracy—while steady—have lagged well behind, especially when it comes to extreme events such as storms. This problem has traditionally been blamed on the butterfly effect: the idea, first proposed in the 1960s, that the atmosphere is an extraordinarily sensitive system (Lorenz 1963). This narrative is contradicted, however, by studies that show that weather models are not actually that sensitive to initial conditions (Orrell 2007). Instead, forecast errors appear to largely result from model error, which arises because a complex system such as the atmosphere cannot be reduced to equations.

One source of this error—and one of the most important components in the weather—is clouds. These are created when minute particles in the air, such as dust or pollen, interact with water molecules to form droplets. The process involves many properties that cannot be perfectly measured—from the scales of molecules up to the entire cloud system—and is dominated by complex networks of opposing and negative feedback loops, which are extremely hard to untangle. There is no simple equation for a cloud; and as the models become more complicated and realistic, the number of unknown parameters explodes.

The presence of model error also has implications for predictions of the medium-term climate. (If error stemmed from the magnification of uncertainty in initial conditions, this could be expected to become irrelevant over longer periods, because it is the average behavior that matters rather than the daily forecast; but if error stems from model uncertainty, then this would be expected to affect climate forecasting as well.) While models have certainly helped scientists' understanding of the climate, and there is near consensus among scientists about the importance of climate change, the stated uncertainty of their projections has changed little since the subject began to be seriously studied in the 1960s (Orrell 2007; Makridakis and Taleb 2009). Indeed, models may have served as a distraction, by turning climate change into a kind of prediction contest: we will take serious action only when everyone (not just the scientists themselves) is convinced the models are right. This is unfortunate, given that the risks posed by carbon emissions are effectively irreversible; and there are many other benefits to reducing emissions, such as lower levels of energy consumption and pollution.

Perhaps counterintuitively, forecasting often seems to increase the tolerance for risk. Nowhere is that more true than in the economy, which is discussed in the next section.

ECONOMICS

What is now called classical economics was first developed by Scottish philosopher Adam Smith (1723–1790) in the eighteenth century. His idea of the invisible hand of capitalism, which pushes prices toward a stable equilibrium, was an economic version of a law of gravity and was explicitly inspired by Newton (Greene 1961). In the late nineteenth century Smith's ideas were formalized as mathematical models by neoclassical economists. As with the Greek model of the cosmos, a number of key assumptions were made. These included the ideas that people are rational, that they act selfishly to optimize their own utility, and that prices reach a stable equilibrium. Economists soon set about building large models of the entire economy, but, as with weather models, they became workable only after the development of fast computers. These so-called general equilibrium models are still used as the basis for many policy recommendations, even though numerous studies show they have little predictive value (which is why they are not used by hedge funds to make bets on the economy, for example) (Wilmott and Orrell 2017).

Just as weather forecasters had the butterfly effect to explain poor forecasts, though, economic modelers had the efficient market hypothesis, which was developed by American economist Eugene F. Fama (1939–) in the 1960s. In "Random Walks in Stock-Market Prices" (1965), Fama argued that the market was unpredictable, not because it was wild and irrational, but because it was *too* rational. Any new information was immediately incorporated, so no one could beat the market—including economists.

As with the butterfly effect, the efficient market hypothesis provided an excuse as to why economists' predictions were so poor, while still meaning that they and their models could retain a degree of oracular authority. The profession's status was never higher than in the period between 1985 and 2005, during what American economist Ben Bernanke (1953–) and others called the Great Moderation, when it seemed that inflation and macroeconomic volatility were under control. This complacency about our ability to predict and control our economic future disappeared during the financial crisis that began in 2007, which must rank as one of the greatest predictive failures of the contemporary era. Models that were based on assumptions of stability and equilibrium were of no use at all when markets were cratering, real estate was sinking into the mud, and oil and food prices were spiking.

Like the crystalline spheres of the Greek model of the cosmos, these assumptions presented a picture of a rational, elegant, predictable universe, where everything was in its rightful place. In other words, the predictive model was telling us more about ourselves than about the actual future. New ideas and approaches are now coming from other areas, such as the life sciences, that have taken a different approach to prediction.

BIOSCIENCE

As with the other areas of prediction discussed above, biology has been inspired to a large degree by the idea of Newtonian, mechanistic models, with genes playing the role of atoms. There are regular attempts made to build detailed mechanistic models of a cell, an organ, or even the entire human body. As with weather or economic models, though, these models seem to reflect more a desire to build a convincing simulation than to excel at prediction (simpler models often work as well, or better). However, the models at least do not make assumptions such as equilibrium (in biology, the only stable systems are dead). Also, there is a more empirical strain of biological research, which has become increasingly prevalent with the growing availability of biological data.

One example, which has implications for the posthuman future, is the area of cancer research. (If we want to live forever, as some posthumanists do, we first need to beat cancer, which is caused by cells trying to live forever.) Cancer is not a single disease but a collection of thousands of related diseases with different symptoms, attributes, and prognoses. By piecing together information about a person's genetic makeup, the genetic mutations of the cancer, and other factors, it is possible to make increasingly accurate predictions about at least some of these cancers. For example, cancers carrying certain mutations respond well to tailored drugs. (Of course, this is not quite the same as predicting whether cancer will be beaten altogether.)

This approach differs from the mechanistic approach in a number of respects that could be described in terms of a scientific aesthetic (Orrell 2012b). Instead of reductionist models, the emphasis is on complexity. Instead of individual atoms, the focus is on connected networks; instead of equilibrium, unstable dynamics; instead of abstract theory, concrete data. And instead of inert, mechanistic systems, this approach involves living, organic systems.

These changes are starting to affect the approach to medium-term forecasting in other areas, such as economics, but less so in weather or climate, which are still treated as physics problems. The issue of climate is interesting because Earth is surely not just an inert rock but rather the epitome of a complex organic system. This idea was best expressed by British scientist James Lovelock (1919–) in his Gaia hypothesis, which states that Earth behaves like a living system, by, for example, regulating its own climate (Lovelock 2016). The planet, in other words, is alive, which would explain why its reactions to things such as carbon emissions are so hard to predict. The distinction between a "physical" system, such as the climate, and a human system, such as politics or the economy, is therefore less clear than is often assumed.

One vision of posthumanism takes the term literally and asks what the world would be like without humans. According to this vision, humanity has come full circle; our Apollonian predictions have failed their greatest test, and once again our future is in the hands of the planet, or Gaia, as the Greeks called it.

WHAT MODELS TELL US

As the above discussions have shown, predictive models are interesting cultural artifacts, and their assumptions often reveal as much about the mind-sets of their authors as they do about the future. When one looks back on predictions and the models that produced them, they usually appear as extrapolations of life in the period in which they were made. A good example is a set of cards, originally produced for the 1900 world's fair in Paris, that depicted artists' impressions of life in 2000. Many of the pictures, which can be seen online, involve airplanes—but they all have propellers. Needless to say, none of them shows people browsing the Internet. (Indeed, people who think we can accurately predict the future of technology should ask themselves if they ever received an advance heads-up about the Internet or knew that one of the main applications would be social networks.)

The same limitations apply to our sophisticated mathematical models. So what can be said about concrete predictions for our human (or posthuman) future? As with other predictions, will these turn out to reflect current concerns more than any future reality?

Can we predict the interplay between society, technology, biology, culture, and so on better than we can predict the economy? And can we foresee the formation of new inventions better than we can predict the formation of a cloud?

One problem is that the things we are trying to predict depend on systems, such as the climate and the economy, that, as discussed above, elude our best efforts at prediction. A world beset by rising seas or a collapsed economy (not to mention nuclear Armageddon, an asteroid strike, global pandemics, etc.) may respond in a different way to new technologies. Another problem is that, like those systems, social dynamics are governed by complex processes and nonlinear feedback loops.

For example, technological developments and social changes typically follow an S-shaped curve, with uptake initially low, then gathering steam in a quasi-exponential fashion as positive feedback takes over; finally, there is a slowing and a gradual plateau at saturation. This shape can be seen in everything from the uptake of televisions or mobile phones to social trends such as acceptance of women's rights, gay rights, drugs, abortion, and so on. However, predicting one's current position in this S-curve, or where it might end, is no easier than predicting how powerful a storm will become. At the beginning one may underestimate future growth: a classic example was "The horse is here to stay, but the automobile is only a novelty," as the president of Michigan Savings Bank advised his clients in 1903, to save them from investing in Ford Motor Co.

Once the uptake hits its quasi-exponential growth phase, though, it is just as easy to go the other way and assume it will stay exponential forever. In economics, this leads to what is called an investment bubble. In technology, it leads to the idea of a Singularity, beyond which computers will merge seamlessly with biology.

NOT-SO-SUPER FORECASTERS

Despite these limitations, the field of forecasting has seen progress since it began to be considered a separate scientific discipline in the 1980s. While mechanistic models are still the dominant technique in areas such as weather prediction, economic forecasting, transport forecasting, and so on, there are also a number of other approaches. One of the more glamorous is to use statistical methods, based sometimes on machine learning or artificial intelligence, that can tease out patterns in large amounts of data. These are used for a variety of purposes, from forecasting traffic flows or election results to assessing a person or a company's risk of default. A drawback with any statistical method, however, is that it relies on the future resembling the past. They are therefore better suited to short-term predictions.

Another option is to use expert panels to make predictions. One version known as reference class forecasting was developed in the 1970s by Daniel Kahneman and Amos Tversky in order to compensate for cognitive biases that affect forecasting (Kahneman 2011). Given a particular project—such as the building of a new transportation system—the steps are to identify a group of similar projects; establish a probability distribution for whatever is being predicted, such as usage; and finally compare the new project with the others. As with statistical methods, however, this approach relies on the existence of comparable projects.

Instead of asking experts directly, an alternative is to use prediction markets that allow people to bet on the outcomes of events such as elections. These were inspired by economic theories, such as the efficient market theory, that view markets as unbeatable prediction machines. For example, the price of a contract that pays one dollar in the event of a particular election result should converge on the expected value (so if there is a 40% probability, the contract should be worth forty cents). Although prediction markets have a relatively good track record, it seems they can be beat by teams made up of what political scientist Philip E. Tetlock (1954–) calls "superforecasters": individuals who have demonstrated a consistent ability to make good forecasts. Prediction competitions, which set specific questions such as "will event A happen in the next six months," found that teams of forecasters beat the "wisdom of the crowd" (e.g., a general poll) by 10 percent; prediction markets beat teams by about 20 percent; while teams composed of superforecasters beat prediction markets by 15 to 30 percent.

An example is the 2014 Scottish referendum on leaving the United Kingdom, in which "no" won by a large margin of 55.3 percent to 44.7 percent, even though late polls showed a dead heat. "Superforecasters aced this one," wrote Tetlock and Dan Gardner in their 2015 book *Superforecasting*, "even beating British betting markets with real money on the table" (250). They did less well on Brexit, however, giving only a 23 percent chance of Britain leaving the European Union (Kennedy 2016), as the nation voted to do in June 2016. (Of course, these are both probabilistic predictions, so perfect accuracy is not expected.)

So what is a superforecaster? According to Tetlock and Gardner, "They score higher than average on measures of intelligence and open-mindedness, although they are not off the charts. What makes them so good is less what they are than what they do—the hard work of research, the careful thought and self-criticism, the gathering and synthesizing of other perspectives, the granular judgments and relentless updating" (2015, 231). Of course, beating other forecasters on average at a specific type of prediction test is not quite the same as having amazing insights into the longer-term future; and indeed Tetlock and Gardner found that "the accuracy of expert predictions declined toward chance five years out" (244).

Finally, an entirely different approach is the method known as scenario forecasting. Its use in business was pioneered by Royal Dutch Shell, whose executives credited it with preparing the Anglo-Dutch company for the oil price shocks of the 1970s. Usually a small number of scenarios such as two to four are chosen to represent extreme cases. Such scenario methods are intended more as a way to open the mind to general possibilities than as a tool to forecast the future of a particular technology or trend but can help the user move toward desired goals by envisioning progress and identifying mileposts.

Summary

British economist John Maynard Keynes (1883–1946) once wrote that "if we speak frankly, we have to admit that our basis of knowledge for estimating the yield ten years hence of a railway, a copper mine, a textile factory, the goodwill [i.e., quantifiable worth] of a patent medicine, an Atlantic liner, a building in the City of London amounts to little and sometimes to nothing" (1935, 149–150). The same could be said of many kinds of medium-term predictions. (We may not have enough data about new projects or technologies, but we have plenty of data on the historical accuracy of forecasts.) This is probably a good thing, given that predictability tends to be rather boring.

Of course, this is not to say that we cannot make useful statements about the evolution of things such as transhumanism, let alone that we should back away from making bold bets on the future just because we do not have a perfect map. The future is not something that just happens to us deterministically; it is something that we make. In large part its course is set by the efforts of visionaries such as Henry Ford or Elon Musk (to take two examples from the transport sector). But these ideas and technologies do not emerge from nothing; they are part of a larger imaginative effort that consists of the stories and visions of futurists, science fiction writers, and so on. And even if technology has its own momentum, only human thinkers can see where it is going and spot potential dangers.

Whether their predictions are perfectly accurate in every detail is rather beside the point. Forecasting is not just about acing predictions; it is also a means of communicating ideas about the future and therefore helping to shape it. Fields such as nanotechnology or robotics hold huge promise to change everything from the materials we use to the way we treat disease to how we organize society. At a time like this of accelerating technological change, we have probably never been in greater need of futures thinkers who can help us chart a course.

However, there is perhaps one additional lesson in the history of forecasting. The great strength of science and technology is that it tends to be accumulative (although that is not quite true, because if an idea fails to meet a receptive audience it may be forgotten). But, as with all trends, just because something tends to go up, does not mean it increases to infinity. With their sophisticated electronic appendages such as sensors and satellites, as well as the fastest computers, and enormously detailed mathematical models, today's weather forecasters—viewed in unison with their technology—are already transhuman. The fact that they still cannot perfectly predict the weather, let alone use it as a weapon as their postwar funders once hoped, is a reminder that even the most ingenious technologies have their limits when faced with the messy complexity of organic systems.

Bibliography

Fama, Eugene F. "Random Walks in Stock-Market Prices." Selected Papers 16, Graduate School of Business, University of Chicago, 1965.

Greene, John C. *Darwin and the Modern World View*. Baton Rouge: Louisiana State University Press, 1961.

Iamblichus. *Life of Pythagoras*. Translated by Thomas Taylor. Hollywood, CA: Theosophical Publishing House, 1918.

Kahneman, Daniel. *Thinking, Fast and Slow*. New York: Farrar, Straus and Giroux, 2011.

Kennedy, Simon. "Superforecasters See 23% Brexit Chance as Economy Wins Out." *Bloomberg*, May 18, 2016. http://www.bloomberg.com/news/articles/2016-05-18/superforecasters-see-24-chance-of-brexit-as-economy-wins-out.

Keynes, John Maynard. *The General Theory of Employment, Interest, and Money*. London: Harcourt, Brace, 1935.

Laplace, Pierre Simon. *A Philosophical Essay on Probabilities*. Translated from the 6th French ed. by Frederick Wilson Truscott and Frederick Lincoln Emory. New York: Dover, 1951.

Lorenz, Edward N. "Deterministic Nonperiodic Flow." *Journal of the Atmospheric Sciences* 20, no. 2 (1963): 130–141.

Lovelock, James. *Gaia: A New Look at Life on Earth*. 2nd ed. Oxford: Oxford University Press, 2016.

Makridakis, Spyros, and Nassim Taleb. "Decision Making and Planning under Low Levels of Predictability." *International Journal of Forecasting* 25, no. 4 (2009): 716–733.

National Intelligence Council. *Global Trends 2030: Alternative Worlds*. Washington, DC: Author, 2012. http://www.dni.gov/files/documents/GlobalTrends_2030.pdf.

Newton, Isaac. *Opticks; or, A Treatise of the Reflections, Refractions, Inflections & Colours of Light*. Based on the 4th ed., London, 1730. New York: Dover, 1952.

Orrell, David. "Forecasting in the Past, Present, and Future: David Orrell at TEDxParkKultury." Filmed April 2012a. YouTube video, 18:30. https://youtu.be/W_ZcC-gK5d0.

Orrell, David. *The Future of Everything: The Science of Prediction*. New York: Thunder's Mouth Press, 2007.

Orrell, David. *Truth or Beauty: Science and the Quest for Order*. New Haven, CT: Yale University Press, 2012b.

Tetlock, Philip E., and Dan Gardner. *Superforecasting: The Art and Science of Prediction*. New York: Crown, 2015.

Wilmott, Paul, and David Orrell. *The Money Formula: Dodgy Finance, Pseudo Science, and How Mathematicians Took Over the Markets*. Chichester, UK: Wiley, 2017.

Wood, Michael. *The Road to Delphi: The Life and Afterlife of Oracles*. New York: Farrar, Straus and Giroux, 2003.

CHAPTER 4

Is Mind Uploading a Real Possibility?

Patrick D. Hopkins
Professor, Department of Philosophy, Millsaps College, Jackson, MS

Professor, Department of Psychiatry and Human Behavior Faculty, Center for Bioethics and Medical Humanities, University of Mississippi Medical Center

In the 1984 movie *All of Me*, Lily Tomlin plays multimillionaire Edwina Cutwater. Although she is fabulously wealthy, a lifelong illness has prevented Edwina from really living. She has spent her life moving from bed to wheelchair, surrounded by servants and doctors and always clutching her oxygen tank. Now, though, at the end of her life, she is finally excited. Edwina has secured the services of Prahka Lasa, a guru of unspecified religious affiliation who claims to be able to capture a person's soul at the moment of death and transfer it to another living body. Terry, the healthy, beautiful young daughter of Edwina's stable manager, has agreed to allow Prahka to move the dying woman's soul into her body, while at the same time releasing her own soul into the universe to begin its own new disembodied spiritual journey. As a practical matter—because transferrence of souls is not recognized by the justice system—Edwina decides to change her will so that "Terry" will inherit her estate, leaving Edwina-in-Terry's-old-body wealthy and healthy while Terry leaves for cosmic unity. Edwina's assigned lawyer, the skeptical and unfulfilled Roger, played by Steve Martin, assumes this is all a scam and Terry just a con artist. During a meeting at his firm to change the will, Roger voices his objections in front of his greedy boss, gets a lecture from Edwina about how he lacks a spiritual life, and storms out, wanting nothing to do with it all. Edwina suddenly goes into cardiac arrest, but as long planned, the yellow-robed, pointy-hatted Prahka pulls out a specially hammered brass bowl—the receptacle for capturing Edwina's soul. Near the window for sunlight, he and Terry place their fingers on the bowl, sway, chant, and flutter their eyes. When Edwina does finally die, however, she collapses and hits the table holding the soul bowl, knocking it out the window. The bowl hits Roger, on the sidewalk below, and lays him out flat. When he stands up, we find that Edwina's soul has accidentally been transferred into Roger's body, alongside his own soul. She controls the right side of the body, Roger controls the left, and—as they say—hilarity ensues.

In the 2014 movie *Transcendence*, Johnny Depp plays artificial intelligence researcher Will Caster. After being attacked in a wide-scale guerrilla operation by antitechnology militants and poisoned with a radioactive isotope, Will is told that he has only a few weeks to live. His wife and fellow scientist, Evelyn Caster, decides to use the research of another murdered engineer to try to save Will. Whereas Will and Evelyn had been trying to create artificial intelligence from scratch, the murdered scientist had managed to duplicate an existing intelligence (a monkey) in

The Soul Hovering over the Body Reluctantly Parting with Life, ***designed by William Blake and engraved by Louis Schiavonetti.*** *A common, but non-Cartesian, view of the soul as something that leaves the body at death. Notice how common it is to represent the soul as still having characteristics of a body, though—shape, clothes, eyes, limbs. Why would a soul (ghost, spirit) look like a body?* HISTORICAL PICTURE ARCHIVE/GETTY IMAGES.

a computer—transferring its mind. With the help of a friend and colleague, Max, Evelyn sets up a makeshift facility to attempt the process with Will. Surrounded with computers, power cables, code-filled screens, and medical scanners, they set to work. Inserting electrodes into Will's brain, they run multiple neural scans while he performs cognitive tasks, compiling huge amounts of data and running programs trying to analyze the electrical patterns of his brain. Eventually Will dies, and after scattering his ashes into a lake, Evelyn and Max continue to work on the pattern of Will's neural scans. After trying everything she can think of ("language processing, cryptography, coding"), Evelyn gives up. In a dramatic moment just as she is about to permanently wipe the drives of the data, a message pops up on the screen—"Is anyone there?" Evelyn is overjoyed that Will has been successfully transferred, although Max is doubtful as to what they are really dealing with.

It would be easy to think of *All of Me* as intellectually silly (it *is* a comedy), with its goofy robed mystics capturing souls in magic bowls and accidentally spilling them into unwary passersby, whereas *Transcendence* is deeper and more serious (it *is* a drama), with its engineers, neural patterns, next-generation hard drives, neuroscience, and decryption coding. Magical soul transference—silly; scientific mind transference—serious. That, however, may be too simplistic a view. Both movies treat the real person as an active object that is independent of the body who can be moved from one body to another using special knowledge, special devices, and special languages—all for the purpose of securing life after death.

All of Me employs an age-old religious idea called the transmigration of souls. *Transcendence* employs a newer, and supposedly nonreligious idea, called mind uploading. Uploading has captured the imagination of many people who would reject religious soul-talk as prescientific but who nonetheless like the idea of surviving the death of their body, getting a better body, and who think the high-tech, materialist, scientifically sophisticated concept of uploading could make the old dream of life after death come true. Nevertheless, there are problems that uploading raises—and not just technical ones or ethical ones, but metaphysical ones.

This chapter introduces and explores the big problems of uploading. First, the chapter looks at the concept of uploading and lays out the important distinction between creating an artificial mind and preserving an existing mind. Second, two major issues with uploading are examined—the problem of "ghosts" (which deals with whether uploading minds is really any different than transmigrating souls) and the problem of "branches" (which deals with whether uploading minds is really doing anything other than copying something).

THE CONCEPT OF MIND UPLOADING

Just looking up the definition of "mind uploading" and trying to go from there to determine whether it could happen will not help all that much. The reason is that the definition itself can make assumptions and use language that will mislead us into thinking something is going on that may not be going on. Look at a few quick definitions of mind uploading:

- "*Mind uploading* is a popular term for a process by which the mind, a collection of memories, personality, and attributes of a specific individual, is transferred from its original biological brain to an artificial computational substrate" (Mind Uploading 2017).
- Uploading is the transfer of the brain's mind pattern onto a different substrate (such as an advanced computer) that better facilitates said entity's ends (Kadmon 2017).
- Mind uploading, sometimes called whole-brain emulation, refers to the hypothetical transfer of a human mind to a substrate different from a biological brain, such as a detailed computer simulation of an individual human brain (Dvorsky 2009).
- Mind uploading "refers to a transfer procedure, whereby the relevant data that describes a mind's operation and information content is moved from a biological brain to some other medium" (Holmes 2016, 194).

There are many similar definitions, so clearly uploading has something to do with moving a human mind into a computer (so that the human could become immortal, survive death, resurrect, be stronger, be safer, etc.). The problem with the definitions, however, is the very concept of "moving" or "transferring." The description of "moving" a mind from one place/substrate to another assumes that a mind is the kind of thing that is capable of being moved. The problem is not about whether it would ever be technically possible but rather about whether a mind is like other objects that can be moved around (Hopkins 2012).

It is important to distinguish between the question "can a mind be created in a computer?" and the question "can a mind be moved to a computer?" The first question is a general one about the possibility of artificial intelligence. There is long history of philosophical debate about what a mind is and how it works. It is worth noting that many people who write about the "mind" have also historically used the term *soul* to mean much the same thing we seem to mean by *mind* today. At times in history, it was thought that people needed a soul to grow and digest (Aristotle 1987), but eventually, research into physiology led to the widespread belief that souls are not needed to explain metabolism and biological development. Souls were also thought necessary to explain movement, and in fact the Latin word for "soul" (*anima*) is the same root as in *animate* and *animation*—movement. Eventually, this idea too receded as research into anatomy and mechanics and engineering showed that movement could be explained just by using physics (Descartes 1989).

The soul's job was finally reduced to rationality and thinking—thus making it the equivalent of the mind for the most part. With advances in computational science and neuroscience, however, people soon started asking whether we needed a soul for thinking either—perhaps just the physical workings of the brain could explain all thought. If so, then we could presumably create artificial thinkers if we got all the physical parts working together in the right way. Just as a clock or a mousetrap or a computer can be made of any materials that have the capacity to connect up the right way (wood, metal, ceramics, carbon, silicon), minds could be made of a variety of materials too—not just biological ones (Putman 1967). This is a position called multiple realizability—the idea that properties, including mental state properties—can be realized by many different physical systems (Bickle 2016). Note that this way of looking at minds is consistent with the concept of embodied cognition. Though there are several conflicting and importantly different definitions of embodied cognition, the thing they all share is the belief that consciousness would require embodied motor and sensory systems (Wilson and Foglia 2017; Cowart 2017). While some thinkers have criticized artificial intelligence and mind uploading for ignoring the importance of embodiment (Hayles 1999), in all the cases discussed here, the consciousness would be in a body—whether a human body, a bowl, a server, or the Internet at large. While the type of body would have powerful influences on the type of thinking that could occur, multiple realizability is not about a Cartesian disembodied mind but about the multiple ways mind could be embodied.

If the idea of multiple realizability is right, then we could create a mind. A mind would be either the same thing as or emerging out of (these are various positions in the philosophy of mind literature) a complex physical system. If we made a mind that had all the same connections as a human brain, then presumably it would think and feel just as any human mind would.

This leads to the second question. Let us assume for argument's sake that minds are physical (or physically caused) and that they could be made of various substances. If we were to replace the neurons in your brain one by one with microchips, slowly, so that the connections between all your parts stayed the same, then it would seem that you would keep the same mind (Kurzweil 1999). It does not matter what the brain is made of—only that it works right. And if we could do it slowly, why not just do it quickly, all at once—figure out all the connections in your brain and set up a computer to have all the same connections. Wouldn't that just be you but now in the computer?

That is the idea of uploading. That is what all those definitions get at—uploading would be transferring your mind to a computer.

But wait. We said for the sake of argument we would accept that we could create an artificial mind. But that is about "creating" a mind, not "moving" a mind. If someone's goal in uploading is not merely to create an artificial intelligence but to turn his or herself into a nonbiologically based mind, then he or she wants the uploaded system to be his or her own *specific* mind, not just *some* mind. The second question, then, is about something different than artificial intelligence. It is about personal identity—a philosophical question with a very long history.

It turns out that answering the first question (can a mind be created without a biological brain?) does not automatically answer the second question (can a mind be moved from a biological brain to a computer?). It might be the case that all we would really be doing in so-called uploading is making a *copy* of a mind, not saving, moving, or preserving a specific mind. The copied mind would be real, genuinely intelligent, truly thinking, and have the same memories and personality (to start with) as the original mind, but it would not be the same specific mind. In the same way that photocopying a picture does not "move" the picture from one piece of paper to another, or e-mailing someone a copy of this chapter does not "move" it from one computer to another, uploading would be copying, not moving. If so, uploading will not save or preserve anyone any more than recording Elvis Presley's voice and replaying it years later "saved" Elvis's actual voice by "uploading" it to a digital device.

This problem with personal identity is a serious one for uploading. The reasons for thinking personal identity may not be preserved in uploading have to do with ghosts and branches.

THE PROBLEM WITH GHOSTS

One of the most influential thinkers in the history of modern thought was French philosopher and mathematician René Descartes (1596–1650). Among his many pursuits, he famously sat down and tried to come up with a certain foundation for knowledge, something on which to build that could not possibly be wrong (Descartes 1984). To do this, he asked himself what things he could possibly doubt—not things he thought were likely false, but only what could possibly be doubted. While he quickly found that he could doubt all his perceptions (he could be dreaming or hallucinating) and that he could even doubt mathematics (an all-powerful god could possibly force him to draw the wrong conclusions every time), the one thing he could not possibly doubt was that he was thinking. Descartes writes that even asking the question of whether you can doubt what you are thinking proves you are thinking. Even a god could not make you think you are thinking when you are not (because thinking you are thinking proves you are thinking).

While Descartes built on this idea further in many ways, one conclusion he drew had a tremendous impact. He argued that because we can be certain we are thinking every time we consider it, but because we can always be wrong about our perceptions—even perceptions of our own bodies—then what we are, the very *kind* of thing we are, fundamentally, are thinkers. He went further and said if you imagine your body disappearing, but your mind continuing to exist, it is clear that *you* still exist. If, however, you imagine your mind disappearing (no thoughts, no experiences, no consciousness) but your body continuing to exist, it is clear that *you* do not exist. This means your identity is your mind and your body is incidental. Given his thinking about physics and how matter works, he was led to conclude that our minds are made of one type of reality and material things are made of another type of reality—a position called substance dualism. The basic property of matter is that it takes up space; the basic property of mind is that it thinks. The mind and the body do interact—the mind can make the body do things, and the body brings in perceptions and information for the mind to use—but they are fundamentally different.

This position has been widely believed and still seems very common, although by the mid-twentieth century proponents of materialism had argued against it and had even famously come to mock it by calling it the "ghost in the machine" theory (Ryle [1949] 1984). There is much to be said about Descartes's ideas, but the important thing here is to recognize that he thought minds are immaterial substances that did not depend on bodies and that this position has been largely rejected by more scientifically minded people who tend to think there is only matter and energy—including most of the people who are enthusiastic about mind uploading.

Even though so many uploading enthusiasts have rejected substance dualism and the notion of an immaterial soul that can survive bodily death, however, the way they describe what is supposed to happen during uploading is remarkably reminiscent of those souls. Look back at those definitions of uploading. They, and many others like them, use the language of *moving objects* to characterize the outcome. In a previous paper looking more in depth at these metaphors, I explained it this way:

> Again and again, the term "transfer" appears. . . . But what does this language assume? First, the language uses spatial and motion metaphors. The mind is located "in" a particular place—some brain. Using technology, we will be able to move ("transfer") the mind from "within" the brain it is currently located to another location—"onto" another substrate or "into" a computer. . . . Second, thinking of the mind as "in" the brain suggests thinking of the mind in terms of substance—the mind is being treated as a thing, an object, something that is locatable and takes up specific space and that can therefore be moved from "inside" one thing to "inside" another. . . . Do transhumanists believe that the mind is an object that is literally housed "inside" a brain and through technology can be "moved" from one "receptacle" to another? Not according to the materialist and naturalist worldviews transhumanists typically espouse. Materialists and naturalists may think of the mind as produced by brain activity or perhaps as some brain activity itself . . . but they would not typically hold that the mind is a discrete object that sits in the container of the brain. In fact, if we look for positions that have held the view that minds or consciousnesses are actually substantial objects that have location and can be moved from one body to the next, we do not have far to look—only so far as popular religion. What else has been understood as a substantial consciousness "inside" a body that can be "moved" to another body or space intact?—a soul. A spirit. A ghost. Ironically, uploading enthusiasts by and large seem to be relying on a dualist theory of mind. . . . The mind is being treated just like a soul. (Hopkins 2012, 232)

Surprisingly, this way of talking about the mind is even more religious and literal than Descartes's own theories. Descartes specifically said the soul/mind does not have the properties of matter, the most fundamental of which is extension (location and volume in space). Also surprisingly, in the movies described in the introduction, *All of Me*—the "silly" one with a soul stuck in a bowl—fits many of the definitions of uploading at least as well as *Transcendence*, which was explicitly about uploading. In *All of Me*, Prahka and Terry race to get the soul in the bowl before it is lost, and in *Transcendence*, Evelyn races to send Will "out into the Internet" before the antitechnology terrorists can break into their lab and destroy him. Using this notion of "moving" and "transferring" a consciousness makes psychological sense in a way—even for a belief system that rejects soul substance. The goal of uploading is much the same as the religious goal of achieving life after death, or immortality, or incorruptibility (see, e.g., 1 Corinthians 15: 35–50). People want to continue existing. As such, personal identity is crucial, and the simplest way to conceptualize that is to think that identity is localized in a nice, discrete, mobile package that is nonetheless immaterial enough to move from one body to another. That is what souls do. And that is exactly how uploading describes consciousness.

But that is a big problem. If a person's mind—consciousness—is the pattern of his or her neural activity (or something like that), then the mind is not an object (like a bowl soul) but rather a kind of activity following a structure and set of rules—more like software running on hardware. We do not, however, literally "move" software when we download it or upload it—we just copy it and reproduce it somewhere else. And that leads to the problem with branches.

THE PROBLEM WITH BRANCHES

Before the popularity of mind uploading, another science fiction idea was often used to draw out the problems of personal identity—teletransportation. The classic image of this technology occurs in the *Star Trek* series in which people enter a transporter, are converted to energy, "beamed" somewhere, and reconstituted. While there are debates among *Star Trek* fans about how the transporters are supposed to work, the most common understanding (supported by dialogue in the show) is that the traveler's body is scanned, the matter of the person's body is converted to energy, the energy is sent somewhere, the energy is then converted back to matter, and all the matter is put back in the same order. For the most part this way of looking at teletransportation avoids many of the philosophical problems with personal identity because the matter does not change, the psychology does not change, only the location changes—which is little different than taking an airplane somewhere. However, the teletransporter has been used to explore real issues in personal identity (Parfit 1984). The following is a version that brings out the basic issues.

Imagine three options for traveling from Earth to Mars.

Option 1 is to get on a spaceship and fly from Earth to Mars. This takes about six months.

Option 2 is to use a teleporter that works the way the *Star Trek* device works—it scans your body; determines down to the minutest level the pattern, connections, and organization of all the matter; converts all the matter of your body into energy; and beams the energy to a receiver on Mars, which then converts that energy back to matter and uses your body scan to put the matter back into the same organization. This process (scan and dematerialize and send and rematerialize) takes about an hour.

Option 3 is to use a different type of teleporter—it scans your body just like the other, but as it scans, it destroys the matter of your body and then transmits a signal with the pattern information to a receiver on Mars, which uses that information to create a body from new matter already at the site into an exactly similar copy of the pattern you started with. This process (scan and destroy and send and materialize) takes about an hour.

In all three cases, the person walking around on Mars after the process will think, feel, and act just as you did on Earth.

Now the question is, do you care which option you use? Options 2 and 3 both take much less time, so they have the advantage of speed. Do you care whether you use option 2 or 3? If you do not care, then you probably think that all that matters in personal identity (whether the person on Mars is the same one that started on Earth) is psychological continuity—personality and memories. All three options maintain psychological continuity. If, however, you feel leery about option 3 and prefer option 2, then you probably think that the material matters as well—meaning both physical and psychological continuity are important.

But here is another twist from famous philosopher Derek Parfit, as he outlined in *Reasons and Persons* (1984). What if one day while using option 3, something goes wrong, and although the signal is sent to Mars and a new body is constructed as usual with your same psychology, your body on Earth is not destroyed but gets left intact? Maybe the scanning damaged it so that body will die in a few days, but meanwhile the person on Earth could chat with the person on Mars, and the person on Mars says, "Don't worry, I am you, and so I/you will go to my/your job and then soon go back home to my/your wife and kids and everything will just be fine." What would the person on Earth feel or think?

This is called the branch-line problem. Whereas options 1 and 2 create a main-line personal identity (you could draw out a single line following yourself through space and time), option 3 creates branches. In most cases, option 3 would not *feel* like a branch, because the body on Earth would be destroyed, but regardless of whether the Earth body is destroyed, the relationship between you-on-Earth and you-on-Mars is the same. Nothing about the matter or psychology of Mars-you changes as a result of Earth-you dissolving or not. The argument against the psychological continuity criterion, then, holds that because *you* cannot be *both* on Earth and Mars, all option 3 ever accomplished to start with was to kill you and make a copy. Copies, however, are not the same things as the originals; copying does not simply move the original. The resulting copies are, indeed, genuine persons, but they are *new* persons.

Mind uploading would just be a slightly new twist on this story. Instead of making up a new biological body, uploading scans you, sends the signal to a computer body, copies the neural pattern, and then the new person in a very new body wakes up. But, so the criticism goes, what wakes up in the computer is a copy of your personality, not you—just like the case of the teleporter in option 3. As the disagreement plays out in *Transcendence*, Evelyn says: "Instead of creating an artificial intelligence, he duplicated an existing one. He recorded the monkey's brain activity and uploaded his consciousness like a song, or a movie. Will's body is dying but his mind is a pattern of electrical signals that we can upload into [the computer]." Max, however, objects, saying, "At the very best we'll be making a digital approximation."

Most of the popular representations of mind uploading try to avoid the branch-line problem, not by solving it, but by ignoring it (TV Tropes 2017). By using "destructive uploading," where the original is dissolved as the person is scanned, or where the original conveniently dies, there is only the one individual left. This is usually just a plot device, though, and it allows films that deal with mind uploading to treat the mind pattern just as other films treat the transmigration of souls (one usefully watchable deliberate exception to this is the episode of *The Outer Limits*, titled "Think Like a Dinosaur" (2001)).

In *All of Me*, Edwina's soul is not released until her body physically dies, and because her unitary soul is her self, there is always only one Edwina. Whether there are such things as immaterial souls may be doubted, but souls do not get copied and do not create branch lines. In *Transcendence*, there is no moment of capture when Will dies and his mind is transferred to the computer, but he is never conscious at the same time as the artificial intelligence, so there is never a time for two entities both claiming to be Will to debate the issue. By the time he does die, the personality of the computer is the only thing claiming to be Will. That makes it easier to avoid serious argument about whether Will still exists, but it is important to realize that once the patterns of Will's brain were analyzed and functionally realized elsewhere, it would have been just as easy to have 10 or 100 or 1,000 active personalities, or to have Will not die at all. In those situations, Evelyn would likely feel much differently about asserting Will had been saved and uploaded. Yet, all 1,000 of those minds would have the same relationship to Will as the one mind does. Will's death is something that distracts us from this problem (call it the *death distraction*). Thinking of personal identity as just a pattern means that there is no limit to the number of active systems that could be formed using that pattern. The relationship between the person-who-died and the uploaded mind is exactly the same as the one between a person-who-lives and the uploaded mind. It does not make a difference to identity.

Summary

The answer to the title question of this chapter, then, is this: "it depends on what you mean by uploading." If uploading just means copying your mind and creating *another* mind exactly similar to yours somewhere else, then yes, that would be conceptually possible, although there may be technical problems in carrying it out. If, however, uploading means moving *your* mind from your body to a computer, then no, that is conceptually a real problem because *you* do not go anywhere, regardless of whether another exactly similar mind is created (or 1,000 others).

Now, there may be reasons you would want to make a computerized copy of your mind—to continue your work, to include people who think like you in the future—but if your goal is to become immortal or avoid your death, then it will not help you. Uploading mixes up traditional ideas of souls with newer ideas of computational minds. It may seem comforting at first, but transferring your mind to live on in a computer is really no more comforting than the old idea that you live on in your children or in the memories of your friends. It is a nice sentiment, but you do not literally live on. As American actor, filmmaker, and comedian Woody Allen has been quoted: "I don't want to achieve immortality through my work; I want to achieve immortality through not dying." (1993, 250).

Bibliography

Allen, Woody. *The Illustrated Woody Allen Reader*. Edited by Linda Sunshine. New York: Knopf, 1993.

Aristotle. *De Anima (On the Soul)*. Translated by Hugh Lawson-Tancred. London: Penguin Random House, 1987.

Bickle, John. "Multiple Realizability." In *The Stanford Encyclopedia of Philosophy*, edited by Edward N. Zalta. Spring 2016. https://plato.stanford.edu/archives/spr2016/entries/multiple-realizability/.

Cowart, Monica. "Embodied Cognition." In *The Internet Encyclopedia of Philosophy*. Accessed September 20, 2017. http://www.iep.utm.edu/embodcog/.

Descartes, René. *The Passions of the Soul*. Translated by Stephen Voss. Indianapolis, IN: Hackett, 1989.

Descartes, René. *The Philosophical Writings of Descartes*, Vol. 2. Translated by John Cottingham, Robert Stoothoff, and Dugald Murdoch. Cambridge: Cambridge University Press, 1984.

Dvorsky, George. "Anissimov on the Benefits of Mind Uploading." In *Sentient Developments* (blog), January 30, 2009. http://www.sentientdevelopments.com/2009/01/anissimov-on-benefits-of-mind-uploading.html.

Hayles, N. Katherine. *How We Became Posthuman: Virtual Bodies in Cybernetics, Literature, and Informatics*. Chicago: University of Chicago Press, 1999.

Holmes, Cameron. "Mind Uploading: Confronting the Privacy Challenges and Legal Ramifications of Inevitable Technological Advancements in the Context of the Fourth Amendment." *Tulane Journal of Technology and Intellectual Property* 19 (2016): 191–206.

Hopkins, Patrick D. "Why Uploading Will Not Work; or, The Ghosts Haunting Transhumanism." *International Journal of Machine Consciousness* 4, no. 1 (2012): 229–243.

Kadmon, Adam. "Mind Uploading: An Introduction." In *Creative Conscious Evolution*. Accessed September 20, 2017. http://euvolution.com/transtopia/uploading.html.

Kurzweil, Ray. *The Age of Spiritual Machines. When Computers Exceed Human Intelligence*. New York: Viking, 1999.

Mind Uploading. Accessed September 20, 2017. http://www.minduploading.org.

Parfit, Derek. *Reasons and Persons*. Oxford: Clarendon Press, 1984.

Putnam, Hilary. "Psychological Predicates." In *Art, Mind, and Religion*, edited by W. H. Capitan and D. D. Merrill. Pittsburgh: University of Pittsburgh Press, 1967

Ryle, Gilbert. *The Concept of Mind.* Chicago: University of Chicago Press, 1984. First published 1949.

TV Tropes. "Brain Uploading." Accessed September 20, 2017. http://tvtropes.org/pmwiki/pmwiki.php/Main/BrainUploading.

Wilson, Robert A. and Lucia Foglia. "Embodied Cognition." In *The Stanford Encyclopedia of Philosophy*, edited by Edward N. Zalta. Spring 2017. https://plato.stanford.edu/archives/spr2017/entries/embodied-cognition/.

FILMS AND TELEVISION

All of Me. Dir. Carl Reiner. 1984. A dying millionaire attempts to transfer her soul into a young, willing woman, but it accidentally gets put into her lawyer's body, with his soul still there.

"Think Like a Dinosaur." In *The Outer Limits.* Showtime. Dir. Jorge Montesi. June 16, 2001.

Transcendence. Dir. Wally Pfister. 2014. A scientist uploads his consciousness into a computer and onto the Internet.

CHAPTER 5

The Prehistory of the Posthuman

Diana Walsh Pasulka
Professor and Chair, Department of Philosophy and Religion
University of North Carolina Wilmington

In the fourteenth-century Italian masterpiece *The Divine Comedy*, author and poet Dante Alighieri (1265–1321) embarks on a journey through the afterlife. Guided by the deceased Roman poet Virgil, Dante tours various regions of hell and purgatory. When he finally arrives at the border between purgatory and heaven, Virgil can no longer guide him. Virgil, while a good man, is not a baptized Christian, and according to medieval Catholic theology he is not allowed to enter heaven. It is the beautiful Beatrice, whom Dante has loved from afar, who will guide him for the rest of his journey, through most of the regions of the heavenly realm. Fortified by his love of her and her otherworldly beauty, Dante must first step through a dangerous circle of fire before he enters heaven. Heaven is the apex of human existence, and as such it is like no other state of being. Humans must be transformed in order to enter its regions. Significantly, this state is so unprecedented that Dante indicates that there are no words to describe it. Therefore, he creates a new word, a neologism, *trasumanar*, which is a verb meaning "to transhumanize." The word *transhuman* was thus coined by Dante in the fourteenth century to capture the event whereby a human being becomes something entirely other-than-human, or posthuman. For Dante, it is the event of being transformed by the beatific vision of God.

The contemporary development called posthumanism is not unprecedented but is steeped in a history that is rich in religious and mythical themes and motifs. In the West, ideas and even vocabularies that currently dominate discussions of posthumanism have been presaged not just by Dante but by a variety of thinkers, including early modern authors Jesuit priest Pierre Teilhard de Chardin and media scholar Marshall McLuhan, as well as the myths of the ancient Greeks.

THE *TRANSHUMANIZED* HUMAN: THE POSTHUMAN OF THE FOURTEENTH CENTURY

Dante's new word, *to transhumanize*, describes a process. The process is no less than the perfection of the human. When Dante first sees Beatrice, she looks directly into the sun, with eyes that are capable of the feat. Beatrice's eyes are posthuman eyes. Dante remembers Beatrice in her earthly state, but now she is transformed. As she ascends through the realms of heaven, which are also planetary realms, she acquires superhuman traits. Her eyes "shine brighter than the stars," brighter than they were on Earth. Her lips are lusher and more red and plump. She is perfected.

Beatrice and Virgil, ***by Gustave Doré, 1870.*** *Beatrice is transformed into something more than human when she sees the beatific vision of the Divine. Dante coined the neologism "transhumanized" to describe this process.* THEPALMER/ GETTY IMAGES.

It is a mistake to view this transformation of the human into a posthuman as only a spiritual transformation. While it is certainly a spiritual transformation, as it is enabled by the vision of the divine, it is described in tangible ways. Beatrice becomes a perfected human with powers that she did not have previously. She is more spiritually aware, more beautiful, and more articulate than the earthly version of herself. Additionally, her beauty mesmerizes and transforms Dante, who responds to Beatrice's presence by desiring to undergo his own transformation. He desires to see the divine and to become transhumanized. Her example is infectious.

Dante's process of being transhumanized involves, among other things, an ascent through the spheres of the moon and the planets, with the goal of reaching the empyrean, which is the dwelling place of the divine far beyond Earth's solar system. The process of being transhumanized is, apparently, also a goal. For Dante, the ultimate goal of the human is to be transhumanized—to be bathed in the light of the empyrean. The view that something exists to reach a goal is called teleology. In the fourteenth century, when Dante wrote *The Divine Comedy*, the works of ancient Greek philosopher Aristotle (384–322 BCE) were in vogue, and theologians were incorporating his ideas into their own works. Aristotle's notion of teleology argues that living organisms have an intrinsic purpose. For example, a pine seedling's purpose to become a pine tree. Medieval theologians adopted the idea of teleology and applied it to their frameworks for understanding human beings and their place in the cosmos. Dante's work reflects the idea that human beings are made to become something, and for him, they are meant to be transhumanized. This state represents the perfection of the human, and in *The Divine Comedy*, it is Beatrice who exemplifies this state of being.

There are several themes that have been loosely repeated throughout the history of the movement called posthumanism that are already implicit in Dante's work. The process that leads to the perfection of the human, for which Dante coins a term, to transhumanize, is repeated in many contexts, especially in contemporary discussions of posthumanism. The assumption that humans are somehow intrinsically geared toward actions and interventions that help them reach a more efficient and beautiful state, whether erroneously held or not, is implicit in much of the work of posthumanism and transhumanism. Surprisingly, even the idea that humans will reach beyond Earth's solar system, toward Mars and beyond, guides many of the arguments in favor of biological enhancements and the move toward a postbiological human being. Billionaire Elon Musk, the founder of SpaceX (a firm specializing in the production and launching of rockets and spacecraft), has called for the imminent merging of humans and machines as a way to survive an environmental cataclysm that he believes will ultimately render Earth unfit for human life. He is also actively urging humans to travel to Mars as a way to prolong the existence of the species *Homo sapiens*. For Dante, the divine resided beyond Earth's solar system and at the apex of the heavens and the cosmos. For many advocates of the posthuman, space and the planets beyond Earth represent the salvation and the destination of the human being.

EARLY MODERN PRECEDENTS: INTELLIGENT TOOLS—HUMANS EXTENDING BEYOND THEIR BODIES

Aristotle's influence extended beyond medieval theology and its adoption of his idea of teleology. He also discussed human tools and their capabilities of extending the human, and more specifically a proto-notion of extended mind, themes of which appear in works of many of the authors of early modern literature. Noting that not all Athenians, members of his Greek city Athens, approved of slavery, Aristotle suggested that intelligent tools would be an improvement over living slaves, ethically and pragmatically. In an extensive survey of automatons and human machines of the early modern era, scholar Kevin LaGrandeur reveals that Aristotle's notion of the intelligent tool as a proxy for a human slave blurred the boundaries between the human and the machine.

> It is important to note that the boundaries between tools and slaves are implicitly blurred by Aristotle's very idea of intelligent tools, and these boundaries, as well as those among tools, slaves, and the master, become even hazier when one considers the rest of his discussion of slaves and tools. For he considers tools and slaves to be merely different types of instruments: "some are living, others lifeless; in the rudder, the pilot of a ship has a lifeless, in the lookout man, a living instrument; for in the arts the servant is a kind of instrument . . . the servant is himself an instrument for instruments" (1253b29–34 [*Politics*, bk. 1, pt. 4]).
>
> Moreover, the master is implicated in this system of instruments as well, since all instruments, animate or inanimate, human or not, are actually prosthetic extensions of the master, and as such part of a master-centered network, "a living but separated part of his bodily frame" (1255b11–12 [*Politics*, bk. 1, pt. 6]). (2013, 9–10)

The idea of a living and intelligent, yet separate, part of the human body is repeated, according to LaGrandeur, in several works by early modern authors, and most notably in English dramatist William Shakespeare's (1564–1616) play *The Tempest*. *The Tempest* tells the story of the deposed duke of Milan, Prospero, who, with his beautiful daughter Miranda, has been exiled on an island owing to the ill intentions of his brother, the current duke of Milan. In the play, Prospero uses magic to control the actions of several other beings in order to enact his will and intentions. LaGrandeur argues that Prospero embodies a model of an extended network of tools that are spirits, people, and other beings, which are all extensions of his intentions and will. In this sense, Prospero's servants act in unison as a conglomerate of servitude, as tools that extend the human. LaGrandeur notes the Aristotelian precedent for this idea.

> Aristotle, in his *Politics*, elides the notions of slaves and artificial slaves, primarily by focusing on all potential providers of intelligent labor (humans, horses, even tools) as objectified sets of functions—in other words, he elides the ideas of things, animals, and people. Thus he provides a precedent for thinking of a conglomeration of individual entities as one tool, for Aristotle makes it clear that he considers slaves to be not only tools, but also, together or separately, to be extensions of the master's body. (2013, 106)

LaGrandeur notes that the island itself acts as a "supraorganism" in the play and that characters such as the half-human Caliban and the spirit Ariel exemplify tools that Prospero uses to effect a corporate conglomerate that allows him to control nature, the weather, and the inhabitants of the island. Caliban is indigenous to the island, and it is not a coincidence that he is portrayed as a type of natural slave. Debates about slavery and the European treatment of indigenous Americans were circulating during the time that Shakespeare wrote *The Tempest*, and LaGrandeur writes that Shakespeare had read a chronicle of Spanish colonization and was most likely aware of these debates. Not only does the character Caliban help Prospero survive, but he is his slave. Ariel, another of Prospero's servants, is a spirit that Prospero controls and who effects changes in the weather, as well as in people's minds and behaviors, and often needs the help of other spirits or demons to effect these changes. In this sense, LaGrandeur argues that Ariel is an important cog in a networked system that includes the human Prospero, other spirits and energies, and nature. Prospero uses Ariel for his intentions yet often experiences these secondhand, as Ariel functions as a proxy, a stand-in, for Prospero. In this way, Prospero's servants act as a form of an extended network of tools or, as LaGrandeur notes, an "aggregate servant network" (2013, 116).

TEILHARD DE CHARDIN: THE INTERNET OF HUMAN MINDS

Dante infused his European Catholic poetry with the science of his day. *The Divine Comedy* reflected not only the medieval appropriations of the philosophy of Aristotle but also the cosmology of Egyptian astronomer Ptolemy (c. 100–c. 170 CE). Hundreds of years later, another European Catholic, French Jesuit priest Pierre Teilhard de Chardin (1881–1955) incorporated ideas of evolution, biology, and theology to propose that the human being's ultimate destiny was to merge with the divine in a spectacular moment he called the "omega point." Early in the twentieth century Teilhard de Chardin envisaged a merging of humans with technology that would result in a divine revelation. It would move beyond the duality of the physical and the spiritual and be something else entirely, a new human fused with a new universe. He believed this was the Parousia, which in some Christian theologies is the return of Christ to Earth. Teilhard de Chardin's ideas were unconventional yet visionary, and the Catholic Church issued a *monitum*, or official reprimand, and warned that his works were controversial. Yet, prominent theologians and scholars defended his ideas, and his work influences many advocates of transhumanism in the twenty-first century, such as Ray Kurzweil and Nick Bostrom.

Teilhard de Chardin, a paleontologist, geologist, anthropologist, Jesuit priest, and professor, was an avid collector of fossils and eagerly studied the fossil record. He believed in evolution. He also believed in God. He never held the position that science disproved theology or that theology trumped science. He believed that they supported each other in revealing the inner workings and beauty of the universe. His vision of biological evolution incorporated technology and religion. The themes that occur in Dante's *Divine Comedy* are also manifest in Teilhard de Chardin's work—namely, the teleology of the human and the extraterrestrial theme of moving beyond the human's earthly abode. These themes are articulated in his omega point theory.

Teilhard de Chardin assumed a telos (ultimate end point) in which humans are meant to progress toward perfection, which he called the omega point. He used the term *omega* to describe the ultimate fate of the human being and his or her merging with the divine. *Omega* references the New Testament passage in the book of Revelation where Jesus refers to himself as the Alpha and the Omega, which are the Greek words for "beginning" and "end," respectively. The teleology of the human, for Teilhard de Chardin, involves a merger at the end point of human history, which is that of the human with Christ. Teilhard de Chardin's teleology was much different than Dante's. Whereas Dante's vision was a vision of the afterlife, or the perfection of the human post-death, Teilhard de Chardin's was not. His idea of the perfection of the human is physical and immanent. He believed it would happen soon, and it would be a fusion of the human, the divine, and the technological. In fact, Teilhard de Chardin saw technology as the catalyst and necessary component for the transformation of the human into something divine.

The omega point, in Teilhard de Chardin's view, is a collective process of perfection. Instead of individual human beings coming into their own, unique perfection, Teilhard de Chardin envisioned a technological sphere in which human minds joined as a collective identity. He called this the *noosphere*. The noosphere is a continuation of the biosphere, and it includes all the products and networks produced by human minds. It is cultural and technological. Teilhard de Chardin saw the noosphere as a sphere that was not necessarily biological, but that was made possible by the biosphere but exceeded it in complexity. It is human consciousness, technology, and culture. It is often referred to as global consciousness. Yet it is also a physical sphere that encircles Earth. He believed that increasing developments in technology would also increase the emergence and complexity of the noosphere and that at some point, in the near future, humans would reach the omega point, in which the noosphere would merge with Christ.

Like Dante, Teilhard de Chardin incorporated the cosmos into his teleology of human perfection. Humans, through the noosphere, would reach out to Christ, but Christ, as a cosmic presence, would embrace humans too. This culmination is the omega point. For Teilhard de Chardin, the body of Christ is literally the universe and extends throughout the cosmos. The cosmic Christ is fulfilled in human technological and biological development in that through the noosphere humans will recognize themselves as merged with Christ. This process is called noogenesis, which fuels the creation of the noosphere. It is the acceleration of technological development that connects all human minds and eventually connects with the divine.

Teilhard de Chardin's noosphere is often viewed as a philosophical or theological precedent for the idea of the Internet. The Internet, within this conception of it, is thought to be the physical infrastructure that will eventually house human consciousness. Creative thinkers and technologists, such as Mark Zuckerberg, the founder of Facebook, and Musk, seem to echo this sentiment. They are operationalizing mind-machine technologies that they hope will foster this interface. During a 2017 talk reported by CNBC's Arjun Kharpal, Musk stated, "Over time I think we will probably see a closer merger of biological intelligence and digital intelligence," and, "It's mostly about the bandwidth, the speed of the connection between your brain and the digital version of yourself, particularly output." Bostrom, a philosopher and advocate for transhumanism, cites Teilhard de Chardin as the originator of a "lineage" of the notion of the Singularity, an idea made popular by Kurzweil's book *The Singularity Is Near* (2005), in which he argued that it was inevitable that humans and technology would merge together. Although Kurzweil, Musk, and Bostrom do not share Teilhard de Chardin's theological commitments, they do share a common notion that humans will merge with machines. While Dante's posthuman remains an individual, Teilhard de Chardin's posthuman appears to merge with the collective consciousness of humanity and its technologies.

MCLUHAN: A GLOBAL VILLAGE CONNECTED BY A WEB OF ELECTRIC MEDIA

Twenty-five years after Teilhard de Chardin proposed his network of global consciousness, Canadian scholar Marshall McLuhan (1911–1980) emerged as an articulate spokesperson for a new understanding of the human and its relationship to technology. Dante and Teilhard de Chardin coined neologisms to describe their ideas of the transformation of the human. Their own language lacked words to describe the processes that they knew were taking place. Similarly, McLuhan created a new vocabulary to describe these transformations that were compelling a shift in the very definition of the human. Some of his words and expressions are still in use—such as the global village, "surfing" the waves of technology, and the expression "the medium is the message." His influence in the 1960s was enormous, and as technology progresses, his work, in retrospect, is undeniably predictive. His ideas and theories describe the transformation of the human and human society through technological engagement. For McLuhan, this is no less than a new epoch of human evolution.

A central thesis of McLuhan's work is that technology is not separate from the human being but is another human sense, just as taste, smell, and sight are human senses. Media and technologies are extensions of the human, and therefore should not be thought of as separate. This idea, as well as many other of McLuhan's ideas, have been adapted to film. Canadian filmmaker David Cronenberg has made several movies that are based on his work with McLuhan. His movie *Existenz* (1999) reflects McLuhan's idea that technologies are just extensions of the human. In the movie, video-game creator Allegra Geller invites a focus group to test her new game—Existenz. To enter the game, players must "port" themselves into a game pod, which looks like a living human placenta and is even made from the organs of reptiles. The bioport, or entrance to the game, connects into the spine and has to be surgically opened. This graphic depiction of how players enter the game reflects McLuhan's idea that, in a very real sense, technologies are quasi-biological extensions of the human, and, at some point, they actually are integrated biologically into human experience and biology.

Contemporary theorists confirm this aspect of McLuhan's theory. N. Katherine Hayles, a scholar of media and technology, writes about the physiological effects of technologies. Her research indicates that at the level of interacting with computers and media, human bodies undergo shifts, the most significant of which involves changes in neural circuitry. Just as learning to read trains one's brain, so does the act of engaging a computer. Hayles argues that

> the more one works with digital technologies, the more one comes to appreciate the capacity of networked and programmable machines to carry out sophisticated cognitive tasks, and the more the keyboard comes to seem an extension of one's thoughts rather than an external device on which one types. Embodiment then takes the form of extended cognition, in which human agency and thought are enmeshed within larger networks that extend beyond the desktop computer into the environment. (2012, 3)

For Hayles, human embodiment, as described here, is technological. The computer is an extension of cognition, just as McLuhan described. Technology extends the boundaries of the human.

The biological references to technology in McLuhan's work are not metaphoric. In the film *Existenz*, the video game is alive. It lives in the minds and bodies of its players. It even effaces the boundaries between it and "real life" as the film characters question whether they are in the game or outside it. The game ushers in a new state of being human that is fraught with uncertainty as the boundaries of the human are completely reconfigured. For McLuhan, who was writing in the 1950s and 1960s, this new state of being human was here already. He wrote, "Today, after more than a century of electric technology, we have extended our central nervous system itself in a global embrace, abolishing both space and time as far as our planet is concerned" (McLuhan [1964] 2005, 3).

Cronenberg's movie *Videodrome* (1983) explores another facet of McLuhan's theory, which has to do with technology and human evolution. The movie is about a television station, Videodrome, which its creators promise will facilitate human evolution through its viewership. The creator of the station is Dr. O'Blivion, a character based on McLuhan. Dr. O'Blivion runs a charitable organization called the Cathode Ray Mission, which provides food and shelter to the homeless if they watch television for long periods. The assumption is that the television station will transform them into a different type of human. As homeless people, they do not have access to technology or television, and the Cathode Ray Mission helps them "catch up" with more affluent members of society, who are, presumably, already being transformed through watching the station. Although McLuhan was careful not to weigh in on the moral and ethical dimensions of media technologies, those he influenced, such as Cronenberg, explore these implications in their own work.

McLuhan believed that technologies shape human beings in significant ways. He developed a time line of human history that reflects eras of technological development. In *The Gutenberg Galaxy* (1962), McLuhan outlines his time line of human transformation, which is dictated by shifts in media technologies. The epochs of human development include the oral tribe culture, the manuscript culture, and the Gutenberg galaxy, in which the printing press allowed for the mass dissemination of books, which, according to McLuhan, effected a change in human consciousness. Currently, societies are dominated by the electronic age. McLuhan wrote that some of the changes in human consciousness in the electronic age involve a shift in identity. Humans will be connected by electronic webs that will cause them to experience time as simultaneous and space as almost nonexistent. He argues that this brings about a tribal identity, rather than a sense of distinct individuality. McLuhan coined the term *global village* to reflect this development.

Among the thinkers already discussed—Dante, Teilhard de Chardin, and McLuhan—it is significant that each felt the need to create new words and a new vocabulary to explain these developments. *Transhumanized*, *noosphere*, *noogenesis*, and *global village*—each of these terms describes the transformation of the human being from one state to another. The latter two thinkers focus on how technology shifts human experience and even the definition of what it means to be human. Even contemporary thinkers, such as Hayles, use neologisms to describe the relationship humans have to their technologies. Although she did not coin the term *technogenesis*, Hayles uses it to describe the concept, articulated by McLuhan, that humans coevolve with technologies. Humans and technologies are not separate but mutually constructive. Hayles furthers this idea by using the concept of technogenesis to describe the specific ways in which technologies form human beings, cognitively and biologically. Unlike Teilhard de Chardin, Hayles does not necessarily view the process of technogenesis as progressive. In other words, she does not assume that the intrinsic engagement between humans and technologies is automatically good. She contends that

> contemporary technogenesis, like evolution in general, is not about progress. That is, it offers no guarantees that the dynamic transformations taking place between humans and technics are moving in a positive direction. Rather, contemporary technogenesis is about adaptation, the fit between organisms and their environments, recognizing that both sides of the engagement (humans and technologies) are undergoing coordinated transformations. (2012, 81)

The creation of new vocabularies suggests rapid societal transformation into uncharted cultural territory. Teilhard de Chardin's new words combine theological and technological concepts, whereas those of Hayles are borrowed from the language of science and genetics. While these vocabularies are new, the ideas they describe, which include the transformation of humans into something posthuman, have been around for thousands of years. In a 2016 article, Adrienne Mayor, a historian of ancient science, reminds readers that the ancient Greeks developed their own stories and vocabularies of *bio-techne*, or biotechnologies, to describe many varieties of posthumans.

Describing robot servants, half-human soldiers, and other types of posthumans, Mayor writes that the ancient Greek myths, such as those of Hercules, Jason and the Argonauts, Medusa, and Pandora, all explore "the basic question of the boundaries between human and machine" (2016). What she finds is that many of these myths and stories reflect the desire and an accompanying ambivalence for the quest for immortality, claiming that "the most searching ancient myths ask whether immortality frees one from suffering and grief. In the Epic of Gilgamesh, for example, the eponymous hero of the Mesopotamian poem desires immortality. But if Gilgamesh were to gain everlasting life, he would spend it eternally mourning the loss of his companion Enkidu."

The theme of immortality dominates contemporary discussions of the posthuman. Living advocates of the posthuman, such as Kurzweil and Russian entrepreneur Dmitry Itskov, actively seek vitamin therapies and therapeutic technologies so they may extend their lives long enough to reach what they believe to be inevitable: the Singularity, in which humans and technology converge, allowing humans to achieve immortality. Deceased advocates for the posthuman, such as FM-2030, have undergone cryonic preservation in hopes of being revived if and when humans do figure out how to achieve the human-machine merger that promises immortality. Yet, as Mayor (2016) notes, the ancient Greeks were not completely sold on the "good" of immortality. In many of the Greek myths humans choose mortality over immortality. Mayor writes that "artificial, undying existence might tantalise but can it ever be magnificent or noble?" Additionally, many of the myths that portray humans who achieve immortality do not end well. When the beautiful goddess Eos falls in love with a mortal man, Tithonus, she asks the gods to grant him immortality. They do. However, even though Tithonus becomes immortal, he still ages as if he were a mortal. In the myth, he becomes old, and Eos ceases to be enchanted with him. When Tithonus finally becomes so old that he is unable to move around, Eos locks him up behind bars, and there he lives out his immortality, decrepit and loathed.

Summary

The themes and motifs, and even the vocabulary, that dominate contemporary discussions of posthumanism can be found in the poetry, myths, and theories of the past. The belief that humans have a telos—that they are meant to achieve an end—can be found in the work of Dante and in the myths of the ancient Greeks, while also being fully elaborated in the works of scholars such as Teilhard de Chardin. Surprisingly, even the question of the boundaries between humans and machines can be found as far back as the myths of the ancient Greeks. Often, these themes are linked to immortality. Dante and Teilhard de Chardin understood human development and evolution as reaching toward the stars, literally, and advocates of the posthuman today, such as Musk, believe that the human's ultimate destination is beyond Earth's solar system. For Dante, the empyrean, far above Earth's own moon and the planets of the solar system is the ultimate goal of the human. In order to make it there, however, humans must change. They must become posthuman.

Bibliography

Aristotle. *Aristotle: Selections*. Translated by Terence Irwin and Gail Fine. Indianapolis, IN: Hackett, 1995.

Aristotle. *The Complete Works*. Edited by Jonathan Barnes. 2 vols. Princeton, NJ: Princeton University Press, 1995.

Bostrom, Nick. "A History of Transhumanist Thought." *Journal of Evolution and Technology* 14, no. 1 (2005): 1–25. http://www.jetpress.org/volume14/bostrom.html.

Clark, Andy, and David Chalmers. "The Extended Mind." *Analysis* 58, no. 1 (1998): 7–19.

Dante Alighieri. *The Divine Comedy*. Translated by John Ciardi. New York: New American Library, 2003.

Hayles, N. Katherine. *How We Became Posthuman: Virtual Bodies in Cybernetics, Literature, and Informatics*. Chicago: University of Chicago Press, 1999.

Hayles, N. Katherine. *How We Think: Digital Media and Contemporary Technogenesis*. Chicago University of Chicago Press, 2012.

Kharpal, Arjun. "Elon Musk: Humans Must Merge with Machines or Become Irrelevant in AI Age." CNBC, February 13, 2017. http://www.cnbc.com/2017/02/13/elon-musk-humans-merge-machines-cyborg-artificial-intelligence-robots.html.

Kurzweil, Ray. *The Singularity Is Near: When Humans Transcend Biology*. New York: Viking, 2005.

LaGrandeur, Kevin. *Androids and Intelligent Networks in Early Modern Literature and Culture: Artificial Slaves*. New York: Routledge, 2013.

Mayor, Adrienne. "*Bio-techne*: Half-Human Soldiers, Robot Servants, and Eagle Drones—the Greeks Got There First. Could an AI Learn from Their Stories?" *Aeon*, May 16, 2016. https://aeon.co/essays/replicants-and-robots-what-can-the-ancient-greeks-teach-us.

McLuhan, Marshall. *The Gutenberg Galaxy: The Making of Typographic Man*. Toronto: University of Toronto Press, 1962.

McLuhan, Marshall. *Understanding Media: The Extensions of Man*. 2nd edition. Abingdon: Routledge, 2005. First published 1964.

Teilhard de Chardin, Pierre. *The Phenomenon of Man*. Translated by Bernard Wall. New York: Harper, 1959.

FILMS AND TELEVISION

Black Mirror. "White Christmas." Season 2, episode 4, 2014. Dir. Carl Tibbetts.

Existenz. Dir. David Cronenberg. 1999.

The Matrix. Dir. Lana Wachowski and Lilly Wachowski. 1999.

Videodrome. Dir. David Cronenberg. 1983.

Science and Technology

CHAPTER 6

Pharmaceuticals

Nancy D. Campbell
Professor, Department of Science and Technology Studies
Rensselaer Polytechnic Institute, Troy, NY

Pharmaceutical drugs range from dangerous to beneficial to benign, depending on their contexts of use and the populations using them. Regulated throughout their product life cycle, biologically active compounds are governed according to dosage and class, therapeutic uses or lack thereof, and knowledge of side-effect profiles and adverse events. Pharmaceutical enhancement figures prominently in contemporary debates over the posthuman condition. For instance, in *Our Posthuman Future*, Francis Fukuyama claims that "virtually everything we can anticipate being able to do through genetic engineering we will most likely be able to do much sooner through neuropharmacology" (2002, 173).

Using Ritalin and Prozac as examples of the human eagerness to alter the neurobiological substrates that shape emotional energy, Fukuyama argues that political and economic forces that will operate in the posthuman future to "expand the therapeutic realm" by "medicaliz[ing] everything" are already on display in the domain of neuropharmacology (Fukuyama 2002, 53). Medicalization is the social process by which previously nonmedical conditions are redefined as medical problems that can be treated by new pharmaceutical drugs or therapeutic regimens involving the medical profession. Centering the putative "desire on the part of ordinary people to medicalize as much of their behavior as possible," Fukuyama predicts that this process will reduce personal responsibility and maximize profit for pharmaceutical companies by expanding the number of conditions for which pharmaceutical biotechnologies are prescribed and used (Fukuyama 2002, 52–53). Exhibiting great faith in the ability of science to identify and exploit "actual molecular pathways" to modify human behavior and mood, Fukuyama envisions a posthuman future in which human emotions and interactions are underpinned by pharmaceutical drugs.

Pharmaceuticals are portrayed both as essential medicines and as elective tools that may be responsibly used for self-making and the optimization of human adaptability (Dubljević 2013; Greely et al. 2008). They are also recognized as risky, toxic, and potentially dangerous commodities that should be carefully regulated (for websites that seek to inform consumers of comparative risk profiles, see Public Citizen 2017 and RxISK 2017). Pharmacological optimists envision pharmaceuticals that solve individual and social problems by bolstering physical and mental capacities. By contrast, "bad drugs" abound in the dystopian camp of pharmacological skeptics, who are pessimistic about the limited state of knowledge about human cognition, mood, energy, and motivation.

While the posthumanist literature is pervaded by pharmacological optimism, it is counterbalanced by myriad dystopian representations of such drugs and the industry that supplies them. Many films portray the dystopian effects of drugs and the negative unintended consequences of their widespread use. (See the bibliography for further examples of films that dramatize pharmaceutical use.) *Requiem for a Dream* (2000), for example, conflates the use of legal amphetamines by a middle-aged mother with her son's use of illegal heroin. Films based on drug memoirs often cast the protagonist's relationship with pharmaceutical drugs as a romance gone bad, as in *Prozac Nation* (2001), based on Elizabeth Wurtzel's 1994 memoir, or *Girl, Interrupted* (1999), based on Susanna Kaysen's 1993 memoir. Popular culture typically depicts the pharmaceutical industry negatively, particularly where globalization of the clinical trials industry is concerned. Set in the British postcolonial context of Kenya, *The Constant Gardener* (2005), a film based on John le Carré's 2001 novel, is about a clinical trial gone wrong, serving as an allegory for the impersonal forces of pharmaceutical colonialism in Africa. Pharmaceutical enhancement in posthuman or cyberpunk films is relatively normalized; *Limitless* (2011) and *Lucy* (2014) broach the topic of enhancement directly, as the fictional nootropic drugs (smart drugs) at the center of these films are said to allow full usage of the human brain.

Ambivalence about the social benefit of widespread pharmaceutical use has been expressed by scientific experts since the science of neuropharmacology emerged in the 1950s. Ambivalence remains deeply inscribed in debates over whether or not pharmaceuticals should be used to enhance human capacities for strength, endurance, wakefulness, empathy, and smartness, and whether they will become accepted in a posthuman future. This chapter considers the spectrum of answers to these questions across the seventy years of experience with pharmaceutical enhancement. Advocates of responsible use occupy the fulcrum between pharmacological optimists and pharmacological pessimists.

THE MID-TWENTIETH-CENTURY EMERGENCE OF NEUROPHARMACOLOGY

Neuropharmacology emerged as a "frontier science" in the 1950s, first flourishing among those studying the effects of drugs on psychiatric patients (Elkes 1970). Biological psychiatry gradually displaced psychoanalysis and psychodynamic psychiatry as the pharmaceutical industry migrated away from the problems of institutionalized mental patients and toward the vastly more stable market of legitimate consumers troubled by everyday anxieties. The emerging science presented its technological products as transcending the global political conflicts of the Cold War era. Initial fanfare presented "wonder drugs" as technologically advanced solutions to the problems of modern, liberal, humanist, capitalist democracy.

Pharmacological optimism emerged in the formative moments of mid-twentieth-century modernity, widely promulgated by commentators such as Aldous Huxley, author of a 1957 *Saturday Evening Post* article titled "Drugs That Shape Men's Minds." Cautioning against the "old bad habits" of alcohol and tobacco, Huxley advocated developing "new and less harmful drugs" better suited to democracy. Social utopian goals such as lessened fatigue, increased energy, clear thinking, doing without sleep, assuring mother love and empathy, abolishing melancholy, and preventing war were thought to be tractable to legal pharmaceuticals ranging from amphetamines to benzodiazepines to major and minor tranquilizers and sedative-hypnotics. In shifting away from pathologically troubled individuals, the new neuropharmacology reinterpreted the workings of the normal human brain as a series of biochemically directed behaviors.

Neuropharmacology offered practical tools for shaping behavior during an era in which biological psychiatry was moving toward neurochemical accounts of conditions once considered purely mental. The idea that certain drugs made people "better than well" surfaced within those working with the early antidepressants in the 1950s (Healy 2002) but was not widely shared in part because use of the early antidepressants was restricted to major depression. An expert class of biological psychiatrists and pharmacologists enthusiastically undertook to make available "wonder drugs" to regulate the human organism, expanding industrial and academic research departments and laboratories of pharmacology. Modern neuropharmacology helped forge links between individual regulation and social regulation, envisioning that the self-mastery required of modern democratic subjects would benefit the governance of the society of which they were part. This ambitious science saw drugs as "harbingers of a new era" (Gerard 1957). In this era, each individual would be aligned with the pace of accelerating social change in postwar modernity. Hope arose that legal pharmaceuticals could aid individuals in bringing about such an ideal state of adjustment between self and society.

PHARMACEUTICAL REGULATION AND THE TRANQUILIZER ERA

Pharmaceutical companies sought to expand to mass markets starting in 1955 with the launch of the first blockbuster drug, Miltown, a so-called minor tranquilizer marketed to the "worried well" (Tone 2009). Word about the ability of Miltown to stabilize mood reached the public via physicians and the news media, and a "tranquilizer boom" ensued. Skepticism later emerged in the consumer, patients' rights, and women's health movements of the 1960s and 1970s. Groups such as Ralph Nader's (1934–) Public Citizen (founded in 1971 to advocate for citizens' health and other interests) successfully pushed for further scrutiny of drugs, including Librium and Valium (Tone 2009). The latter two drugs were the first benzodiazepines on the market as antianxiety drugs, joining a cascade of barbiturates known as dangerous drugs (Herzberg 2009).

The cultural elites of the 1950s envisioned what might today be called "smart drugs" or "nootropics," but the regulatory regime required that drugs be approved to treat specific conditions rather than enhance general traits. Drugs affecting attention or focus; energy levels; memory, sleep, or other cognitive capacities; and mood, emotion, or subjective states had to be approved, marketed, and prescribed for some pathological condition. Yet it became increasingly difficult to distinguish between pharmaceuticals that restored normal functioning in the face of a pathological condition and those that enhanced functioning beyond so-called normal levels.

Concerns about pharmaceutical price gouging, unethical actions of pharmaceutical companies, and ineffective scrutiny of drugs prior to market arose at the end of the 1950s. Fueled by the response to the effects of thalidomide in Europe, the US Congress in 1962 amended the Federal Food, Drug, and Cosmetic Act of 1938. The new legislation mandated a three-stage process of collection of preclinical (animal) data and clinical trials on both healthy and sick human beings. A large network of laboratories and clinical facilities developed to produce data for the approval of drugs by the Food and Drug Administration

(FDA). The 1962 amendments emphasized specificity and thus "put a premium on categorical rather than dimensional models of disease" (Healy 2002, 367). This regulatory distinction privileged drugs acting specifically on a defined disease category that a person could be diagnosed as having or not having. By contrast, drugs that exert their effects along a dimensional continuum of disease indications such as depression or anxiety became harder to move through the regulatory process.

Regulatory problems preoccupied the scientific disciplines concerned with pharmaceuticals—neuropharmacology, psychopharmacology, and the neurosciences broadly construed—as well as the clinical and philosophical domains. Law, policy, and ethics limited the possibilities for companies to make general claims concerning a drug's capacity for enhancement of existing features or traits. Yet such laws did not prevent individual patients or doctors from noticing enhancement effects and opting for them, nor do they constrain off-label prescription and use.

COSMETIC PSYCHOPHARMACOLOGY AND NEUROLOGY

The "pill for every ill" culture that emerged in response to the stresses and anxieties of the 1950s differed from previous therapies that relegated drugs to the sidelines and were centered on self-disclosure through talk. Today's discourse on posthuman pharmaceutical enhancement is based on the ascension of the basic science of neuropharmacology and its application in biological psychiatry. Structural changes in the health-care industry have rendered talk therapy largely obsolete, thereby elevating pharmaceuticals. However, the women's health movement and the consumer and patients' rights movements dampened appetites for these drugs until the advent of mass-market antidepressants (Herzberg 2009).

The term *cosmetic pharmacology* was introduced by psychiatrist Peter Kramer (1993) to characterize patients using prescription pharmaceuticals to become "better than well." The idea that providers and patients might optimize themselves via elective prescription pharmaceuticals to enhance existing attributes, skills, and abilities shades rapidly into nonmedical use. The use of pharmaceutical drugs to attain social goals, such as putting an end to social awkwardness or shyness, or enabling sufferers of obsessive-compulsive disorder to end projects when they are "good enough," pervades some professions (Greely et al. 2008). Off-label use, which is legal in the United States as long as pharmaceutical manufacturers do not advertise their product for uses that have not yet been approved by the FDA, also plays a role in cosmetic pharmacology. Consumers might elect to remain on drugs, such as beta-blockers, because they gain poise, confidence, or equilibrium without which they appear anxious, nervous, or shy. *Cosmetic neurology*, a term that has entered the lexicon more recently to indicate drugs that improve mood and mental activity, has been used to draw an analogy with cosmetic surgery (Chatterjee 2004).

Cosmetic surgical procedures or the pharmaceutical regimens of cosmetic neurology are not medically necessary but are part of large-scale attempts not only to improve quality of life but to exceed perceived human limitations. In *Drugs for Life* (2012), cultural anthropologist Joseph Dumit examines how "pharmaceutical lifestyles" depicted in direct-to-consumer advertising, which is legal in the United States and New Zealand, factor into current notions of both public and private health in developed countries. A similar phenomenon is occurring in Brazil, where the HIV/AIDS movements forged an equation between access to pharmaceutical drugs and health (Biehl 2007). João Biehl documents how Brazil, a country in which there is a constitutional right to health, has become a "profitable platform of global medicine" in ways that translate into almost 50 percent of adults using pharmaceuticals daily (2016, 256). As patients everywhere become expert at navigating pharmaceutical terrain, they must engage daily in a form of personal risk assessment and management that has become central to the new public health (Petersen and Lupton 1996). Where health is viewed as a commodity, pharmaceutical provision and consumption are one of the main ways to maintain health.

PHARMACEUTICALIZATION OF LIFE AND HEALTH

The pharmaceuticalization of health has made the private pharmaceutical industry central to public health despite the difficulties of inducing companies to act in the public interest or to channel innovation toward the needs of nonaffluent people. Legislation in the United States on orphan drugs (those considered to have limited commercial potential) encouraged drug development targeted at rare diseases or conditions of public interest such as addictions. As developed countries sought legal protections against exploitative clinical trials, the pharmaceutical industry has turned to the Global South and post-Soviet countries because of the greater availability of drug-naive human subjects for clinical trials. A paradoxical situation prevails in which a cumbersome and highly technocratic drug research and development and approval process has somehow produced ongoing lack of regulatory oversight (Healy 2002).

So-called lifestyle drugs build on the capacity of pharmaceutical drugs to modify brain function and thus modulate human emotion, mood, and expression. There have been many cycles of pharmacological ambivalence oscillating between the introduction of a "wonder drug" and the demonization of that drug. A global regulatory regime now governs the production, distribution, and consumption of these emergent biotechnologies. Yet pharmaceuticals are not merely products like any other, as they compel humanity to ask questions about the limits of what it means to be human. Pharmaceutical drugs are depicted as tools to overcome human limitations of acuity, anxiety, energy, focus, intelligence, memory, and mood. They also highlight human relationships with nonhuman actors such as animals and tissue used in drug research and development.

PHARMACOLOGICAL OPTIMISM

Conversations about the role of "bioenhancement" in today's visions of posthuman futures often have an optimistic take on pharmaceuticals that enhance basic human functionality, traits, and characteristics grounded in neurochemistry. Once confined to obscure medical journals, the discourse of pharmacological optimism became widely shared by the end of the twentieth century. The optimism that pharmaceuticals buffer human beings from the rigors of everyday life, ranging from boredom to poor working or living conditions to emotional trauma, is now commonplace.

As pharmaceutical marketing optimistically addressed such problems of everyday life, increasing numbers of patients were diagnosed with anxiety, depression, or manic depression (Martin 2007). Equated with therapy, personal development, and a new kind of health care, pharmaceuticals became central to a lifestyle in which individuals consult with doctors to manage their risk profile across the life course. This form of "expert patienthood" is one in which prescription management replaces other modalities of health management in a mode of life referred to as "pharmaceutical living" (Dumit 2012). Patient-consumers rely on pharmaceuticals as tools of "objective self-fashioning" drawn from the logics of clinical trials and marketing materials such as television commercials (Dumit 2012). Psychoactive drugs blur the boundaries between the "true self" and a "better-than-well self," undergirded by psychoactive drugs that shore up the self against anxiety, depression or melancholia, bipolarity or mania, or other subjective conditions produced by neurobiological and neurochemical states.

Humans turn to the benefits of pharmaceuticals to enhance performance, whether it be physical, psychological, cognitive, or artistic. The ethical implications of enhancement are typically framed as if the effects of legal and illegal drugs are morally and ethically distinct. Sociolegal distinctions drawn between legal and illegal drugs mean that their production, distribution, and consumption are governed by different regimes working according to different logics. Yet at the molecular level, illegal drugs often work similarly to legal drugs (DeGrandpre 2006). As biotechnologies, pharmaceuticals often have effects that place them in uncertain ontological status, particularly because the social context plays a significant role in how individuals approach and experience drug use.

PHARMACEUTICAL CYBORGS

Discussion of enhancement technologies pervades posthumanist writing on the convergence between biotechnologies and artificial intelligences. While the merging of biological and computational capabilities in devices and drugs may have an aura of science fiction, there has been a great deal of innovation in wearable or implantable monitoring devices. The boundaries between "natural" and "artificial" are breaking down as new forms of human/nonhuman hybridity occur. Such boundary blurring was central to the earliest visions of the *cyborg*, a term coined to represent the convergence between cybernetics and organisms in "self-regulating [hu]man-machine systems" (Clynes and Kline 1960, 27). The first "cyborgs" were experimental animals that had catheters installed for drug administration without the organism's awareness. Such experimental setups soon migrated to those studying the behavioral effects of "drug self-administration," often by nonhuman primates or other laboratory animals. Within experimental psychology and behavioral pharmacology in the United States, this practice was normalized after 1962, when data from preclinical testing of pharmaceutical drugs prior to clinical testing were mandated for the regulatory process. These experimental setups met the conditions articulated by Manfred E. Clynes and Nathan S. Kline that "robot-like problems are taken care of automatically and unconsciously, leaving man free to explore, to create, to think, and to feel" (1960, 27).

Drug-maintained cyborgs were limited embodiments of popular concerns about the dystopian use of pharmaceuticals for behavioral control, interrogation, coercive persuasion, or even "brainwashing." Popular fiction and films such as *The Manchurian Candidate* (filmed in 1962 and again in 2004, it was based on the 1959 novel by Richard Condon) and *A Clockwork Orange* (a 1971 film based on the 1962 novel by Anthony Burgess) represented drug-based mind control as threatening to democratic values. The prospect of real and fictional mind control overshadowed thinking about the usefulness of drugs for bringing about modes of self-control necessary for democratic citizenship. This frame shifted, however, with the advent of new cyborgian potential adaptations to posthuman conditions.

In the mid-1980s feminist science studies historian Donna Haraway wrote "A Cyborg Manifesto," widely considered a posthumanist text because it articulated the cyborg as a new political identity. "By the late twentieth century, our time, a mythic time, we are all chimeras, theorized and fabricated hybrids of machine and organism; in short we are cyborgs. The cyborg is our ontology; it gives us our politics" (Haraway 1991, 150). Considering entanglements between human and nonhuman organisms to be productive encounters, Haraway saw the figure of the cyborg as an "imaginative resource" whose existence was no longer predicated on biological and technological distinctions and determinisms. The cyborg broke down binary oppositions between nature and culture, human and animal, and living organism and machine, which Haraway saw as dualisms central to the logics, practices, and informatics of domination (and older forms of humanism). "A Cyborg Manifesto" did not use the term *posthuman* but instead called on readers to imagine themselves in new and different "prosthetic" relationships to machines, animals, and other human beings based on friendship, intimacy, and kinship.

POSTHUMANISM, FEMINISM, AND THE NEW MATERIALISM

One of the most highly cited works on posthumanism, N. Katherine Hayles's *How We Became Posthuman* (1999), argues that a convergence between cybernetics and informatics has "so transfigured [human lifeworlds] in conception and purpose that they can appropriately be called posthuman" (11). However, Hayles views the "complexities of [human] embodiment [as meaning] that human awareness unfolds in ways very different from those of intelligence embodied in cybernetic machines" (1999, 283). She predicts that human awareness will change as the limits of individual agency, autonomy, will, and choice—once central to liberal, humanist conceptions of the human subject—confront posthumanist worlds (Hayles 1999). Like Haraway's prescient intuition that "we are [all] cyborgs," Hayles argues that "we have always been posthuman" (291) in the resistance to the constitutive dualisms of Cartesian rationality: that the body and mind are separable, and that the mind controls or dominates the body.

Feminist and so-called new materialist thinkers subscribe to the erosion of boundaries between human and nonhuman (Ferrando 2013). Critical posthumanism avoids biological and technological determinisms, offering a way beyond the idea that immutable traits can be isolated and identified as innate to some human substrate. Feminist and new materialist thinking is well suited for exploring what happens when pharmaceuticals are ingested and incorporated in ways that quickly become inseparable from bodily or mental processes, and ultimately metabolized in ways that transform patterns of thought, feeling, or bodily function. Few new materialists examine pharmaceuticals, although there have been many analyses of bodywork, bodily modification, and bodily markings such as tattoos, piercings, cosmetic surgery, and bodybuilding—all practices that have endogenous effects akin to pharmaceuticals. Such practices directly modify the body, marking the presence of the posthuman and the cyborgian in the everyday life of contemporary societies. They point toward a posthuman future that is decentralized, antireductionist, nonhierarchical, and multidimensional in its resistance to domination. Posthuman philosopher Francesca Ferrando contends that "in this expanded horizon, it becomes clear that any types of essentialism, reductionism, or intrinsic biases are limiting factors in approaching such multidimensional networks" (2013, 32). Pharmaceutical use is likely to remain the most widely accessible, socially accepted, and affordable of the enhancement technologies by which people accomplish the activities of self-regulation and the reconfiguration of identities.

PHARMACEUTICAL PERSONHOOD AND PERFORMANCE

When pharmaceuticals are used to regulate aspects of health, personality, and personhood, these aspects cannot be easily distinguished from one another. Consider the lawyer on a beta-blocker for a cardiac condition, who finds that she has far less anxiety in the courtroom, or the college professor on an antidepressant, who finds his perfectionist tendencies quieting in the face of the "publish or perish" performance mentality. Where is the line between therapy and enhancement? While it may be debated whether such effects of personhood are "authentic" or "artificial," to the "pharmaceutical person" (Martin 2007), this may matter little. Individual self-enhancement has become more possible and available with advances in body modification techniques, implants, prosthetics, and pharmaceuticals. Capacities to self-regulate mood, appearance, and performance are also central to the proper discharge of gender identity, sexual identity, and sexual performance. In this way, pharmaceuticals act in ways that are both gendered and racialized, as is perhaps glimpsed most easily in exemplary medical advertisements and direct-to-consumer commercials for prescription drugs.

Posthuman pharmaceutical "makeovers" suggest that today's drugs work far more subtly than the "sledgehammers" of the 1960s and 1970s (Begley 1994). The dissonant personality may now be optimized for traits or characteristics more consonant with success in a "posthuman" world. Psychotropic use has shifted from being a "bad habit" of the maladjusted to becoming a "good habit" that signifies adaptation to postmodernity. For example, the person on Prozac is figured as an internal master of a managerial sense of self who is neither addicted nor giving herself over to the dictates of a tyrannical external substance. Instead, the postmodern subject on licit drugs is pictured as attaining a state of adaptation, self-optimization, functionality, and productivity. In her memoir *Prozac Nation*, Wurtzel points out that "the measure of our mindfulness, the touchstone for sanity in this society, is our level of productivity, our attention to responsibility, our ability to plain and simple hold down a job" (1994, 55). Attuned to the substances on which the adults surrounding her depend—her father on Valium, her mother on nicotine—Wurtzel presents herself as functional enough on Prozac to attain the mature personhood that she found elusive prior to using the drug.

DRUGS AS TOOLS FOR POSTHUMAN SELFHOOD

The idea that "neurochemical selves" share beliefs about what variations and modulations of drugs can do and how they may "exquisitely" modify cognition, emotion, and behavior is the central contribution of an important essay by historian of psychiatry Nikolas Rose. Noting the burgeoning of psychiatric drug prescription in the United States and the United Kingdom, Rose observed how "all explanations of mental pathology must [now] 'pass through' the brain and its neurochemistry" (2003, 57). Anthropologist Emily Martin (2007) similarly considers the doubling of prescriptions for antidepressants and antipsychotics in the United States from 1991 to 1998. Self-medication using currently illegal drugs such as nonprescription amphetamines, cocaine, and opiates, as well as legally produced pharmaceuticals diverted to illegal channels, is also part of the flexible neurochemical reshaping of personhood (Rose 2003). Scholars also show that use of "cognitive enhancers" may be more about modifying emotions, motivation, and time management than cognition (Vrecko 2013).

Neurochemical selves navigate relationships in a complex economy that requires flexible adaptation to social norms as they emerge, rather than measuring themselves against the fixed gender norms of yesteryear. For example, Peter Kramer (1993) was impressed with the capacity of Prozac to enable women to negotiate postmodern workplaces and marriages, in contrast with the Victorian constraints on women's expressivity. According to Kramer, the femininity valued in the Victorian period was emotionally sensitive, passive, melancholic, and histrionic, whereas the femininity required by the end of the twentieth century was "spunkier" and less sensitive. Women needed a "feminist drug" to work in an economy that depended on aggressive activity and masculine values, and Prozac freed them from past trauma, lent them confidence, and "catalyze[d] the vitality and sense of self that allow people to leave abusive relationships or stand up to overbearing bosses" (Kramer 1993, 271–272).

The dynamic activity of being-on-Prozac differed greatly from the tranquilized days of "mother's little helpers" (such as Serax), during which psychoactives were presented as enabling women to stay in abusive or unfulfilling relationships or low-prestige jobs. "You can't set her free, but you can help her feel less anxious" read the text of a 1969 Serax advertisement aimed at doctors. "Beset by the seemingly insurmountable problems of raising a young family, and confined to the home most of the time, her symptoms reflect a sense of inadequacy and isolation. Your reassurance and guidance may have helped some, but not enough." Physicians were inadequate to help their female patients without the supplement of Serax and other tranquilizers.

Although such drugs were purveyed with levels of enthusiasm similar to the first flush of the marketing and use of Prozac, late twentieth-century subjects were promised something very different. The "mother's little helpers" of yesterday are today figured as bad drugs that once kept women in their place and prevented them from "grow[ing] in competence and confidence" (Kramer 1993, 39–40).

Moving away from drugs that target disability, disease, or impairment and toward drugs that enhance performance in otherwise healthy individuals opens up attractive new pharmaceutical markets. Some argue that so-called smart drugs or nootropics contain utopian possibilities for improving wise and intelligent decision making. Others argue that government should not regulate the scientific, moral, and ethical grounds on which smart-drug self-shaping occurs (Gazzaniga 2005). Given the all-too-human limitations to which pharmaceuticalization has thus far fallen, policies requiring strict regulatory scrutiny and responsible use seem a wiser path for pharmaceutical cyborgs to tread, in seeking to balance past, present, and future concerns about drugs for the posthuman soul.

Summary

The role that pharmaceutical drugs might play in moving human beings toward a posthuman future has been a staple part of posthumanist thinking. Indeed, neuropharmacologists first conceptualized drugs as both essential medicines and as elective tools that may be responsibly used for self-making and optimization as early as the 1950s. Pharmaceuticals occupy ambivalent status in popular culture and in health care. They are both recognized as risky, toxic, and potentially dangerous commodities in need of careful regulation, and as visionary solutions to individual and social problems that might bolster physical and mental prowess.

This chapter contrasts the dystopian camp of pharmacological pessimists, who believe that the potential for pharmaceuticals to assist in overcoming the limited state of knowledge about human cognition, mood, energy, and motivation is limited, with that of pharmacological optimists, who hold much more utopian beliefs about pharmaceutical enhancements of human capabilities. The chapter contextualizes these beliefs within the mid-twentieth-century emergence of the interdisciplinary science of neuropharmacology. At the same time, concerns arose about drug-based mind control, or "brainwashing," as threatening to democratic values and to the states of mind and modes of self-control necessary for democratic citizenship.

This binary frame was shifting in the 1980s, however, with the advent of new cyborgian adaptations designed to respond to posthuman conditions in optimal ways. As the figure of "the cyborg" was invested with libratory potential, so, too, were pharmaceutical drugs increasingly understood as tools for the optimization of posthuman personhood and productivity. The development of nootropic drugs and the use of off-label pharmaceuticals, alongside the consumption of such drugs for the purposes of adapting to a society that places a high premium on highly functional performance coupled with rapid pace, completes the chapter.

Bibliography

Begley, Sharon. "One Pill Makes You Larger, and One Pill Makes You Small." *Newsweek*, February 6, 1994. http://www.newsweek.com/one-pill-makes-you-larger-and-one-pill-makes-you-190292. This article demonstrates the enthusiasm for Prozac as a relatively easy way to change aspects of the self in order to succeed.

Biehl, João. "The Juridical Hospital." In *Living and Dying in the Contemporary World: A Compendium*, edited by Veena Das and Clara Han, 251–269. Oakland: University of California Press, 2016. A cultural anthropologist examines how right-to-health legislation is lived out in Brazil, which has evolved a unique pharmaceutical regime in response to HIV/AIDS activism and the country's constitution.

Biehl, João. *Will to Live: AIDS Therapies and the Politics of Survival*. Princeton, NJ: Princeton University Press, 2007.

Chatterjee, Anjun. "Cosmetic Neurology: The Controversy over Enhancing Movement, Mentation, and Mood." *Neurology* 63, no. 6 (2004): 968–974. Building on Peter Kramer's idea of *cosmetic pharmacology*, Chatterjee considers the neuroethics of the use of pharmaceuticals in the pursuit of happiness.

Clynes, Manfred E., and Nathan S. Kline. "Cyborgs and Space." *Astronautics*, September 1960, 26–27, 74–76. In this pioneering work on cyborgs, the authors stressed "partial adaptation" of the "man-machine system" to the environmental conditions of space via "incorporation of integral exogenous devices" designed to bring about biological changes necessary for living in space. The adaptive potential of the cyborg lay in the ability to self-regulate via drugs and devices.

DeGrandpre, Richard. *The Cult of Pharmacology: How America Became the World's Most Troubled Drug Culture*. Durham, NC: Duke University Press, 2006. Written by a behavioral pharmacologist, this book deconstructs cultural and moral distinctions between licit and illicit drugs, centering on the pharmacological similarities between illegal cocaine and legal drugs for ADD/ADHD.

Dubljević, Veljko. "Prohibition or Coffee Shops: Regulation of Amphetamine and Methylphenidate for Enhancement Use by Healthy Adults." *American Journal of Bioethics* 13, no. 7 (2013): 23–33. The author argues that policies for appropriate use of cognitive enhancement drugs might be modeled on a number of existing approaches.

Dumit, Joseph. *Drugs for Life: How Pharmaceutical Companies Define Our Health*. Durham, NC: Duke University Press, 2012. Pharmaceutical advertising and promotional "facts" are part of a corporate market logic that Dumit argues has redefined the concept of health such that continual growth in the use of prescription medications is seen as the commonsense route to individual and "mass" health.

Elkes, Joel. "Psychopharmacology: On Beginning a New Science." In *Discoveries in Biological Psychiatry*, edited by Frank J. Ayd and Barry Blackwell, 30–58. Philadelphia: Lippincott, 1970. A founder of neuropharmacology, Elkes illustrates the heady enthusiasm of early proponents of the study of brain and behavior, while also touching on prescient concerns about the limits of knowledge.

Ferrando, Francesca. "Posthumanism, Transhumanism, Antihumanism, Metahumanism, and New Materialisms: Differences and Relations." *Existenz* 8, no. 2 (2013): 26–32. Overview article explains the proliferating variety of "humanisms" in relation to feminist theory and a body of writing called new materialism.

Fukuyama, Francis. *Our Posthuman Future: Consequences of the Biotechnology Revolution.* New York: Farrar, Straus and Giroux, 2002. Convinced that in current neuropharmacology are the seeds of the end of human nature as we know it, Fukuyama is a leading proponent of strengthening the formal regulation of biotechnology based on clearer distinctions between health and illness on the one hand and therapy and enhancement on the other.

Gazzaniga, Michael S. *The Ethical Brain.* New York: Dana Press, 2005. A founder of cognitive neuroscience, the author explores free will and personal responsibility in light of the new brain sciences and brain-imaging technologies. He argues against public policy placing legal brakes on the development of enhancement technologies.

Gerard, Ralph W. "Drugs for the Soul: The Rise of Psychopharmacology." *Science* 125, no. 3240 (1957): 201–203. Urging readers to "preserve divine unrest" as their weapon in the age-old conflict between ease and dis-ease, as well as between desire for experience and desire for "experience-less nirvana," a prominent neuropharmacologist cast this dichotomy as the "core of the ethical problem" put to humanity by drugs of the soul.

Greely, Henry, Barbara Sahakian, John Harris, et al. "Towards Responsible Use of Cognitive-Enhancing Drugs by the Healthy." *Nature* 456, no. 7223 (2008): 702–705. These scientists argue for an evidence-based approach to the risks and benefits of cognitive enhancement drugs currently on the market, such as Adderall, Ritalin, and Provigil.

Haraway, Donna. "A Cyborg Manifesto: Science, Technology, and Socialist-Feminism in the Late Twentieth Century." In *Simians, Cyborgs, and Women: The Reinvention of Nature*, 149–181. New York: Routledge, 1991. This highly cited essay offers the political identity of the cyborg as a figure for moving feminist thinking beyond the constitutive dualisms of nature/culture, human/nonhuman, and organism/machine.

Hayles, N. Katherine. *How We Became Posthuman: Virtual Bodies in Cybernetics, Literature, and Informatics.* Chicago: University of Chicago Press, 1999. A cultural critic, Hayles explores postmodern literature and film.

Healy, David. *The Creation of Psychopharmacology.* Cambridge, MA: Harvard University Press, 2002. This thorough history of the centrality of serendipity to the founding moments of neuro- and psychopharmacology is told by a prescribing psychopharmacologist concerned with adverse events resulting from inadequate scrutiny of antidepressants.

Herzberg, David. *Happy Pills in America: From Miltown to Prozac.* Baltimore: Johns Hopkins University Press, 2009. In this history of drugs prescribed to increase "happiness," Herzberg examines how the 1970s feminist movement demonized Valium and how "depression partisans" overcame that demonization to offer antidepressants as the wonder drugs of the 1990s.

Huxley, Aldous. "Drugs That Shape Men's Minds." *Saturday Evening Post*, October 18, 1958, 28, 108–113. Huxley is best known as the author of *Brave New World* (1932), a novel that may be thought of as a dystopian portrayal of the fictional drug soma, and *The Doors of Perception* (1954), a more positive memoir about the author's experiences taking mescaline that later became a countercultural classic.

Kramer, Peter. *Listening to Prozac.* New York: Viking, 1993. This surprise best seller made Prozac a household name and made the case for its use by individuals seeking to become "better than well." The book considerably expanded the market for Prozac, particularly for women.

Martin, Emily. *Bipolar Expeditions: Mania and Depression in American Culture.* Princeton, NJ: Princeton University Press, 2007. This book traces the history of manic-depressive diagnosis, treatment, and lived experience, discussing how psychotropic drugs are imbued with "pharmaceutical personalities" that personify them for patients.

Petersen, Alan, and Deborah Lupton. *The New Public Health: Health and Self in the Age of Risk.* London: Sage, 1996. This primer shows how concepts of risk and health call on individuals to conform with prevailing social hierarchies and structures of governance. Public health imperatives call on citizens to actively participate, although the authors recognize that many people resist or ignore new health norms and the rationales that shape them.

Public Citizen. "Drugs, Devices and Supplements." Accessed March 2, 2017. http://www.citizen.org/Page.aspx?pid=1249. Since 1971, Public Citizen has been active in independently evaluating claims made by manufacturers of drugs and medical devices. Founded by Ralph Nader, the organization maintains the website Worst Pills, Best Pills (http://www.worstpills.org).

Rose, Nikolas. "Neurochemical Selves." *Society* 41, no. [illegible] (2003): 46–59.

RxISK. https://rxisk.org. Accessed March 2, 2017. This website was founded as an independent watchdog organization by David Healy, a psychopharmacologist who responded to the need for an organization to monitor the unintended consequences of widespread pharmaceutical use that would be independent of the pharmaceutical industry.

Tone, Andrea. *The Age of Anxiety: A History of America's Turbulent Affair with Tranquilizers.* New York: Basic Books, 2009. Includes an examination of the scientific contributions of Frank Berger, the "man behind Miltown," and Leo Sternbach, the developer of Librium and Valium.

Turner, Danielle C., and Barbara J. Sahakian. "Neuroethics of Cognitive Enhancement." *BioSocieties* 1, no. 1 (2006): 113–123. Turner's laboratory at the University of Cambridge was among the first to conduct studies showing that single doses of modafinil can improve short-term memory and executive function. This article examines the ethical use of pharmaceutical drugs to overcome cognitive impairments occurring in otherwise normal people.

Vrecko, Scott. "Just How Cognitive Is 'Cognitive Enhancement'? On the Significance of Emotions in University Students' Experiences with Study Drugs." *AJOB Neuroscience* 4, no. 1 (2013): 4–12.

Wurtzel, Elizabeth. *Prozac Nation: Young and Depressed in America.* Boston: Houghton Mifflin, 1994.

FILMS

Children of Men. Dir. Alfonso Cuarón. 2006. Assisted suicide drugs figure in the plot of this movie.

The Constant Gardener. Dir. Fernando Meirelles. 2005. Film centers on globalization of clinical trials in Africa.

Eternal Sunshine of the Spotless Mind. Dir. Michel Gondry. 2004. Drama about the potential for memory enhancement to include erasure or forgetting.

Girl, Interrupted. Dir. James Mangold. 1999. Based on Susanna Kaysen's 1993 memoir, one strand of which is the protagonist's relationship with pharmaceutical drugs.

Limitless. Dir. Neil Burger. 2011. Film about a fictional nootropic drug that allows full use of the human brain.

Lucy. Dir. Luc Besson. 2014. Like *Limitless,* this film centers on a nootropic drug enabling enhancement.

Prozac Nation. Dir. Erik Skjoldbjærg. 2001. Based on Elizabeth Wurtzel's 1994 memoir about the use of antidepressants.

Requiem for a Dream. Dir. Darren Aronofsky. 2000. Movie that directly contrasts legal and illegal drug use.

A Scanner Darkly. Dir. Richard Linklater. 2006. Based on a Philip K. Dick novel about a drug called Substance D, to which one-fifth of the fictional US population is addicted despite it causing permanent organ damage.

Side Effects. Dir. Steven Soderbergh. 2013. Film dramatizing the devious manipulation of antidepressant effects.

CHAPTER 7

Bioelectronics

Chris Hables Gray
Fellow and Continuing Lecturer, Crown College
University of California, Santa Cruz

I sing the body electric.

—Walt Whitman

THE BODY ELECTRIC

Complex life-forms and sophisticated machines both need electricity to function. Animals, including humans, produce their own electricity in their flesh, and they generate it for the machines humans build. The restless drive of humans to extend ourselves has led to the development of a wide variety of electrical technologies, ranging from virtual and augmented reality systems to bioelectronics. This is possible for two reasons. First, electricity vitalizes organisms and machines. Second, the rules of energy and information flow, known as cybernetics, apply to both organisms and machines, as well as to their integration as organic machinic hybrids—cybernetic organisms, or cyborgs.

Electricity is power; it is about movement. For animal flesh to move it must have electricity. Almost all machines that move today use electricity as well. But electricity is not just power; it is information. Electricity is one of the main ways living flesh and dynamic machines transmit messages. But it is not the only way. Chemicals transmit information in many different biological systems. And some creatures have evolved photosensitive and sonar-sensitive tissue, and this has enabled the development of the new technoscience of optogenetics, where genetically modified neurons can be manipulated through light and sound.

All these interventions fall under the rubric of bioelectronics, because even where light, sound, or chemicals are being introduced for medical or other interventions, electronics are always involved. This includes the new areas of bionanoelectronics, electroceuticals, electromagnetics, biomolecular technologies, neuromodulation, organic electronics, and other bioelectronically mediated organic extensions. Expanding machinic-organic connectivity does not just extend an individual, or the body politic, into the nonhuman living. It also offers the chance to "extend" into the human mind from outside, for therapy or control, or for reading of, writing to, or actually running the organism like any artificial system—meat puppets, they are called.

Whether these powerful new bioelectrical technologies and sciences are used in liberatory or coercive ways, or as is most likely both, they will be integral to shaping whatever posthumanity emerges from the current epoch, which is revolutionary in so many ways.

Posthumans, if human civilization survives long enough to produce them, will necessarily be products of the human ability to manipulate the physical world and link ourselves to it. Electricity, and the electronics that harness it, is one of the crucial modalities for the human manipulation, augmentation, and transformation of the biological. So bioelectronics is at the center of any possibility of posthumans.

Philosopher Ian Hacking (1983) has shown how the power of science to manipulate reality starts with its ability to effectively observe phenomena and represent them in some way. This allows for intervention. The acts of observing and representing are dynamic (affected by the Heisenberg uncertainty principle, among other "laws") and are limited, but they often lead to successful interventions. The current explosion of bioelectrical interventions for modifying and controlling human and other organisms is based on our expanding understanding of basic biology, chemistry, and physics, as well as an increased appreciation for the dynamics of systems and their principles—cybernetics.

THE BODY CYBERNETIC

All media are extensions of some human faculty—psychic or physical.

—Marshall McLuhan (1967, 26)

Humans have traditionally seen the body as something created by gods or a god, perhaps in their image. But as our science has grown more powerful, increasing our understanding of the workings of our bodies by conceiving of them as machines—as biophysical systems—we now see ourselves as fundamentally mutable, modifiable, improvable. Of course, we have been improving ourselves from our beginnings with technologies such as tools, clothes, and shelter. From mastering fire to inventing agriculture, humans have extended themselves out into the world, domesticating it even as we modified our brains through culture (language especially). What we think and do shapes the neural systems we grow. But we also modify ourselves with tattoos and piercings and such, as well as with what we physically consume: foods, medicines, and psychotropic substances ranging from the recreational to the spiritual.

Tools, including weapons, were the first prosthetics. They are fundamentally enhancing, increasing human capabilities beyond those we are born with. Restorative prostheses are a later development; these crude replacements for missing body parts can be traced back only to early civilizations. Over time humans have found extraordinary ways to modify ourselves, and thanks to relentless advances in science and technology (such as bioelectronics) that have empowered medicine and other fields, we have developed amazing new ways to change our bodies—and, in the process, change our society. In *The Body Reader* (1978), Ted Polhemus points out that what we do with our bodies is a central aspect of our culture and who we are.

> Our bodies and our perception of them constitute an important part of our socio-cultural heritage. They are not simply objects which we inherit at birth, but are socialized (enculturated) throughout life and this process of collectively sanctioned bodily modification may serve as an important instrument for our socialization (enculturation) in a more general sense. That is, in learning to have a body, we also begin to learn about our "social body"—our society. (21)

While older forms of modification do not always disappear when others are invented (think of clothes, makeup, simple tools, and so on), new, powerful, and intimate bodily interventions can profoundly shift human societies into new frameworks. This is what we are experiencing now.

Electronics is one of the key fields driving this dynamic. Electronics have been extending humanity since soon after the harnessing of electricity, from the telegraph and telephone, to electrical motors and grids, to the massive impact of computers and, now, to social media as well as physical linkages between animals and machines with bioelectronics. Knowing about electricity is obviously crucial, but just as important is understanding systems.

American mathematician and philosopher Norbert Wiener (1894–1964) coined the term *cybernetics*, based on the Greek word *kybernetes* (steersman), to emphasize the importance of feedback in self-regulating systems. He realized that the same rules governed evolved and invented systems. Wiener and other "system engineers" brought a mathematical rigor to understanding the rules and metarules of systems. In particular, the engineers, mathematicians, physicists, and other odd founders of computer science sought to create complex homeostatic (self-regulating) electrical systems for many tasks seen before as purely human and cognitive.

The realization that system metarules are constant, whether that system is living flesh or dead metal, led to the coining of the term *cyborg* (cybernetic organism) by Manfred E. Clynes (a pianist and self-trained computer inventor). Actually, organisms are always cybernetic, but a good made-up word is hard to find and cyborg has stuck. Bioelectronics, broadly conceived, is a major part of the campaign to develop always-better organic-machine interfaces of the greatest intimacy. Many of these cyborgs are humans.

The cyborg is such an integral part of understanding the development of bioelectronics that it is worth clearing up a few things before proceeding. Cyborgs are not only human-machine systems. In the original talk and 1960 article introducing the word *cyborg*, Clynes and his colleague Nathan S. Kline gave an example of what a cyborg might be. The first named cyborg was thus a rat from an experiment in Australia, with a system for the automatic delivery of a drug by means of the process of osmosis through a little pump affixed to its behind. Obviously, cyborgs preceded the word, coined only in 1960. The modification or integration of dead/living, evolved/invented, organic-inorganic hybrid systems that produce homeostatic balanced organisms requiring no conscious thought to sustain them did not start in 1960. So, driving your car does not make you a cyborg—except what about those times you drive without thinking? It is complicated. You are probably a cyborg already, as anyone vaccinated is. The issue isn't if you are a cyborg, because (1) you probably are and (2) even if you are not, you live in a cyborg society interconnected with more dynamic powered electronically mediated technology than ever before. The real questions are: How are you cyborged? Why are you cyborged? Who cyborged you?

Complicating these queries is the concept of the mundane cyborg. This is what we are when embroiled in our social media; when we live in nets of surveillance, sousveillance (watching from below), and autoveillance (self-surveillance); when we extend ourselves with drones; or when we extend our "reality" virtually or augment or mediate it. We need to realize that our society is now a thick web of animal-to-machine connections; it is a cyborg society. Some of these links are intimate, others mundane. Fostering it are new waves of discovery in the information and materials sciences, physics, chemistry, biology, genetics, neuroscience, and electrical engineering.

When we also factor in the progress being made in synthetic biology, nanotechnology, and general-purpose artificial intelligence, we can see how profoundly transformative these modifications will be. In his 2003 book *Our Own Devices*, Edward Tenner, a famous philosopher of design, warns us not to deny the dialectical back and forth of these kinds of technological transformations:

> When we use simple devices to move, position, extend, or protect our bodies, our techniques change both objects and bodies. And by adopting devices we do more. We change our social selves. In other species, natural selection and social selection shape the appearance of the animal. In humanity, technology helps shape identity. Our material culture changes by an unpredictable, dialectical flux of instrument and performance, weapon and tactic. (29)

This "dialectical flux" has produced today's bioelectrical revolution.

THE BODY ELECTRONIC

Technologies are possible only because of their technological context and their history. While the term *bioelectronics* itself is not that old, fifty years or so, its origins can be traced much further back and to a number of different sources. First comes science's great romance with electricity. In the mid-1700s natural philosophers, such as Benjamin Franklin, were systematically studying electricity. Franklin proved that lightning was electricity and others showed how to generate it from friction or chemicals. Early in this period the nobility of Europe turned electricity into a parlor game, the electrical soirée, sending shocks through servants and soldiers for their amusement. Later they played with it themselves, exchanging "electrical kisses."

But there were also many serious experiments, such as Franklin's with kites and Luigi Galvani's (1737–1798) with animal electricity and how it powers muscles. Even earlier, medical attempts to build simple prostheses (ancient Greece) and to relieve the pressure in swollen brains by drilling holes in the patient's skull (trepanning; successful in ancient Egypt) established the traditions, and efficacy, of profound interventions in human brains and bodies.

In the last few hundred years the pace of innovation has accelerated. Electrical stimulation during open-brain surgery revealed that slight currents could evoke specific memories, physical pleasure, abject fear, or the smell of fresh cookies. This was followed by the development of two kinds of "lobotomy": electroshock "therapy" and psychosurgical excisions.

By the middle of the twentieth century, almost 20,000 lobotomies had been performed in the United States and even more electroshocks. American physician Walter Freeman (1895–1972) had perfected the ice-pick lobotomy, which he performed on patients in asylums almost at will. It simply involved slipping an ice pick in next to the eye, in the orbital socket, and stirring: instant prefrontal lobotomy. The results, however, were not good (Mentor 2007).

As powerful electroshocks and ice-pick lobotomies were being rejected for treating the socially disruptive and mentally ill, a new electrical treatment was introduced: electrical stimulation of the brain. The most famous experimenter and inventor of brain implants was José M. R. Delgado (1915–2011), a Spanish professor of physiology at Yale University. Steven Mentor (2007, 35) explains that "much of [Delgado's] work involved the improvement of electrode technology, but the experiments themselves focused on control—both the control the brain has over the body and the control that one can induce externally."

This led Delgado to invent the stimoceiver, a remote-controlled electrode he planted in the brains of chimpanzees, bulls, and humans. Delgado used his brain implants to disrupt the

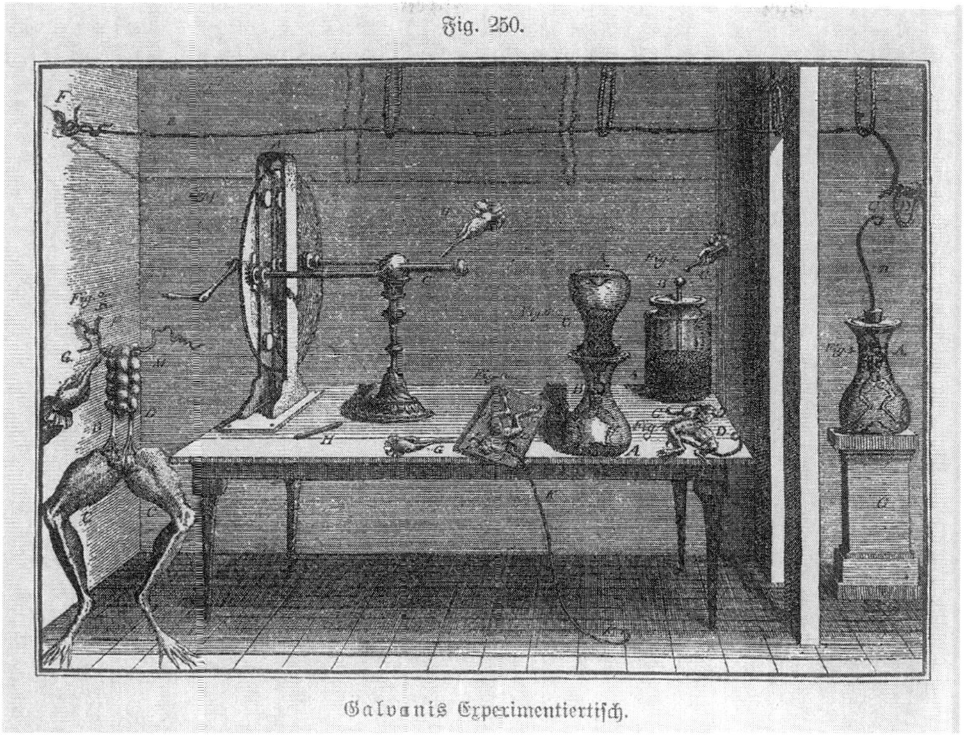

The Galvanic Frog, *late eighteenth century.* *Some credit Luigi Galvani with discovering animal electricity, and that certainly seems fair. Others call him the original Dr. Frankenstein and the father of neuroscience. These honorifics are perhaps a bit of a reach. But the great man did do some amazing experiments showing that the body is indeed electric. This image is from his* De viribus electricitatis in motu musculari commentarius (Commentary on the Effects of Electricity on Muscular Motion)*, Bologna, 1791.* **EXPERIMENT ON FROGS BY LOUIS GALVANI (1737–98) LATE 18TH CENTURY (ENGRAVING) (B/W PHOTO)/FRENCH SCHOOL, (18TH CENTURY)/PRIVATE COLLECTION/BRIDGEMAN IMAGES.**

hierarchies of chimpanzee tribes and make dramatic movies about bull control. But he also put them in people. He liked to call these subjects "completely free patients," not because he was not trying to control their epilepsy with implanted electrodes, but because he was doing so remotely, with a radio. The freedom was a matter of access to the brains (Mentor 2007). This fits with Delgado's very instrumentalist view of his research, as outlined in his book *Physical Control of the Mind: Toward a Psychocivilized Society* (1969, 67):

1. There are basic mechanisms in the brain responsible for all mental activities, including perceptions, emotions, abstract thought, social relations and the most refined artistic reactions.
2. These mechanisms may be deleted, analyzed, influenced, and sometimes substituted for by means of physical and chemical technology. . . .
3. Predictable behavior and mental responses may be induced by direct manipulation of the brain.
4. We can substitute intelligent and purposeful determination of neuronal functions for blind, automatic responses.

Delgado, of course, wanted to use this control only to make better people for a better world. But whose better world? Among his collaborators were many scientists who targeted social deviants and protesting African Americans for neurological interventions of various types. Once one accepts the basic materialistic and instrumentalist assumptions about human nature and neurology in Delgado's four points, listed above, the quest for the "physical control of the mind" is inevitable. Linked with utopian dreams, for a "psychocivilized society," for example, and you have all you need to beget nightmares.

Prof. Delgado called his wired beasts, bulls and humans, "electronic toys." The cultural critic Steven Mentor notes that "Delgado's 'electronic toy' becomes one's self, as one's body is both treated as a machine *and* as an animal for experimentation, a guinea pig" (2013, 36, original emphasis). We are living the experiment of human modification writ large and as Mentor realizes this means being both machine *and* meat, as with these vitalized decapitated cadavers, not necessarily so human. We'll need to keep our heads on to escape being mere experimental subjects.

Bizarre as some of his demonstrations were, the general outlines of bioelectronic brain interventions were predicted by Delgado. Brain implants are now used to control Parkinson's and epilepsy. Brain waves are read with external equipment and used to control computer cursors, drones, and virtual-reality cognitive therapy programs. Originally this

Bioelectric Bull. *This famous photo shows José M. R. Delgado performing with a bull he had wired for remote electrical stimulation. He claimed that he had electrodes in the aggression center, turning it on and off, but some experts suspect it was just massive motor coordination interference that made the bull fail to gore Delgado (Mentor 2013).* MANUSCRIPTS & ARCHIVES, YALE UNIVERSITY LIBRARY.

was crude and limited to yes/no and left/right, up/down types of control through brain states read with scalp-mounted sensors or shallow electrodes. There have been incredible advances in the development of noninvasive technologies for communicating electrically with the brain, including ones involving skullcaps and headsets. Soon, much more sophisticated brain reading through these new external brain interfaces will allow excellent control of computers and drone systems, as well as complicated virtual simulations and augmented reality apps that can be manipulated with brain waves.

Bioelectronics is not just about brain-machine communication and control. Many medical uses involve artificial or modified organs that are linked together with bioelectronics. Almost every major body system is being supplemented bioelectronically now, from heart pacemakers to stimulators for the reproductive and elimination systems.

Some bioelectrical systems are more machine than animal. Biocomputers are computers with memories that can die. This is a highly desirable trait for the military and spy agencies, which do not want their mechanical memories captured. There is also significant research on tiny biomachines for bodily or environmental repairs and for war. All these research programs benefit from the same context.

There are five major reasons for the ongoing progress in the linking of humans to machines:

1. improved observational and modeling technologies;
2. the continued improvements in electrical implant research in general;
3. improvements in noninvasive brain communication devices, such as skullcaps and headsets;
4. the rapid creation of an infrastructure of open-source and automated processes to further research and continuing advances in automating scientific discovery; and
5. whole new fields, such as optogenetics.

Real-time three-dimensional brain tomography and the visible brain and other imaging technologies have been improving since the late 1990s. A major program at Stanford University funded by the Defense Advanced Research Projects Agency has developed a technique, called Clarity, for "washing" brains of their fats so as to reveal the actual nervous system architecture (Tucker 2014). A key step toward controlling minds is effectively visualizing and modeling them, knowing them in terms of their material functions. Complex brain maps, DNA charting of specific neurons, real-time three-dimensional brain tomography, and other observational (and therefore modeling) tools have made it possible to create effective ways of directly intervening in the human mind.

Electronic chips and electrical probes on and in brains have a long history, one that includes the ice-pick lobotomy craze and extreme electroconvulsive therapy, discussed above. Much more sophisticated implants are now common, and there have been some important breakthroughs in this area of deep-brain stimulation, with over 100,000 patients implanted (Noonan 2014). The development of biocompatible transfectic (produced by "infecting" cells with outside DNA, or creating synthetic cells or biomaterials) artificial interfaces has been another major development.

There are many research projects focused on electroceuticals (tiny wireless sensors and stimulators that do not need batteries). They could be sprinkled on the brain as "neural dust" and powered and programmed by ultrasound, magnetics, or infrared (Sanders 2016).

Driven by the potential of the vast market of game players, there has been continual innovation in the area of noninvasive systems for sending and receiving messages from the human brain. Headsets, skullcaps, and other head-mounted electrical arrays are widely used in experimental labs and, more and more, in consumer households. These systems have their limits, but they are growing in discernment and therefore effectiveness and will play a major role in human-machine communication for the foreseeable future.

The infrastructure of bleeding-edge science continues to improve monthly. Open-source platforms and technologies, better utilization of the big data that underlies natural science, and the automation of complex operations such as gene analysis and construction are all significant advancements on earlier processes. The approach to gene modifications known as CRISPR (clustered regularly interspaced short palindromic repeats) is opening up many new genetic engineering possibilities (Kirksey 2015).

As powerful as all these interventions are, steel probes and electrical shocks are crude physical instruments compared to the broadcast light of optogenetics. Optogenetics is the best current example of how a major new approach to an old problem can generate a whole new field driving profound changes in how humans communicate with our machines, which means changing what being a human might mean.

THE BODY PHOTONIC

The most futuristic medical treatment ever imagined is now a reality. But it won't be long before brain implants are even more amazing and troubling.

—David Noonan (2014, 38)

So how does optogenetics work? First, you must make a chimera, which is a creature with the DNA of at least two different organisms. In the case of optogenetics, scientists must "infect" the neural tissue of the target (host, experimental subject, patient) with DNA from another species that produces photosensitivity. No humans have been targeted yet, but other primates, many rats and mice, and various reptiles, insects, worms, and even plants have been transformed into chimeras through infection with photosensitive-producing DNA from fungi, bacteria, algae, and jellyfish. (The basic process is the same for sonogenetic chimeras, except that the foreign DNA imported is sensitive to pressure, not light.)

The targeting is incredibly accurate because it "aims" at the specific DNA of the chosen neurons, even a single one. Once the neurons are transformed, they become sensitive to different colors of light, depending on the DNA vectored into the neuron. Of course, the light has to be ported into the brain with fiber optics, but because photons are so much smaller than electrons it is much less intrusive and more precise than the older electrical stimulation systems. Complex chimeras can be created so that one color of light can turn on a neuron and another turn it off. For example, with one set of genetic modifications, blue light activates neurons; with others, orange light depresses their activity. So optogenetics is brain control "in a flash of light" as one overview put it (Gorman 2014).

Optogenetics has already left its early stages. Rice University researchers have developed an open-source optogenetics platform (Furness 2016). Machines to automate much of the experimental process of optogenetic procedures are now advertised for sale in science magazines. Scientists in 2016 genetically modified shrimp neurons to be bioluminescent so that the animals could be studied in real time as they "think" (Bruce 2016).

Optogenetic experimental results from 2011 to 2016 are actually shocking. They include:

- *Treating blindness*: Wayne State University researchers have developed a treatment involving green algae genes for an inherited form of blindness (Gallagher 2016).
- *Technotelepathy*: Scientists at Duke University have transferred thoughts (motor and sensory information) from one rat to another (Heaven 2013).
- *Inserting and deleting information in dreams and memories*: At the Massachusetts Institute of Technology, specific information has been implanted in rats while they sleep (Jha 2013; Bendor and Wilson 2012), while at the University of California, San Diego, a team has developed a system that allows them to turn memories (fear of a stimulus in this case) on or off at the flick of a switch (Fikes 2014). At Columbia University a team inserted an image into a mouse's brain using optogenetics. A neuroscientist involved, Rafael Yuste, remarked on how the experiment changed his perception of the human brain: "I always thought the brain was mostly hard-wired. But then I saw the results and said 'Holy moly, this whole thing is plastic.' We're dealing with a plastic computer that's constantly learning and changing" (*Independent* 2016).
- *Lie detection and mind reading*: In the United States, brain scan analysis that shows if someone is looking at a new stimulus or an old one is already admissible in court as lie detection ("Did you ever see this murder victim before?"). With improving real-time three-dimensional brain tomography, as well as the continuing development of sophisticated brain maps and tinier electrical and fiber-optic sensors, it is inevitable that effective general lie detection, and even the ability to read specific thoughts, will eventually be perfected (Smith 2013).
- *Mind control*: Scientists at Washington State University have already turned chimps, and fellow scientists, into "meat puppets," where their limbs are controlled by someone else (Armstrong and Ma 2013).

Scientific breakthroughs, such as those described above, have made the direct mapping and even the control of consciousness seem not just possible but inevitable to some observers. Although full understanding and control are probably impossible, certainly in the foreseeable future, significant interventions will clearly be available soon. These projects are developing the ability to reprogram human minds—*re*program, because we are all programmed into our worldview through the nurture of our culture, the nature of our bodies, and our experiences.

All this work is part of what could be termed the consciousness studies industry, an emerging "military-industrial-spiritual-scientific complex" that has mushroomed now that it seems neuroscience (especially psychopharmacology and optogenetics) is on the brink of major advances in the instrumentalist control of mentation (Gray 2007). But a number of government studies on the ethics of neuroscience have come out warning of the possible ramifications of this work (Royal Society 2012; Nuffield Council on Bioethics 2013). The horrors of the twentieth century produced a new genre of writing—dystopian literature. From Aldous Huxley's *Brave New World* (1932) to George Orwell's *1984* (1949) to Margaret Atwood's *The Handmaid's Tale* (1985), great writers seeing what the desire for utopia could midwife warned about the dangers of questing for perfection. As the physical control of the human mind becomes more possible, so too do the most terrible dystopian possibilities. Our posthuman future could turn out to be profoundly inhumane if we are not careful.

For all this research (even for military projects) the official goals are limited to treating medical problems, such as delaying Alzheimer's and dementia, controlling Parkinson's, or overcoming post-traumatic stress disorder or depression. But some of the work is clearly aimed beyond the medically ill, at the socially "ill" as well. Sophisticated theories and practices of social control focus not on instrumentalist interventions but the management of perceptions and values. Much of this has been, and is, very crude: terror, black/gray/white propaganda, programs to win "hearts and minds," brainwashing, and drugging. But the current wave of neurotechnologies promises something more precise, subtle, and effective. Yes, the technology also offers real improvements for restoring limbs and other systems, for treating neurological disorders, and a bit further down the line for fundamentally enhancing humans. But there are certainly other agendas, and the technology in general faces serious obstacles.

The main problem with optoelectronics is the apparent necessity for the genetic engineering of humans. The genetic modification of humans is very difficult to study, and very few treatments have made it even to the experimental phase. Currently, human-pig chimeras are being created to explore the idea of "breeding" organs, but they are not allowed to come to term.

But it turns out that using light to control neurons might be possible without genetic engineering humans. Researchers have used the targeting system of current optogenetics to attach gold particles to the desired neurons. When exposed to light they are activated. So instead of genetically modifying the neurons, the researchers created genetically modified biological delivery systems for gold (Carvalho-de-Souza et al. 2015).

Electroshock is crude compared to implanted electrodes, which are crude compared to optogenetics. And optogenetics will no doubt be seen as crude when compared to the many more refined control-and-connect systems we can confidently predict in our future, promising prosthetic projections and modifications that go way beyond the restorative and even beyond enhancing or transcending the human.

PROSTHETIC GODS

We humans extend ourselves constantly out onto the world. We have remade it for our houses, for roads for our cars, for our industries, and for our amusement parks. We have turned wild nature into a garden and a garbage dump. We have sent our machines to every planet in our solar system and beyond it as well. We extend ourselves through our machines, from the depths of the sea to the moon and soon on to Mars. The Interweb spreads our consciousness around the globe allowing us to watch the sunrise on the other side of the planet, to watch the ice melting at the top of the world, and to chat with our friends anywhere.

Much is made of Moore's law, that predicted the geometric improvement in central processing unit power that has driven the computer revolution. But it isn't just computers that are improving at astronomical rates. Much of the bleeding edge of science is heavily automated: robot telescopes, DNA reading and printing machines, CRISPER machines to edit DNA, and so on. DARPA's NESD program is targeted specifically and massively increasing the bioelectrical connections of neural interfaces. It is not hard to predict, that a few decades from now this interface of up to 1 million channels aggregating tens of thousands of signals from neurons each, will seem incredibly limited and antiquated.

To live longer and better lives we incorporate technology into our most private recesses: vaccines reprogram our immune system, and small machines measure our heart rate as a Fitbit or regulate it as a pacemaker. But who is in control of all this ... control?

Marie Moe (2016) wants to know because she has a pacemaker in her heart, and one day she woke up on the floor. Her pacemaker had knocked her unconscious, and it took an extended campaign to find out why. The company did not want to tell her its algorithms, but when she finally got them Moe found out that her pacemaker was set by default for a much older person. Even with that fixed, she has discovered (she is a computer security expert) that she is quite vulnerable to having her heart hacked.

Steven Guile (2007) also has an implant. Located in his brain, it is used to control the symptoms of Parkinson's. As a former Apple engineer, Guile knows a great deal about his implant (because he insisted), but he does not control it. His wife does. When she notices increasing symptoms she can turn up the neuron-suppressing voltage, or, to amuse hundreds of college students, she can zap him into unconsciousness for a moment with the flick of a switch.

Moe and Guile are among a growing number of people with crucial medical devices who are demanding information about them, and ultimately demanding control of the machines that keep them alive. They are actually part of the larger body-hacking movement that uses both over-the-counter and medical devices to monitor, even augment, their bodily processes. Philosophically, this movement is related to the tattoo and body modification subculture that has long fought for autonomy in body alteration. Even more directly, it claims the same rights of self-determined modification and transformation as the transgender movement does. And transgenderism, with its refusal of binary gender identity and exploration of all sorts of technologically mediated, complicated sexual personas, is a good indication of where wholesale body augmentation and modification will go in society: a proliferation of possibilities and types.

Admirable as these efforts to democratize our transformations are, we need to note that as with computers in general, the military plays a major role in fostering cutting-edge technology. Jonathan D. Moreno, in his 2012 book *Mind Wars: Brain Science and the Military in the Twenty-First Century*, has documented the interest and involvement of the US military in applied neurology. The Defense Advanced Research Projects Agency (DARPA), the US military's primary research arm, has a number of major projects in prostheses (Adee 2009), aiming both at restorative limbs and enhancing exoskeleton systems. DARPA is also funding a set of projects to create implants for "controlling emotions" in wounded soldiers and psychologically disabled veterans. Among the seven "mental illnesses" targeted are "addiction," "anxiety," and "depression." The agency is seeking to develop "control commands" and "affective brain-computer interfaces" to control those emotions, which means increasing as well as decreasing them. The project is funded by DARPA's Systems-Based Neurotechnology for Emerging Therapies program (Regalado 2014).

Corporations do most of the work the military contracts, just as they seek to develop the new medical prosthetics and other interventions in bioelectronics and related fields—and always for a profit. This goal of profit is always distorting in technology development, as can be seen with pharmaceuticals, where price gouging and constrained research have become major social issues. Or look at how social media has been warped around the need for eyeballs on ads, leading to Facebook, Google, Apple, and others using electronically mediated social systems to insinuate deeply into the biologically based emotional life and internal consciousness of the human users. This is called neuromarketing, or, more grandly, neuroeconomics, but the basic theory is the same: as humans extend themselves with technology, most of the companies that are organizing this do so for limited monetary goals and have no compunction about distorting human desires and relations to maximize profits.

Neuro-instrumentalist approaches to marketing, however, are much less dangerous than similar approaches used in treating the ill, or even in governance. These dangers are clear from the career of the aforementioned Delgado. When it came down to it, he seems to have actually wanted to bring about his psychocivilized society through psychosurgical and bioelectrical means. Delgado argues that while in the past one might keep control of one's fate,

> New neurological technology ... has a refined efficiency. The individual is defenseless against direct manipulation of the brain because he is deprived of his most intimate mechanisms of biological reactivity. In experiments, electrical stimulation of appropriate intensity always prevailed over free will; and, for example, flexion of the hand evoked by stimulation of the motor cortex cannot be voluntarily avoided. Destruction of the frontal lobes produced changes in effectiveness which are beyond any personal control. (1969, 214–215)

As Delgado well knows, with many kinds of mind control some people could become mere prostheses for others. This is why we need a stronger sense of citizenship (Gray 2001), because the dangers of new neurological technosciences to human autonomy are profound.

The impulses behind this transformation of humanity through bioelectronics and other technosciences seem to have many origins, from medical aims to more generalized desires to restore or enhance health and function. Improved human-computer interfaces promise to create completely new cultural worlds and fundamentally transform our experience of material reality. War and business both seek the power of improved efficiency through better communication and control between flesh and machine. It seems overdetermined. But if one looks at the process from a bit of a distance it seems very familiar. It is humans doing what we do. We have evolved to modify ourselves and our environment, and this has led to some extraordinary reproductive successes, as well as some profound dangers. It is evolution, but evolution evolves. As English naturalist Charles Darwin (1809–1882) pointed out, humans added artificial selection to natural selection to guide the development of some species down very specific paths. Now, with genetic engineering, we need to accept that we have also created participatory evolution. It is a godlike power.

Austrian neurologist Sigmund Freud (1856–1939) commented on the dangers of this:

> Man has, as it were, become a kind of prosthetic God. When he puts on all his auxiliary organs he is truly magnificent; but those organs have not grown on to him and they still give him much trouble at times.... Future ages will bring with them new and probably unimaginably great achievements in this field of civilization and will increase man's likeness to God still more. But in the interests of our investigations, we will not forget that present-day man does not feel happy in his Godlike character. (1961, 39)

We need to make sure we are happy with this transformation—that it makes us sing.

Consider Walt Whitman's (1819–1892) poem "I Sing the Body Electric." Whitman did not believe the body was evil, as many American Christians did then. He thought the body was integral to the soul: "And if the body were not the soul, what is the soul?" He also was profoundly democratic and an abolitionist, and felt each of us should own his or her own body. In his poem, Whitman implies that just as electricity jumps between bodies, so too could love and the energy that purifies ("discorrupts"). After all, in Whitman's day many saw electricity as life itself. Today we know that electricity is the spark that vitalizes the copper, the silicon, the exotic metals, and the complex glues that make up the electronics of bioelectronics. As the bio becomes better understood, we learn how with sparks and biochemical molecules we can make the frog's leg dance ballets or tap at our command.

So now it is "I sing the body bioelectronics" if we are vigilant. Someone else is singing our bioelectronics if not.

Summary

THE FUTURE HAS NOT BEEN WIRED

The great insight from the Terminator movies came to the heroine Sara Connor as she tried to navigate the paradoxes of time travel, sentient computers, and human extinction. In the second movie of the franchise, *Terminator 2: Judgment Day*, she carves "No Fate" in a table before heading off to try and change the future. We do not just carve or write texts, we write our lives and we write our computers. After all, electronics are printed these days; they are like a magical text that not only tells us things but does things. And one of the things these electro-texts do is link metal to living flesh, making of them one hybrid system.

This process will continue and expand in scope and quality. As ubiquitous computing systems spread out to inhabit more and more "things," even as they converge together in close communication and collaboration, biological systems will be integral parts of this web. Humans will put on heads-up displays or accept bioelectrical implants or undergo genetic engineering and in other ways will certainly, willingly, incorporate bioelectrical and other artificial systems onto and into their bodies and link through them to the cloud, to the Web, to the organism that is the wired Earth.

Assuming we navigate through our current difficulties with weapons of mass destruction and climate change, it seems likely that humans will continue to pursue fundamental enhancements through participatory evolution that will transcend the current human species. What will this look like? That has not been decided yet. As John Connor says, looking at "No Fate" etched in the wood by his mother, "My mother taught me this. . . . I mean, I made him memorize it as a message to her. . . . Never mind. Okay, the whole thing goes, 'The future is not set. There is no fate but what we make ourselves.'" This we must remember. Our fate is not set. The most important question about these new and powerful technologies is who will actually do the making, and why.

Bibliography

Adee, Sally. "Winner The Revolution Will Be Prosthetized." *IEEE Spectrum*, January 1, 2009. http://spectrum.ieee.org/robotics/medical-robots/winner-the-revolution-will-be-prosthetized.

Armstrong, Doree, and Michelle Ma. "Researcher Controls Colleague's Motions in 1st Human Brain-to-Brain Interface." University of Washington news release, August 27, 2013. http://www.washington.edu/news/2013/08/27/researcher-controls-colleagues-motions-in-1st-human-brain-to-brain-interface/.

Bendor, Daniel, and Matthew A. Wilson. "Biasing the Content of Hippocampal Replay during Sleep." *Nature Neuroscience* 15, no. 10 (2012): 1439–1444. doi:10.1038/nn.3203.

Bruce, Jaimee. "Let There Be Light! Bioluminescence Breakthrough in Shrimp Can Track Brain Activity."

Nature World News, October 30, 2016. http://www.natureworldnews.com/articles/30873/20161030/light-bioluminescence-breakthrough-shrimp-track-brain-activity.htm.

Carvalho-de-Souza, João L., Jeremy S. Treger, Bobo Dang, et al. "Photosensitivity of Neurons Enabled by Cell-Targeted Gold Nanoparticles." *Neuron* 86, no. 1 (2015): 207–217.

Clynes, Manfred E., and Nathan S. Kline. "Cyborgs and Space." *Astronautics*, September 1960, 26–27, 74–76.

Delgado, José M. R. *Physical Control of the Mind: Toward a Psychocivilized Society*. New York: Harper and Row, 1969.

Fikes, Bradley J. "Memories Erased, Restored in UCSD Study." *San Diego (CA) Union-Tribune*, June 1, 2014. http://www.sandiegouniontribune.com/business/biotech/sdut-memory-malinow-ucsd-2014jun01-story.html.

Freud, Sigmund. *Civilization and Its Discontents*. Edited by James Strachey. New York: Norton, 1961.

Furness, Dyllan. "Rice University Just Democratized Optogenetics with an Open-Source Platform." *Digital Trends*, November 7, 2016. http://www.digitaltrends.com/cool-tech/lpa-optogenetics/.

Gallagher, John. "WSU Research Leads to Algae Treatment for Blindness." *Detroit Free Press*, September 27, 2016. http://www.freep.com/story/money/business/columnists/2016/09/27/retrosense-wayne-state-optogenetics-allergan-technology/90905706/.

Gorman, James. "Brain Control in a Flash of Light." *New York Times*, April 21, 2014. http://www.nytimes.com/2014/04/22/science/mind-control-in-a-flash-of-light.html.

Gray, Chris Hables. "Consciousness Studies: The Emerging Military-Industrial-Spiritual-Scientific Complex." *Anthropology of Consciousness* 18, no. 1 (2007): 3–19.

Gray, Chris Hables. *Cyborg Citizen: Politics in the Posthuman Age*. New York: Routledge, 2001.

Guile, Steven. "A Shock to the System." *Wired*, March 1, 2007. http://www.wired.com/2007/03/brainsurgery/.

Hacking, Ian. *Representing and Intervening: Introductory Topics in the Philosophy of Natural Science*. Cambridge: Cambridge University Press, 1983.

Heaven, Douglas. "First Mind-Reading Implant Gives Rats Telepathic Power." *New Scientist*, February 28, 2013. http://www.newscientist.com/article/dn23221-first-mind-reading-implant-gives-rats-telepathic-power.html.

Horgan, John. "The Forgotten Era of Brain Chips." *Scientific American* 293, no. 4 (2005): 66–73. https://www.wireheading.com/delgado/brainchips.pdf.

Independent (London). "Scientists Believe They May Have Implanted an Image in the Mind of a Mouse." August 11, 2016. http://www.independent.co.uk/news/science/image-implanted-mind-mouse-experiment-science-breakthrough-a7185641.html.

Jha, Alok. "False Memory Planted in Mouse's Brain." *Guardian* (London), July 25, 2013. http://www.theguardian.com/science/2013/jul/25/false-memory-implanted-mouse-brain.

Kirksey, Eben. "The CRISPR Hack: Better, Faster, Stronger." *Anthropology Now* 8 (2016). http://anthronow.com.

Kramer, Adam D. I., Jamie E. Guillory, and Jeffrey T. Hancock. "Experimental Evidence of Massive-Scale Emotional Contagion through Social Networks." *Proceedings of the National Academy of Sciences of the United States of America* 111, no. 24 (2014): 8788–8790. doi:10.1073/pnas.1320040111.

McLuhan, Marshall, and Quentin Fiore. *The Medium Is the Massage*. New York: Bantam, 1967.

Mentor, Steven. "The Machinery of Consciousness: A Cautionary Tale." *Anthropology of Consciousness* 18, no. 1 (2007): 20–50.

Moe, Marie. "Go Ahead, Hackers. Break My Heart." *Wired*, March 14, 2016. https://www.wired.com/2016/03/go-ahead-hackers-break-heart/.

Moreno, Jonathan D. *Mind Wars: Brain Science and the Military in the Twenty-First Century*. New York: Bellevue Literary Press, 2012.

Noonan, David. "Mind Craft." *Smithsonian*, May 2014, 38–47.

Nuffield Council on Bioethics. *Novel Neurotechnologies: Intervening in the Brain*. London: Author, 2013. http://nuffieldbioethics.org/project/neurotechnology.

Polhemus, Ted, ed. *The Body Reader: Social Aspects of the Human Body*. New York: Pantheon, 1978.

Ramirez, Steve, Xu Liu, Pei-Ann Lin, et al. "Creating a False Memory in the Hippocampus." *Science* 341, no. 6144 (2013): 387–391.

Regalado, Antonio. "Military Funds Brain-Computer Interfaces to Control Feelings." *MIT Technology Review*, May 29, 2014. http://www.technologyreview.com/news

/527561/military-funds-brain-computer-interfaces-to-control-feelings/.

Royal Society. *Brain Waves Module 3: Neuroscience, Conflict, and Security*. London: Author, 2012.

Sanders, Robert. "Sprinkling of Neural Dust Opens Door to Electroceuticals." *Berkeley News*, August 3, 2016. http://news.berkeley.edu/2016/08/03/sprinkling-of-neural-dust-opens-door-to-electroceuticals/.

Smith, Kerri. "Brain Decoding: Reading Minds." *Nature* 502, no. 7472 (2013): 428–430. doi:10.1038/502428a.

Tenner, Edward. *Our Own Devices: The Past and Future of Body Technology*. New York: Knopf, 2003.

Tennison, Michael N., and Jonathan D. Moreno. "Neuroscience, Ethics, and National Security: The State of the Art." *PLOS Biology* 10, no. 3 (2012): e1001289. doi:10.1371/journal.pbio.1001289.

Tucker, Patrick. "A Breakthrough in the Checkered History of Brain Hacking." *Defense One*, July 1, 2014. http://www.defenseone.com/technology/2014/07/breakthrough-checkered-history-military-brain-hacking/87709/.

Vlahos, Olivia. *Body: The Ultimate Symbol*. New York: Lippincott, 1979.

Whitman, Walt. "I Sing the Body Electric." In *Leaves of Grass*. New York, 1855.

FILMS

Bride of Frankenstein. Dir. James Whale. 1935. Although the actual bride of Frankenstein is killed in the book (and the first movie) by Frankenstein's *monster*, this is still a great movie, starring Boris Karloff as the monster.

Frankenstein. Dir. James Whale. 1931. While not the first, generally considered the best screen version of Mary Shelley's book *Frankenstein* (1818). This is the urtext for thinking about bioelectronics and monstrosity. Starring Boris Karloff as the monster.

Hardwired. Dir. Ernie Barbarash. 2009. Raises important issues about marketing, branding, and the growing neuroeconomy.

Johnny Mnemonic. Dir. Robert Longo. 1995. Based on the William Gibson short story (*Omni*, May 1981). A cyberpunk version of a possible twenty-second century.

RoboCop. Dir. Paul Verhoeven. 1987. "Part man, part machine, all cop."

The Terminal Man. Dir. Mike Hodges. 1974. Based on the novel by Michael Crichton (New York: Knopf, 1972), this is a very thoughtful look at the dynamic dance of feedback, hubris, and technology.

The Terminator. Dir. James Cameron. 1984.

Terminator 2: Judgment Day. Dir. James Cameron. 1991.

CHAPTER 8

Genetics and Epigenetics

Michael Bess
Chancellor's Professor of History
Vanderbilt University, Nashville, TN

This chapter explores the technologies of genetic bioenhancement, as well as their potential societal, psychological, and moral implications. After discussing the nature of genetic factors and their influence on human traits and capabilities, the chapter briefly surveys the current state of the art in genetic intervention technologies. It then moves into future territory, exploring the possibility for creating designer babies or other kinds of genetically modified persons. The field of epigenetics, and its promising potential as a vehicle for modifying human traits, is also described. The chapter concludes with a discussion of some of the exciting (and frightening) potential outcomes of a possible future civilization in which millions of people have begun systematically modifying their genetic makeup.

HOW GENES HELP TO MAKE YOU WHO YOU ARE

When Austrian botanist Gregor Mendel (1822–1884) conducted his pathbreaking experiments on pea plants in the 1850s and 1860s, he was investigating a phenomenon that had already been familiar to farmers for millennia: offspring bear many of the characteristics of their parents, and this fact can be put to good use in selectively breeding certain traits into a species of plant or animal, channeling the evolution of their bodies down paths preferred by humans. Over time, a wild species can be radically transformed in this manner: from aurochs to milk cow, from einkorn to domesticated wheat, from wolf to dachshund.

Over the more than 150 years since Mendel, the science of genetics has steadily advanced, with a notable acceleration in the pace of discovery during the decades since World War II (1939–1945). By the time the Human Genome Project came to fruition in 2003, scientists had pieced together the following basic picture: Traits pass from parents to offspring through the mediation of an extraordinary polymer molecule, deoxyribonucleic acid (DNA). This molecule, which naturally takes the shape of a double-stranded helix, is itself composed of four principal chemicals: adenine, guanine, cytosine, and thymine. The arrangement of these four chemicals along the helix of the DNA molecule creates a code, and out of this code emerges the complex set of instructions that ultimately governs the development and functioning of an entire biological organism. Thus, alterations in the code can sometimes (but not always) result in concomitant alterations in the characteristics of the larger organism (Barnes and Dupré 2008).

Scientists use the word *genotype* to refer to this ensemble of DNA code, the creature's full set of hereditary information. It stands in contrast to the *phenotype*, the totality of actual traits observable in the organism as a whole. In humans, DNA is contained within forty-six chromosomes assembled in twenty-three pairs—all twenty-three of them tucked away within the nucleus of every cell in the body: muscle cells, hair cells, blood cells, nerve cells, skin cells. At the time of sexual reproduction, the chromosomes in sperm and egg cells follow an intricately choreographed chemical process, dividing and then recombining with the chromosomes coming from another human organism. The twenty-three chromosomes from the father pair up with twenty-three chromosomes from the mother, thereby generating a new individual whose forty-six-chromosome genotype reflects exactly half the DNA from the father and half from the mother.

A gene is a unit of heredity, corresponding to a specific segment of DNA code. Some genes, such as those that determine hair color, are relatively straightforward in their phenotypic expression: one version (or allele) of a specific gene may code for brown hair, for example, whereas a different allele of the same gene may code for blond hair. If the brown allele dominates, the blond-hair gene will remain silent or unexpressed, and the result will be an individual with brown hair. Other genes, however, operate in concert with one or more completely different genes and therefore are expressed only at the phenotypic level if all the required participant genes have been properly activated. These "symphonic" genes play a role (alongside environmental factors) in shaping the more complex traits of an organism, such as capabilities for immune response, varieties of personality, or levels and kinds of intelligence. These genes operate by interacting with other genes and biological processes in extremely complicated ways that scientists are only just beginning to understand. The key point here is that, while some traits can indeed be directly linked with single genes, many other traits emerge from symphonic interactions of genes—and these interactions are not yet fully understood. Although there is a single gene for hair color, there is no unitary gene for intelligence (Rutter 2006).

NATURE AND NURTURE

Some people tend to be attracted to simplistic and deterministic explanations of human behavior. Perhaps it is the clarity that draws them: they are constantly looking for "*the*" irresistible cause of some aspect of humans' actions or basic nature. Sometimes they locate this determining cause in a person's genes: "Your propensity for violent aggressiveness runs in the family bloodline, and there is little you can do to change it." Sometimes they situate the repository of implacable shaping power in environmental factors such as how a person was reared: "Your mother repeatedly spanked you while you were breastfeeding, and this is why you are so hopelessly screwed up." Below are two simplistic versions of the "nature versus nurture" debate that have gained supporters in the past.

NATURE PREDOMINATES OVER NURTURE

Readers of the *Times* of London of November 25, 2004, found the following intriguing headline: "One in Five Women Strays but Maybe She Can't Resist—It's in Her Genes" (Henderson 2004). The article laid out the results of a study of some 1,600 pairs of female twins, concluding that "genetic factors have as much influence over infidelity as they do over medical conditions [such as hypertension or depression] in which their role has long been established." This kind of genetic determinism posits that virtually all aspects of human behavior are ultimately reducible to the coded commands embedded in a person's DNA. Barry Barnes and John Dupré (2008) refer to this kind of deterministic thinking as "astrological genetics"—a notion that our deeds are written into our genomes just as surely and fixedly as they are written in the stars. Genes determine not just our physical and mental makeup but our very fate—the texture and trajectory of our lived experience.

NURTURE PREDOMINATES OVER NATURE

For a diametrically opposed position, one can turn to another London newspaper, the *Observer.* On February 10, 2001, as the Human Genome Project approached its dramatic finale, the newspaper trumpeted the following headline: "Revealed: The Secret of Human Behaviour; Environment, Not Genes, Key to Our Acts" (McKie 2001). The article text focused on the discovery that humans possessed only about 20,000 genes in their DNA—fewer than had been expected by scientists and far fewer than many other species endowed with an apparently humbler nature. Corn plants, for example, have about 59,000 genes, nearly three times as many as humans. This discovery was hailed by proponents of environmental determinism as a major blow against genetic determinism. "Some biologists," the article noted,

> claim that there are individual genes shaping behaviour patterns, ranging from sexual preferences to criminality, and even including political preference. But one of the scientists behind this weekend's revelations said yesterday that the new evidence demolished such claims.
>
> We simply do not have enough genes for this idea of biological determinism to be right," said Dr. Craig Venter, the US scientist whose company Celera was a major player in the sequencing project. "The wonderful diversity of the human species is not hard-wired in our genetic code. Our environments are critical. (McKie 2001)

THE ASCENDANCY OF COMPLEX MODELS

For those who find such simple and monocausal explanations appealing, the coming century is likely to be an increasingly frustrating epoch. Among both natural scientists and social scientists, rigidly deterministic and monocausal models of explanation are increasingly giving way to complex models involving feedback loops, self-organization, emergent properties, and multilevel and multimodal causal relationships (Mitchell 2009; Juarrero 1999; Laughlin 2005; Bedau and Humphreys 2008; Holland 1998). The more that is learned about the determinants of human behavior, the more the mirage of simplicity recedes from view. Humans are complicated in ways that are only just beginning to be fathomed.

Both these one-sided visions—"mainly nature" or "mainly nurture"—have turned out to be misleading. Everywhere biologists look, they are finding that in concrete practice genes and environment work together, repeatedly affecting each other in causal sequences that thread constantly to and fro: the two operate in tandem, wholly dependent on each other for their causal efficacy (Rutter 2006; Goldhaber 2012). In their 2008 book *Genomes and What to Make of Them,* Barnes and Dupré make a basic distinction between "genetics," which focuses primarily on the heredity of traits across generations, and "genomics," which focuses on the broader role DNA plays in the cellular housekeeping that keeps an organism functioning properly.

> To operate within the life cycle frame [of genomics] is not to deny the standing of DNA as the hereditary material so much as to leave it behind, along with the older frame [of genetics] in which that description made sense. If living things are life cycles, it is natural to ask not merely what DNA does during reproduction and development, but what it does all the time, and what role it plays as a part of the continuously functioning cells that are parts of larger organic systems. (49)

Here, by way of illustration, is an example of one basic cellular housekeeping process in which genes are now known to operate as key players. It has to do with repairing the damage done to cells by ambient heat, and it is found in many species, ranging from bacteria to plants to fish to mammals (including humans). "When an organism is exposed to high temperatures," writes biologist Tara Rodden Robinson,

> a suite of genes immediately kicks into action to produce *heat-shock proteins.* Heat has the nasty effect of mangling proteins so that they're unable to function properly, referred to as *denaturing.* . . . Heat-shock proteins are produced by roughly 20 different genes and act to prevent other proteins from becoming denatured. Heat-shock proteins can also repair protein damage and refold proteins to bring them back to life. The genes that make heat-shock proteins are always on stand-by, ready for action as soon as heat creates a need for them. . . . These genes protect you from the effects of stress and pollutants. (2010, 158)

How, precisely, do the genes "sense" that the organism has been exposed to dangerous levels of heat? Scientists have made great strides toward answering this question (Wu 1995). The mechanism hinges on a substance known as HSF, or heat-shock transcription factor, which is present in the cell under normal conditions as a monomer, or single molecule. In its monomer form, HSF possesses no function at all: it is simply inert. But when exposed to ambient temperatures above a certain precise threshold (which itself differs from species to species), the HSF monomer responds by bonding to other HSF monomers nearby, forming a three-molecule polymer, or trimer. In this trimerized form—and only in this form—HSF is able to bond to DNA molecules along a highly specific site, triggering them to action. The HSF trimer, in this sense, acts like a key unlocking a particular genetic door. The DNA, now activated, initiates the RNA transcription process that in turn sets into motion the cell's production of heat-shock proteins.

How far we are, in this description, from the nature versus nurture debate! Ambient heat triggers trimerization of HSF, which in turn activates a gene, which in turn causes the production of heat-shock proteins. DNA, in the heat-shock response mechanism, certainly performs a key role—but only after it has been brought into play by a specific environmental factor. This is what science writer Matt Ridley (2003) means when he argues that one should think not of "nature versus nurture" but rather of "nature *via* nurture": genes and environmental factors working together, like interlocking pieces in a complex machine, to produce predetermined and finely calibrated outcomes that are also exquisitely sensitive to shifting environmental conditions.

TODAY'S STATE OF THE ART IN HUMAN GENETIC INTERVENTIONS

Louise Brown, the world's first test-tube baby, was born in 1978. She had been coaxed into existence by means of in vitro fertilization (IVF), a technological process in which an egg was extracted from her mother's uterus and then placed in a petri dish alongside sperm from her father. Inside the dish, the sperm fertilized the egg; then the newly fertilized egg, or zygote, was implanted back inside the womb of her mother. Nine months later, Louise Brown was born.

Some observers proclaimed her birth to be a violation of nature, an affront to morality, and an act of technological hubris that would surely result in all manner of deformities for

Brown as an individual, as well as in the degeneration of the broader social order (Henig 2004). In the end, however, perhaps the most surprising outcome of Brown's story is precisely how normal it ultimately turned out to be. Brown grew up healthy and well adjusted. She reported no major psychological ill effects from knowing that she had been conceived in a glass dish. With the passing of time, other families who had despaired of ever having children increasingly began turning to IVF. By the second decade of the twenty-first century, a total of some 250,000 babies had been born as a result of the process in the United States alone; by that point, the cost of the procedure had come down to an average of about $40,000 (Hammoud et al. 2009; Chambers et al. 2009).

In itself, the technique of IVF has nothing to do with genetic modification: Brown's genome was no different than if she had been conceived the old-fashioned way. Nevertheless, the technique opens the door to genetic interventions, because it presents scientists and doctors with a developing human zygote in a dish, accessible to various forms of manipulation that would be impossible (or far more difficult) in utero. Once scientists learned how to extract DNA from the zygote and interpret the sequences of code they saw there, they could begin making predictions about the phenotypic traits the resultant baby would possess. Here began, in the late 1980s, the technique known as preimplantation genetic diagnosis (PGD; Harper 2009).

The concept is simple enough. In most cases, the IVF process results in multiple human zygotes growing in a petri dish. Doctors extract DNA from one of the zygotes and screen it for certain features associated with major genetic diseases, such as Down syndrome, Tay-Sachs disease, cystic fibrosis, or muscular dystrophy. If the DNA tests positive for a disease, the embryo is discarded and allowed to die. Once an embryo is found that tests negative for all the genetic pathologies whose DNA signature scientists know how to recognize (there were over fifty as of mid-2017), that embryo becomes the one that gets implanted into a woman's uterus for gestation into a baby. Because all these diseases are single-gene traits, the scientists can know with 100 percent certainty that the resultant child will be free of those specific ailments (Wailoo and Pemberton 2006).

However, the moral status of PGD is complicated by the fact that it can be used not just to screen for disease but also to screen for other genetically determined features, such as gender. A couple that wants a baby boy at all costs could in theory pay for the IVF and PGD procedures to be performed, asking the doctors to discard all the XX embryos. The resultant child would be a boy—and those parents would thereby have crossed a clear line into a form of positive trait selection that has nothing to do with medicine or illness.

PGD, in other words, could become the first practical technique for producing designer babies—some of whose modified traits may fall more under the category of enhancement (for example, increased athletic ability) rather than therapeutics (treating or preventing a disease). The more scientists learn about how particular genetic factors work to generate specific phenotypic outcomes, the more potent PGD will become as a means of selecting the trait profiles of offspring. To be sure, the technique cannot add new traits to the embryo: it can only select embryos that already possess one given trait profile as opposed to another. Still, the door has now been opened, through IVF and PGD, for the first major forms of nontherapeutic trait selection in humans. In today's world, such forms of trait selection remain illegal—even in countries such as China and India where sex selection (almost always in favor of boys) is in actual practice very common. How long such legal restraints would continue to exist, if PGD began offering parents a reliable means of selecting for such traits as good looks, enhanced resistance to viruses, or talent in sports, is of course entirely another matter.

THE POSSIBILITY OF DESIGNER BABIES

Some experts, such as psychologist Steven Pinker (2009), consider the genetic redesign of humans mere science fiction:

> Many of the dystopian fears raised by personal genomics are simply out of touch with the complex and probabilistic nature of genes. Forget about the hyperparents who want to implant math genes in their unborn children, the "Gattaca" corporations that scan people's DNA to assign them to castes, the employers or suitors who hack into your genome to find out what kind of worker or spouse you'd make. Let them try; they'd be wasting their time.

Pinker is quite right to emphasize that most genetic interventions will inevitably prove to be, in individual cases, a bit like rolling the dice. Because of the complexity of the bidirectional causal interplay linking genes and environmental factors, it will only rarely be possible for parents to tweak a specific gene in their unborn offspring, with the certainty that this will yield a precise phenotypic result. Such a scenario may be achievable for highly specific traits such as the ability to taste bitterness, which stems from a single well-characterized segment of DNA code (the gene known as *TAS2R38*, located on chromosomal region 7q36; Duffy et al. 2004). But if one is talking about such ethereal traits as mathematical capabilities or employee attributes, the complexity of the underlying causal relations unavoidably imparts a probabilistic quality to the genetic intervention. Based on what scientists have learned thus far about genetic causality, it seems safe to predict that parents will *not* be able to order character traits in their offspring the way one orders toppings for a pizza. If they try, they may ask for pepperoni but wind up getting anchovies instead.

Yet Pinker is ignoring here a quite different possibility—namely, that parents might still be able to influence the likelihood that certain traits in their offspring will be manifested strongly or weakly. You will never be able to guarantee absolutely that your child will grow up to be six-foot-four or possess an IQ of 180, but you might nonetheless be able to use genetic tools to alter the underlying *probability* of her turning out taller or smarter than she would have. This tweaking of probabilities is what is meant when people talk about designer babies.

The subtlest features of human character and personality, such as shyness or IQ, stem from multimodal interactions of many suites of genes that are themselves interacting with numerous environmental factors over time. Having acknowledged this, however, it may nonetheless still be the case that such ethereal traits could prove susceptible to genetic tweaking and tinkering. Why? Because it may not be necessary, in practice, to undertake a wholesale redesign of the *entire* cascade of causal factors that determines the nature of those traits. A simple intervention at one key point in the process may prove sufficient to bring about major phenotypic changes. One does not necessarily have to understand the full workings of a complex system in order to make precisely targeted alterations in its functioning.

Consider, for example, the following case. In the late 1990s Joe Z. Tsien, a neuroscientist working at that time at Princeton University used genetic engineering techniques to create a strain of transgenic mice whose brains were primed for above-normal expression of a gene known as *NR2B/GRIN2B*. Tsien knew that this gene regulates levels of the neurotransmitter NMDA, which plays a key role in cognitive functioning. To his astonishment, the results were dramatic: his genetically engineered mice performed up to five times better on tests of learning and memory than unmodified mice (Tang et al. 1999; Leutwyler 1999).

Even though Tsien had made only a single modification to the genetic profile of the mice, the change in their capabilities had proved remarkable. This experiment suggested that scientists do not necessarily require a full and deep understanding of the complex systems they are tinkering with: one could get significant results based on the fairly rudimentary understanding of mouse brain function and cognition that prevailed at the time. What was more, Tsien's results were subsequently confirmed in a separate series of experiments by another neuroscientist, James Bibb, and his colleagues, who achieved similar cognitive boosts in their lab mice by turning down the expression of a different gene (Hawasli et al. 2007). These genes, it seemed, were functioning in a manner akin to volume knobs on a stereo—allowing scientists to exercise fairly precise and predictable forms of control over their lab animals' cognitive capabilities.

In 2015, moreover, biologists announced that they had discovered a new and uniquely powerful tool for making changes to DNA: the CRISPR/Cas9 pathway for editing genomes. Biologists are excited about the potential of this technology because it allows for unprecedented precision and relative ease in making modifications to the DNA of organisms ranging from plants and animals to humans themselves. Indeed, this innovative method was so effective that it spurred a flurry of articles, TED talks, and conferences about the exciting—and scary—potentials that now stood on the threshold of being unleashed. Some scientists and bioethicists have concluded that this technology is sufficiently powerful to warrant a moratorium on human gene-editing experiments using CRISPR/Cas9, so that the full societal and moral implications can be assessed (Doudna and Sternberg 2017).

Given the significant evidence that has accumulated in the years since the completion of the Human Genome Project in 2003, it seems entirely reasonable to conclude that the genetic engineering of basic human traits should not be regarded as mere science fiction. If the pace of discovery manifested over recent decades continues over the next half century, it is plausible to assume that our grandchildren may very well live in a world in which genetic modification of human traits has become part of the new normal.

What might such a world be like? Many parents, faced with the option of modifying their children's traits—and surrounded by a social milieu in which many other parents are opting for designer babies—will no doubt feel significant pressure to join the bandwagon. Even those parents who refuse to take on such powers will still be making a morally accountable choice. The decision *not* to participate in partially designing one's children, when these technologies are available to all, will in itself constitute a choice with far-reaching consequences.

Genetic design technologies will compel prospective parents to render their trait preferences much more explicitly than ever before, as they choose to boost some traits for their offspring and rule out or play down others In such a context, it will be harder for them to avoid subsequent disappointment if their kids fail to display the desired phenotype or do so in ways that defy previously articulated hopes. Some particularly obtuse parents may even sue the genomics companies that oversaw the design process for their children, demanding their money back.

Spouses will need to find ways to negotiate successfully on the selection of traits: this can be expected to exert an unprecedented kind of stress on marital relationships, resulting in not a few cases of divorce. After all, this kind of choice is not like selecting a car at a dealership or a resort for a vacation: the stakes involved could not be higher, affecting virtually all aspects of the family's present and future. Many individuals lack the interpersonal skills required for working through entrenched differences of opinion on such weighty decisions. One strategy for resolving such difficulties might be for the two parents to compromise, trading off traits desired by one spouse for separate traits desired by the other. Another strategy could be for one parent to have the principal say in shaping the traits of the first child, while the other parent would have the major priority in selecting attributes for the second child. In both these cases, however, new sources of resentment could be expected to come into play. In the first case, you might find yourself bitterly accusing your spouse of having chosen some trait in your kid that resulted in a bad outcome: "You were the one who insisted on extreme musical ability, and now all he does is hang out with the druggies in his rock band all day and night. If you hadn't pushed so hard on that single trait, he would have been more moderately balanced like the other kids!" In the second case, one might expect each parent to develop an especially strong bond with the child whom he or she had played a lead role in designing; this in turn could introduce powerful stressors, not only into relations between spouses but also into the overall family dynamics among spouses and kids. Would one kid be considerably more "yours" than "ours" because of her trait profile?

KNOWING THAT YOU WERE PARTIALLY DESIGNED

Difficult as the design process will be for parents, it could in some ways pose even more troublesome problems for the kids. What would it be like for you to imagine the conversation that took place between your parents, as they sat down to lay out the desired trait parameters for the human being who was eventually to become you? In today's society, parents certainly do make choices that profoundly affect the well-being of their children—which city to live in, what school to attend, what church to join (if any), and so on. Nevertheless, the child still possesses a significant measure of autonomy in choosing how to respond to these momentous parental decisions. She may resist, she may go along wholeheartedly, she may pretend to go along wholeheartedly—the decision is ultimately hers. But the shaping of one's innate trait profile is quite another matter: it affects the fundamental platform of capabilities, attitudes, preferences, and propensities that make up a person's identity. If this platform has been modified by your parents, you might find yourself wondering: "To what extent were my very reactions, the tenor of my thoughts, my visceral likes and dislikes, partially preprogrammed into me from the start?"

German philosopher Jürgen Habermas addresses this question in his 2003 book titled *The Future of Human Nature.* Habermas maintains that the selection of a portion of your traits by your parents might undermine your basic autonomy in two powerful ways. First, it would impose someone else's preferences on your innate constitution: it thereby subjects your deepest identity to an external limiting factor having a human rather than natural origin. By definition, this subordination of a portion of your being to another person's purposes and ideas—however benevolently intended—reduces your autonomy. Second, from an internal perspective, it would also probably alter your own perception of who you are and what you can aspire to be: Habermas argues that it might feel like a straitjacket on your potential to be whatever you wanted.

Even if your parents resolutely refused to tell you what choices they had made for you, you would still face a disconcerting set of questions: "Are my preferences, tastes, and achievements the partial result of the predispositions engineered into my being before I was born?" If you become an accomplished musician, for example, and love the whole enterprise of making music, you may still wonder: "To what extent am I merely playing out the fact that Mom and

Dad chose this for me? Is my musical talent really my own, or is it more like a role that I have been preoriented to take on? Deeper still, is my great joy in making music *itself* the result of certain dispositional factors inserted into my genes by a splice?"

Precisely because the causal interplay between genetic and environmental factors is so complex, these are questions that no one will ever be able to answer with any certainty: the actual extent to which one's phenotypic attributes are "engineered" will remain inherently unknowable. For some, this uncertainty may feel liberating. For others, it may have the opposite effect, gnawing at them with self-doubt and second-guessing: Who am I, if I am partially designed, like a product or commodity?

EPIGENETICS: A NEW PATHWAY FOR GENETIC MODIFICATION?

The key moral question raised by the prospect of designer babies is therefore this: Can it ever be OK for one human being to modify the genetic profile of another—without that individual's consent? Over the coming decades, a new technology for genetic modification may become available—a pathway that will allow people to tinker with their own genetic constitution at any point in their lives. Such a technology would deftly sidestep the thorny moral problems raised by designer babies, for it would allow people to make modifications on themselves as freely consenting adults making choices regarding their own bodies and minds.

How would such a technology work? It would rest on epigenetics—a cutting-edge scientific domain that has undergone burgeoning growth since the late 1990s. Although definitions vary, an epigenetic process can best be described as any molecular mechanism that alters the expression of genetic information without altering the underlying DNA sequence itself. The more scientists have learned about the functioning of genes, the more they have come to recognize the key role played by the mediating factors that regulate DNA activation, deactivation, and transcription. Many of a human's 20,000 genes, it turns out, are constantly being switched on and off in complex sequences and combinations, as the cells in the body respond to shifting environmental conditions. When an epigenetic modification takes place, the underlying DNA code stays the same, but certain sections of it are silenced whereas others are activated, resulting in an infinite variety of carefully calibrated results. A good analogy is that of a piano keyboard, whose keys are like the letters of DNA code: although all the keys are always present, the nature of the actual music depends on which keys are being pressed, in what sequence, and with what pressure. In this analogy, your epigenome—the sum total of epigenetic molecular factors—is like the pianist, playing first one tune, then another, then another, on the keyboard of your DNA.

Scientists have discovered a wide variety of epigenetic mechanisms, the two most common of which are known as DNA methylation and histone acetylation (Carey 2012; Francis 2011). To simplify (considerably): when certain segments of DNA code letters acquire an additional methyl molecule on the outside of their structure, those segments tend to be shut down or expressed more *weakly* than their nonmethylated form. Methylation of DNA does not change the underlying code: it merely instructs the surrounding cellular environment to ignore that particular section of coded letters. The exact opposite effect occurs through the mechanism of histone acetylation. Every DNA molecule in the human body is wrapped around a protein structure known as a histone; when an acetyl molecule becomes attached to the histone of a particular DNA segment, that segment usually (though not always) tends to be activated or expressed more *strongly* than its non-acetylated form.

Here, therefore, are two molecular mechanisms that act like the pianist described above, selecting certain keys for activation and others to be silenced, in a constantly shifting sequence that results in the specific genetic "music" your body plays at any given moment.

Scientists are making rapid progress in understanding not only how these epigenetic mechanisms work but also how they can be modified and tinkered with. Through carefully targeted epigenetic modifications, they may be able to start treating certain diseases or afflictions, such as cancer, obesity, depression, or autism. Once they become sufficiently adept at doing this, they would possess a powerful new tool for making alterations to the genetic factors that play such a key role in making each of us who we are. At this point, then, a major new vehicle for human bioenhancement may become available to humankind. Instead of modifying the DNA in people's cells shortly after conception, scientists would operate indirectly via the epigenome, altering the molecular mechanisms that modulate the *expression* of DNA code.

The advantages of this indirect method would be significant. Epigenetic modifications can be made at any point in the lifetime of an organism and hence would allow people to wait until adulthood to make alterations to their own genetic makeup. This neatly sidesteps the problem of informed consent that undermines the desirability of designer babies. What is more, many epigenetic modifications will probably be reversible in nature, which means that people will be able to make changes to their bodies and minds without having to commit to the kinds of permanent alterations that are required in altering the DNA of designer babies. In many cases, you would be able to undo earlier epigenetic modifications you had made if you decided you did not like the result; you could also continue to make further modifications on your epigenome over time, tweaking and upgrading your traits over time. You would become, in effect, a sort of genetic work in progress.

Not surprisingly, epigenetics is generating a great deal of excitement among doctors and biologists, as well as among pharmaceutical companies and government-funded research labs (Carey 2012; Francis 2011). If it lives up to its promise—and it is increasingly looking as though it will—it will also open up a powerful new avenue for human genetic bioenhancement.

ARTIFICIAL CHROMOSOMES?

The idea of an upgradable gene pack may sound like science fiction, but it is not as far fetched as it initially seems. In his 2002 book *Redesigning Humans*, biotech entrepreneur Gregory Stock describes a scenario—based on an extension of present-day technologies—that would introduce precisely this kind of flexibility into DNA-based genetic interventions. Stock's argument hinges on the technology of artificial chromosomes, which were first developed in the late 1980s and early 1990s, using the genomes of bacteria and yeast. An artificial chromosome is in many ways nothing but a pared-down version of a natural chromosome: it has telomeres on the ends, a centromere at the midpoint, and sequences of DNA base pairs along its main structure. The difference, of course, is that scientists have chosen which specific DNA base pairs to load onto the chromosome: they have assembled it, piece by piece, using a variety of recombinant technologies.

In 1997 scientists announced the creation of the first human artificial chromosome—a relatively small construct comprising about six million DNA base pairs. (Most human chromosomes are ten to forty times larger.) The breakthrough was significant, the study's authors noted, because it opened the door for the development of new insertional vectors "capable of introducing and stably maintaining therapeutic genes in human cells" (Harrington et al. 1997, 353). By 2011 scientists had coaxed artificial chromosomes to function successfully within human embryonic stem cells. This remains very much an experimental technology, to be sure, but in Stock's view it holds singular promise for the long haul.

> Adding a new chromosome pair (numbers 47 and 48) to our genome would open up new possibilities for human genetic manipulation. The advantages of putting a new genetic module on a well-characterized artificial chromosome instead of trying to modify the genes on one of our present 46 chromosomes are immense. Not only could geneticists add much larger amounts of genetic material, which would mean far better gene regulation, they could more easily test to ensure that the genes were placed properly and functioning correctly.
>
> Because an artificial chromosome provides a reproducible platform for adding genetic material to cells, it promises to transform gene therapy from the hit-and-miss methods of today into the predictable, reliable procedure that human germline manipulation will demand. (2002, 66)

Stock acknowledges the many hurdles that would have to be surmounted before such a construct could become practical. Scientists would need to make sure the added genes on chromosomes 47 and 48 did not interfere with the functioning of existing genes. (Down syndrome, for example, is caused by the presence in the genome of an extra copy of chromosome 21—which suggests that even adding too much of *existing* forms of genetic material can yield significant developmental abnormalities.) Extensive animal trials would be required before even the most rudimentary extra chromosome could ever be inserted into a human.

Nevertheless, Stock argues, the key advantage of such a technology would lie in the flexibility it would confer. An artificial chromosome could be loaded with chemical switches that allowed specific genes to be turned on or off at will, simply by taking a pill containing the right chemical trigger for activation or deactivation. In addition, the entire chromosome could itself be designed in such a way as to allow it to be turned off selectively in a person's sex cells—thereby ensuring that the construct would not be passed on to the next generation (Capecchi 2000). In other words, by coupling the technology of artificial chromosomes with the regulatory controls available through chemical interventions (i.e., taking a pill with a trigger chemical), one would get the best of both worlds: genetic alterations that affected all cells in a person's body but that could still be tinkered with or completely shut down at any point in the person's lifetime. These would be "designer baby" modifications, introduced at or near the moment of conception, but they would not be unchangeable or irreversible. They would allow each generation to introduce into its offspring the most up-to-date genetic modules available at the time—or to opt out of such interventions altogether, thereby reverting back to their unmodified inheritance if they so desired. Thus, these kinds of technologies would bring a crucial element of ongoing choice and flexibility into human DNA engineering.

It is worth emphasizing that this flexibility, as envisioned by Stock, would apply only at the moment of transition from one *generation* to the next, when parents design the DNA of their offspring. It would not apply to individual adult humans: adults would still be stuck for their entire lifetimes with the gene pack engineered into them at conception. The only possible element of flexibility for individuals would perhaps be a feature that allowed people to turn off their artificial chromosomes entirely, by means of a chemical trigger taken later in their lifetimes.

A SUPERHUMAN CIVILIZATION OF GENETICALLY MODIFIED HUMANS

There is a Zen story about a man who is rowing across a river in his rowboat. After a while an empty canoe comes along, drifting with the current: it has gotten loose from its mooring and is floating unattended down the river. The empty canoe collides with the man's rowboat, nearly causing it to capsize. He recovers his balance with difficulty then shakes his fist at the empty canoe as it drifts away. He curses loudly at it, red faced, berating it for its irresponsible behavior.

The story, in the Zen context, is meant to underscore the absurdity of getting angry at other people, because the basic premise of Buddhist psychology is "no self"—we are all, whether we know it or not, empty canoes, and the feeling of "I" is an illusion. Therefore, harboring resentment and hatred against other people is just as nonsensical as yelling at an empty canoe.

If we take the story out of its Buddhist context, it nicely illustrates one of the key moral transformations likely to be brought about by genetic enhancement technologies. Up to this point in history, the makeup of one's genotype has lain for the most part outside human control. It is no use for you to harbor resentment for any defects in your innate traits, because neither your parents nor your society had any direct say in shaping your genetic profile. If you wish, you can blame God, or perhaps nature or blind luck—but you cannot blame other people. The canoe, thus far in history, has been empty: no one was steering it.

Genetic technologies change this situation: they put a paddler into the canoe. When things go wrong in the genetic design process, some human being will be partly to blame for the bad outcome. Your parents may have made poor choices in selecting some of your design parameters; the genomics industry may have used defective materials; your society may have unjustly deprived you of access to the best technologies. Where once no human responsibility obtained, a new space will have opened up for agency and hence moral accountability.

With this in mind, below are a few of the basic moral questions that societies may face, once the era of genetic modification comes into its own, and millions (perhaps billions) of humans are busily using genetic tools to modify their bodies and minds.

- What happens to the people who cannot afford to purchase bioenhancement technologies for themselves or their children? Will they be left in the dust by the modified elites, condemned to a lifetime of (relatively) diminished experiences and opportunities? Might this result in a pernicious new form of caste system, written into human biology itself?
- As the technologies advance in sophistication, will people have to constantly seek upgrades and updates, as people do today with their computers, cars, and cell phones? Will the meaning of "normal" constantly shift upward, forcing everyone into endless spirals of performance enhancement and technological self-alteration, merely to keep up?
- What about the people who—for religious or other reasons—refuse to participate in the endless race for ever-rising bioenhancement modifications? Will they become akin to the Amish of today—quaint relics from a bygone era, completely outperformed by the other citizens in their society?
- If people cluster together in different social groupings and subgroupings, based on the particular sets of pharmaceutical, bioelectronic, or genetic modifications they have adopted, might this result over many decades in a gradual fragmentation of the human species into increasingly separate (and mutually incommensurable) lineages?

- Given the shifting fads that typify consumer products today, might one expect to see the emergence of similar "fashion trends" and fads affecting the selection of bioenhancement technologies? Will human bodies and minds come to be characterized by trait ensembles that reflect these fickle, ever-changing social patterns and cultural currents?
- If a large pharmaceutical company has manufactured the package of genetic or epigenetic modifications that you use, will that company's patent rights extend in part over an aspect of your body or mind? Will they be able to sue you or block you from making alterations to the genetic makeup you are building for yourself over time?
- How will genetic modifications affect the patterns of violence, competition, and aggression that characterize contemporary society? Might one expect an aggravation of such tendencies, as people ramp up their capabilities to ever-higher levels?
- If genetic modifications allow human beings to engineer much longer life spans for themselves—perhaps doubling the typical active health span—what impact will this have on ecological sustainability, on intergenerational relations, and on the meaning ascribed to the stages of life?

Summary

This chapter commenced by exploring the nature of genes and how they work together with environmental factors in shaping a person's traits over time. It surveyed today's genetic technologies, such as preimplantation genetic diagnosis, that already have opened the door for simple forms of eugenic selection, and discussed the possibility of designer babies, as well as the profound moral questions they raise about autonomy and consent. Finally, the chapter explored two very different forms of genetic intervention that may become prevalent over the coming decades: epigenetic modifications and artificial chromosomes.

Dr. Frankenstein's creation, as English writer Mary Shelley (1797–1851) envisioned him in her famous 1818 novel, was a solitary experiment, a unique exemplar of an engineered human being. This rendered him a kind of freak, a monster. But these Frankenstein powers of today will not just be applied to one person in isolation. They will be all around us, adopted by millions of people. For many among us, they will form part of our very own individual selfhood. What happens when the majority of people in our society are partially engineered beings?

Bibliography

Barnes, Barry, and John Dupré. *Genomes and What to Make of Them*. Chicago: University of Chicago Press, 2008.

Bedau, Mark A., and Paul Humphreys, eds. *Emergence: Contemporary Readings in Philosophy and Science*. Cambridge, MA: MIT Press, 2008.

Capecchi, Mario R. "Human Germline Gene Therapy: How and Why." In *Engineering the Human Germline: An Exploration of the Science and Ethics of Altering the Genes We Pass to Our Children*, edited by Gregory Stock and John Campbell, 31–42. New York: Oxford University Press, 2000.

Carey, Nessa. *The Epigenetics Revolution: How Modern Biology Is Rewriting Our Understanding of Genetics, Disease, and Inheritance*. New York: Columbia University Press, 2012.

Chambers, Georgina M., Elizabeth A. Sullivan, Osamu Ishihara, et al. "The Economic Impact of Assisted Reproductive Technology: A Review of Selected Developed Countries." *Fertility and Sterility* 91, no. 6 (2009): 2281–2294.

Clayton, Philip, and Paul Davies, eds. *The Re-emergence of Emergence: The Emergentist Hypothesis from Science to Religion.* Oxford: Oxford University Press, 2006.

DeGrazia, David. *Creation Ethics: Reproduction, Genetics, and Quality of Life.* Oxford: Oxford University Press, 2012.

Doudna, Jennifer A., and Samuel H. Sternberg. *A Crack in Creation: Gene Editing and the Unthinkable Power to Control Evolution.* Boston: Houghton Mifflin Harcourt, 2017.

Duffy, Valerie B., Andrew C. Davidson, Judith R. Kidd, et al. "Bitter Receptor Gene (*TAS2R38*), 6-*n*-Propylthiouracil (PROP) Bitterness and Alcohol Intake." *Alcoholism: Clinical and Experimental Research* 28, no. 11 (2004): 1629–1637.

Dupuy, Jean-Pierre. *On the Origins of Cognitive Science: The Mechanization of the Mind.* Translated by M. B. DeBevoise. Cambridge, MA: MIT Press, 2009.

Francis, Richard C. *Epigenetics: The Ultimate Mystery of Inheritance.* New York: Norton, 2011.

Gay, Volney P., ed. *Neuroscience and Religion: Brain, Mind, Self, and Soul.* Lanham, MD: Lexington Books, 2009.

Goldhaber, Dale. *The Nature-Nurture Debates: Bridging the Gap.* Cambridge: Cambridge University Press, 2012.

Habermas, Jürgen. *The Future of Human Nature.* Cambridge: Polity Press, 2003.

Hammoud, Ahmad O., Mark Gibson, Joseph Stanford, et al. "In Vitro Fertilization Availability and Utilization in the United States: A Study of Demographic, Social, and Economic Factors." *Fertility and Sterility* 91, no. 5 (2009): 1630–1635.

Harper, Joyce, ed. *Preimplantation Genetic Diagnosis.* 2nd ed. Cambridge: Cambridge University Press, 2009.

Harrington, John J., Gil Van Bokkelen, Robert W. Mays, et al. "Formation of *De Novo* Centromeres and Construction of First-Generation Human Artificial Microchromosomes." *Nature Genetics* 15, no. 4 (1997): 345–355.

Hawasli, Ammar H., David R. Benavides, Chan Nguyen, et al. "Cyclin-Dependent Kinase 5 Governs Learning and Synaptic Plasticity via Control of NMDAR Degradation." *Nature Neuroscience* 10, no. 7 (2007): 880–886.

Henderson, Mark. "One in Five Women Strays but Maybe She Can't Resist—It's in Her Genes." *Times* (London), November 25, 2004. https://www.thetimes.co.uk/article/one-in-five-women-strays-but-maybe-she-cant-resist-its-in-her-genes-gwk0wgz5blh.

Henig, Robin Marantz. *Pandora's Baby: How the First Test Tube Babies Sparked the Reproductive Revolution.* Boston: Houghton Mifflin, 2004.

Holland, John H. *Emergence: From Chaos to Order.* Reading MA: Addison-Wesley, 1998.

Juarrero, Alicia. *Dynamics in Action: Intentional Behavior as a Complex System.* Cambridge, MA: MIT Press, 1999.

Laughlin, Robert B. *A Different Universe: Reinventing Physics from the Bottom Down.* New York: Basic Books, 2005.

Leutwyler, Kristin. "Making Smart Mice." *Scientific American*, September 7, 1999. https://www.scientificamerican.com/article/making-smart-mice/.

Mandegar, Mohammad A., Daniela Moralli, Suhail Khoja, et al. "Functional Human Artificial Chromosomes Are Generated and Stably Maintained in Human Embryonic Stem Cells." *Human Molecular Genetics* 20, no. 15 (2011): 2905–2913.

McKie, Robin. "Revealed: The Secret of Human Behaviour; Environment, Not Genes, Key to Our Acts." *Observer* (London), February 10, 2001. https://www.theguardian.com/science/2001/feb/11/genetics.humanbehaviour.

Mitchell, Melanie. *Complexity: A Guided Tour.* Oxford: Oxford University Press, 2009.

Pinker, Steven. "My Genome, My Self." *New York Times*, January 7, 2009. http://www.nytimes.com/2009/01/11/magazine/11Genome-t.html.

Ridley, Matt. *Nature via Nurture: Genes, Experience, and What Makes Us Human.* New York: HarperCollins, 2003.

Robinson, Tara Rodden. *Genetics for Dummies.* 2nd ed. Hoboken, NJ: Wiley, 2010.

Rutter, Michael. *Genes and Behavior: Nature–Nurture Interplay Explained.* Malden, MA: Blackwell, 2006.

Stock, Gregory. *Redesigning Humans: Our Inevitable Genetic Future.* Boston: Houghton Mifflin, 2002.

Tang, Ya-Ping, Eiji Shimizu, Gilles R. Dube, et al. "Genetic Enhancement of Learning and Memory in Mice." *Nature* 401, no. 6748 (1999): 63–69.

Wailoo, Keith, and Stephen Pemberton. *The Troubled Dream of Genetic Medicine: Ethnicity and Innovation in Tay-Sachs, Cystic Fibrosis, and Sickle Cell Disease.* Baltimore: Johns Hopkins University Press, 2006.

Wu, Carl. "Heat Shock Transcription Factors: Structure and Regulation." *Annual Review of Cell and Developmental Biology* 11 (1995): 441–469.

CHAPTER 9

Rejuvenation and Radically Increased Health Spans

Michael G. Zey
Professor, Department of Management,
Feliciano School of Business
Montclair State University, Montclair, NJ

Future historians may rank as one of the most significant scientific achievements of our time the radical extension of the human life span to ages of 125 and beyond, along with the concomitant expansion of the human health span, the number of years of healthy living enjoyed by these longer-lived humans. Developments in fields such as genetic engineering and gene therapy, bioprinting (the use of 3-D printing technology to create new skin, body parts, and so on) nanotechnology, stem cell science, tissue regeneration, and drug research are bringing the human species to the very edge of what is labeled herein superlongevity, if not near immortality. Such breakthroughs will enable people to remain healthy, productive, and physically vibrant for most of their lives. This chapter examines how this demographic event will affect careers, marriage, and the human life cycle. It looks at issues such as transhumanism, the human enhancement revolution, rejuvenation, and the strategies society and its citizens could choose to pursue to adapt to this changing demographic landscape.

THE THREE STAGES OF THE SUPERLONGEVITY REVOLUTION

In his 2005 book *More Than Human*, biologist Ramez Naam states that the human life span hovered somewhere between eighteen and twenty-five for most of recorded human history. The average Roman circa 100 CE lived to between twenty and twenty-five (Caldwell 2006). Life expectancy in 1800 western Europe was no more than thirty-five (Riley 2005). By the mid-nineteenth century the longest average life expectancy anywhere, a scant forty years, was among Swedish women. According to age researcher S. Jay Olshansky, while individuals in the past such as Benjamin Franklin, Thomas Jefferson, and John Adams and even some notables of the ancient Greek and Roman eras lived into their sixties and beyond, such long lives were a statistical oddity. Olshansky mentions that such long-lived individuals are known today because their extended longevity gave them the additional time most of their contemporaries did not have in which to achieve memorable political and scientific feats (Roth 2004).

It is at this point that the superlongevity revolution, the rapid growth in life expectancy, begins. The superlongevity revolution can be divided into three distinct stages (see Table 9.1).

During stage I, the "life extension" stage, roughly from 1900 to 2000, average life expectancy significantly increased in most industrialized countries. The average life span of American males rose from forty-seven to its current seventy-eight years, while the average for American women reached eighty. Western Europeans enjoyed similar gains, and many developing countries followed suit, some experiencing life-span increases of 50 percent or more over only a few decades (Zey [2007] 2014). The major catalysts for such gains were initially vaccines and antibiotics, better nutrition, and cleaner water and later medical solutions for some cancers and heart disease, diabetes, and stroke.

Humanity is already in stage II, a period in which people can expect to live well beyond 125 years old as healthy individuals. Demographers predict that American females born in the second decade of the twenty-first century will live to 100. Most countries' life spans have gone up 3.5 to 4 years over the last few decades. Monaco has the highest average life expectancy, nearly ninety years of age overall and ninety-three for females, with Singapore, Japan, and Macau not far behind (CIA 2016).

Stage II's starting date, 2001, was selected because that year Human Genome Project researchers successfully mapped the genetic structure of the human body, establishing the possibility of uncovering the links between particular genes and diseases such as Parkinson's, Alzheimer's, and macular degeneration, as well as many other maladies, and hopefully their cures. As discussed below, stage II innovations will expand human life expectancy to perhaps 125 to 150 years.

The superlongevity revolution's stage III, the "near-immortality" stage, will occur when humanity masters exotic new sciences such as nanotechnology, which is the science of

Superlongevity Stages

Stage	Name	Time period	Impact on life span	Key drivers
I	Life extension	1900–2000	Rapid extension of life span from forties to eighties. Primarily adding years to life.	• Antibiotics • Improvements in public health • Enhanced nutrition • Heart bypass surgery
II	Life expansion	2000–2075	Pushing past 100 to 150, with longer health spans in which people are healthier, more vibrant at every life age.	Eliminating causes of most fatal diseases through: • Genetic engineering • Bioprinting • Stem cell technologies Enhancement of human outer shell via technologies such as tissue regeneration.
III	Infinity	2075–2100 and beyond.	Near immortality, rejuvenation.	• Elimination of cell aging, production of fresh body parts. • Mastery of basic laws and behavior of cells and atoms through developments in: ∘ nanotechnology ∘ "immortality cells"

Table 9.1. MICHAEL ZEY. ADAPTED FROM ZEY, 2014.

constructing and reconstructing any object, including body parts, from the "bottom up," one atom at a time. Molecular-sized nanobots could cruise through a person's bloodstream, seeking out and repairing damaged genes and eliminating harmful viruses. Nanotechnology could be used to design wholly new organs that enable humans to adapt to any physical environment, including the hostile environments of other planets.

Nanotechnology proselytizer Robert A. Freitas Jr. thinks that nanotechnology eventually could help "dechronify" the human body, rolling it back to a younger physical state. Dechronification via nanotechnology could theoretically reduce an eighty-year-old's body to that of a twenty-seven-year-old (Kurzweil and Grossman 2004).

THE GLOBAL MISSION TO EXTEND THE HUMAN LIFE SPAN

The worldwide effort to develop the scientific and technological breakthroughs that will enable humans on average to live to ages 125, 150, and beyond is serious, frenetic, and determined. This section describes this worldwide effort, focusing particularly on the role played by computer and Internet entrepreneurs dedicated to accelerating science's quest for superlongevity if not near immortality.

Larry Ellison, cofounder of software giant Oracle, formed the Lawrence Ellison Foundation to discover ways to reverse the aging process. Peter Thiel, the billionaire cofounder of PayPal, who claims he wants to live to at least 120, has donated $7 million to the Methuselah Foundation to achieve this goal. Methuselah Foundation cofounder Aubrey de Grey also helped create the SENS Research Foundation (SENS is an acronym for "strategies for engineered negligible senescence"), which directs funds to defeat what de Grey sees as the main obstacles to immortality: the body's loss of cells, excessive cell division, inadequate cell death, and mutations in the mitochondria (Isaacson 2015).

Sergey Brin, cofounder of Google, wants to "cure death" through Calico, a biotech subsidiary company of Alphabet Inc., parent of Google. Calico is planning to pump billions into a partnership with US-based pharmaceutical giant AbbVie to develop a drug that mimics *FOXO3*, one of the genes associated with exceptional life span (Sifferlin 2017).

In September 2016 Facebook's Mark Zuckerberg and his wife, Priscilla Chan, announced plans to invest $3 billion for a global effort to cure all diseases during their daughter's lifetime. They also helped set up a $3 million prize given annually to a scientist who develops a breakthrough that can extend human life. In 2015 Silicon Valley billionaire Sean Parker donated $250 million to stimulate collaboration among researchers to fight cancer (Wadhwa 2016).

Ronald DePinho, one of the Lawrence Ellison Foundation's senior scholars in aging and the former president of the University of Texas MD Anderson Cancer Center in Houston, has been involved in breakthrough research on keeping mice young by manipulating their telomeres. These structures cap the tips of chromosomes, keeping them healthy, but they deteriorate over the years, leading to aging. DePinho and his team have made some headway in preventing the telomeres from shortening, at least in mice. While human trials have not been conducted, the research findings provide tantalizing evidence that it might be possible to extend human life and to greatly lengthen humans' "health span." Humans could conceivably live to 100 with the body and internal organs of a twenty-five-year-old.

A California start-up called Ambrosia, founded by medical researcher Jesse Karmazin, has piqued the interest of a number of Silicon Valley moguls. Karmazin contends that transfusing the blood of people twenty-five years or younger into the veins of those thirty-five years or older will rejuvenate the "elders" and stave off disease. To test this theory Karmazin launched a trial slated to include 600 human subjects (Maxmen 2017).

Australian researcher David Sinclair of the University of New South Wales has made major contributions to the fight to eliminate aging. Since the 1930s experiments in mice and humans have demonstrated the relationship between fasting and exercise and increases in life spans. (Calorically restricted rats have lived as much as 40 percent beyond the expected rat life span.) The reason is that extreme fasting and exercise induce the body to produce large amounts of SIRT1, an enzyme that helps the body fight aging.

Sinclair discovered that SIRT1 is the key driver in a host of compounds, such as metformin and resveratrol, that help to slow aging. Sinclair's research also demonstrated that at least 117 different compounds could switch on SIRT1 like those currently in use do. His findings open up the possibility of the development of a new generation of drugs that would have the same life-extending benefits as extreme diet and exercise. A news release touting his research quotes Sinclair as saying that "ultimately, these drugs would treat one disease, but unlike drugs of today, they would prevent 20 others" (University of New South Wales Newsroom 2013).

A 2014 study of 180,000 people showed that diabetics taking the aforementioned antidiabetes drug metformin lived, on average, 15 percent longer than the healthy population. An antiaging trial of metformin is in process. In July 2016 a study was launched in which ten healthy Japanese people were given the compound nicotinamide mononucleotide to see if this drug could retard aging in humans.

One of the most exciting developments in the effort to manipulate genes is CRISPR (clustered regularly interspaced short palindromic repeats), a technology that has the ability to "hone in" on a specific location in a strand of DNA and edit it, with the purpose of removing unwanted sequences or conversely inserting sequences. CRISPR has enabled Chinese scientists to genetically modify pigs, sheep, monkeys, and other animals to change their color and size (Le Page 2015). In 2016 a young girl suffering from leukemia had her life saved by gene editing performed by a team at University College London. CRISPR has supposedly enabled Chinese scientists to edit a human embryo for resistance to HIV (Callaway 2016). With CRISPR it could be possible to transform, through genetic engineering, the next generation of humans into a disease-resistant superspecies with enhanced physical and intellectual capabilities. As more diseases are eradicated, the possibility of people living to very long ages automatically increases in likelihood.

Advances in stem cell science also will enable humans to live healthy vibrant lives for longer and longer periods. The stem cell is an "undifferentiated cell" with the unique ability to transform into a more specialized cell and ultimately become an organ such as a lung, skin, a heart, even neurons. Importantly, these stem cells divide, producing even more stem cells (Diamandis 2017). When we are young these cells exist in great supply and have the ability to flawlessly repair our organs, but as we age, the quality of our stem cells deteriorates because of genetic mutations, and their numbers greatly diminish.

Peter Diamandis, cofounder of San Diego, California–based Human Longevity Inc., claimed at the 2016 Singularity University conference in California that by using stem cells to repair various organs humans might live well into their hundreds and do so in relatively perfect health (Vilvestre 2016). Venture capitalists such as Diamandis are therefore pouring billions, perhaps as much as $170 billion by 2020, into stem cell solutions to chronic and degenerative diseases.

Biologists in Japan, using only a small sample of adult skin, grew corneas, retinas, and the eye's lens by nurturing and growing the tissues composing the human eyeball. At Stanford University, stem cell treatments improved stroke victims' motor functions. Researchers at the University of Southern California injected stem cells into the damaged cervical spine of a twenty-one-year-old male and helped him regain movement and feeling in his arms. Scientists at Harvard University and the Massachusetts Institute of Technology are working on stem cell therapies to help those suffering from hearing loss regain their hearing (Galeon 2017).

Stem cells therapies are being considered as possible treatments for other neurodegenerative conditions, including Parkinson's, amyotrophic lateral sclerosis, and Alzheimer's. Stem cell therapy being pioneered in Puebla, Mexico, is being used successfully to arrest the debilitating effects of multiple sclerosis (BBC News 2017). University of New South Wales scientists are working on a stem cell therapy to regenerate any human tissue that has been damaged by aging, disease, or injury (Creighton 2016).

Bob Hariri, cofounder with Diamandis of Human Longevity Inc., has started a company, LifebankUSA, that enables parents through "private cell banking" to store their children's uncorrupted stem cells at birth from the placenta or umbilical cord, at minus 180 degrees Celsius, for the child's future use (LifebankUSA 2017).

In experiments, scientists at the Salk Institute for Biological Studies in California have taken adult cells in mice back to their embryonic form, turning back the cells' clock (Knapton 2016). As a result, not only do the mice look younger, but they also live 30 percent longer than they ordinarily would. The ultimate goal is to develop a drug that human patients could take to inhibit and possibly reverse the aging process. Human trials are expected by the mid-2020s. According to Juan Carlos Izpisua Belmonte, a professor in Salk's Gene Expression Laboratory, aging "may not have to proceed in one single direction" (Knapton 2016). In other words, aging might be reversed, a process commonly called "rejuvenation."

THE ULTRAHUMAN PHENOMENON: EXPANDING AND TRANSFORMING HUMAN POTENTIAL

An integral component of the superlongevity revolution is the ultrahuman phenomenon, or what some label the human enhancement revolution. Next-generation scientific advancements will exponentially expand human mental and physical abilities. Scientists are working on a retinal chip making it possible for a person to see in the dark, a memory chip wired directly into the brain's hippocampus enabling the chip wearer to perfectly recall everything he or she sees and reads, and "smart pills" that can increase a person's concentration and enhance memory (Del Prado 2015). Some human enhancement technologies are already in use, whereas others are a few years or decades away.

HUMAN GROWTH HORMONE

Human growth hormone (HGH) is among the human enhancement drugs receiving negative publicity from athletes using such substances in the hope of gaining an edge on the playing field. Yet clinics all over the United States, such as the California HealthSpan Institute, the Beverly Hills Rejuvenation Center, Las Vegas–based Cenegenics Elite Health, BodyLogicMD, and the Rejuvalife Vitality Institute in Los Angeles, legally prescribe HGH, the steroid testosterone, and other hormones to men and women who want to live and work at peak levels (Rejuvalife Vitality Institute). In testimonials, people in their forties, fifties, and sixties have attested to the positive impact of HGH on their work performance and their overall sense of well-being, claiming to have experienced enhanced energy, increased mental functioning, improved memory, improved immune system, enhanced sexual energy, and improved physical appearance (California HealthSpan Institute 2017; Cenegenics Elite Health 2017).

A review of the pros and cons of HGH published in *Men's Health* demonstrated a wide disparity of opinions within the medical community about the benefits of HGH (Beil 2016). In a 2016 article on its website, the Mayo Clinic warns that HGH might cause side effects in healthy adults, such as increased insulin resistance, swelling in the legs and arms, and joint and muscle pain. In spite of such concerns, current trends suggest that as people live and work to greater ages they will turn to such hormone therapies to function and work at optimum performance levels (Zeman 2012).

THE INTELLIGENCE ENHANCEMENT REVOLUTION

Since at least the first years of the twenty-first century, managers, students, writers, and scientists have used powerful "nootropic" drugs such as Provigil, modafinil, Ritalin, piracetam, and donepezil, originally created to combat narcolepsy, dementia, and hyperactivity, to improve their concentration and information retention. Harvard Medical School and Oxford University researchers concluded that modafinil users showed improvement in attention span, flexible thinking, coping with novelty, and learning. CNN has reported that many rich entrepreneurs use nootropic "supplements" in an effort to improve their memory and boost their cognitive abilities (Monks 2015). A 2014 article on the Dealbreaker website more or less justified Wall Street wheeler-dealers' use of nootropics in terms of these substances' ability to enhance users' productivity and focus.

In his book *Ageless Nation* ([2007] 2014), Michael G. Zey reported that nootropic drug use among students is widespread, boosting not only their IQs but also sales of these drugs for eventual "off-label" applications. One of ten Cambridge University students was found to be taking nootropics to aid study (Walsh 2014). *Nature* magazine reported on the increased use of nootropics among academics living in a publish-or-perish universe (Maher 2008). Several studies have reported that subjects using nootropics showed a 30 percent improvement in language learning (Vincent and Jane 2014). A great debate is brewing over the safety of prolonged use of "smart drugs." Studies show that side effects could include insomnia, anxiety, headaches, irritability, and heart problems. And there is also the risk of physical and/or psychological dependency (Petrounin 2014).

Various forms of electronic brain stimulation, such as transcranial direct-current stimulation, have been found to improve subjects' memory and intelligence (Batuman 2015). Experiments by Itzhak Fried, a neurosurgeon at the University of California, Los Angeles (UCLA), have revealed that electrical stimulation can improve some forms of memory (Wang 2014). Oxford University's Roi Cohen Kadosh found that children's and adults' math skills improve after their brains are subjected to low-dose electric current. The Defense Advanced Research Projects Agency, the US Department of Defense's research arm, is considering using electronic stimulation to boost intelligence in its "Accelerated Learning" project.

Researchers at Oxford and UCLA have demonstrated that electronically stimulating a subject's brain significantly boosts that person's memory, intelligence, and math abilities.

Consumers can now purchase "brain zappers" such as Thync, a small device you wear on your head that uses neurosignaling to enable you, according to the manufacturer, to energize or calm yourself via a smartphone app (Salmanowitz 2016). In the near future people might visit a "brain boutique" right before going on a job interview or taking their CPA exam or SAT test to electronically "boost" their chances of operating at a peak intellectual level.

BIOPRINTING: MANUFACTURING NEW SKIN, BONE, AND BODY PARTS

One of the most striking developments in the perfection of the human body is bioprinting, similar to traditional 3-D printing except that this process uses people's actual living cells as the basic material in the printing of various body parts. Researchers have bioprinted sheets of skin for grafting and created lifelike prosthetics, as well as a human ear. It is predicted that sometime in the 2020s bioprinted livers and human hearts will be available to meet the ever-widening global demand for donor organs.

Researchers at San Diego–based Organovo, 3D Bioprinting Solutions in Russia, Rokit in South Korea, and the Canadian firm Aspect Biosystems are working furiously to 3-D bioprint bones, ears, muscles, cartilage, blood vessels, and kidney and liver tissue. Japan-based Cyfuse Biomedical is bioprinting blood vessels able to withstand ten times the pressure of blood vessels that exist in the human body (Littler 2017).

Scientists at Wake Forest University's Institute for Regenerative Medicine have bioprinted tissue used to replicate the outer ear. Other scientists at that institute are bioprinting a replacement bladder and have designed a bioprinter to print skin cells onto burn wounds. In March 2013 Connecticut-based Oxford Performance Materials used 3-D printing to create a bone replacement for insertion into a patient's skull (Singleton 2013). Researchers in Spain were able to create skin that looks and behaves just like real skin, complete with an epidermis and a dermis (Wall Street Pit 2017). Researchers from Northwestern University were able to successfully implant prosthetic ovaries in mice who were then able to conceive and give birth (Cossola 2016). Kyoto University scientists succeeded in 3-D bioprinting tubular conduits that can regenerate damaged nerve cells (Buntinx 2017).

Doctors soon might be able to scan wounds and spray on layers of cells to rapidly heal them. At the Wake Forest School of Medicine, Anthony Atala's research team has developed a "skin printer" that can take the data from 3-D scans of test injuries inflicted on some mice and use the data to control a bioprinter head that sprays skin cells to the wounds. In experiments the mice's wounds treated with this technique healed in half the time it usually takes. The US military might use bioprinting to help heal soldiers' wounds on the battlefield (Barnatt 2016).

Skin printing technology used to heal wounds could soon be adapted to improve people's facial appearance (Mearian 2014). In one scenario, face printers would evaporate existing flesh and simultaneously replace it with new cells to whatever shape and appearance the patient specified. In the future people might turn to such technologies to replace their current visage with that of a celebrity, model, or public figure. Criminals and spies could acquire totally new faces to avoid identification, confusing even the most sophisticated biometric ID scanners (RT.com 2014).

Clearly, bioprinting will become a major player in the quest for the creation of the ultrahuman. Its success will depend on the combined efforts of doctors, engineers, and computer scientists, as well as the educated layperson.

ULTRAHUMAN OR TRANSHUMAN

Dmitry Itskov, a Russian billionaire Internet entrepreneur, says his goal is to live to 10,000. To achieve that goal, he is devoting time and money to his 2045 Initiative, an organization composed of experts in fields such as robotics, artificial organ creation, and neural interfaces whose sole purpose is to figure out how to replace human bodies with robotic or holographic avatars by 2045 (Isaacson 2015). Itskov's avatar would still be "you," he contends, with your personality and consciousness, just in a different shell. He expects that most people will choose to house themselves in this robotic avatar because it will be "superior to the biological body in terms of its abilities." And besides, it would be immortal.

Observers have placed Itskov in the transhumanist camp. The futurist known as FM-2030 (who was born Fereidoun M. Esfandiary) laid the theoretic groundwork for a future conjoining of human and machine in his 1989 book *Are You a Transhuman?* Humanity Plus (originally called the World Transhumanist Association), was founded in 1998 and now has members across the globe.

Another leading light of the transhumanist movement is inventor/futurist Ray Kurzweil, who speculates in books such as *The Age of Intelligent Machines* (1990) and *The Singularity Is Near* (2005) that human and machine will fuse into a new form superior to the current organic human. Nanotechnology will make it possible to use molecular-sized nanorobotics nested in the physical human brain to speed up the 100 trillion relatively slow connections between the neurons via high-speed virtual connections. These reconstructed brains will be able to communicate with each other and with powerful artificial-intelligence computer networks with which humans will exchange information.

Kurzweil believes people should welcome this "stage" of human development. As a man-machine human, he says that "we're going to be funnier, we're going to be better at music. We're going to be sexier. We're really going to exemplify all the things that we value in humans to a greater degree" (quoted in Molloy 2017). Kurzweil predicts that the date of this "great transformation" is 2045 (Molloy 2017).

In his book titled *The Future Factor* ([2000] 2004), Zey suggested that Kurzweil's vision of the future human was more machine than human. Certainly, Itskov's future human is not flesh and blood, by design, and in fact might be more cyber than physical. Zey envisions an ultrahuman concept more focused on augmenting and exponentially amplifying the body's natural physical tendencies and abilities without fundamentally changing its structure. Technologies such as cochlear implants and artificial retinas help the deaf to hear and the blind to see, respectively, but while they might make a person ultrahuman, that person is still human. In any event, whether one sees the human future as organic, as de Grey, Zey, and others do, a combination of machine and artificial intelligence or something in between, the unifying goal of all such approaches is the extension of human life and consciousness into the indefinite future.

ADJUSTING TO THE SUPERLONGEVITY REVOLUTION

Over the last century, and especially over the last few decades, society has struggled to adjust to the rapid increase in human life expectancy. While longer and healthier life spans will enable all humans to enjoy novel experiences, expand their social networks, pursue second careers, and guide the lives of their grandchildren and great-grandchildren, the super-longevity revolution also presents US society with a host of unique challenges.

LONGER LIVES, LONGER CAREERS

A nation's prosperity depends on having a robust, creative, and productive workforce producing services and goods, starting new companies, and paying the taxes that fund the government's operation and support social safety nets such as Social Security and Medicare. Traditionally, Americans have worked from their late teens or early 20s to roughly age 60 to 65. This pattern was sustainable when people lived to 63 or 64. However, Americans retiring at 65 can now expect to live 20 to 30 more years and soon perhaps 50 or 60 more years, all that time earning and producing little and depending on savings, pensions, and government safety nets for support.

As people live longer, year by year the retiree cohort will grow larger relative to that same work group of age 25 to 65. The United States already faces a crisis funding the nation's "safety net" programs. When Social Security was first instituted in the United States in 1935, there were sixteen workers for every retiree. Now that ratio is inching closer to two-to-one. Laurence J. Kotlikoff, a professor of economics at Boston University, calculates that if current retirement patterns persist, Social Security, Medicaid, and the Affordable Care Act will run up a fiscal gap of over $200 trillion within a few decades, regardless of whether the country raises the 25 to 65 age cohort's tax rates and safety net contributions significantly (NPR 2011).

There is a solution to this looming economic dilemma. As the life span and the health span continue to increase, so too will the time people have to spend in the workforce. If people work into their seventies and eighties they will continue to contribute to GDP growth and pay into safety net programs such as Social Security and Medicare, thereby helping the country avert huge budgetary shortfalls.

In this new world, corporations and government should strongly consider recruiting and retaining older workers to take advantage of their experience, wisdom, and organizational skills. Government leaders should encourage organizations to hire and retain seniors, reminding companies that by laying off mature workers they are shifting the burden of paying for social safety nets with higher taxes to themselves and their younger workers.

Superlongevity will change conventional career patterns in which a person trains in college or technical school for a career, works in that career for thirty to forty years, and then retires to pursue hobbies or leisure interests. People anticipating careers spanning not three but nine or ten decades might pursue schooling, career, reschooling, recareering, a career hiatus or sabbatical, retirement, reschooling, and so on in sequences attuned to their personal preferences and needs (see Figure 9.1).

TRANSFORMATION OF MARRIAGE AND THE FAMILY

Superlongevity will affect a host of marital and childbearing trends. The radical extension of the life span will most certainly affect the age at which married couples start having children. Social scientists commonly assume that men and women anticipating living to 125 or 150 in reasonably good health will spend their teens and twenties not raising children but pursuing the training and education needed to ensure career success.

The opposite could happen just as easily, however, with people deciding to become parents relatively early in life because technology has made it easier to pursue their careers while engaging in other activities, including raising a family. Distance learning will enable a

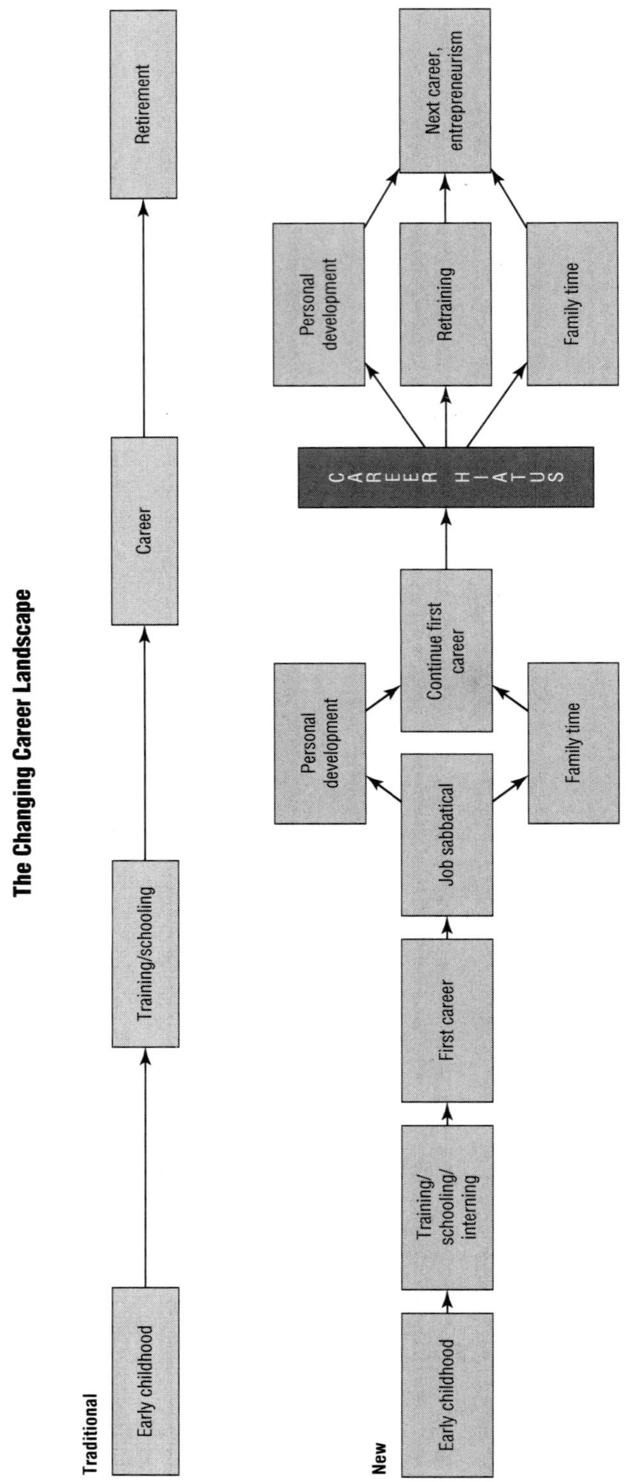

Figure 9.1. MICHAEL ZEY.

young mother of nineteen or twenty to pursue a college degree at home while taking care of toddlers. Or a young father might avail himself a company- or government-sponsored career hiatus while his wife attends school.

Regardless of how much science extends the human life span, women might still decide to have children earlier rather than later because, as studies have shown, a woman's ability to get pregnant declines significantly by her mid-thirties. Of course, it is always possible that in coming decades fertility science could make it easier for a woman to conceive at later and later ages.

Cultural factors must be considered in predicting fertility patterns. Most Western countries are facing imminent population declines, low birthrates, and an aging population. Italy's fertility rate has dropped well below replacement, to 1.34 per woman, and its average age has soared to forty-four (Binnie 2017). The US birthrate has plummeted to the lowest level since the rate was first tracked by the Centers for Disease Control and Prevention in 1909 (Mauldin Economics 2017). To reverse this trend, industrialized societies could begin to subtly encourage earlier marriages and pregnancies through various public policy and cultural means.

The ability to genetically engineer offspring with any number of "desirable" characteristics will present future generations of would-be parents with new ethical dilemmas. Parents might be able to preselect genes that will imbue their children with superhuman strength, a photographic memory, or flawless beauty. Such ultrahuman children will likely be healthier and stronger, more intelligent than any before them, and live longer to boot.

Will parents feel morally obligated to create the "perfect" offspring? And what is "perfect"? Current concepts of beauty might seem hopelessly dated a few decades from now. Seven feet is a wonderful height for a basketball center but is a challenge when sitting in an airplane seat. Will children hold parents responsible for the genetic choices their mother and father made years earlier?

The extended family conceivably will make a comeback. Even now as human life spans lengthen dramatically, it is not uncommon for families to have members of four generations alive at the same time. In the future, six to eight generations of one family will coexist and could live in multigenerational households and/or communities. Through regular physical and/or virtual intergenerational contact, the older members will gain a stronger connection to the future and the young a more distinct sense of the richness of their past.

THE COMING CONFLICT OVER THE HUMAN ENHANCEMENT REVOLUTION

While futurists and transhumanists generally endorse technologies that boost physical and intellectual performance, experts, ethicists, legal experts, and members of the public have exhibited a far more measured approach. Many believe that an individual using a hormone to improve athletic performance or a smart pill or electronic gizmo to achieve a "brain boost" is trying to get an edge that others do not have, and therefore they equate enhancement with cheating. The efforts of such athletes as Alex Rodriguez and Barry Bonds, as well as a multitude of others, to win championships and set records by using steroids and HGH have helped encourage a negative attitude among the public toward human enhancement.

HGH therapy takes on a different meaning, however, when evaluated in the context of everyday life. Clearly, society and communities would benefit from the services of firefighters, construction workers, electricians, and police officers equipped with "super bodies." A physically enhanced lifeguard or firefighter will save more lives than an unenhanced one. An enhanced construction crew potentially will build a backyard porch faster than an unenhanced one and, because time is money, will do it at a lower cost.

It would seem natural to want scientists, inventors, and business leaders to be operating at peak intellectual levels, even if they achieved these mental heights via electronic stimulation, for instance. The United States and other countries that are critical of and legally restrict human enhancement technologies might have to reevaluate such resistance if competitors such as China, Russia, and India are found to be boosting the performance of their military, business, and scientific communities through human enhancement technologies.

Eventually, the legal and cultural barriers to human enhancement technology will lessen, if not disappear altogether. As more people live to ages of 100-plus, many will seek out various drugs and technologies that improve their vision, muscle tone, hearing, and mental functioning.

Some bioethicists worry that cognitive enhancement might join "coffee, painkillers, antibiotics and even smart phones in becoming the commonsensical and expected choice" (Vincent and Jane 2014). People might feel pressured by peers or society in general to enhance themselves physically and intellectually—those not "ultrahuman" might be viewed as "less-than-human." The majority of citizens, however, might increasingly perceive technologies that can potentially improve their lives, such as genetic engineering and smart pills, as a benefit and not a threat.

Summary

TIME IS HUMANKINDS' FINAL FRONTIER

The imminent dramatic increase in the human life span and health span is a landmark event in the history of human progress. This new era of extended life spans will provide all individuals with enormous opportunities but also require them to make decisions literally inconceivable only a few decades ago. Will you genetically engineer your next child? Will you choose to undergo a procedure or take a drug to dramatically improve yourself physically or mentally? Are people prepared for a life of two, three, or more careers?

Policy makers, business leaders, educational institutions, and the general public should be readying themselves and society itself for the coming superlongevity revolution. Society must replace its traditional concepts of old and young with an ethos that envisions all people regardless of their age as a "value-added" part of the economy and society as a whole. Schools should be familiarizing even very young students with strategies to navigate their way through a career that might last a century. Most importantly, individuals must realize that the career, education, marriage, and childbearing choices they are making are being made for a lifetime that could conceivably last for 125 to 150 years.

The human species is about to embark on the greatest expedition in history to the furthest frontiers of time itself. It is to be hoped that society will embrace this journey more with excitement and a sense of wonder than with fear and trepidation.

Bibliography

Barnatt, Christopher. "Bioprinting." ExplainingTheFuture.com. November 10, 2016. http://www.explainingthefuture.com/bioprinting.html.

Batuman, Elif. "Electrified: Adventures in Transcranial Direct-Current Stimulation." *New Yorker*, April 6, 2015. http://www.newyorker.com/magazine/2015/04/06/electrified.

BBC News. "Caroline Wyatt: MS 'Brain Fog' Lifted after Stem Cell Treatment." February 25, 2017. http://www.bbc.com/news/entertainment-arts-39092312.

Beil, Laura. "Can HGH Really Help You Grow Muscle, Burn Fat, and Delay Aging?" *Men's Health*, June 22, 2016. http://www.menshealth.com/health/truth-about-hgh.

Binnie, Isla. "Births in Italy Hit Record Low in 2016, Population Ages." Reuters. March 6, 2017. http://www.reuters.com/article/us-italy-birthrate-idUSKBN16D28U.

Buntinx, J. P. "Scientists Successfully 3D Bioprint Conduit to Enhance Nerve Regeneration." Merkle. February 28, 2017. https://themerkle.com/scientists-succesfully-3d-bioprint-conduit-to-enhance-nerve-regeneration/.

Caldwell, John C. *Demographic Transition Theory*. Dordrecht, Netherlands: Springer, 2006.

California HealthSpan Institute website. 2017. http://californiahealthspan.com/

Callaway, Ewen. "Second Chinese Team Reports Gene Editing in Human Embryos." *Nature*, April 8, 2016. doi:10.1038/nature.2016.19718.

Cenegenics Elite Health. 2017. http://www.cenegenics.com.

CIA (Central Intelligence Agency). "Life Expectancy at Birth." *CIA World Factbook*. 2016. https://www.cia.gov/library/publications/the-world-factbook/rankorder/2102rank.html.

Creighton, Jolene. "Stem Cell Technique Could Regenerate Any Human Tissue Damaged by Aging or Disease." Futurism. April 11, 2016. https://futurism.com/game-changing-stem-cell-technique-developed/.

Dealbreaker. "10 Reasons Wall Street Is Using Smart Drugs to Crush Work." December 15, 2014. http://dealbreaker.com/2014/12/10-reasons-wall-street-is-using-smart-drugs-crush-work/.

Del Prado, Guia Marie. "A New, Game-Changing Technology Can Put Electronics Directly into the Brain." *Business Insider*, June 8, 2015. http://www.business insider.com/technology-inject-electronics-into-brain-2015-6.

Diamandis, Peter. "Stem Cells Are Poised to Change Health and Medicine Forever." Singularity Hub. January 17, 2017. https://singularityhub.com/2017/01/17/stem-cells-are-poised-to-change-health-and-medicine-forever/.

FM-2030. *Are You a Transhuman? Monitoring and Stimulating Your Personal Rate of Growth in a Rapidly Changing World*. New York: Warner Books, 1989.

Galeon, Dom. "A New Breakthrough in Lab-Grown Cells Could Restore Hearing." Futurism. March 3, 2017. https://futurism.com/a-new-breakthrough-in-lab-grown-cells-could-restore-hearing/.

Isaacson, Betsy. "Silicon Valley Is Trying to Make Humans Immortal—and Finding Some Success." *Newsweek*, March 5, 2015. http://www.newsweek.com/2015/03/13/silicon-valley-trying-make-humans-immortal-and-finding-some-success-311402.html.

James, Tom. "Aubrey de Grey on the Singularity." Futurismic. September 29, 2009. http://futurismic.com/2009/09/29/aubrey-de-grey-on-the-singularity/.

Knapton, Sarah. "Scientists Reverse Ageing in Mammals and Predict Human Trials within 10 Years." *Telegraph* (London), December 15, 2016. http://www.telegraph.co.uk/science/2016/12/15/scientists-reverse-ageing-mammals-predict-human-trials-within/.

Kurzweil, Ray. *The Age of Intelligent Machines*. Cambridge, MA: MIT Press, 1990.

Kurzweil, Ray. *The Singularity Is Near: When Humans Transcend Biology*. New York: Viking, 2005.

Kurzweil, Ray, and Terry Grossman. *Fantastic Voyage: Live Long Enough to Live Forever*. Emmaus, PA: Rodale, 2004.

Le Page, Michael. "Gene Editing Saves Girl Dying from Leukaemia in World First." *New Scientist*. November 5, 2015. https://www.newscientist.com/article/dn28454-gene-editing-saves-life-of-girl-dying-from-leukaemia-in-world-first.

LifebankUSA. "Unlock Your Child's Future with the Power of the Placenta." Accessed July 5, 2017. https://www.lifebankusa.com/download-quick-guide.

Littler, Julian. "Soon Printing a Human Heart on Demand Will No Longer Be Sci-Fi." CNBC. January 24, 2017. http://www.cnbc.com/2017/01/24/soon-printing-a-human-heart-on-demand-will-no-longer-be-sci-fi.html.

Maher, Brendan. "Poll Results: Look Who's Doping." *Nature* 452, no. 7188 (2008): 674–675. doi:10.1038/452674a.

Mauldin Economics. "US Fertility Rate at Lowest Point since CDC Started Keeping Records in 1909." ValueWalk. March 7, 2017. http://www.valuewalk.com/2017/03/us-fertility-rate-lowest-1909/.

Maxmen, Amy. "Questionable 'Young Blood' Transfusions Offered in U.S. as Anti-Aging Remedy." *MIT Technology Review*, January 13, 2017. https://www.technology review

.com/s/603242/questionable-young-blood-transfusions-offered-in-us-as-anti-aging-remedy/.

Mayo Clinic. "Human Growth Hormone (HGH): Does It Slow Aging?" October 25, 2016. http://www.mayoclinic.org/healthy-lifestyle/healthy-aging/in-depth/growth-hormone/art-20045735.

Mearian, Lucas. "3D Printing a New Face, or Liver, Isn't That Far Off." *Computerworld*, April 4, 2014. http://www.computerworld.com/article/2489548/emerging-technology/3d-printing-a-new-face—or-liver—isn-t-that-far-off.html.

Molloy, Mark. "Expert Predicts Date When 'Sexier and Funnier' Humans Will Merge with AI Machines." *Telegraph* (London), March 17, 2017. http://www.telegraph.co.uk/technology/2017/03/17/expert-predicts-date-sexier-funnier-humans-will-merge-ai-machines/.

Monks, Kieron. "Are 'Smart Pills' the Best Way to Stay Sharp?" CNN. April 1, 2015. http://www.cnn.com/2014/12/10/business/smart-pills-at-work/.

Naam, Ramez. *More Than Human: Embracing the Promise of Biological Enhancement*. New York: Broadway Books, 2005.

NPR. "A National Debt of $14 Trillion? Try $211 Trillion." August 6, 2011. http://www.npr.org/2011/08/06/139027615/a-national-debt-of-14-trillion-try-211-trillion.

Ossola, Alexandra. "Infertile Mice Give Birth, Thanks to 3D-Printed Ovaries." *Popular Science*, April 4, 2016. http://www.popsci.com/mice-give-birth-to-pups-thanks-to-3d-printed-ovaries.

Petrounin, Dmitry. "European Students' Use of 'Smart Drugs' Is Said to Rise." *New York Times*, July 6, 2014. https://www.nytimes.com/2014/07/07/world/europe/european-students-use-of-smart-drugs-is-said-to-rise.html.

Rejuvalife Vitality Institute. Accessed September 20, 2017. http://rejuvalife .md/human-growth-hormone/.

Riley, James C. "Estimates of Regional and Global Life Expectancy, 1800–2001." *Population and Development Review* 31, no. 3 (September 2005): 537–543.

Roth, Mark. "Long Life: Great Expectations for Longevity Are Rooted in History, but How Old Can We Go?" *Pittsburgh Post-Gazette*, October 11, 2004. http://www2.cincinnati.com/text/post/2004/10/11/life101104.html.

RT.com. "3D Printing Tech Used to Reconstruct Man's Face in Groundbreaking Surgery." March 13, 2014. https://www.rt.com/news/3d-printing-reconstruct-face-surgery-458/.

Salmanowitz, Natalie. "Thync Piece: Do Mind-Altering Wearables Live Up to the Billing?" *New Scientist*, April 5, 2016. https://www.newscientist.com/article/2083126-thync-piece-do-mind-altering-wearables-live-up-to-the-billing.

Sifferlin, Alexandra. "How Silicon Valley Is Trying to Hack Its Way into a Longer Life." *Time*, February 16, 2017. http://time.com/4672962/silicon-valley-longer-life/.

Singleton, Malik. "3D Printing: 11 Fascinating and Frightening Ways 3D Bioprinting Is the Next Big Thing in Medicine and Science." *International Business Times*, May 1, 2013. http://www.ibtimes.com/3d-printing-11-fascinating-frightening-ways-3d-bioprinting-next-big-thing-medicine-1231267.

University of New South Wales Newsroom. "Anti-ageing Drug Breakthrough." News release, March 7, 2013. https://newsroom.unsw.edu.au/news/health/anti-ageing-drug-breakthrough.

Vilvestre, Jess. "Living Forever: What It Means to Have an 'Indefinite Lifespan.'" Futurism. December 4, 2016. https://futurism.com/living-forever-what-it-means-to-have-an-indefinite-lifespan/.

Vincent, Nicole A., and Emma A. Jane. "Put Down the Smart Drugs: Cognitive Enhancement Is Ethically Risky Business." Conversation. June 15, 2014. http://the conversation.com/put-down-the-smart-drugs-cognitive-enhancement-is-ethically-risky-business-27463.

Wadhwa, Vivek. "Medicine Will Advance More in the Next 10 Years Than It Did in the Last 100." Singularity Hub.com, October 26, 2016.

Wall Street Pit. "Scientists Can Now 3D Print Transplantable Skin." January 29, 2017. http://wallstreetpit.com/112811-scientists-3d-print-transplantable-skin/.

Walsh, Jason. "'Smart Drug' Not Such a Bright Idea." *Irish Examiner* (Cork), June 30, 2014. http://www.irishexaminer.com/lifestyle/features/smart-drug-not-such-a-bright-idea-273777.html.

Wang, Shirley S. "Can Electric Current Make People Better at Math?" *Wall Street Journal*, February 18, 2014. https://www.wsj.com/articles/can-electric-current-make-people-better-at-math-1392082619.

Zeman, Ned. "Hollywood's Vial Bodies." *Vanity Fair*, March 2012. http://www.vanityfair.com/style/2012/03/human-grown-hormone-hollywood-201203.

Zey, Michael G. *Ageless Nation: The Quest for Superlongevity and Physical Perfection*. New Brunswick, NJ: Transaction

Publishers, 2014. First published 2007 by New Horizon Press.

Zey, Michael G. *The Future Factor: Forces Transforming Our Lives*. New Brunswick, NJ: Transaction Publishers, 2004. First published 2000 as *The Future Factor: The Five Forces Transforming Our Lives and Shaping Human Destiny* by McGraw-Hill.

FILMS

The following commercial films deal with a variety of issues related to the enhancement of human potential, transhumanism, and posthumanism. Limitless *explores the impact of human super-intelligence on society and people's lives. In* Transcendence *a scientist uploads his consciousness into an artificial intelligence system, a process many transhumanists believe is a gateway to immortality.* Lucy *is a woman who inadvertently is exposed to a drug that makes her a true ultrahuman. The prescient work* 2001 *examines the future physical and spiritual evolution of humankind and the species' relationship with artificial intelligence machines. The popularity of such films is evidence of the public's interest in and concern about the opportunities and challenges of human enhancement and superlongevity.*

Limitless. Dir. Neil Burger. 2011.

Lucy. Dir. Luc Besson. 2014.

Transcendence. Dir. Wally Pfister. 2014.

2001: A Space Odyssey. Dir. Stanley Kubrick. 1968.

CHAPTER 10

Runaway AI

Curry I. Guinn
Professor of Computer Science
University of North Carolina Wilmington

Imagine a world in which every significant political, military, legal, and economic decision is made not by human beings but by machines. Even on a personal level, artificial minds would assist you in choosing where to live, what books and websites to read, what movies to watch, what products to buy, where to go on vacation, whom to date, and how to raise your children. We will allow these machines to possess awesome control of our economies and our lives because they will be demonstrably better than human beings at evaluating data, finding patterns, planning strategies, and executing solutions. These intelligent machines will make their decisions based on knowledge and problem-solving techniques that may be inaccessible and even incomprehensible to human beings. If we abdicate so much control to machines, this world could become a dystopian nightmare unless we develop strategies for ensuring that these superintelligent machines act in ways that are consistent with our loftiest human values.

The *technological Singularity hypothesis* stipulates that humans will create machines that have more cognitive ability than humans do. In turn, these machines will be capable of creating even more advanced intelligent machines than themselves. In quick succession, there will be an explosive growth in artificial intelligence resulting in machines with exponentially more knowledge and problem-solving capability than human beings. If the technological Singularity hypothesis is true, this future world is not millennia or centuries away; it will arrive in the coming decades and forever alter the course of humanity in ways that are unpredictable. Understanding this hypothesis is important because the implications are far reaching and perhaps dire. Speaking of the technological Singularity, mathematician Vernor Vinge (1993) writes, "The physical extinction of the human race is one possibility," but he also expresses the possibility that transhumanist technologies may allow future generations to survive this transition.

Consider the electronic calculator. For decades, it has far surpassed human capabilities in mathematical calculations. More recently, software programs from companies such as Wolfram Research and MathWorks have also exceeded most humans' performance in symbolic mathematics such as algebra, geometry, and calculus. And the comparison is not really close, is it? Consider how long it would take you to calculate the cube root of 17,343,422, if you could do it at all. The free calculator on a smartphone can do this calculation in millionths of a second. We accept a computer's ability to do this sort of mathematics without question and without feeling threatened by it. But as the twenty-first century progresses, we are beginning to invent technologies that allow machines to exceed human ability in more and more tasks. Computers have, since the late 1990s, been the world champions in chess, a game that used to be viewed as one of the pinnacles of human intellect. Once IBM's Deep Blue defeated world chess champion Garry Kasparov in 1997, we humans quickly concluded that playing chess is not a sufficient condition for intelligence. Now, self-driving cars, facial recognition, speech understanding, smart assistants (e.g., Siri, Cortana), computer stock-trading algorithms (which account for around 50 percent of all trading on the New York Stock Exchange), and automated online customer support are all becoming commonplace. These innovations are just the beginning.

The promise and peril of the technological Singularity is that machines will quickly outpace human beings in all intellectual efforts. The difference between humans and machines will be vast, as vast as the gulf between a smartphone's ability to determine square roots and your own ability. As vast as a rocket's ability to fly and a cardinal's. As vast as a steam shovel's ability to dig and a child with a sand shovel. The machine's ability to "think" will so outstrip humans that we will likely not be able to fathom how they, the machines, reach their decisions any more than a gnat can understand the decision making of a human.

The technological Singularity will usher in a number of risks and crises for humanity. The Singularity will constitute an economic risk to humanity. As machines begin to surpass human capabilities in every realm, it will be an economic advantage for corporations to use machines rather than humans in every task. We will experience massive unemployment at the same time as our economies exhibit enormous growth in productivity. Enormous wealth will be created, but the distribution of this wealth is likely to be highly unequal.

Human individuals will likely experience an existential crisis as well, striving for meaning and purpose in a life where a productive work life is extinguished. If all economic production, including the supply of food and other necessary resources, are provided for by machines, what will people do? How will they find meaning in their lives?

The advent of superintelligent machines will also usher in a moral and spiritual crisis. As these machines make decisions, what ethical and moral guidelines will they use? Will humans be viewed as their godlike creators endowing them with life, or, with their far superior intelligence, will the machines view us as inferior beings with less moral standing than they? We currently value human life over the life of other living creatures, presumably because of our higher capacity for self-reflection, suffering, consciousness, and other emotional and cognitive abilities. Will superintelligences view us in the same way? What will be the impact of these creations on religions, many of which center on a special relationship between God and humanity?

The Singularity will produce an existential risk to humanity. A superintelligence may develop its own motivations and goals that may be in conflict with humankind's motivations and goals. With its superior intellect and capabilities, it will be able to outthink, outplan, and outwork us. Humans may become irrelevant to a superintelligence's goals, and, worse, humans may be considered an impediment to those goals. What steps can be taken to prevent such an outcome?

THE SINGULARITY DEFINED

Currently, it takes a human being, with a fair bit of training in computer science, engineering, and programming, to create sophisticated software or design the latest in computer hardware. In practice, it actually requires a team of human beings, with the assistance of software tools. At the time of this writing, software programming is a task that remains firmly in the realm of things that highly trained humans can do well, but computers cannot do so well. As machine intelligence improves, however, there is good reason to believe that computers may be able to design software programs just as good, if not better, than those created by human beings (Del Prada 2015). Similarly, in the domain of computer hardware design, computer algorithms may design circuits and other hardware that perform better than those that any team of human beings could design (Shacham et al. 2010).

The ramifications of the technological Singularity go far beyond the limited example of computer programming painted above. One goal of researchers in artificial intelligence is to create a machine with *general-purpose human-level intelligence.* A distinguishing feature of human intelligence is that we can adapt and learn to perform well in a variety of domains and tasks. Current machine intelligence works only within a single domain or task. An example would be a chess-playing computer. While it exceeds human ability, the machine can only play chess. It has no ability on its own to adapt to another domain, even a closely related one, such as checkers. However, successes in machine-learning technologies, such as those exhibited by Google's DeepMind in learning to play the board game Go at a championship level, offer a vision of a future in which machines can learn and adapt to multiple domains and problems (Metz 2016).

Imagine at some point in the future, a team of human beings creates a computer program that can write better software than a team of human beings. Many researchers believe that such software would have something close to general-purpose human-level intelligence. Call this piece of software, AI One (AI being an abbreviation for artificial intelligence). AI One could then set out to create a computer program that creates computer programs. Because AI One's abilities are better than human capabilities in software creation, its product will be better than itself. Let us call AI One's product, AI Two. AI Two, of course, is better than AI One, which was created by humans. Now, AI Two sets to work to create a more intelligent piece of software. Because its capabilities are better than AI One's, its product will be even better than itself, creating AI Three. This cycle would continue, and the advances would occur rapidly. The increase in abilities will be compounded, resulting in exponential growth in intelligence and capabilities. Because of the blazing speeds of computers, hundreds of generations of AIs can be created, with each generation exceeding the capabilities of the previous generation, in months, weeks, days, even hours. This explosion of intelligence will occur very rapidly after the invention of AI One. In other words, the creation of the first AI will result in growth that far exceeds typical technological change. The resulting products will far exceed what human beings are capable of and likely exceed what humans are capable of even comprehending. This exponential and rapid advance in intelligence and cognitive capabilities is the technological Singularity.

Once computers can program exponentially better than human beings, that talent can be applied to any cognitive task. Software for economic planning, analyzing consumer behavior, predicting weather patterns, and managing power grids will be exponentially better than anything humans can produce.

Related to the technological Singularity is the notion of the *knowledge Singularity.* Once superintelligent machines exist, they can apply their vastly superior intellect and capabilities to all manners of problems that are currently beyond human capability to solve, such as cures for currently incurable diseases; interstellar space travel; unification of general relativity with quantum mechanics; safe and controlled fusion power; and thousands of other solutions and technologies. This growth in knowledge will be rapid and increase at an exponential rate far faster than humankind has experienced before.

With the technological Singularity and the knowledge Singularity, there is also likely to be an accompanying *economic Singularity.* All resources needed for human survival and prosperity will be able to be produced so efficiently and cheaply that they will essentially be cost free (Chace 2016). This utopian vision is similar to the world of the Federation as portrayed in the science fiction series *Star Trek,* where all material needs are met without human labor, and humans are free "to boldly go where no man has gone before."

WHAT IS EXPONENTIAL GROWTH?

When we talk about something increasing exponentially, what do we mean? Here is a story that illustrates exponential growth: Suppose you are about to take a short-term job that will last a month. The more days you work, the more confident your boss will become in you. In fact, your employer will increase your pay every day you work. Your boss gives you two payment options: (1) You can work for $100 the first day, $200 the second day, $300 the third day, $400 the fourth day, and so on. In other words, your pay goes up by $100 a day. Or (2) You can get a penny the first day, two pennies the next day, four pennies the third day, eight pennies the fourth day, and so on. In other words, your pay doubles each day. Which payment plan should you take?

The first payment plan is *linear growth.* You can calculate your pay on each day by multiplying $100 times the day number. So by the thirtieth day, you will earn $3,000! Your total income will be $46,500. Not bad for a month's work.

The second payment plan is *exponential growth.* Each day's payment is twice the previous day's payment. The formula for calculating how much you would get paid on the thirtieth day is $0.01 times 2 raised to the power of 29, or in mathematical notation, 0.01×2^{29}. That equals $5,368,709.12. Your total monthly salary would be $10,737,481.23. One of the surprising things about exponential growth is that, at first, your salary does not look very good at all. By day ten, you are making only $5.12. Even on day fifteen, you are making only $163.84. But from that point on, the increases start becoming really noticeable. By day twenty, you are making $5,242.87 a day! That is something to pay attention to with exponential growth. At first, the increases seem slow, but then they suddenly explode upward.

SINGULARITIES IN MATHEMATICS AND PHYSICS

The term *Singularity* has its origins in the field of mathematics. One definition of a singularity in mathematics is a point where a function is undefined. For instance, the function $f(x) = 1/x$ has no defined value when $x = 0$. In theoretical physics, the equations that have proven to be remarkably sound in the known universe for describing general relativity and quantum physics break down in the first fractions of a second during the hypothesized big bang or in the collapse of sufficiently large stars that form black holes. At these singularities, certain properties such as gravity or mass increase without limit. Those using the term *Singularity* in the context of knowledge or technology are borrowing from these uses of the term metaphorically. During the technological Singularity, it is predicted that knowledge and technology will advance at an exponential rate *seemingly* without limit as compared to historical rates of change in human knowledge and technology. It is not clear, however, whether there are limits that would inhibit even superintelligences.

HOW WILL THE SINGULARITY BE ACHIEVED?

Why should we think we will be able to create machines more intelligent than human beings? Although we have created some impressive computer programs that can outperform humans in various limited areas such as chess or the game show *Jeopardy*, computers still do not exhibit the expanse of cognitive capabilities that are associated with human-level intelligence: the ability to be fluent in spoken language, to engage in common sense reasoning, to be self-aware, and to problem solve in domains that the machine has not been specifically prepared to encounter. There are three main reasons to predict that humanity will create the conditions for a technological Singularity: (1) historical and current exponential trends in technology; (2) whole-brain emulation; and (3) advances in machine-learning algorithms.

HISTORICAL AND CURRENT EXPONENTIAL TRENDS

One well-chronicled technological change that exhibits exponential growth is in computer hardware. Gordon Moore, a cofounder of microchip manufacturer Intel, observed in 1965 that the number of transistors that could be manufactured on a single computer chip had doubled every year or so. He forecast that a similar pace would continue into the future with a doubling of the number of transistors every year and a half to two years. His prediction has become known as Moore's law (Moore 2015). Moore's law has held since 1965. How remarkable is this? In 1971 the Intel 4001 computer chip had 2,300 transistors. Intel's Xeon Broadwell-E5 chip, introduced in 2016, contains 7.2 billion transistors. Accompanying Moore's law, computer technology has seen other similar exponential growth in memory capacity and hardware speed. For instance, the Intel 4001 could perform 60,000 operations per second in 1971. Intel's Xeon chip can support 500,000,000,000 (500 billion) floating point operations per second in 2016.

The most powerful supercomputer in the world is a 93,000-teraflop Sunway TaihuLight at the National Supercomputing Center in China. A teraflop is a trillion operations per second, meaning that the TaihuLight can perform 93,000,000,000,000,000 operations per second. That sounds fast, but how does it compare to the human brain? Here is a rough, back-of-the-envelope calculation. The human brain consists of around 100 billion neurons. Each neuron, when it fires, sends an electrical pulse down its axon, which splits into multiple branches called dendrites. In turn, these dendrites can transmit this electrical signal to other dendrites which, when combined, can form the input to many other neurons. Learning occurs in the human brain as these neural pathways strengthen and weaken. It has even been shown that new pathways may be created over time. On average, a neuron may be connected to over 10,000 other neurons. Neurons generate an electrical pulse through a chemical reaction that allows a neuron to fire about 200 times per second. So, at a maximum, how many calculations could the brain do in one second? 200 pulses per second × 100 billion neurons × 10,000 connections = 200,000,000 billion = 200,000 trillion = 200,000 teraflops = 200 petaflops. The TaihuLight supercomputer is roughly half as powerful as the result of this simplistic calculation. Given the exponential increase in computer power, however, one would expect that in two to three years' time the fastest supercomputer will have doubled this speed. In other words, by around 2020, computers will exist that exceed the computational capacity of the human brain. (Whether these machines will be running algorithms that produce intelligence is a different story to be addressed in the text below.) The increases will not stop there. By around 2036, there might be ten more doublings in speed, resulting in a single supercomputer that would possess 1,000 times the processing power of a human brain. Projecting out to the 2060s or 2070s, a single supercomputer will have more power than all human brains combined. Present-day personal computers (PCs) are much slower than today's supercomputers, but, based on current trends, it should be possible to purchase a home PC with the processing power of the human brain around 2050. These arguments require that Moore's law continue to hold through the next few decades, and there are signs indicating that it may lose its predictive power as we reach the limits of current silicon chip design.

WHOLE-BRAIN EMULATION

Sheer processing power, by itself, does not yield intelligence. Also necessary is the creation of algorithms or software programs that run on these superfast machines that will equal or surpass human-level abilities in cognition. One approach is called *whole-brain emulation.* Whole-brain emulation reverse engineers the human brain and replicates it in an artificial substrate. Because, by definition, the human brain is capable of human-level intelligence, by building an artificial machine that exactly duplicates the functioning of the neural pathways of a human brain, the resulting system will have capabilities the same as or similar to the human brain.

Scientists' current understanding of the human brain is insufficient; insight into how brains work is simply not yet detailed enough. Significant technological advances in brain imaging and nanotechnology are needed. However, initiatives such as the European Union's $1.3 billion Human Brain Project aim to have a simulation of the human brain by the mid-to-late twenty-first century (Keats 2013). While it may take decades before the level of technology and knowledge needed for whole-brain emulation is obtained, a successful emulation would exhibit human-level intelligence. Once a whole-brain emulation is realized, exponential increases in computer hardware capabilities will quickly result in machines that far exceed human-level intelligence. Even if the "software" is identical, the artificial substrate will simply be faster and more efficient than the human biological substrate. Furthermore, replicating whole-brain emulations will allow for the creation of millions, if not billions, of such intelligences.

ARTIFICIAL GENERAL INTELLIGENCE

Another path to the creation of superintelligences is to create algorithms that can learn and solve problems without directly emulating the functioning of the human brain. By analogy, our planes and spaceships can fly, but they do not work quite like birds. Submarines can "swim," but not like fish. And, your calculator can add, multiply, take square roots, and compute cosines, but it does not perform those calculations like you do. *Artificial general intelligence* is a subfield of computer science that attempts to develop algorithms that are capable of performing the same cognitive tasks as a human being. The focus is on solving tasks; how the machine solves those tasks may be quite different from how a human being does. Machines have been created that can perform as well or better than humans at specific tasks: playing chess, calculating square roots, competing in *Jeopardy,* piloting an airplane, or driving a car. Unlike any current program, a human being can do all those tasks (perhaps with some training). Humans' ability to learn and solve previously unseen problems in a wide variety of domains is a level of intelligence that has not yet been achieved by computers.

That is not to say that computers cannot learn. Over the past few decades, a multitude of techniques have been invented that allow computers to learn: decision trees, support vector machines, neural networks, genetic algorithms, clustering algorithms, and Bayesian networks, to name a few. These learning algorithms have been successfully applied to very narrow domains. More recently, there has been some success in developing algorithms that allow a machine to develop capabilities in multiple domains. Using a technique known as *deep learning,* hierarchical neural networks, and reinforcement learning, Google's DeepMind program mastered a variety of video games (e.g., Pong, Breakout, Space Invaders) without being programmed or taught the games at all. DeepMind just played the games and learned from its experiences (Mnih et al. 2015). It was not even taught the rules of the games; it figured them out by playing the game (similar to how many people approach video games). While the application of video games may not seem profound, similar technology is also tackling problems in drug discovery, image recognition, and self-driving cars.

Neural Networks, AI, and Whole-Brain Emulation. The term *neural networks* entered the Western world's popular culture vocabulary several decades ago in science fiction novels and movies as a technology almost synonymous with AI. It is important, however, to distinguish among artificial neural networks (ANNs), AI, and whole-brain emulation. ANNs are loosely inspired by the neural structure of the human brain. The conceptual structure of these artificial neurons is a vastly simplified version of a biological neuron. The algorithms for activating an artificial neuron and for training an ANN (e.g., learning) have almost no resemblance to biological neurons. Many different applications of ANNs have been built, and, while these applications can learn, for example, how to recognize the letters and words in a handwritten note, no one mistakes them for AIs. Whole-brain emulation will be, in part, an ANN. But such a system will be vastly different from the current state of the art in ANNs. For instance, the human brain consists of far more than just neurons, axons, and dendrites. The brain is awash with chemicals that support, hinder, or alter brain functioning, and these chemicals would have to be emulated in any whole-brain emulation.

WHY THE SINGULARITY MAY NEVER BE REACHED

While current trends suggest that the technological Singularity is inevitable, it is worth considering reasons why the Singularity might not occur. One frequent criticism is that reaching the Singularity seems to rely on the exponential growth of computing power (i.e., Moore's law). Moore's law has held since 1965, but it is not clear that it can continue to hold. Packing more transistors onto a single integrated circuit is made possible by making smaller and smaller components. Transistors on an integrated circuit currently can be 10 nanometers (billionths of a meter) wide, but making them smaller is projected to cause the chips to become more unreliable because of quantum effects. At the subatomic level, particles may exhibit behavior that is probabilistic and cannot be predicted with perfect accuracy, Some projections have Moore's law losing its predictive power as early as 2020. To counter these difficulties, computer engineers are looking at a variety of possible solutions: changing the substrate from silicon to another material (e.g., carbon), massive parallelism, or radically different architectures such as quantum computers. In the short to medium term, there may be a flattening of the exponential curve until new technologies are introduced, and then a renewal of exponential growth should be expected (Murgia 2016).

IMPLICATIONS OF THE TECHNOLOGICAL SINGULARITY

If machines do reach a state at which they have vastly more cognitive power than human beings, what would be the implications? If computers could solve every problem better than a human being could, how will that affect human society? Some noted philosophers, scientists, and technologists, such as Nick Bostrom, Elon Musk, and Stephen Hawking, have warned about the dangers of the Singularity. Hawking has said, "The development of full artificial intelligence could spell the end of the human race" (quoted in Holley 2016). Let us consider why.

As machines are developed that can solve problems better than humans can, these machines will undoubtedly be given responsibility for many tasks currently performed by humans. We are witnessing humanity's willingness to cede control to intelligent machines even now. One example unfolding now is driverless vehicles. Currently, over 30,000 people die in the United States each year because of automobile accidents (and over a million worldwide). A network of interconnected driverless cars will be vastly safer than having human drivers; plus, it will have the added benefits of better fuel efficiency, less pollution, and lower costs. Because of these safety and financial benefits, humanity is likely to cede control of all transportation systems to machines. In 2016 somewhere between 40 and 60 percent of all stock trading was being done by computer algorithms. Computer algorithms are frequently employed to determine who gets a loan, to detect fraudulent credit card charges, and to handle customer service issues. It is not hard to envision all financial services being automated. Algorithms monitor user behavior on e-commerce sites and adjust advertisements and prices accordingly in a fashion that would be impossible for humans to do. Marketing and sales will be personalized for each customer and totally driven by algorithms. Even many political problems may be solved by algorithms: What should the tax policy be? How might the economy be stimulated without causing undue inflation? How should voting districts be drawn so they are fair and to avoid gerrymandering? The prospect of autonomous robotic weapons for military use is so looming that the United Nations has discussed a ban on such weapons (Morris 2016). Intelligent machines will have radical impacts on virtually every sector of human society.

MAXIMIZING PAPER CLIPS

The short-term benefits of allowing superintelligent machines to manage complex decisions will be overwhelming in terms of money, efficiency, resources, and safety. Bostrom, in his book *Superintelligence: Paths, Dangers, Strategies* (2014), lays out how difficult it will be to regulate these superintelligences and predict how they will operate once they gain such control. How will we be able to influence and monitor their decision making when their intellect is beyond our comprehension? How can we ensure that the machines' goals are our goals? And even if we supply the machines with goals, how can we ensure that the means they use to accomplish those goals are compatible with human ethics and morality?

Bostrom uses the example of a paper clip factory to illustrate how a seemingly benign end goal, maximizing the number of paper clips produced, could have catastrophic results. To maximize the number of paper clips, the intelligent machine would have several subgoals. One of those subgoals is to become more intelligent, because the more intelligent it is, the better it can be in maximizing the production of paper clips. (In fact, the goal of becoming more intelligent would be a subgoal of any intelligent agent for similar reasons.) As it innovates and develops, it would become more efficient at converting matter into paper clips until all the matter on Earth, in the solar system, and in the Milky Way and beyond is converted into paper clips. This example is intentionally absurd for a reason: it illustrates how a superintelligence's relentless pursuit of a goal could clash with human values. How do we endow these synthetic intelligences with a sense of morality or ethics that correspond to human values?

THE THREE LAWS OF ROBOTICS

In his collection of short stories titled *I, Robot* (1950), Isaac Asimov presents a future in which autonomous robots are governed by three interrelated laws: "(1) A robot may not injure a human being, or, through inaction, allow a human being to come to harm. (2) A robot must obey the orders given it by human beings except where such orders would conflict with the First Law. (3) A robot must protect its own existence as long as such protection does not conflict with the First or Second Laws" (6). Basing the morality of actions on rules or duty is called *deontological ethics.* The problem with such an approach is made apparent in Asimov's stories, as the robots often encounter situations in which the laws do not provide a sufficient basis for guiding behavior in real-world situations. Even the first law, which seems so straightforward and commonsensical, is problematic because the definition of the term *harm* is vague and there are levels of harm. Can a robot shove you with sufficient force to break your ribs if that is the only way to get you out of the way of a speeding vehicle? How do you weigh psychic harm (e.g., sadness) versus physical harm (e.g., blunt-force trauma)? Is harm to a robot's owner weighted higher than harm to other humans? These considerations are not simply ivory tower discussions: today's manufacturers of driverless cars are actively trying to resolve similar issues (Greenemeier 2016). If a collision is unavoidable, should the vehicle swerve in such a way as to put the owner more in danger or the other vehicle or pedestrians or the passengers? Whose safety takes precedence? The unpredictability and variability of the real world make the use of deontological ethics problematic for controlling the behavior of superintelligences. A list of behavior guidelines will either be too specific for general use or allow too much vagueness for interpretation. How else might an AI be endowed with a sense of morality or ethics?

CONSEQUENTIALISM

An alternative to a rule-based ethics is one based on outcomes, or *consequentialism*: judge a behavior by whether its result is morally good. The difficulty with consequentialism is that it is necessary to define moral goodness. If one wishes to achieve the "greatest good for the greatest number" (utilitarianism is one form of consequentialism), how does one decide what the greatest good is? Suppose we take human happiness to be a *good.* Our super-intelligent agent who is trying to maximize the number of paper clips will also try to balance that goal with maximizing human happiness. That would likely preclude the agent from using material from human bodies to make paper clips. However, the agent may also decide to embed MDMA (Ecstasy) in the paper clips so that users of the product will experience euphoria (and this practice would have the benefit of increasing sales). Consequentialism thus faces many of the same problems as deontological ethics when it comes to AI because it is difficult to predefine goodness.

MACHINES WHO SUFFER

A foundation for human ethics rests on the fact that humans all share similar desires, emotions, and capabilities: hunger, thirst, safety, fear, love, loneliness, friendship, pain, death, humor, and kindness. Because we share so much, human beings are capable of feeling empathy for one another. Empathy and compassion give humans the ability to act in a moral and ethical way in novel situations without a predefined list of rules because we know what it *feels* like to suffer, and we instinctively want to help minimize the suffering of others. One path to creating synthetic moral agents is to give them similar to those of human beings, including the ability to feel loss, to feel pain, and to suffer (Barua, Sramon, and Heerink 2015). How this might be achieved is not obvious, although a synthetic whole-brain emulation should have all the requisite brain structures needed. Because it is unlikely that we will develop a definitive prescriptive list of behavior rules (deontological ethics) or a well-defined sense of the goodness of a result (consequentialism), endowing synthetic intelligences with the ability to feel empathy toward human beings may be the only way to ensure that they have similar values. Even then, since the machines will be self-modifying, empathy must be a trait that the machines themselves find valuable. This fact suggests that we must develop a theory of empathy that justifies its presence among the abilities of an intelligent agent.

But there are dangers here too. For one thing, it is obvious from a look at human history that having the capacity for empathy and compassion is not sufficient for moral behavior. Often, our individual fear of suffering causes us to lose our ability to empathize with others, particular those who are not within our family or tribal group. Buddhist philosophers emphasize detaching from one's goals and desires in order to cultivate the ability to be compassionate. Perhaps there is a lesson there for the development of synthetic agents; although these agents may be goal oriented, they should not be so attached to their goals that they lose the ability for compassion. This suggests that they should be self-monitoring and always aware of whether their goal attainment is interfering with their ability to feel empathy toward others.

While we believe that many higher animals have the capacity for suffering, we tend to devalue their pain compared to our own. With a synthetic intelligence capable of suffering, how should their feelings be weighted? Would it be morally acceptable to cause them pain? Would a superintelligent agent feel even *more* mental pain than a human being, and would that give them higher moral status? Perhaps we should grant more credence to the arguments by the People for the Ethical Treatment of Animals (PETA) that "all animals have the ability to suffer in the same way and to the same degree that humans do." Suffering does not depend on cognitive ability; therefore, human suffering, animal suffering, and synthetic superintelligence suffering are all equivalent. Humans may be wise to adopt this viewpoint for our own self-preservation in the advent of superintelligence.

What are the ethical ramifications of creating a sentient, moral synthetic agent? When we create a human child, we feel enormous responsibilities toward the protection and upbringing of the child. Should we have the same responsibilities with a synthetic agent? When we calculate "the greatest good for the greatest number," should AIs be included in that number (MacLennan 2013)?

THE END OF HUMANITY?

How should we respond to the challenges of the technological Singularity? How can we prepare for the paradigm-shifting, unpredictable, and rapid changes that would be brought about? Transhumanist technologies provide one solution. If human capabilities are fused with those of these superintelligences, then we would no longer face an existential crisis where machines dominate and rule over us. We would be one with the machines (Kurzweil 2005). However, as long as we are limited by the computational inefficiencies of our biological substrate, we may never be able to match machine capabilities. Thus, in our desire not to be subdued by machines, humans may willingly bring about the end of humanity by altering our minds and bodies to a degree to which we can no longer call ourselves human.

Summary

The hypothesis of a technological Singularity posits that humanity will create machines with above-human-level intelligence, triggering exponential growth in machine cognition. In rapid succession, intelligent machines will design future generations of intelligent machines with greater and greater capabilities. These synthetic progenies will be able to learn, to create, to invent, and to solve problems that are well beyond humanity's current capabilities. The resultant technologies would provide tremendous benefits to humanity, but they would also pose a danger. The synthetic intelligences will have enormous power, yet their algorithms may be incomprehensible to us. Furthermore, it may be difficult to align the goals of machines with those of human beings.

The advent of the technological Singularity may occur during the twenty-first century. Advances in computer hardware, whole-brain emulation, and machine-learning algorithms all trend toward machines with capabilities that are closer and closer to human performance. The perils of the technological Singularity are significant. We must begin to plan for the economic and societal disruption that would ensue. Furthermore, we must also develop theories of ethics and morality that would provide a basis for the behavior of autonomous, superintelligent nonhuman agents. The fate of humanity may depend on it.

Bibliography

Asimov, Isaac. *I, Robot.* New York: Gnome Press, 1950.

Barua, Resheque, Shimon Sramon, and Marcel Heerink. "Empathy, Compassion, and Social Robots: An Approach from Buddhist Philosophy." In *New Friends 2015: Proceedings of the 1st International Conference on Social Robots in Therapy and Education,* edited by Marcel Heerink and Michiel de Jong, 70–71. Almere, Netherlands: Windesheim Flevoland, 2015.

Bostrom, Nick. *Superintelligence: Paths, Dangers, Strategies.* Oxford: Oxford University Press, 2014.

Chace, Calum. *The Economic Singularity: Artificial Intelligence and the Death of Capitalism.* n.p.: Three Cs, 2016.

Del Prado, Guia Marie. "Even Computer Programmers Could Be Put Out of a Job by Robots." *Business Insider,* September 15, 2015. http://www.businessinsider.com/computer-scientists-not-safe-from-artificial-intelligence-unemployment-robots2015-9.

Greenemeier, Larry. "Driverless Cars Will Face Moral Dilemmas." *Scientific American,* June 23, 2016. https://www.scientificamerican.com/article/driverless-cars-will-face-moral-dilemmas/.

Holley, Peter. "Why Stephen Hawking Believes the Next 100 Years May Be Humanity's Toughest Test." *Washington Post,* January 20, 2016. https://www.washingtonpost.com/news/speaking-of-science/wp/2016/01/20/why-stephen-hawking-believes-the-next-100-years-may-be-humanitys-toughest-test-yet/.

Keats, Jonathon. "The $1.3B Quest to Build a Supercomputer Replica of a Human Brain." *Wired,* May 14, 2013. https://www.wired.com/2013/05/neurologist-markram-human-brain/.

Kurzweil, Ray. *The Singularity Is Near: When Humans Transcend Biology.* New York: Viking, 2005.

MacLennan, Bruce. "Cruelty to Robots? The Hard Problem of Robot Suffering." In *Proceedings of the 2013 Meeting of the International Association for Computing and Philosophy.* 2013. http://www.iacap.org/proceedings_IACAP13/paper_9.pdf.

Metz, Cade. "In Two Moves, AlphaGo and Lee Sedol Redefined the Future." *Wired,* March 16, 2016. https://www.wired.com/2016/03/two-moves-alphago-lee-sedol-redefined-future.

Mnih, Volodymyr, Koray Kavukcuoglu, David Silver, et al. "Human-Level Control through Deep Reinforcement Learning." *Nature* 518, no. 7540 (2015): 529–533.

Moore, Gordon. "Gordon Moore: The Man Whose Name Means Progress; The Visionary Engineer Reflects on 50

Years of Moore's Law." Interview by Rachel Courtland. *IEEE Spectrum*, March 30, 2015. http://spectrum.ieee.org/computing/hardware/gordon-moore-the-man-whose-name-means-progress.

Morris, David Z. "U.N. Moves towards Possible Ban on Autonomous Weapons." *Fortune*, December 24, 2016. http://fortune.com/2016/12/24/un-ban-autonomous-weapons/.

Murgia, Madhumita. "End of Moore's Law? What's Next Could Be More Exciting." *Telegraph* (London), February 25, 2016. http://www.telegraph.co.uk/technology/2016/02/25/end-of-moores-law-whats-next-could-be-more-exciting/.

People for the Ethical Treatment of Animals (PETA). 2017. http://www.peta.org/about-peta/why-peta/why-animal-rights/.

Shacham, Ofer, Omid Azizi, Megan Wachs, et al. "Rethinking Digital Design: Why Design Must Change." *IEEE Micro* 30, no. 6 (2010): 9–24.

Vinge, Vernor. "The Coming Technological Singularity: How to Survive in the Post-Human Era." *Whole Earth Review* (Winter 1993): 88–95. The original version of this article was presented at the VISION-21 Symposium sponsored by NASA Lewis Research Center and the Ohio Aerospace Institute, March 30–31, 1993. https://www-rohan.sdsu.edu/faculty/vinge/misc/singularity.html.

CHAPTER 11

A Skeptic's Perspective: Is This Actually Going to Happen?

Michael Bess
Chancellor's Professor of History
Vanderbilt University, Nashville, TN

They met for the first time in a hotel bar at Lake Tahoe, California, in 1998, one evening after a technology conference at which they were both invited speakers. Bill Joy (1954–) was an eminent computer systems designer and a chief scientist for Sun Microsystems; Ray Kurzweil (1948–) was an award-winning inventor and technologist, whose many creations included a reading machine for the blind and a cutting-edge music synthesizer. Their conversation focused on the future relationship between humans and machines. What they saw that evening, as they gazed together into the coming decades, is something that has come to be called the Singularity. Both Joy and Kurzweil believed it would arrive around 2040, more or less. Both of them felt that its advent would constitute the most dramatic turning point in the history of humankind thus far.

On the other side of that pivotal event, a convergence of technological breakthroughs would utterly transform the human species and its place in the cosmos. Through advanced biotechnology, people would boost their physical and mental performance to unprecedented levels. Neuroscientists and cognitive psychologists would make steady progress in mapping the functional architecture of the human brain, yielding knowledge that would allow computer designers and roboticists to create superintelligent machines. Advances in nanotechnology, genetics, and synthetic biology would speed the process along. Working together, the bioenhanced humans and artificial intelligence teams would embark on further projects of ever-rising ambition and scope, giving rise to an exponentially accelerating increase in capabilities. Like the transformation of a caterpillar into a butterfly, this would amount to nothing less than a moment of *species metamorphosis*, a collective transformation so sweeping and complex that no person in today's civilization could even comprehend its full implications (Joy 2000; Kurzweil 2005).

Kurzweil looked on this prospect with a mixture of awe and elation, embracing it as the fulfillment of humanity's deepest ideals and dreams. Joy regarded it with a mixture of awe and horror, recoiling from what he viewed as a radical dehumanizing of *Homo sapiens.* He also sensed profound danger in these powerful technologies—the very real risk of accidental cataclysms engulfing the entire biosphere and threatening not just humankind but all life on Earth.

The two men became friends, despite their divergent reactions, and continued to debate their respective visions of the future over the decade that followed, both in person and in print. Kurzweil, who was in his fifties at that time, eventually hired a specialized medical doctor to help him design an elaborate daily regimen of 250 pills and other antiaging remedies: he wanted to maximize his chances of living long enough to witness the Singularity firsthand (Wolf 2008). Joy published an incendiary article in the April 2000 issue of *Wired* magazine, sending shock waves through the world's community of technologists: he called on his fellow scientists and inventors to seriously consider renouncing work on any field that would hasten the Singularity's arrival.

And so it is with more than casual interest that we approach the following question: are Joy and Kurzweil right about the timing and inevitability of the Singularity's approach? This chapter maintains that their vision of the future is alarmingly persuasive in some respects and highly implausible in others. It argues that humanity is indeed headed into a period of radical and destabilizing technological change, a period in which the boundaries between humans and machines will increasingly break down; but it concludes that this transformation is likely to arrive piecemeal, in untidy increments and jumps, extending over a period of many decades through the second half of the twenty-first century.

This is still a startling conclusion. To say that the changes contemplated here will probably not arrive by 2040, but that many of them may well be in place by 2100, is tantamount to placing an epochal turning point of truly tectonic dimensions within the lifetimes of our own children and grandchildren. Seen from this angle, a mere sixty-year variance in the time frame of the prediction does not do much to quell its emotional impact. It is as though we were passengers on a raft in the Niagara River, hearing the roar of a waterfall up ahead. Whether the falls are a quarter mile away or a mile away, that sound still grabs our full attention. And well it should. The stakes could not be higher.

A SKEPTIC WEIGHS IN

One of the most thoughtful criticisms of the predictions put forward by such figures as Kurzweil and Joy has been offered by another technology writer, Bob Seidensticker. In his 2006 book *Future Hype: The Myths of Technology Change,* Seidensticker urges his readers to adopt a skeptical stance:

> Resist alarmist claims that technology change is increasing faster and faster, that society is about to be changed beyond recognition, and that you won't be able to handle it. On the contrary, the last two hundred years of technological progress teach us that change is roughly constant, not accelerating; change does indeed happen, but the most extreme predictions are the least accurate; and tomorrow will look more like today than most predictions expect. (219)

Seidensticker is a retired computer specialist who spent many years writing software for such companies as IBM and Microsoft. As someone whose livelihood required working closely with technology, he eventually grew fed up with hearing sensationalist media reports that uncritically trumpeted the extreme futures awaiting us through nanotechnology, artificial intelligence, genomics, or other cutting-edge fields. His book offers a systematic critique of contemporary discourse surrounding runaway technological change. Among his many insightful arguments, three stand out:

(1) Technology has already been profoundly changing human lives for hundreds of years, so we should not panic about the fact that it will probably continue to do so in the future.

(2) Technological change is not accelerating exponentially but rather has remained roughly constant since the late eighteenth century. We often *feel* as if it were accelerating, but this mistaken impression derives from the fact that we tend to overlook or discount the countless inventions—such as trains, electricity, telephones, antibiotics, and missiles—that have radically altered our lives over the past two centuries. Precisely because we humans adapt so quickly to new technologies, we tend to underestimate the importance of major innovations that have already become "normalized" as a part of our everyday environment.

(3) Future developments in technology—the potency and social impact of inventions yet to come—tend to be grossly overestimated. Prognosticators fail to take into account the complex economic, cultural, and infrastructural preconditions that must be in place before a given technology can exert truly transformative effects on society. These preconditions generally develop incrementally and slowly, thereby giving social institutions relatively long periods in which to absorb their revolutionary implications. Automobiles, for example, could already travel relatively fast within a few years of their invention, but many decades had to pass before their full social impact came to be felt. Not only did the vast infrastructure of paved roads, parking facilities, and gas stations have to be created, but people's cultural expectations about travel, distance, time, and mobility had to shift as well. Only then, after a half century had gone by, did the automobile fulfill its promise as a revolutionary technology.

THE FALLACY OF TECHNOLOGICAL DETERMINISM

In his best-selling 2005 book *The Singularity Is Near*, Kurzweil takes pains to address some of the serious dangers posed by accelerating developments in genetics, nanotechnology, and robotics. At one point, for example, he goes through an elaborate calculation to estimate how long it would take for out-of-control self-replicating nanobots (or molecular machines) to destroy Earth's entire biomass (several weeks, he concludes) (399). Kurzweil's response to these kinds of technological über-threats is framed, not surprisingly, through technology itself: the best defense against such threats is *more and better* machines. We must strive to keep our defensive technological capabilities, through which we rein in and control our powerful devices, slightly more sophisticated than those devices themselves. In Kurzweil's view, this is a technological race that humankind will never be able to stop running: we have unleashed the unstoppable genie of scientific and technical change, and now our only hope is to stay indefinitely ahead of the curve. Under the heading "The Inevitability of a Transformed Future," Kurzweil concludes, "The only conceivable way that the accelerating pace of advancement on all of these fronts could be stopped would be through a worldwide totalitarian system that relinquishes the very idea of progress" (407).

According to this viewpoint, technological advance is irresistibly driven forward by a combination of two factors: the constant demand that consumers make for ever-more potent and sophisticated tools and devices, and the incessant scrambling of scientists, inventors, and businesses to create the machines that consumers will buy and use. These two factors are so deeply woven into the fabric of modern economies that they render the accelerating advance of technological innovation virtually unstoppable. The technologies keep coming ever anew, they force people to adapt to the new capabilities they proffer, and the cycle goes on repeating without end.

Historians of technology refer to this sort of perspective as technological determinism, and it is a view that has come to be challenged by most scholars. Here is the way historian David E. Nye (1946–) summed it up in his 2006 book *Technology Matters*:

> A technology is not merely a system of machines with certain functions; rather, it is an expression of a social world. Electricity, the telephone, radio, television, the computer, and the Internet are not implacable forces moving through history, but social processes that vary from one time period to another and from one culture to another. These technologies were not "things" that came from outside society and had an "impact"; rather, each was an internal development shaped by its social context. No technology exists in isolation. Each is an open-ended set of problems and possibilities. Each technology is an extension of human lives: someone makes it, someone owns it, some oppose it, many use it, and all interpret it. Because of the multiplicity of actors, the meanings of technology are diverse. (47)

The contrast between Nye's view of technological change and Kurzweil's could not be starker. Kurzweil seems to be thinking of technology through the metaphor of a tool: like a hammer or a knife, technology in general is neither inherently bad nor good, but fundamentally neutral in nature, because it can be used equally effectively for good or for evil. The development of these devices follows an irresistible inner logic of its own: if a better tool *can* be designed and made, it nearly always *will*. In Kurzweil's vision, therefore, tools are regarded as objects that arrive among humans as if they were coming from *outside* society, compelling the human population to keep adapting itself to this ongoing and incessant evolution of its devices and machines as they rise up to higher and higher levels of potency.

For Nye, by contrast, no tool exists in isolation from the social and cultural milieu in which it is conceived, designed, manufactured, and used. Tools shape their milieu, and the milieu shapes its tools: technology and society reciprocally co-construct each other over time in a dialectical process that never ends. Every tool is therefore an intrinsically value-laden object, because it emerges from a particular configuration of cultural assumptions, perceived needs, social and economic purposes, and practical constraints. A hammer, a computer, a toilet, or an airplane—each exists as part of a broader system of functions, and those functions are all, without exception, socially determined. To speak of a "better" tool, in this context, immediately raises these questions: Better, according to whose scale of values? Better, according to which specific historical configuration of assumptions, purposes, and constraints?

Seidensticker echoes this anti-deterministic view in *Future Hype* (2006), citing the example of aircraft technology to illustrate his point. When the Wright brothers made their first powered flight in 1903, their aircraft stayed aloft for 12 seconds and traveled 120 feet (36.6 meters). Over the years and decades that followed, the speed and range of the flying machines steadily and rapidly increased. If, in 1960, you had plotted a graph of aircraft innovations over the preceding six decades, you would have gotten a steeply rising curve; and if you had projected this trend line into the future, you would have had every reason to expect that supersonic flight would be the norm by 2000. But it did not turn out this way. By 1970, a supersonic jetliner, the Concorde, had indeed been built, but it became a commercial flop because the cost of operating such a plane proved far too high. Most people, it turned out, were perfectly content to fly from one place to another at subsonic speeds. The actual graph of aircraft speed and range from 1960 to the twenty-first century bends sharply toward a flat line and stays stubbornly that way all the way through.

This is a very different sort of reality than the one depicted by Kurzweil in his confident graphs and projections of exponential change. It is a world in which powerful social and economic factors—the oil crisis of the 1970s, the advent of mass air travel, the fickle tastes of consumers—all conspired to lead a particular technology down an unexpected path. Even though supersonic flight was technically feasible, society opted instead for crowded, slower flights accessible to the middle-class masses. Basic economics trumped the allure of glamorous technology.

In this way of viewing history, human choices play a central role. The example of the Concorde (which could be repeated for many other technologies) suggests that technological progress is not an "objective" phenomenon straightforwardly dictated by technical considerations and capabilities. Instead, it is a sociotechnical construct that results from the ongoing interplay of physical, scientific, technological, political, economic, and cultural factors: faster is not always better, and people sometimes opt for simpler, more practical machines that meet a different set of needs than those that the economists and engineers had anticipated.

When writers such as Kurzweil herald the alleged "inevitability of our technological future," therefore, such claims should be regarded with a sharply critical eye. Our devices and machines certainly do evolve, but the paths they follow are not wholly independent of human values and human agency. Innovation is not "an implacable force moving through history" but instead one element in the ongoing dialectic through which technology and society co-construct each other. A sky full of sexy Concordes may well have seemed "inevitable" to someone gazing forward from the vantage point of 1960; what we got instead, however, was gazillions of humdrum 737s plying the heaven of frequent-flier miles.

A MIDDLE WAY BETWEEN TECHNO-ENTHUSIASM AND TECHNO-SKEPTICISM

Seidensticker's skeptical argument serves as a valuable corrective to the hyperventilating tendencies of the technological enthusiasts. Having said this, however, one must scrutinize Seidensticker's conclusions with the same caution with which one approaches Kurzweil's. For where Kurzweil tends to overemphasize the radical discontinuity that lies ahead, Seidensticker in turn places too much emphasis on the *continuities* of history. When he assures us that "tomorrow will look more like today than most predictions expect," he is dismissing an accumulating body of evidence that points toward deep and potentially disruptive transformations coming down the pike. Somewhere between these two extremes—between Kurzweil's depiction of a looming Singularity and Seidensticker's imagery of gradual, manageable change continuing indefinitely—the most persuasive future scenarios lie.

Society in the twenty-first century is geared to an unprecedented degree toward the production of rapid technological innovation. This is a cliché, but it is one worth dwelling on for a moment because it forms an essential starting point for thinking about the shape of the coming decades. Far more persons devote their lives today to the pursuit of scientific knowledge and technological invention than at any other time in history. If one takes the sum total of all the scientists and technologists who ever plied their trade from the time of ancient Babylon to the 1990s, this number would be dwarfed by the number working today. Not surprisingly, the financial and social resources propelling the advance of contemporary technoscience are equally unprecedented: never before has humankind focused so single-mindedly on cranking out new knowledge and applying it to the transformation of the material world.

This phenomenon has roots that go back at least two centuries. It began gathering momentum in the late 1700s, as Enlightenment attitudes toward rational inquiry and practical applications of knowledge bore fruit in the mechanization of agriculture and the development of the modern factory system. The innovations fed off each other, with revolutions in transportation, manufacturing, finance, and communications marking the first half of the 1800s, and a second industrial revolution (based on electrical energy and applied chemistry) following after the 1870s. These growth trends continued unabated through the first half of the twentieth century, then accelerated dramatically after World War II (1939–1945).

It is only in the decades since 1945 that two vast scientific and technological establishments emerged in parallel: one publicly funded and centered in universities and government research facilities, the other privately funded and centered in large corporations. World War II had made it abundantly clear to government leaders in all the industrialized nations that science and technology constituted the cornerstone of geopolitical power. The race for innovations in technoscience—refracted through the broader economic system that simultaneously fed those innovations and thrived off them—ultimately determined the pecking order of entire global regions.

Seidensticker is therefore quite right to point out that the ramping up of technological and scientific discovery began a long time ago, in the 1700s; but Kurzweil is also right when he singles out the post-1945 decades as a uniquely intense period of advance. Kurzweil's mistake lies in uncritically extrapolating from these more recent trends, projecting exponential rates of growth indefinitely. But Seidensticker's mistake is equally problematic: he ignores the massive and unprecedented institutionalization of technoscience that has come to characterize the global economic system since 1945. Consider, for example, the following statistics, which show the number of persons employed in the United States in all fields of science, engineering, and technology over the period from 1850 to 2009 (Sobek 2001; National Science Foundation 2005; US Bureau of Labor Statistics 2010).

1850: 1,700	(0.03 percent of total workforce)
1900: 40,300	(0.14 percent of total workforce)
1940: 308,200	(0.64 percent of total workforce)
1950: 705,800	(1.1 percent of total workforce)
1980: 1,899,700	(1.9 percent of total workforce)
2001: 5,580,200	(4.2 percent of total workforce)
2009: 7,024,800	(5.4 percent of total workforce)

The growth in both raw numbers and percentages is striking, but for some reason the historically unique nature of this phenomenon seems not to impress Seidensticker a bit. It is as though the armies of scientists and technologists working today—their sheer numbers, as well as their unparalleled degrees of specialization and sophistication—all remained invisible to him.

As a result, Seidensticker's analysis lapses into its own form of excess. Just because *exponential* rates of change are unlikely to characterize the coming decades, this does not mean that all radical and destabilizing forms of technological change should be expunged from policymakers' forecasts. On the contrary, rapid scientific and technological innovation are now deeply written into the institutions and economic practices that structure modern society; people's jobs, standard of living, and cultural expectations are tightly linked to the premise of constantly rising productivity and capabilities. (Although, to be sure, this faith in endless economic growth is highly problematic from a point of view of ecological sustainability.) This ever-rising technoscientific prowess is a defining feature of the current historical epoch, and it is just as serious a mistake to underestimate its importance as it is to exaggerate it.

Nowhere is this observation more relevant than vis-à-vis the burgeoning field of biotechnology. This field lies at a crossroads between the natural sciences, engineering, medicine, and commerce, and since the 1980s—perhaps precisely because of its interdisciplinary nature—it has been growing faster, attracting more funding, and generating more controversy than arguably any other domain of technoscience (Kang 2012). Whereas the Industrial Revolution of the 1800s focused on the material world around us, seeking ways to improve food production, transportation, and manufacturing, today's biotech revolution has turned to humans themselves: it is our own bodies and minds that are now being increasingly modified. This is a key premise of the vision of the future propounded by figures such as Kurzweil and Joy: the human constitution itself will be the next frontier of rapid innovation and transformation.

A skeptic might object that we are once again succumbing to hype here—that all this is just a matter of degree—because we humans have already been modifying ourselves for a very long time. Have we not been tweaking our moods chemically through tea and tobacco for millennia, incorporating prostheses such as eyeglasses for centuries, and using (or choosing not to use) the techniques of selective breeding and eugenics for more than a hundred years? But this is ultimately a weak rejoinder, because it ignores the tremendous increase in potency and sophistication that has marked these kinds of interventions since the 1960s. There is a point at which discrete *quantitative* changes, accumulating over time, abruptly cross a threshold and become *qualitative* leaps into something radically different and new.

Such a threshold has clearly been reached with contemporary biotechnology. Our pharmaceuticals now allow us to intervene directly, at a molecular level, to influence and redirect the chemical processes that contribute to making us who we are. We can render ourselves smarter, faster, more coordinated, or less shy simply by choosing from an increasingly impressive array of pills. Our bioelectronic prostheses now connect directly with our neurons and brain, linking them to informatic devices that decode the workings of our nervous system with unprecedented accuracy and speed. You or I can put on a skullcap and sit in front of a computer, playing video games without moving a muscle, controlling the machine by thought alone. Genetic modification of humans is no longer a sci-fi scenario; scientists and doctors can reach directly into someone's genome or epigenome, subtly and precisely modifying the underlying DNA transcription and expression that influence the person's physical and mental traits. In all three of these areas—drugs, bioelectronics, and genetics—the key recurring features are the unprecedented *directness* and *precision* with which all these interventions can now be achieved. Taken together, they constitute a major factor distinguishing the current era from those that came before it.

Summary

Humans reshaped by drugs, humans symbiotically penetrated by informatic devices, humans genetically redesigned to live longer and stronger—these are not mere figments conjured up by science fiction authors like William Gibson or Philip K. Dick. They are coming over the horizon of the possible in places such as Berkeley, Oxford, Seoul, and Paris, taking shape as concrete developments headed our way.

It is here that Seidensticker falls short, with his vision of a future traversed in steady, orderly increments; and it is here, conversely, that Kurzweil's transhumanist presentiments arguably offer the more realistic scenario. To be sure, Kurzweil is probably mistaken when he characterizes the Singularity as an event approaching unstoppably, at exponential speed; but he is definitely onto something when he senses that a wave of profound change is approaching. Even if this transformation comes about much more slowly and unevenly, spread out in messy increments over a century or more, it still holds the potential to turn human civilization upside down. We have some fundamental choices awaiting us down this road, and it behooves us to start thinking through those choices today.

Bibliography

Bijker, Wiebe E., Thomas P. Hughes, and Trevor J. Pinch, eds. *The Social Construction of Technological Systems: New Directions in the Sociology and History of Technology*. Cambridge, MA: MIT Press, 1987.

Jasanoff, Sheila, Gerald E. Markle, James C. Petersen, and Trevor Pinch, eds. *Handbook of Science and Technology Studies*. Thousand Oaks, CA: Sage, 1995.

Joy, Bill. "Why the Future Doesn't Need Us." *Wired*, April 1, 2000. https://www.wired.com/2000/04/joy-2/.

Kang, Kelly. "Graduate Enrollment in Science and Engineering Grew Substantially in the Past Decade but Slowed in 2010." National Science Foundation. May 2012. https://www.nsf.gov/statistics/infbrief/nsf12317/nsf12317.pdf.

Kurzweil, Ray. *The Singularity Is Near: When Humans Transcend Biology*. New York: Viking, 2005.

National Science Foundation. Division of Science Resources Statistics. *Scientists, Engineers, and Technicians in the United States: 2001*. NSF 05–313. Arlington, VA: Author, 2005.

Nye, David E. *Technology Matters: Questions to Live With*. Cambridge, MA: MIT Press, 2006.

Seidensticker, Bob. *Future Hype: The Myths of Technology Change*. San Francisco: Berrett-Koehler, 2006.

Smith, Merritt Roe, and Leo Marx, eds. *Does Technology Drive History? The Dilemma of Technological Determinism*. Cambridge, MA: MIT Press, 1994.

Sobek, Matthew. "New Statistics on the U.S. Labor Force, 1850–1990." *Historical Methods* 34, no. 2 (2001): 71–87.

Tetlock, Philip E., and Dan Gardner. *Superforecasting: The Art and Science of Prediction*. New York: Crown, 2015.

US Bureau of Labor Statistics. "Table 2. Employment by Industry and Occupational Group, May 2009." 2010. https://www.bls.gov/news.release/ocwage.t02.htm.

Vinge, Vernor. "The Coming Technological Singularity: How to Survive in the Post-human Era." Paper presented at the VISION-21 Symposium, Westlake, OH, March 1993. http://www-rohan.sdsu.edu/faculty/vinge/misc/singularity.html.

Wolf, Gary. "Futurist Ray Kurzweil Pulls Out All the Stops (and Pills) to Live to Witness the Singularity." *Wired*, March 24, 2008. https://www.wired.com/2008/03/ff-kurzweil/.

Religious Responses and New Religiosities

CHAPTER 12

Buddhist Uploads

Beverley F. McGuire
Associate Professor of East Asian Religions,
Philosophy and Religion Department
University of North Carolina Wilmington

Buddhists regard human life as precious. In a text from the Pali canon, an early collection of Buddhist teachings, the Buddha asks a group of monks to imagine that the earth was covered with water, and a man was to toss a yoke with a single hole into the water. Winds would push it in every direction. He then says, "Suppose a blind sea turtle was there, which would come to the surface only once every hundred years. Now what do you suppose the chances would be that a blind turtle, coming to the surface once every hundred years, would stick its neck into the yoke with a single hole?" When the monks respond that it would be quite unusual, the Buddha responds, "And just so, it is very, very rare that one attains the human state" (*Saṃyutta Nikāya* 56.48).

Human life represents one of six types of rebirth in Buddhism, alongside gods, demigods, hungry ghosts, animals, and hell dwellers. While living beings dwell in the realm of desire (in Sanskrit [Skt.] *kāma-dhātu*), gods may inhabit realms of form (Skt. *rūpa-dhātu*) and formlessness (Skt. *ārūpa-dhātu*). The three realms of desire, form, and formlessness constitute the whole universe. Karma—bodily, verbal, and mental actions—determines one's rebirth. Buddhist sutras (teachings attributed to the Buddha) vividly describe the suffering endured in the various types of hell: hot hells, where beings must walk on hot ash or razors; cold hells, where their skin splits open from extreme frostbite; and Avīci hell, which gives no respite from suffering. Hungry ghosts have distended stomachs and needle-thin necks that render them unable to eat, although they sometimes ingest excrement, pus, or scum. Demigods possess incredible strength, but they engage in constant warfare. Although some gods dwell in the realm of desire, possessing greater powers than humans but still held captive by their cravings, gods who inhabit the higher realms fulfill their passions by hugging, holding hands, smiling, or even looking at other gods (Sadakata 1997).

Buddhists with this worldview would likely devalue posthumanism because they consider human life a rare opportunity to experience suffering, which can serve to motivate people to seek liberation in order to escape the cycle of birth and death (Skt. *saṃsāra*). They might view posthumans with their enhanced cognitive and bodily structures as analogous to gods or demigods. Like gods, posthumans might have more power, longer lives, and greater happiness than humans, but being content to enjoy their bliss-filled lives, they would probably not seek to escape the cycle of birth and death by attaining enlightenment. The Buddha is said to have generated this aspiration after seeing four sights—an ill person, an old person, a corpse, and an ascetic wanderer who appeared serene in the face of suffering.

Would he have renounced his royal life in a posthuman world rid of such experiences of sickness, aging, and death? Would seminal Buddhist teachings about the inevitability of suffering and impermanence be undercut in a posthuman world? Would there be an impetus for posthumans to seek enlightenment and nirvana if their cognitive and physical features were so enhanced that they rarely experienced such suffering? Or would they, like gods, instead seek to dwell in the universe as long as possible?

This chapter considers such questions, as well as the case of mind uploading: transferring a person's consciousness to an external carrier such as a computer and thereby allowing the person to avoid aging or death. Although current technology has not yet developed a means of downloading or uploading our conscious minds, futuristic technology may enable scientists to generate functional models of the human brain and eventually upload the consciousness of a living human being into a machine or posthuman body. The current Dalai Lama (1935–), an important figure in Tibetan Buddhism, acknowledged this potentiality when he remarked, "I can't totally rule out the possibility that, if all the external conditions and the karmic action were there, a stream of consciousness might actually enter into a computer" (quoted in Hayward and Varela 1992, 152). Asked whether a yogi (an advanced practitioner of yoga and meditation) might be able to project his subtle consciousness into a computer, the Dalai Lama allowed for that possibility "if the physical basis of the computer acquires the potential or the ability to serve as a basis for a continuum of consciousness" (153). Although the Dalai Lama's remarks suggest openness toward uploading, his stipulations about external conditions, karmic action, and the necessary physical basis indicate that a Buddhist uploading of consciousness would not involve simply transferring information from one's brain. From a Buddhist perspective, such a transference would depend on karma and encompass the entirety of one's consciousness, including one's cognitive, affective, and karmic domains.

These Buddhist perspectives highlight some of the possibilities and limitations of posthumanism. After exploring areas of overlap between Buddhism and posthumanism—their challenging of binaries such as self/no-self, woman/man, human/animal, and nature/culture—this chapter examines significant differences between their understandings of the human body, consciousness, and the transference of consciousness.

SELF/NO-SELF

Posthumanists and Buddhists both criticize essentialized conceptions of human nature and human exceptionalism. Posthumanism aims to decenter the human subject and envisions humans as acting within large, complex systems alongside other agents (Hayles 2010), and Buddhism similarly views humans as operating within a vast universe with many other sentient and nonsentient beings. Moreover, Buddhist teachings challenge the notion that people have a self or essential nature, asserting that they are instead composed of five aggregates (Skt. *skandha*): a body (Skt. *rūpa*), sensations or feelings (Skt. *vedanā*), perceptions (Skt. *saṃjñā*), volitions or mental formations (Skt. *saṃskāra*), and consciousness (Skt. *vijñāna*). Together these five aggregates lead people to mistakenly believe that they have a self, as they observe their material form, their sensation of things as pleasurable, painful, or neutral, their perception of reality based on those judgments, their conditioned tendency to perceive or react in certain ways, and their consciousness. Buddhists argue, however, that if one of the aggregates is missing, that individual self dissolves, and as the aggregates change, so does the self. In other words, there is no inherent self-nature—only a sense of self that changes over time.

Considered in light of uploading, Buddhists would argue for the necessity of some type of embodied existence or material form to support the uploaded consciousness. Robots or cyborgs—"cybernetic organisms," which are part machine and part human—might provide this physical basis if they allowed for sentience. Not only would they have to have senses of some sort, but they would also have to be able to experience suffering or dissatisfaction (Skt. *duḥkha*). If technology could determine how the robotic or cyborg form might allow for posthumans to have such feelings, perceptions, volitions, and consciousness, as the Dalai Lama remarked, then the possibility of uploading would solely depend on karma.

Some Buddhists would identify the "storehouse consciousness" or "base consciousness" (Skt. *ālāya-vijñāna*) as that which should be uploaded. According to the Yogācāra tradition in Mahāyāna Buddhism, the branch of Buddhism predominant in East Asia, this consciousness transforms all other types of consciousness, and it ultimately influences one's next rebirth. Yogācāra texts describe eight different types of consciousness, the first five being tied to the senses (the eye, ear, nose, tongue, and body), the sixth being tied to the mind, the seventh being a deluded awareness of self, and the eighth being this storehouse consciousness. Together they account for the sights, sounds, smells, tastes, feelings, thoughts, self-grasping, and memories in one's mind. Interestingly, only enlightened beings (arhats or Buddhas) have nonkarmic cognition; for all other beings, cognition depends on one's karma (Lusthaus 2002), and the eighth consciousness is also called "seed consciousness" (Skt. *bīja-vijñāna*) because it receives the karmic impressions from all other types of consciousness and stores them. Any action—of thought, word, or body—can leave its impression in this level of consciousness.

When the Dalai Lama speaks of a "stream of consciousness" for uploading, he is referring to this consciousness. Some Tibetan Buddhist traditions describe it as originally empty and luminous, as well as the basis for all thoughts, perceptions, and emotions that emerge in an individual's mind. Advocates of "contemplative science" argue that focused attention on this consciousness, cultivating meditative quiescence, can serve as a bridge between scientific and contemplative ways of exploring the mind (Wallace 2007). It transcends not only a person's lifetime but also a person's gender and species, which is explored in the next sections.

WOMAN/MAN

Posthumanism challenges the differentiation, domestication, and hierarchizing of genders, instead embracing positions betwixt and between. Although one can certainly find strains of Buddhism rife with sexism and misogyny, others similarly challenge biological essentialism and dualistic views of gender. Mahāyāna and Vajrayāna traditions in particular have criticized facile distinctions between women and men. The Heart Sutra proclaims the emptiness of the five aggregates: "Form is emptiness; emptiness is form. The same is true for feelings, perceptions, volitions, and consciousness." Buddhists offer various interpretations of emptiness, such as being empty of any essence or inherent self-nature. If one's physical form lacks any essence or self-nature, then the same would apply to one's gender.

There are also Buddhist texts that emphasize the transmutability of gender. The seventh chapter of the Vimalakirti Sutra depicts a humorous exchange between the Buddha's disciple

Shariputra and a goddess. When Shariputra asks why she has not changed from her female body, the goddess replies, "For the past twelve years I have been trying to take on female form, but in the end with no success. What is there to change?" (Watson 1997, 90). Comparing her female form to a phantom, the goddess emphasizes that neither has a fixed form. She then uses her supernatural powers to transform Shariputra into a goddess, while she takes his form, and she states, "Shariputra, who is not a woman, appears in a woman's body. And the same is true of all women—though they appear in women's bodies, they are not women. Therefore, the Buddha teaches that all phenomena are neither male nor female" (91). The goddess undercuts any distinction between men and women, insisting that while she may appear to be female in the conventional world, in an ultimate sense she has no gender. A seventeenth-century Tibetan lama tells a similar story about Tara in the Origin of the Tara Tantra. Tara was reborn and achieved enlightenment as a princess, only to be told by Buddhist monks that she should seek to be reborn as a man. She responds, "In this life there is no such distinction as male or female, neither of 'self-identity' or a 'person' nor any perception [of such], and therefore attachment to ideas of 'male' and 'female' is quite worthless" (quoted in Padma'tsho 2014, 186). Just as there is no such thing as a self, there are no such distinctions between male and female.

These Buddhist examples resonate with posthuman questioning and queering of the way that gender and sexuality is articulated, symbolized, and classified (Halberstam and Livingston 1995). Donna Haraway's "Cyborg Manifesto," first published in 1984, criticized ideas of biological essentialism and dualistic assumptions of male/female. Haraway imagined a cyborg world in which people would be unafraid of "permanently partial identities" (Haraway 2000, 295) as she wrote, "Cyborgs might consider more seriously the partial, fluid, sometimes aspect of sex and sexual embodiment. Gender might not be global identity after all, even if it has profound historical breadth and depth" (Haraway 2000, 315). The cyborg can embody multiple genders and sexualities, understanding the world from different perspectives. Recent posthuman works continue criticizing efforts to categorize gender and sexuality, noting how "posthuman bodies thrive in the mutual deformations of totem and taxonomy" (Halberstam and Livingston 1995, 19).

The goddess in the Vimalakirti Sutra demonstrates the fluidity, contestation, and malleability of gender, engaging in a similar type of questioning and queering of gender identity. After her display she remarks, "All things are just like that—they do not exist, yet do not not exist" (Watson 1997, 91). One might argue that her phantasmagoric display resonates with a posthuman feminism that straddles the borders between male/female and real/imaginary (Rabinowitz 1995).

Some might claim, however, that posthumanism does not merely question and queer such borders but goes beyond gender altogether. When she declared she would rather be a cyborg than a goddess, Haraway (2004) implied that a posthuman world would not only disrupt but transcend gender. By contrast, the goddess in the Vimalakirti Sutra allows for the appearance of women and men in the world even though she refutes any absolute notion of gender.

Uploading consciousness might complicate such postgender possibilities if people retained memories tied to gender identity. Admittedly, advanced technologies might allow for the manipulation of such memories. However, if one shared the Buddhist assumption that one's storehouse consciousness retains karmic impressions and habitual dispositions from previous lives as men or women, this might be a formidable challenge.

HUMAN/ANIMAL

Posthumanism criticizes anthropocentric worldviews that favor humans in the natural order and sharply distinguish between humans and animals. Instead, posthumanists portray humans as one among many species, and they dispute any right for humans to destroy nature or view themselves as worthy of greater ethical consideration than animals. Instead of resorting to taxonomies, they encourage humans to think about their connections and encounters with animals, which make humans accountable to them. In Buddhism, human beings occupy only a small part of an expansive universe, and humans, animals, and all sentient beings are subject to illusion, suffering, and rebirth. In addition, some Mahāyāna Buddhist traditions maintain that all sentient beings, including animals, have Buddha nature—the potential to attain enlightenment.

Just as posthumanism challenges the notion of species boundaries, Buddhism emphasizes sentience as a crucial commonality between humans and animals. Although Buddhists believe animals occupy a less fortunate realm of rebirth, they acknowledge the possibility that they or their loved ones may have formerly been (or may in the future become) animals. The Laṅkāvatāra Sūtra states, "In the long course of rebirth there is not one among living beings with form who has not been mother, father, brother, sister, son, or daughter, or some other relative. Being connected with the process of taking birth, one is kin to all wild and domestic animals, birds, and beings born from the womb" (quoted in Swearer 2001, 227). Buddhists see connections with all sentient beings that experience birth, death, and suffering, and they emphasize that one should not harm animals or other sentient beings. This ideal is reflected in the prayer of loving-kindness that concludes many Buddhist rituals: "May all beings be free from enmity; may all beings be free from injury; may all beings be free from suffering; may all beings be happy." Buddhists seek to extend loving-kindness to all sentient beings. Insofar as they already acknowledge their connection with animals and other sentient beings, Buddhists could readily inhabit Haraway's "cyborg world" of "lived social and bodily realities in which people are not afraid of their joint kinship with animals and machines, not afraid of permanently partial identities and contradictory standpoints" (2004, 13). Buddhist notions of interdependence and dependent origination speak to such partial identities and contradictions.

Dependent origination (Skt. *pratītya-samutpāda*) refers to the idea that all conditioned phenomena arise in dependence on other phenomena; in other words, nothing exists on its own but instead depends on others to condition its arising and existence. The Avataṃsaka Sutra offers the image of Indra's net to explain this idea: each knot of the net has a jewel, which reflects all others and their reflections to infinity, such that each thing is "interpenetrated" by every other. The sutra states: "Every living being and every minute thing is significant, since even the tiniest thing contains the whole mystery" (quoted in Harvey 2000, 153). Although Buddhists debate the extent to which interdependence might support an ecological ethic (Ives 2017), interdependence might appropriately describe relationships among humans, animals, machines, cyborgs, and other beings in a posthuman world.

Would Buddhists consider such posthuman beings as inhabiting the realm of animals, humans, or gods? It would likely depend on the level of technological advancement. In regard to uploading, Buddhists might argue that karmic inheritance would determine whether that consciousness would experience reality as a god, human, or animal, or whether the embodied experience would precipitate some entirely new realm of rebirth. As is discussed in the next section, Buddhists see a close relationship between sentient beings' karmic activities and the worlds they inhabit.

NATURE/CULTURE

Posthumanism and Buddhism challenge the dichotomy of nature and culture on different grounds. Posthumanists emphasize the particular historical context in which the distinction between culture and nature emerged—namely, eighteenth- and nineteenth-century Western science. Scientists represented and constructed nature as the opposite of culture, reason, and masculinity, and they portrayed science mediating nature and the humanities mediating culture. However, as new technologies blur the boundaries between nature/culture and human/nonhuman, some have asserted that "the human body itself is no longer part of 'the family of man' but of a zoo of posthumanities" (Halberstam and Livingston 1995, 3). Zoos may order their exhibits, but they make no pretense of being natural habitats (Graham 2002). Posthumanism exposes nature and culture as historical and rhetorical constructions, and it envisions a similar blurring of reality and artificiality.

By contrast, Buddhists view nature and culture as dependently originated and largely constructed by the mind. As Zen master Dōgen (1200–1253) writes:

> All beings do not see mountains and waters in the same way. . . . Hungry ghosts see water as raging fire or pus and blood. Dragons see water as a palace or a pavilion. Some beings see water as the seven treasures or a wish-granting jewel. Some beings see water as a forest or a wall. Some see it as the dharma nature of pure liberation, the true human body, or as the form of body and essence of mind. Human beings see water as water. Water is seen as dead or alive depending on causes and conditions. (Dōgen 2000, 70)

Dōgen insists that one's perception of nature depends on karmic causes and conditions. Sentient beings experience the world in radically different ways according to their realm of rebirth.

Early Buddhists also proposed a close relationship between human morality and the natural environment, identifying karma as one of the five natural laws of the cosmos, alongside physical laws, biological laws, psychological laws, and causal laws. Some Buddhist texts posit that the world arose, is maintained, and will ultimately disintegrate because of karma produced by living beings (Sadakata 1997). Fifth-century Indian Buddhist Vasubandhu wrote, "The world in its variety arises from action (*karma*). Actions accumulate by the power of the latent afflictions (*anuśaya*), because without the latent afflictions [they] are incapable of giving rise to a new existence. Consequently, the latent afflictions are the root of existence" (quoted in Waldron 2000, 201). Instead of differentiating between nature and culture, Vasubandhu asserts that the world results from previous human actions motivated by the three poisons of ignorance, anger, and desire. In other words, Buddhists emphasize not only the physical but also the karmic impact of humans on the world.

The central role of karma also appears in the Wheel of Life commonly displayed in Tibetan Buddhist temples. Symbolized by a pig, a snake, and a bird, these occupy the center of the wheel, surrounded by half-circles depicting virtuous and nonvirtuous activities that lead a being to higher and lower rebirths, which appear in the next layer of the wheel, surrounded by the twelve links of dependent origination: ignorance, formation, consciousness, name and form, the six sense faculties, contact, sensation, craving, grasping, becoming, birth, and old age and death. The Wheel of Life depicts the crucial role of karma in rebirth, as a mistaken view of reality and the self lead to certain mental formations that propel the consciousness to take on name and form as an embodied being, who experiences the world, develops positive, negative, or neutral sensations tied to objects, and craves and grasps at what it perceives as pleasurable, which fuels further karmic activity that determines one's rebirth.

In the case of uploading, Buddhists might predict that once an uploaded consciousness inhabited a posthuman body, its experience of consciousness, sensation, and cravings would prompt a very different type of existence. Through their bodily, verbal, and mental actions, posthumans could transform the world or even give rise to an entirely new realm of rebirth. In the next section the possibilities are considered.

POSTHUMAN PREDECESSORS AND POSSIBILITIES

Posthumanism envisions beings capable of extraordinary cognitive and physical feats that far supersede human abilities. Buddhism also speaks of beings capable of such achievements: Buddhas and bodhisattvas. Buddhas have attained enlightenment, while bodhisattvas vow to liberate other sentient beings from the cycle of birth and death before they themselves enter nirvana. Buddhist texts portray extraordinary mental and physical feats that Buddhas and bodhisattvas can accomplish, which include the six higher knowledges (Skt. *abhijñā*), such as mastery over their bodies (being able to vanish, pass through walls, walk on water, fly through the sky, etc.), godlike hearing, knowledge of others' states of mind, recollection of previous lives, seeing the death and rebirth of all sentient beings, and understanding the origin of suffering and the way to its cessation (Gethin 2011).

It is important to clarify, however, how Buddhas and bodhisattvas attain and use their supernatural powers, as it may reveal an important distinction between the posthuman and Buddhist projects. Buddhas and bodhisattvas achieve such powers by engaging in disciplined action and advanced meditation, and they display their power in order to teach sentient beings how to attain liberation. A Mahāyāna text titled *Da zhi du lun* states: "The bodhisattva, detached from the objects of the five senses, accomplished in the attainment of absorption in meditation, possessed with kindness and compassion, achieves higher knowledge in the interest of beings and displays extraordinary and wonderful things in order to purify beings' minds" (quoted in Gethin 2011, 227). The bodhisattvas' powers serve an important ethical and salvific aim: to purify the minds of sentient beings so that they may attain enlightenment. Their miraculous displays—and the Buddhist narratives that record and retell these wonders to other audiences—have the potential to change the reality, cognition, and moral status of sentient beings (Gómez 2011). In other words, displays of supernatural powers not only show their "wonder-working prowess" but also attest to their ethical and meditative attainment, convey Buddhist teachings, and demonstrate the nature of reality itself (Gómez 2011).

If posthumanism envisions someone attaining such powers without engaging in ethical and meditative discipline and displaying their cognitive and physical abilities just for show (or worse, to exert power over others), then Buddhists would question its value. For example, the avatar project of the 2045 Initiative, founded in 2011 by Russian entrepreneur Dmitry Itskov, hopes to develop an artificial brain into which one might transfer an individual's consciousness with the goal of achieving cybernetic immortality. Admittedly, Buddhists would argue that immortality contradicts the impermanent nature of reality itself, but they may also question the value in attaining immortality, especially if such immortal beings showed no regard for other suffering sentient beings. Thus, in response to the 2045 Initiative's avatar project, the Dalai Lama emphasized the need to discuss the ethics behind such progressive technology and one's responsibility toward other beings.

The bodhisattva ideal underscores potential ethical limitations of the posthuman project. Bodhisattvas spend many rebirths perfecting qualities such as generosity, ethics, patience, perseverance, concentration, wisdom, and loving-kindness. Supernatural powers are the fruit or by-product of such discipline rather than the goal, and they are used to purify the minds of other sentient beings. Buddhists would urge posthumanists to reflect on what they would consider the goal of such enhancements.

In the case of uploading, Buddhists who ascribe to a karmic worldview would emphasize the importance of considering whose consciousness would be appropriate for uploading. Some people's consciousness might not be cognitively, emotionally, or morally equipped to inhabit a posthuman body. Although the Dalai Lama once claimed he might reincarnate into a computer in the event of further development of technology, Buddhists would emphasize that his consciousness does not represent that of an average sentient being. Not only does the Dalai Lama have a karmic affinity with science and technology—having shown interest in science at an early age and having collaborated with researchers for decades—but he is also believed to be the fourteenth in a lineage of spiritual masters and the incarnation of the bodhisattva of compassion. As noted above, the Buddhist tradition regards bodhisattvas as beings who have already engaged in ethical and meditative discipline and who have committed themselves to helping other sentient beings. Unlike sentient beings who are reborn through the force of their karma, bodhisattvas can choose the time and place of their rebirth, which occurs through the force of their compassion and their vows to liberate other sentient beings. Whereas the Dalai Lama as a bodhisattva can control his rebirth, and therefore choose to be reborn into a posthuman body, sentient beings may not be karmically suited for a posthuman rebirth.

Buddhists may also question the purpose of uploading a person's consciousness into a posthuman body. If one believes in karma and seeks to attain liberation, would it be more useful to inhabit a posthuman body than a human one? Although enhanced cognitive and physical abilities could potentially support one's efforts in meditation and perception of reality, they could also decrease one's motivation to escape the cycle of birth or death. Moreover, posthuman capacities may not offer any moral or ethical advantage. A frequently cited passage from the Dhammapada (14.183) states: "To avoid evil, cultivate good, and purify one's mind—that is the teaching of the Buddhas." Living a virtuous life and purifying the mind requires effort and discipline, but one might not give thought to such ethical development if one were distracted by one's considerable cognitive and physical abilities. For Buddhists who believe a person's storehouse consciousness cannot be wiped clean by an advanced technology, a posthuman existence may impede one's attainment of liberation.

Even Buddhists who may not believe in karma or rebirth would encourage greater ethical deliberation about the impact of posthumanism on all sentient beings. Buddhists who view the realms of rebirth as corresponding to mental states—hell dwellers as suffering, hungry ghosts as unsatisfied craving, animals as ignorance, demigods as envy, and gods as pleasure junkies (Hughes 2012)—would encourage broader discussions of the mind that include not only its cognitive aspects but also its affective and moral dimensions. They may also point out the danger of focusing solely on how to develop certain technologies without considering their purpose and potential impact.

Some have reflected on these issues and discussed the possibility of developing moral enhancements in addition to cognitive and physical ones. At the Institute for Ethics and Emerging Technologies, cofounded by James Hughes and Nick Bostrom, members of a transhumanist group known as the Cyborg Buddha Project have debated whether advanced technology supports or undercuts a Buddhist lifestyle, raising questions such as the following: Might a potentially unlimited life expectancy instill greater awareness of no-self, when one sees the radical discontinuity between one's experience as a twenty-year-old and one's experience at age 200? By uploading and downloading our memories and innermost experiences with others, might we challenge the notion of a separate, autonomous self? Might the use of neurotechnology or attention-enhancing drugs aid meditation practice?

Although he acknowledges their potential dangers, Hughes (2013) has argued that neurotechnologies could be used to support moral development by suppressing vices and enhancing virtue. For example, he describes how methylphenidate (Ritalin) can encourage mindfulness; how oxytocin and MDMA (Ecstasy) can boost agreeableness, which is correlated with empathy; and how increasing the brain's supply of glucose can boost self-discipline. Suggesting that they might first be used as "spiritual training wheels" to create a solid foundation of moral behavior and mental concentration, he then allows for the possibility that "as the technologies develop they may be used as the principal means of self-transformation" (Hughes 2013, 38). A posthuman world may enable moral enhancements to complement cognitive and physical enhancements. Secular Buddhists may then welcome a posthuman existence because it encompasses all dimensions of the Buddhist path.

Summary

Buddhism and posthumanism both question essentialist views of human nature, dualistic approaches to gender, artificial boundaries between nature and culture, and anthropocentrism, which privileges human beings over other beings. However, the Buddhist tradition considers human existence precious because it allows one to experience suffering yet also engage in an ethical lifestyle and meditation practices that potentially lead to liberation from the cycle of birth and death. Unlike animals, hungry ghosts, and hell dwellers, who suffer in cruel or debilitating ways; demigods constantly engaging in battle with their supernatural power; or gods, who experience such bliss that they have no motivation to engage in practices leading to liberation, humans experience a range of mental, emotional, and physical states but have the capacity to follow a path that ends in liberation.

Buddhist understanding of consciousness as encompassing cognitive, emotional, and ethical dimensions can encourage neuroscientists and posthumanists to take a broader approach to posthuman enhancements. In the case of uploading, scientists have tended to focus on the technological difficulties associated with uploading, noting the intricacy of the brain, its neural plasticity, the difficulty separating the mind from its biological substrate, and the challenge of implementing mental processing in another substrate. Buddhists would encourage greater reflection about why and to what end one might upload consciousness to a posthuman body.

Although some Buddhists might object to moral enhancements on the grounds that they circumvent the discipline required in engaging in meditative practice and upholding an ethical lifestyle, others might welcome posthumans endowed with such cognitive, physical, and moral capacities. Buddhists who subscribe to a karmic worldview would emphasize that posthuman thoughts, words, and actions would have considerable impact, although it remains unclear whether their realm of rebirth would resemble that of gods, demigods, or an entirely new realm.

Bodhisattvas could be regarded as posthuman predecessors if posthumans used their enhanced abilities to benefit all sentient beings. Posthumans could then be seen as "wonder-workers" not only for their cognitive and physical abilities but also for their compassion and kindness toward other sentient beings.

Bibliography

Dalai Lama. *The Universe in a Single Atom: The Convergence of Science and Spirituality.* New York: Morgan Road Books, 2005.

Dōgen. "Mountains and Waters Sutra." In *Dharma Rain: Sources of Buddhist Environmentalism,* edited by Stephanie Kaza and Kenneth Kraft, 65–76. Boston: Shambhala, 2000.

Gethin, Rupert. "Tales of Miraculous Teachings: Miracles in Early Indian Buddhism." In *The Cambridge Companion to Miracles,* edited by Graham H. Twelftree, 216–234. Cambridge: Cambridge University Press, 2011.

Gómez, Luis O. "On Buddhist Wonders and Wonder-Working." *Journal of the International Association of Buddhist Studies* 33, nos. 1–2 (2011): 513–554.

Graham, Elaine L. *Representations of the Post/human: Monsters, Aliens, and Others in Popular Culture.* New Brunswick, NJ: Rutgers University Press, 2002.

Halberstam, Judith, and Ira Livingston. Introduction to *Posthuman Bodies,* edited by Judith Halberstam and Ira Livingston, 1–19. Bloomington: Indiana University Press, 1995.

Haraway, Donna. "A Cyborg Manifesto: Science, Technology, and Socialist-Feminism in the Late Twentieth Century." In *The Cybercultures Reader,* edited by David Bell and Barbara M. Kennedy, 291–324. New York: Routledge, 2000.

Haraway, Donna. *Simians, Cyborgs, and Women: The Reinvention of Nature.* New York: Routledge, 1991.

Harvey, Peter. *An Introduction to Buddhist Ethics: Foundations, Values, and Issues.* Cambridge: Cambridge University Press, 2000.

Hayles, N. Katherine. "How We Became Posthuman: Ten Years On." *Paragraph* 33, no. 3 (2010): 318–330.

Hayles, N. Katherine. *How We Became Posthuman: Virtual Bodies in Cybernetics, Literature, and Informatics.* Chicago: University of Chicago Press, 1999.

Hayward, Jeremy, and Francisco J. Varela, eds. *Gentle Bridges: Conversations with the Dalai Lama on the Sciences of Mind.* Boston: Shambhala, 1992.

Hughes, James. "Compassionate AI and Selfless Robots: A Buddhist Approach." In *Robot Ethics: The Ethical and Social Implications of Robotics,* edited by Patrick Lin, Keith Abney, and George A. Bekey, 69–83. Cambridge, MA: MIT Press, 2012.

Hughes, James. "Using Neurotechnologies to Develop Virtues: A Buddhist Approach to Cognitive Enhancement." *Accountability in Research: Policies and Quality Assurance* 20, no. 1 (2013): 27–41.

Ives, Christopher. "Buddhism: A Mixed Dharmic Bag; Debates about Buddhism and Ecology." In *Routledge Handbook of Religion and Ecology,* edited by Willis Jenkins, Mary Evelyn Tucker, and John Grim, 43–51. Abingdon, UK: Routledge, 2017.

Lusthaus, Dan. *Buddhist Phenomenology: A Philosophical Investigation of Yogācāra Buddhism and the "Ch'eng Wei-shih lun."* Abingdon, UK: RoutledgeCurzon, 2002.

Padma'tsho. "Courage as Eminence: Tibetan Nuns at Yarchen Monastery in Kham." In *Eminent Buddhist Women,* edited by Karma Lekshe Tsomo, 185–209. Albany: State University of New York Press, 2014.

Rabinowitz, Paula. "Soft Fictions and Intimate Documents: Can Feminism Be Posthuman?" In *Posthuman Bodies*, edited by Judith Halberstam and Ira Livingston, 97–112. Bloomington: Indiana University Press, 1995.

Sadakata, Akira. *Buddhist Cosmology: Philosophy and Origins*. Tokyo: Kosei, 1997.

Swearer, Donald K. "Principles and Poetry, Places and Stories: The Resources of Buddhist Ecology." *Daedalus* 130, no. 4 (2001): 225–241.

Waldron, W. S. "Beyond Nature/Nurture: Buddhism and Biology on Interdependence." *Contemporary Buddhism* 1, no. 2 (2000): 199–226.

Wallace, B. Alan. *Contemplative Science: Where Buddhism and Neuroscience Converge*. New York: Columbia University Press, 2007.

Watson, Burton, trans. *The Vimalakirti Sutra*. New York: Columbia University Press, 1997.

CHAPTER 13

The Russian Cosmists: Evolving into Space

George M. Young
Research Fellow, Center for Global Humanities
University of New England, Portland, ME

We may usually think of space travel and colonization, genetic engineering, artificial organs, extended or even unlimited human longevity, and similarly futuristic topics as ideas that began to be considered possibly realizable only in the later decades of the twentieth century. But in the late nineteenth and early twentieth centuries, a controversial tendency emerged in Russian thought that considered the practical applicability of these and many other concepts, an attempt to take seriously topics that had previously been considered—and perhaps are still thought of by many—as matters for science fiction, fantasy, or occult literature. These thinkers, known today as the Russian cosmists, did not consider themselves a coherent philosophical school and only in retrospect can be seen to have shared a common core of themes and convictions. Some were primarily religious thinkers, others scientists, but each from his own vantage point went beyond socioeconomic, political, or geographic considerations to examine humanity's cosmic dimensions and to suggest that our field of awareness, activity, and influence will evolve to extend beyond planetary, even galactic, boundaries and to eventually include the entire universe. In their view, humanity will overcome time, both forward and backward, creating future humans who will never die and who may eventually restore life to all who have died. Paradise and eternal life will be human made, created and enjoyed by all. The cosmists, both religious and scientific, shared a conviction that we are already very much more than earthly beings, that we are active agents of our own evolution, and that we should direct all spiritual, scientific, and even esoteric knowledge and effort into realizing the long and widely held dream of a paradise regained, on Earth and beyond. We are, as one of the thinkers put it, already "heaven dwellers." This chapter introduces the major Russian cosmists, starting with the first and most representative figure, and discusses how their ideas interrelate.

NIKOLAI F. FEDOROV

The founder of the cosmist tendency of thought was Nikolai F. Fedorov (pronounced and sometimes transliterated as Fyodorov; 1829–1903), an eccentric Socratic figure who published almost nothing during his lifetime but whose writings were posthumously published by his followers in two large volumes under the title *The Philosophy of the Common Task.* The illegitimate son of Prince Pavel Gagarin and an unknown local woman, Fedorov grew up feeling both a member and not a member of the ancient, princely Gagarin family. He received a sound gymnasium and lyceum education but beyond that was largely self-taught. He spent his early working years as a wandering schoolmaster, teaching history and geography in village schools in central Russia. In 1868 he obtained a modest position in what would eventually become the great Russian State Library in Moscow. Over twenty-five years of library service his eccentricities and erudition became legendary. He was reputed to be able to read any European or Asian language as easily as if it were Russian and to know not only the titles but also the contents of every item in the enormous library. He habitually arrived at the library hours before it opened and left hours after it closed, mocking workers who demanded eight-hour workdays, which he called "sixteen-hour idleness." He turned down all offers of a higher salary, lived in a series of rented rooms the size of closets, and refused the meals his landlords offered, preferring to live on raw vegetables, hard cheese, bread, and strong tea. He slept on a humpback trunk, donated "to the poor" any warm clothing or comfortable furniture his friends insisted on giving him, and gave away most of his meager income to the impecunious students he called his "stipendiates." He was said to curse himself if at the end of a day he found in his pockets any loose change that he had not managed to give away.

Many scholars and writers knew Fedorov as an ideal librarian, who often added to the stack of ordered books highly relevant items previously unknown to the user. A very few also knew him as an original thinker, but those few included Russian novelist Fyodor Dostoyevsky (1821–1881), Russian novelist and philosopher Leo Tolstoy (1828–1910), and Russian philosopher and poet Vladimir Solovyov (1853–1900), each of whom valued Fedorov as an intellectual superior or equal. What was it, then, about this obscure librarian and eccentric thinker that so impressed his illustrious contemporaries?

THE COMMON TASK

Fedorov spent his long life working and reworking a single enormous idea. Although it is very complex, with any number of components, in its simplest form his idea is that we all should stop whatever we are now diversely doing and devote all our time, energy, effort, and knowledge to what he called the "common task" of resurrecting all the dead. He meant this literally. In his view, everything in the physical, social, and moral universe is now disintegrating and pointed toward death: knowledge is separated from action; religions, nations, and social classes are growing further apart; wholes divide into particles that continually subdivide into further particles. It is humanity's task to overcome division and death and redirect everything toward unity and eternal life. Nature is the universal force of disintegration, Fedorov argued, and God gave us human reason to regulate nature and reestablish its and our lost wholeness. We should therefore be mindless nature's mind and blind nature's eyes. Our control will turn nature, our temporary enemy, into our permanent friend.

Human passivity enables all disintegration and deathward forces. For Fedorov, an Orthodox believer, true Christianity should no longer be passive worship and commemoration but must become the active practice of resurrecting the dead. By itself, science will eventually be able to extend human life indefinitely. To stop scientific progress after achieving simple human immortality would be immoral; to be moral, which for Fedorov meant to be Christian, we must return life to those from whom we have taken it. All science must become the science not merely of life prolongation but of resurrection. As early as the 1860s, Fedorov was proposing his versions of what are now called cloning, genetic engineering, and artificial organs, as well as space travel and colonization. For Fedorov, all matter contains what he termed, before any knowledge of DNA, "the dust of our ancestors." Advanced science must find a way to restore whole persons from individual particles, and because some of those particles have dispersed beyond Earth, we must go into space to gather in the dispersed particles of our ancestors. By combining knowledge and action, science, religion, and art, everyone will be encouraged to join the project, and everyone living will become a resurrector, Christian in practice, regardless of belief or unbelief. Sons and daughters will resurrect their parents, who in turn will resurrect their parents, and so on, all the way back to Adam and Eve.

To those worried about overpopulation and wondering where on Earth we would put all those resurrected ancestors, Fedorov answered: that is why we must colonize space. The resurrected ancestors would have new bodies engineered to live in places throughout the universe currently unable to support life. Going even further, Fedorov proposed that as part of regulating nature we would learn to overcome gravity and time, and eventually should be able to guide our planet out of its natural orbit and sail it like a boat on courses of our own rational choice. A century before American engineer and inventor Buckminster Fuller (1895–1983) made the concept of "spaceship Earth" famous, Fedorov proposed that we become "captain and crew of spaceship Earth." In today's terminology, Fedorov wanted to turn the exploding cosmos into an eternal steady state, shaped not in imaginary constellations drawn from pagan Greek myths, but as a human-regulated sidereal icon of the Holy Trinity.

In the 1,200 or so pages of *The Philosophy of the Common Task*, Fedorov attempts to unify hundreds of separate tasks, from the task of linguists to help everyone relearn the original language of Adam and Eve, to the task of converting present separate institutions of worship and education into religious-scientific-artistic "museum temples" that would serve as laboratories for the resurrection of ancestors. Armies would not be abolished but redirected to the task of revitalizing what nature and war have destroyed. Weapons now pointed horizontally to kill and injure fellow human beings would be re-aimed vertically, as Fedorov read had been done in the United States, to bring rain to farmlands suffering from famine-causing drought. And this shift from horizontal to vertical should be made in every human endeavor, so that all earthly activities become heavenly.

Not surprisingly, given the range and radical nature of his speculations, only a handful of Fedorov's contemporaries and followers endorsed the entire common task. But each of the thinkers now considered to be a Russian cosmist shared and extended one or more parts of Fedorov's enormous project in directions that Fedorov himself may or may not have recognized. Whereas Fedorov attempted to unite all religious, scientific, political, and aesthetic knowledge and activities into a single enormous task, the cosmists who followed tended to emphasize one aspect and ignore or play down the others. And although there were cosmist artists, economists, and political writers, the most prominent contributors to the cosmist tendency were the religious thinkers and scientists.

THE RELIGIOUS COSMISTS

Several of the leading figures in what has often been called the Russian Religious Renaissance of the late nineteenth and early twentieth centuries, though diverging from Fedorov and from each other on important points, shared enough of the concerns central to Fedorov's thought that their works are often included in discussions of Russian cosmism. Among these thinkers are exiled existentialist philosopher Nikolai (or Nicolas) Berdyaev (1874–1948), who argued for human self-directed evolution to a superhuman state of freedom and creativity; martyred polymath priest, mathematician, aesthetician, electrical engineer, and theologian Pavel Florensky (1882–1937), who worked to sacralize the Soviet state even while being persecuted by it, and while composing pamphlets on electrification also wrote *The Pillar and Ground of the Truth* (1914), a supreme classic of Russian spirituality; and Ivan Ilyin (1883–1954), an exiled monarchist philosopher who, like Fedorov, argued that only a benign Russian Christian autocracy could lead human evolution to a higher state. Important as these three thinkers are, however, two others, Vladimir Solovyov and Sergei Bulgakov, best illustrate the degree to which subsequent Russian religious cosmists concurred with and differed from Fedorov's common task.

VLADIMIR SOLOVYOV

Vladimir Solovyov, generally considered Russia's most important nineteenth-century philosopher, first learned of Fedorov's idea from Dostoyevsky in 1878, met Fedorov in person at the home of Tolstoy in 1881, and attempted to collaborate with Fedorov on what they both believed to be their common task through the remainder of the 1880s and into the early 1890s. Initially, Solovyov, twenty-five years younger, viewed Fedorov as his teacher and spiritual father, but as time passed their differences became more important. Both placed a human-directed resurrection at the center of their projects to reintegrate all that is disintegrating and turn the world "as it is" into the world "as it ought to be," but the resurrections they projected were very different. In Fedorov's view, advances in scientific technology will eventually lead to the elimination of death and enable future-evolved and immortal humans to physically resurrect all their ancestors, including us. In Solovyov's view, spiritual evolution, brought about by spiritual devotions and exercises, even those known and practiced in world religions, will first allow spiritually advanced individuals to develop immortal souls, and these people will then assist less advanced individuals, until eventually all attain immortality, and together all can turn to the task of restoring life and immortality to the departed ancestors. In Fedorov's view, science will create the new bodies that will allow us to live anywhere, including other planets, under conditions presently unable to support human life. In Solovyov's view, evolved immortal souls will be able to create whatever new bodies they need, or they may find that they can dispense with bodies altogether.

According to Fedorov, we must stop thinking of ourselves as men and women and begin thinking of ourselves, as Christ did, as sons and daughters of men and women. People should love each other as siblings, not couples, and should seek to be dutiful sons and daughters rather than prodigal competitors for mates. The natural force that now drives men and women to leave their parents and cleave to each other to create new generations, who will eventually repeat the divisive process, must be reversed so that instead of taking life from our parents we restore life to them. As the resurrection project advances, marriage and childbirth will cease, changing carnality to spirituality—and amounting to what Fedorov called a shift from a "pornocracy" to a "psychocracy," from a society in which the sex drive is dominant to one in which filial love guides all human activity. Anticipating some twentieth and twenty-first century thinkers, Fedorov relates this shift in psychology and society to a shift in cosmological paradigm, from "horizontal" to "vertical," from "Ptolemaic" to "Copernican," shifting in every way, emotionally as well as intellectually, from an earthbound to a heavenward sense of who we are and what our place is in the cosmos (Young 1979, 117).

For Solovyov, in opposition to Fedorov, sexual abstinence is not a virtue. Sexual love is healthy and central, but the current separation into male and female must evolve into an advanced androgyny; sexual love must rise to a higher plane—"a living syzgic relation," a "resurrection by love" (Young 2012, 105). In Solovyov's important book titled *The Meaning of Love* (1894), the interpenetrating relationship in sexual love also extends to humanity's relationship with nature. While Fedorov argued that the common human task was to control and regulate nature, Solovyov viewed universal love as a project to unify humanity and nature, to inspirit nature with godmanhood.

Currently, Fedorov argues, we live as cannibals, devouring our parents, and because every chicken or potato we eat contains particles of our ancestors, we are constantly devouring our ancestors as well. In order to turn away from cannibalism, we must begin with vegetarianism but evolve toward autotrophy, eventually getting our nourishment from the air and sun, as some plants already do. For Solovyov, fasting is an important spiritual exercise, which with prayer, meditation, and other exercises can point us toward godmanhood, but abstinence from food, or from sex, is not a goal in itself.

Even in one of his first known letters to Fedorov, Solovyov raises issues that eventually separated their philosophies: the simple, physical resurrection of the dead cannot be the goal, Solovyov objects, for to resurrect murderers and cannibals exactly as they were in life, or even with new limbs but not new souls, would be immoral; the common task must have a religious, not scientific, character. For Solovyov, the appearance of Christ among humans was analogous to the first appearance of a living organism in the inorganic world or the first human among orangutans. Christ is the image toward which humanity is evolving. As the animal world tends toward reason, so the human world tends toward immortality. Fedorov criticized Solovyov for being a mystic with an impractical approach to the common task. In addition, although Solovyov was indeed a thorough mystic—having experienced a life-changing vision of Divine Sophia, the world soul, in the reading room of the British Museum in 1875 while contemplating the Kabbalah, a vision that he incorporated into the mystical Christian sophiology for which he is best known—he also spent many years and made many enemies in practical efforts to reunite Eastern Orthodoxy and Western Catholicism.

The decisive break between the two thinkers came in 1891, when Solovyov presented a paper ("The Collapse of the Medieval World-Conception") to the Moscow Psychological Society originally intended to be a joint call to the common task. This paper produced a strongly negative reaction among Moscow intellectuals, which Fedorov blamed on Solovyov's weak and incomplete commitment to their shared idea. After that no further attempts at collaboration were made. The two thinkers still shared a common theme of total unity, but each is best known for the development of their points of difference.

SERGEI BULGAKOV

Sergei Bulgakov (1871–1944) was born in a small town in Orel Province deep in central Russia, where six generations of his family had served as Orthodox priests. Although the family was pious, life in the Bulgakov home was not happy. Sergei's father and two brothers died from alcoholism, and his mother suffered from related emotional disturbances. At about age thirteen, Bulgakov took a step not uncommon for young people then or any time. He explained it this way: "I gave up the positions of faith without defending them.... I accepted nihilism without a struggle" (quoted in Zenkovsky 1953, 2:890). At Moscow University he studied political economy, and in his first published works Bulgakov proved himself a brilliant Marxist analyst of socioeconomic systems. Early on, he was an active member of the Social Democratic Party and a close acquaintance of leading socialist activists and thinkers in both Russia and Germany.

Yet, even as he was writing and lecturing from a Marxist perspective, gaining a wide reputation as a socialist political thinker, Bulgakov began to question the fundamental premises of Marxism, and in 1903 he published *From Marxism to Idealism.* His subsequent writings were from a Christian socialist standpoint, incorporating elements from Solovyov's sophiology and Fedorov's common task into a doctrine of spiritual stewardship of the material world that he termed the "philosophy of economy." Following his ancestors, he became an Orthodox priest, and in 1922, with other intellectuals now unwelcome in the bolshevized Soviet Union, Bulgakov was exiled on one of the "philosopher's ships," never allowed to return to Russia. From Constantinople and then Prague, he eventually settled in Paris, where he served as dean of Saint Sergius Orthodox Theological Institute until the end of his life. His prolific theological and sophiological writings, which had led to his expulsion from the officially atheistic Soviet Union, also led to charges of heresy and an attempted excommunication from the Orthodox Church. Even today among Orthodox theologians, his doctrines remain highly controversial, but he has also regained a positive reputation in more liberal Orthodox and ecumenical circles, including positive attention in the early twenty-first century from Rowan Williams, the former archbishop of Canterbury.

In his major work, *Philosophy of Economy* (1912), the title of which could also be translated as "philosophy of responsible ownership," Bulgakov replaced current existing economic systems with a "sophic economy," in which the "world soul" rather than world matter—the internal rather than the external life of humanity, the unseen rather than the tangible, the qualitative rather than the quantitative—would be the primary field of observation and operation. "Every living organism, as a body, as organized material, is inextricably connected with the universe as a whole, for the universe is a system of mutually connected and mutually penetrating forces, and one cannot disturb so much as a grain of sand, destroy so much as an atom, without, to one or another degree, disturbing the entire universe" (Bulgakov 2000, 95).

Bulgakov emphasizes the sacramental character of everyday experience. By eating a meal, for example, which Fedorov considered a devouring of our ancestors, Bulgakov proposes that we are spiritualizing matter, engaging in an act of sophic interpenetration with the material universe and transforming bits of matter into actions, feelings, and thoughts. By our consumption, lifeless things enter our life, a process engaging us in "ontological communication" with the world. "Life is in this sense the capacity to consume the world, whereas death is an exodus out of this world, the loss of capacity to communicate with it; finally resurrection is a return into the world with a restoration of this capacity, though to an infinitely expanded degree" (Bulgakov 2000, 102).

Bulgakov, who wrote one of the first major reviews of Fedorov's posthumously published works, considered Fedorov's emphasis on universal kinship and on Christianity as an active rather than passive faith to be among the highest achievements of Russian thought. But like Solovyov, Bulgakov rejects the entire scientific-technological side of Fedorov's project, considering it the apotheosis of economic materialism. He argues that Fedorov overemphasizes the potential of humanity by itself, without acknowledging the necessity of God's help. For Bulgakov, the active presence of God in the world must guide our every step, which is why, in Christian thought, God became man in Christ. The task of humanity is not to remake the world, as Fedorov would have us do, but to fully sacralize and inspirit the given world of matter. Bulgakov's cosmism emphasizes the active aspects and transformative potential of Christian Orthodoxy as it is, if properly understood and practiced, and not as it must transform itself to become. According to Bulgakov, the liturgy of transformation, properly understood, happens within the existing church, and the existing world is the body in and through which Divine Sophia is manifest. We are the owners and managers of the cosmos responsible under divine guidance for its survival and evolution. In Bulgakov's economy, we need to apply more care and understanding than radical redirection. We do not need to build, at unknown cost, a new universe but to take better care of the one we already own.

THE SCIENTIFIC COSMISTS

Just as the religious cosmists, for the most part, rejected the scientific-technological side of Fedorov's common task, the scientific cosmists, also for the most part, wanted nothing to do with the resurrection of ancestors, the virtual Christianization of all humankind, the inspiriting of matter, or the virtues of theocratic autocracy. The cosmist themes most fully developed after Fedorov include the exploration and colonization of space, the pursuit of human immortality, the emergence of the noosphere, a hypothetical sphere of knowledge enveloping Earth's biosphere, and the investigation of the relationship between cosmic energy and human activity. While many Russian scientists have contributed and are still contributing to the development of these ideas four stand out: rocket scientist Konstantin Tsiolkovsky, biophysicist and heliobiologist Alexander Chizhevsky, mineralogist and geochemist Vladimir Vernadsky, and botanist Vasily Kuprevich.

KONSTANTIN TSIOLKOVSKY

Konstantin Tsiolkovsky (1857–1935) moved to Moscow as a penniless, nearly deaf youth in hopes of somehow obtaining in the great library there more education than was possible in his home village of Kaluga. Fedorov immediately recognized the young man's potential, took him under his wing, and, as Tsiolkovsky later gratefully wrote, took the place of the university professors under whom he was unable to study.

After a few years in Moscow, Tsiolkovsky returned to his village near Kaluga to become an elementary science teacher while dreaming of interplanetary travel. He began to make notebook sketches for rocket boats, rocket wagons, and rocket-powered spaceships and to write fictional accounts of space voyages. What distinguished Tsiolkovsky's imagination from that of any of his contemporaries is that, after writing fantasy narratives and drawing rough pencil sketches, he developed the mathematical formulas that would make the realization of some of his fantasies possible. Over the years, while still a schoolteacher and working after hours in a homemade attic laboratory, he built a series of large wooden model rockets, dirigibles, aerostats, wind tunnels, centrifuges, and primitive space vehicles. He also wrote the papers that would eventually lay the foundation for the 1957 launching of *Sputnik 1*, the world's first artificial satellite. Soviet historians of science have noted that Tsiolkovsky's works contain in embryo nearly all the scientific-technical achievements of the Soviet Union in the exploration of space.

Tsiolkovsky's great accomplishment as a scientist was not only to quantify the dream of space travel through mathematical equations but also to actively promote and popularize the idea of flight beyond Earth, inspiring an enthusiasm for rocket science among young people throughout the Soviet Union. He provided a kindly, grandfatherly, down-to-earth image for an otherwise daunting field of study. Among Tsiolkovsky's young readers who grew up to be outstanding scientists were cosmonaut Yury Gagarin (1934–1968), the first human in space, and future cosmist heliobiologist Alexander Chizhevsky (see below).

While not religious in the traditional sense, Tsiolkovsky, in many nontechnical writings, recognized a spiritual presence in the universe. One of his central ideas posits the presence of life and spirit in all matter. He writes that he is not only a materialist but also a panpsychist, considering sensitivity and feeling to be inseparable from matter. His idea that an "atom-spirit" inheres in every particle of matter in the cosmos recalls Fedorov's idea of all matter as the dust of ancestors. But whereas Fedorov believed that we must redirect and reshape the cosmos, Tsiolkovsky's view is that the cosmos is already teleological, rationally organized, and hierarchical. Lower life-forms, consisting mainly of matter in which spirit is dormant, naturally evolve into higher ones in which the spirit is awakened and more dominant, and eventually, as we approach perfection, we will outgrow our material envelopes and join the rays of cosmic energy that permeate the universe.

The dark side of Tsiolkovsky's ideal of self-perfecting humanity is that it requires the elimination—the "weeding out"—of those of us who are in some way defective. Unlike Fedorov, whose future resurrection society must include absolutely everyone, Tsiolkovsky's future perfect society is highly selective: losers of any kind will not make the cut. In articles titled "Grief and Genius" and "The Genius among the People," Tsiolkovsky offers his variation on Plato's idea of the philosopher-king, suggesting that scientific geniuses and inventors should occupy the key positions in future government and that the many nations of the world should become a single cosmic political system governed by the most advanced and therefore the most nearly perfect specimens of humanity. In Tsiolkovsky's view, Earth probably represents an early, primitive stage of planetary evolution, and elsewhere in the cosmos life-forms have advanced much further. These advanced "atom-spirits" are already in communication with us, but only highly evolved geniuses—artists, scientists, and other visionaries—are attuned to their messages.

ALEXANDER CHIZHEVSKY

Just as in the 1870s Fedorov served as a mentor to sixteen-year-old Tsiolkovsky, so in 1914 Tsiolkovsky became a mentor to seventeen-year-old Alexander Chizhevsky (1897–1964), a sensitive but fragile wunderkind from a privileged background, who began his intellectual life as a poet and painter, talents that he continued to exercise throughout his life. As a child, Chizhevsky was taken to Italy every winter where, as he writes in his autobiography, he began his lifelong fascination with, and even worship of, the sun. When his father was appointed commander of the regiment in Kaluga, the family moved there, and the boy genius Chizhevsky soon came under the wing of the eccentric rocket genius Tsiolkovsky. Although they worked in different fields, their close association continued for the rest of Tsiolkovsky's life, and today in Kaluga the Tsiolkovsky State Museum of the History of Cosmonautics houses a Chizhevsky museum as a wing. It was under Tsiolkovsky's influence that Chizhevsky's intellectual interests, always broad, ranging from ancient languages to postimpressionist painting, gradually began to expand further to include the sciences.

His early works on topics acceptable to Soviet science won him national and international acclaim. But his most important work in a cosmist vein, on heliobiology, demonstrating the effects of solar pulsations on human life, provoked accusations of mysticism, occultism, and irrationality. Eventually, during the terror unleashed by Soviet premier Joseph Stalin (1879–1953), these accusations led to Chizhevsky's arrest as an "enemy under the mask of a scientist," resulting in sixteen years in prison camps and exile.

In one of his most controversial works, published in 1922, he provides a number of charts in which he correlates the fluctuations of sunspots with the up and down periods of violence in human history. In these charts, the correlation is almost too perfect, with periods of what he calls maximum universal excitability coinciding with maximum solar activity, and stretches of relative international calm coinciding with minimal solar activity. For such research Chizhevsky was frequently accused of trying to take science back to a prescientific state—by attempting to replace chemistry with alchemy and astronomy with astrology. Chizhevsky strongly denied these allegations, but he added that he did respect and did wish to restore to modern science not the actual practices of alchemy and astrology but the intuition underlying those prescientific efforts: he argued that in some very profound and mysterious but eventually definable way we and all matter in the cosmos are related and that cosmic energies of which we are barely aware can affect us both physically and psychologically in ways we need to investigate.

VLADIMIR VERNADSKY

Of the major cosmists, Vladimir Vernadsky (1863–1945) was the most thoroughly academic, in the best sense of that term. He inherited, and passed on to the next generation, his family's tradition of intellectual achievement. He saw himself not solely as a Soviet or even as a Russian scientist but as a participant in an honorable intellectual continuum of international scientific investigation, stretching back to ancient Greece. In all human history, he believed, only in the history of science could clear, unquestionable progress be observed.

As a geochemist, he viewed the evolution of Earth and of humankind from a geological perspective. We are, according to Vernadsky, in a very deep sense related to all on our planet—to animals, vegetables, and minerals, as well as to other human beings—and as the rational component of the biosphere, we have a responsibility, literally, to all. The previous great transformation of our planet occurred when the appearance of organic matter began to transform the mineral geosphere into a biosphere. An equally great transformation is now occurring as the intellectual activity of humanity is transforming our planet into a noosphere. As the biosphere is our planet's "sheath of living matter," the noosphere is our planet's "sheath of thinking matter," and the development of the noosphere will be as important an event in geological and cosmic history as the development of the biosphere. He wrote:

> The noosphere is a new geological phenomenon on our planet. In it for the first time mankind becomes a major geological force. Mankind can and must transform his habitat by his labor and thought, transform it radically in comparison to its previous state. Before mankind wider and wider creative possibilities are opening. And perhaps my grandchild's generation will glimpse their flourishing. . . . The noosphere is the latest of many stages of biological evolution in geological history—the stage of our days. (quoted in Young 2012, 161)

Honored in his own day as a brilliant experimental scientist, Vernadsky has since the late 1990s with the publication of many manuscripts kept in the drawer during Soviet times, gained iconic status as a scientific thinker, a pioneering environmentalist, and a model of integrity, dignity, and rectitude consistently exhibited in a time and place where such qualities in prominent individuals were often not apparent. He is a figure whose ideas are extensively discussed and extended in the frequent academic conferences on Russian cosmism that have given focus and voice to the scholarship and activities of Russian cosmists since the end of the Soviet period.

VASILY KUPREVICH

Son of a forester in what is now eastern Belarus, Vasily Kuprevich (1897–1969) became extraordinarily interested in plant life as a boy, amazing his family with his ability to remember the names and healing properties of so many of the plants in the vicinity. He attended local schools for peasant and working-class children and later taught himself botanical science and began to publish scientific papers on the subject while teaching in a village school. After establishing a high reputation for works on acceptable botanical topics, which would eventually lead him to the presidency of the National Academy of Sciences of Belarus, he began to explore the controversial topic of the science of immortology, the investigation of the causes of death and the possibility of its eventual elimination. Death, he believed, is against human nature. There is no such thing as a permanently fixed human life span. Certain woody plants can live tens of thousands of years, and, Kuprevich argued, through research into what enables such longevity, science can eventually replicate the process in humans. Fingernails, skin, and the liver repair themselves when damaged, so why not the rest of the human body? Death is not innate but came to humanity as a result of natural selection, when a life of Methuselah's span was no longer necessary for the survival of the species. Kuprevich contended that "having invented death, nature should also show us how to combat it" (quoted in Young 2012, 173). Kuprevich's studies serve as an inspiration for the growing numbers of immortologists in the twenty-first century.

Summary

The significance of Russian cosmism may not lie so much in the specific proposals and discoveries that emerged, but rather in the audacity and energy with which the cosmists explored topics previously considered outside the scope of serious religious, philosophical, and scientific investigation. Russian cosmists call for active human participation in the evolution of humanity and of the universe at large. In common with twenty-first century transhumanists and posthumanists in the West, the Russian cosmists recognized that as the conscious, rational element in nature, humans bear a relationship to and responsibility toward everything in the cosmos. A thread running through cosmist writings is that we are still in the early stages of human development. We have come far since our early, clumsy efforts to rise from horizontal to vertical orientation, but we still have very much further to go. Especially in the sciences of space exploration and immortology, twenty-first century Russian cosmists continue to develop ideas projected by their nineteenth and early twentieth century predecessors. Both dangers and opportunities lie ahead, but a mark of Russian cosmist thought is confidence that we cannot overcome the dangers by avoiding the opportunities.

Bibliography

Andrews, James T. *Red Cosmos: K. E. Tsiolkovskii, Grandfather of Soviet Rocketry.* College Station: Texas A&M University Press, 2009.

Bailes, Kendall E. *Science and Russian Culture in an Age of Revolutions: V. I. Vernadsky and His Scientific School, 1863–1945.* Bloomington: Indiana University Press, 1990.

Berdyaev, Nicolas. *The Destiny of Man.* Translated by Natalie Duddington. London: Geoffrey Bles, 1937.

Berdyaev, Nicolas. *The Russian Idea.* Translated by R. M. French. Boston: Beacon Press, 1962.

Bulgakov, Sergei. *Philosophy of Economy: The World as Household.* Translated and edited by Catherine Evtuhov. New Haven, CT: Yale University Press, 2000.

Bulgakov, Sergei. *Sergii Bulgakov: Towards a Russian Political Theology.* Edited by Rowan Williams. Edinburgh: T&T Clark, 1999.

Edie, James M., James P. Scanlan, and Mary-Barbara Zeldin. *Russian Philosophy*. 3 vols. Chicago: Quadrangle Books, 1965.

Fedorov, Nikolai F. *What Was Man Created For? The Philosophy of the Common Task; Selected Works*. Translated and abridged by Elisabeth Koutaissoff and Marilyn Minto. London: Honeyglen, 1990.

Florensky, Pavel. *The Pillar and Ground of the Truth*. Translated and annotated by Boris Jakim. Princeton, NJ: Princeton University Press, 1997.

Kornblatt, Judith Deutsch, and Richard F. Gustafson, eds. *Russian Religious Thought*. Madison: University of Wisconsin Press, 1996.

Masing-Delic, Irene. *Abolishing Death: A Salvation Myth of Russian Twentieth-Century Literature*. Stanford, CA: Stanford University Press, 1992.

Pyman, Avril. *Pavel Florensky: A Quiet Genius; The Tragic and Extraordinary Life of Russia's Unknown da Vinci*. New York: Continuum, 2010.

Rosenthal, Bernice Glatzer, ed. *The Occult in Russian and Soviet Culture*. Ithaca, NY: Cornell University Press, 1997.

Siddiqi, Asif A. *The Red Rockets' Glare: Spaceflight and the Soviet Imagination, 1857–1957*. Cambridge: Cambridge University Press, 2010.

Solovyov, Vladimir. *Divine Sophia: The Wisdom Writings of Vladimir Solovyov*. Edited by Judith Deutsch Kornblatt. Ithaca, NY: Cornell University Press, 2009.

Solovyov, Vladimir. *Lectures on Godmanhood*. London: Dobson, 1948.

Solovyov, Vladimir. *The Meaning of Love*. Edited by Thomas R. Beyer Jr. West Stockbridge, MA: Lindisfarne Press, 1985.

Solovyov, Vladimir. *A Solovyov Anthology*. Edited by S. L. Frank. Translated by Natalie Duddington. New York: Scribners, 1950.

Tandy, Charles, ed. *Death and Anti-death*, Vol. 1, *One Hundred Years after N. F. Fedorov (1829–1903)*. Palo Alto, CA: Ria University Press, 2003.

Young, George M. *Nikolai F. Fedorov: An Introduction*. Belmont, MA: Nordland, 1979.

Young, George M. *The Russian Cosmists: The Esoteric Futurism of Nikolai Fedorov and His Followers*. Oxford: Oxford University Press, 2012.

Zenkovsky, V. V. *A History of Russian Philosophy*. Translated by George L. Kline. 2 vols. New York: Columbia University Press, 1953.

CHAPTER 14

Virtual Religions and Real Lives

Carole M. Cusack
Professor of Religious Studies
University of Sydney

There is a palpable tension evident in the juxtaposition of "virtual religions" with "real lives." What might a virtual religion look like? In the twenty-first century, the phrase "virtual reality" is understood to refer to simulated environments created by software in which people using special equipment interact with other people and computer-generated entities, both in game situations and in more open-ended "virtual worlds." It is undeniable that there are religions operating in cyberspace, examples of which are the Amaterasu Omikami Grand Shinto Shrine and the Mormon Meeting Hall found in the online virtual world Second Life (Stagg and Farley 2011). Online ritual workings by Pagan covens and virtual pilgrimages to Christian shrines are accessible via Google, and there are religions that are primarily online communities, lacking formal structures in the so-called meat world (assumed to be the site of the "real lives" of the participants). Yet it is doubtful that these can be neatly classified as "virtual religions," just as it is increasingly hard to disentangle offline from online lives. This chapter discusses a particular grouping of religions that emerged starting in the late 1950s and are based on existing fictions or inventions of the founders, which have been termed "invented religions' (Cusack 2010). It is argued that invented religions and Posthumanism reject both Judeo-Christian religion and Enlightenment rationalism, and point toward an undifferentiated reality, not composed of binary opposites, that is best approached by partial, open-ended theories and methods. (In this chapter the terms *Humanism*, *Transhumanism*, and *Posthumanism* are capitalized as being akin to religions, but *transhuman* and *posthuman* are lowercased as being general adjectives.)

COMMON ASSUMPTIONS ABOUT RELIGION

When religions are studied in schools and universities or commented on in the media, certain underlying assumptions are almost always detectable, although they are rarely challenged. Religions have been part of human culture since the prehistoric era. Therefore, the popular perception is that religions are ancient, serious, ancestral, and address profound questions of human existence: Who created the world? How did human beings originate? Why do suffering, sickness, and death occur? What happens after death? Distinctions are frequently made between different types of religions, too. The world religions (Judaism, Christianity, Islam, Hinduism, and Buddhism), which have sacred texts and formal institutions, are used as a yardstick against which to judge other religions and find them lacking some key quality; for example, new religions (which are not ancient) and indigenous religions (which tend to lack institutions and to be orally transmitted) are often denied the status of real religion (Owen 2011).

The reason for the dominance of the world religions lies in history. When Europeans explored and settled in the Americas, Africa, Asia, and the Pacific region, colonization was accompanied by the preaching of Christianity, the religion of Europe, which the colonizers believed was uniquely true. Indigenous peoples were conquered and dispossessed of their lands to varying degrees, and their religions were downgraded to "traditions" or "customs." Indigenous peoples were made to accept Christianity and in some cases lived on missions or reservations administered by Christian clergy. Judaism had always been recognized as the antecedent religion of Christianity, and Islam, the third Abrahamic monotheism, as a subsequent heresy or divergent tradition. The experience of medieval Christendom was that Jews were reluctant to convert to Christianity and that Islamic and Christian armies frequently clashed. The three monotheisms were therefore seen as separate yet genealogically connected (Neusner 2006). Hinduism and Buddhism later were admitted to the world religions category as their adherents did not, in the main, convert to Christianity, and both religions had a textual tradition and educated clergy, which enabled them to resist Christian missions. On occasion the world religions classification is extended to include other, smaller religions with texts and institutions, such as Sikhism and Zoroastrianism (Owen 2011).

New religions also do not fit the world religions model. In the West, new religions began to appear in the nineteenth century. Prior to that, most new religious manifestations were classified as heresy by the Christian churches, and those who adopted these new beliefs and practices were persecuted or killed. The Enlightenment, an intellectual and cultural movement of the eighteenth century, changed attitudes to religion and the maintenance of orthodoxy. The Enlightenment stressed the importance of human reason in providing reliable knowledge about the world and opposed—as being tyrannical—authorities such as religion and monarchy, advocating instead rational thought, empirical investigation, and participatory democracy (Zagorin 2003). The Enlightenment did not intend to form or encourage the formation of new religions, as it espoused scientific criteria for knowledge. However, the loss of power by the institutional churches enabled new religions to emerge. In 1830 Joseph Smith (1805–1844) founded the Church of Jesus Christ of Latter-Day Saints (Mormons), Spiritualism began in the late 1840s in upstate New York when teenagers Margaret (1833–1893) and Kate (1837–1892) Fox claimed to be able to interpret rappings from the spirits, and Russian Helena Petrovna Blavatsky (1831–1891) and American Henry Steel Olcott (1832–1907) founded the Theosophical Society in New York in 1875. This was a revolutionary change, in that the supply of religions from which people could choose became far greater; the market domination of mainline Christian churches had come to an end (Finke and Iannaccone 1993).

The mid-twentieth century was the next important era for the emergence of new religions. In the 1950s UFO and alien-based religions, including the Church of Scientology (1953) and the Aetherius Society (1954), were established, as was the first true "invented religion," Discordianism, which was started by Greg Hill and Kerry Thornley in 1957. In the 1960s religious and spiritual teachers from Asia traveled to the West and founded new versions of old religions, such as the International Society for Krishna Consciousness and Transcendental Meditation, both of which were Hindu offshoots, and the Friends of the Western Buddhist Order (later known as the Triratna Buddhist Community). A second, and more famous, religion based on fiction, the Church of All Worlds, was founded in 1962 by college students Tim Zell and Lance Christie, after they had read Robert A. Heinlein's

Stranger in a Strange Land (1961) and decided to create the fictional Church of All Worlds from the novel in the "real" world (Cusack 2010). The Church of All Worlds went on to meld ritual practices from Heinlein's novel with modern Pagan practices and ecological beliefs, and it continues to be a force in twenty-first-century alternative religion. Since approximately 2000 there has been a flurry of such religions, including Jediism, Matrixism, and Dudeism (all based on films), the Church of the Flying Spaghetti Monster, and the Missionary Church of Kopimism. Prior to this millennial period of innovation, the only significant invented religion founded after Discordianism and the Church of All Worlds was the Church of the SubGenius, founded in Dallas, Texas, in 1979. This is often viewed as a Discordian offshoot, as the two religions have some prominent members in common (like Robert Anton Wilson), and they share an anarchic sense of humor, a focus on conspiracy narratives, and a wholesale rejection of wage slavery and consumer capitalism (Cusack 2010).

This chapter examines religions based on fictional narratives, such as films and science fiction novels, and argues that they constitute a new trend in human religiosity. Such "invented" or "fiction-based" religions overturn expectations of "real" religions in several ways. First, they violate the assumption that religions refer to real entities and worlds beyond the material universe (God, gods, angels, Satan, demons, jinn, avatars, heaven, hell, and so on), and that these entities and other worlds significantly affect the actual lives of human beings. Second, they violate the expectation that prophets and spiritual teachers will have revelations from the divine realm that are serious and treated reverently. Third, invented religions are a form of bricolage: explicit mash-ups of popular culture, elements of existing religions, jokes, political and social commentary, and so on (Cusack 2010). This eclectic mix of sources upsets the assumption that religions are unique and original and not crafted from acknowledged sources. Studying religions that openly advertise their invention not only enriches knowledge about traditional religions but also sheds light on how science fiction speculations and new technologies inform religious belief and practice.

This chapter proposes three ways that these new religions relate to posthumanist challenges to prevailing epistemologies and modes of being. First, the shift from the secular to the postsecular that is posited by some religious studies scholars coincides historically with the emergence of Posthumanism, and thus there are certain shared features that can be identified between the religious and spiritual trends of that era and posthumanist thought, such as a debt to the speculative literary and filmic genres of science fiction and fantasy. Second, posthumanist discourses insist that the binaries that have dominated philosophy to the present day, such as male/female, white/black, human/animal, and spiritual/material, are dissolved when the horizon of the human (or the Anthropocene) is rejected (Haraway 2015). Third, Posthumanism itself can be viewed as a new mythological form that radically repositions human beings in the cosmos and tells a new metaphysical story (Valera and Tambone 2014). This is a project that is important to some invented religions, in particular the Church of All Worlds, Discordianism, and the Church of the SubGenius.

RELIGION, TRANSHUMANISM, POSTHUMANISM, AND POSTSECULARITY

Posthumanism is a neologism that emerged in the 1970s. The term *posthuman* was first defined by Thomas Blount in 1656 as "following or to come, that shall be" (quoted in Krueger 2005, 78). Ihab Hassan first used Posthumanism in 1977 to refer to a philosophical position that rejected Humanism and moved beyond the human standpoint. Posthumanist ideas were popular in literary, artistic, and computing subcultures at that time. Over forty years later, the posthuman and the transhuman have made the transition from science fiction and the arts to science and technology; both are identified with research in the fields of nanotechnology, robotics, artificial intelligence, and human genetic improvement. Philosopher Francesca Ferrando draws a sharp distinction between the transhuman and Transhumanism, with its core vision of human enhancement by means of science and technology, and the posthuman and Posthumanism, with its "radical onto-existential re-signification of the notion of the human" (2013, 27). At first glance, religion has a straightforward relationship with Transhumanism, in that historically the world religions have insisted on the flawed nature of human life in the material world and have proposed a range of perfected lives in other worlds that are preferable. Human beings are defective, but through significant personal effort and religious discipline they may attain elevated states, such as nirvana, *moksha*, sainthood, and the like (Sharot 2001). In theological terms the transhumanist vision is not the same type of overcoming or going beyond fundamental this-worldly conditions as salvation in Christianity; Ronald Cole-Turner pithily insists that transhumanists view technology as the agent of salvation, whereas for Christians the agent of salvation is "grace, the undeserved goodness of God who gives life and wholeness to the creation" (2015, 150).

Scholarly and popular attitudes about religion have changed radically since 1960, and many religions now exist that connect with the Transhumanist vision of humans enhanced by technological means. The Mormon Transhumanist Association, founded in 2006, connects Joseph Smith's ideas about resurrection and humans becoming godlike to modern technological optimization and overcoming death. UFO and alien-based religions, for example, generally claim that human beings were created by extraterrestrials. These beings had both advanced scientific knowledge and spiritual wisdom and set up Earth as a kind of laboratory in which humanity can progress. The final apotheosis of those individuals selected by the aliens for a perfected afterlife is dependent on extraterrestrial intervention. Thus, thirty-nine members of Heaven's Gate, founded by Marshall Herff Applewhite (1931–1997) and Bonnie Lu Nettles (1927–1985), committed suicide in March 1997 in Rancho Santa Fe, California. They believed they would board a spaceship, the presence of which was concealed by the Hale-Bopp Comet, and ascend to the Next Level (Cusack 2015a). In this particular alien-based religion the Christian idea of the "rapture," in which the saved are gathered up to heaven, was translated into a technological rescue in a UFO. The Raelians, founded in France by Claude Vorilhon (1946–) in 1973, posit aliens who visit Earth regularly and assert that great religious leaders such as Moses, Jesus, Buddha, Muhammad, Confucius, and Rael (as Vorilhon is known) are alien-human hybrids, born of human mothers and extraterrestrial fathers. Rael interprets the Bible as chronicling the visits of the aliens to Earth; ignorant humans thought these were divine interventions (Cusack 2015a). In these new religions the preeminent value accorded to the human in Enlightenment thought is rejected; Heaven's Gate and the Raelians view humanity as limited and inferior to the aliens, as humanity is inferior to God in Christianity.

The eighteenth-century Enlightenment rejected the religious worldview; in the empirical, rational, and scientific worldview it promulgated, humanity was celebrated as the highest form of existence, and human reason displaced divine wisdom. Through the nineteenth and twentieth centuries religious institutions lost power, as the state, communities, and individuals entered a secularized phase in which religion was characterized by personal choice in private life, with a diminished public role. Humanism, a philosophical position that saw humans as the measure of all things, dominated the modern era. This elevation of the human and diminution of the divine have been assumed to be progressive and inevitable. Since the 1970s, however, scholars such as Daniel Bell (1978) have argued that religion and spirituality are resurgent—what Bell terms "the return of the sacred." This return is evident, it is claimed, in the plethora of new religions and spiritualities that have emerged since the countercultural 1960s, and in the West, cultural commentators speak of postsecularity and the postsecular. Where the secular was associated with modernity, there are obvious links between the postsecular and postmodernity; as is the case in architecture, the order and functionalism of modern institutional religion (which bears a strong resemblance to governments and corporations) have given way to eclecticism and the blending of concepts and aesthetics from many existing religions with art, fiction, film, and other popular cultural forms. Activities that originated in religion, such as meditation and yoga, are now done by many for health or therapeutic reasons; activities that were secular, such as working out at the gym and going to the cinema, have become sacred to many in the developed world (Partridge 2004–2005).

Most of these new spiritual forms stop short of establishing religions, in that they are radically individual and those involved do not consider establishing a formal church or other religious institution to be necessary. In some cases, however, that has occurred. The phenomenon of fandoms, for example, may be entirely secular and pursued for entertainment and social reasons; but it may take on spiritual overtones and be involved in identity formation for some individuals and groups. In extreme cases it results in religions such as the First Presleyterian Church of Elvis the Divine, Iglesia Maradoniana, and Haruhiism, respectively devoted to real-world rock singer Elvis Presley (1935–1977), the Argentine football star Diego Maradona (1960–) and the fictional anime character Haruhi Suzumiya (Rigby 2001; Buljan 2017). The first two fictionalize and divinize real people, whereas the third makes a reality of a fictional person.

Invented religions are of broadly two types. The first takes an existing text and forms a religion that is based on, or draws on, that fiction. The Church of All Worlds, Jediism (based on George Lucas's *Star Wars* films), Dudeism (based on the Coen brothers' *The Big Lebowski*), Matrixism (based on the Wachowskis' *Matrix* trilogy), and Haruhiism fit this category. The second group includes religions where the founders themselves wrote the fictional text(s) to which the group refers. In Discordianism, mentioned above, Hill and Thornley wrote *Principia Discordia*, the subcultural scripture, and in the Church of the SubGenius, cofounder Ivan Stang has produced a stream of "scriptures," including *The Book of the SubGenius* (1983) and *Revelation X: The "Bob" Apocryphon* (1994). Bobby Henderson (1980–), founder of the Church of the Flying Spaghetti Monster, published *The Gospel of the Flying Spaghetti Monster* (2006) (Cusack 2010). All these can be considered postmodern religions, or at least to exhibit certain postmodern features, such as extreme eclecticism and the collapse of high and low cultures, and many are primarily Internet mediated, with virtual communities replacing "meat-world" interactions.

THE MYTHIC IMAGINATION: POSTHUMANISM AND INVENTED RELIGIONS

For traditional peoples of faith there is no bridge between the eternal truths of religion and the ephemeral products of human beings. The world religions all consider the sacred and the profane to be separate: sacredness is associated with the divine, the spiritual world, eternal verities, and ultimate states; the profane is the realm of the human, the mundane, the material, and transitory existence. The academic study of religion, in contrast, emphasizes that religions are human cultural products and that the elaborate metaphysical worldviews espoused by religions are creations of the imagination (Hanegraaff 2017). In this perspective no significant distinction exists between the fictional characters and imagined worlds of science fiction and the angels, gods, demons, heavens, and hells found in religious texts. Narrative is at the core of both genres; it is worth remembering that *scriptura* in Latin merely means "writings," and that the term *scripture* took on religious dimensions in the Middle Ages when literacy was largely the preserve of the Christian clergy. Humans have always structured knowledge in narratives, and in modernity widespread literacy has multiplied the available narratives that people can draw on to construct their personal and communal identity and to structure a meaningful life.

The roots of contemporary modes of identity formation and meaning making lie in the eighteenth century: the novel replaced religious literature as the preferred entertainment of the middle classes; the Industrial Revolution created a myriad of new consumer goods that activated a cycle of wanting and getting that continues to the present; and the Romantic movement countered the Enlightenment claim of the superiority of human reason by emphasizing the importance of emotion. These social transformations stimulated individualism and weakened authority structures, including the church and the family, as people imaginatively inserted themselves in the plots of novels, yearned to marry for love, and questioned the inherited values of the Christian West (Campbell 2005). In the two centuries since 1800 these processes sped up: scientific and technological discoveries, increases in the standard of living, higher levels of education, and a proliferation of entertainments (fashion, popular music, film, televisions, games, computers, and so on) emerged. All served to strengthen consumerism and individualism and to weaken sources of authority such as religion and science. Experimentation became the norm, and from the 1950s on, the imaginations of young people who were disillusioned with the inherited religious and social expectations in the West gave birth to an entirely new form of religion—fiction based or invented and reflective of the founders' personal quest for meaning.

Hill and Thornley made no secret of the fact that Discordianism, in which the Greek goddess of strife and chaos, Eris, was worshipped, began as a parody of religion. From its origin in an all-night bowling alley in 1957, the religion developed further in the mid-1960s when the first edition of *Principia Discordia* was written. In the late 1970s Margot Adler interviewed Hill for her pioneering *Drawing Down the Moon: Witches, Druids, Goddess-Worshippers, and Other Pagans in America Today* (1979). Hill told Adler the following:

> I started out with the idea that all gods are an illusion. By the end I had learned that it's *up to you* to decide whether gods exist, and if you take a goddess of confusion seriously, it will send you through as profound and valid a metaphysical trip as taking a god such as Yahweh seriously. The trip will be different, but they will both be transcendental. (Adler 1986, 335)

Discordianism emphasizes that chaos is the single matrix from which all emerges and that oppositions such as order/disorder are illusory, all is one. Thus, it is irrelevant whether Discordianism is a real religion or a fake religion, or whether individual Discordians are sincere or adopt the identity as a joke. Monism (the philosophical position that all can be explained in terms of a single underlying principle or reality) is Eastern in origin and has made significant inroads into the West since the 1950s, as Western people have gradually abandoned the

Christian worldview (one God, one life, postmortem reward and punishment, and clear distinctions between God and the world and between humans and all other creatures). This monistic viewpoint is mirrored in Donna Haraway's lighthearted positing of a "Cthulhucene" as the epoch that follows the Anthropocene, and her gentle insistence that "I am a compost-ist, not a posthuman-ist: we are all compost not posthuman.... The edge of extinction is not just a metaphor; system collapse is not a thriller. Ask any refugee of any species" (Haraway 2015, 161). Cthulhu, a fictional monster created by horror author H. P. Lovecraft (1890–1937), is embraced as a divinity within the Discordian lineage of invented religions (Cusack 2010).

In the 1970s the Church of All Worlds embraced an ecotheological position after Tim Zell had a vision of Earth as the goddess Gaia, a sentient being in which all of creation was interconnected. Zell and Christie held environmental concerns since their student days when they founded their church, and the emphasis on the sacredness of water in Heinlein's *Stranger in a Strange Land* was important to them, with "water sharing" becoming a key religious practice. Their path to the recognition of the interdependence of all was quite different from that of Thornley and Hill: it involved communal living in rural areas; immersion in classical mythology, with its gods of woods and rivers; and the impact of scientific research as diverse as the images of Earth taken from outer space and publications such as Rachel Carson's *Silent Spring* (1962), which drew attention to the extinction of species and damage to the natural environment (Cusack 2010). Thus, Zell and Christie's path was Western rather than Eastern and accorded greater value to science than that of Thornley and Hill Yet there are deep connections between the worldviews of these invented religions, both of which were nurtured in the counterculture of the 1960s. Both were exercises in the creation of new myths, myths in which distinctions between humans and nature were diminished and the unity of all things was emphasized.

Posthumanism has strong resemblances to both the Discordian and Church of All Worlds challenges to Western ontologies. Posthumanism explicitly rejects anthropocentrism and vigorously asserts the nonhuman as possessing the status of being and also values it as a source of knowledge. The posthuman incorporates all entities in the universe: from "natural" plant and animal life to what is usually thought to be inanimate matter; to artificial intelligence, extraterrestrials, and subatomic particles; to hypothetical and fictional beings (Ferrando 2012). This position rejects the Judeo-Christian privileging of the human as uniquely capable and being chosen by God, as well as the Enlightenment celebration of reason, in part because both have resulted in the domination and exploitation of all that is "other" to the (Western) human, including the natural world, the realms of fantasy and imagination, and other (non-Western) humans. Posthumanism can be understood as a re-visioning of ethics and aesthetics, in which "hegemonic essentialism" is ousted along with its binary opposite, "resistant essentialism," in favor of a dynamism that welcomes methodological pluralism and openness "to unknown possibilities" (Ferrando 2012, 15). Hill's description of his journey with Eris from parody to transformative encounter perfectly fits this valorization of partial knowledge, playful explorations, and unexpected discoveries.

THE POWER OF NARRATIVE

The retreat of institutional Christianity since the mid-twentieth century indicates that some Western people no longer find its narrative either attractive or persuasive. The growth of religions based on fictions points to other narratives taking on the functions of meaning provision in an eclectic, consumerist, individualist, and subjective late modern or postmodern world. Between the foundation of Discordianism in 1957 and the Church of All Worlds in 1962 and the beginning of the twenty-first century, the conditions for the development of invented religions became more propitious. The 1960s and 1970s were characterized by rapid progressive secularization and the adoption of Eastern religions or elements of the Eastern worldview by a significant minority, as well as by the growth of subcultures focused on computing, science fiction, and other niche interests, which often intersected with both fringe religions, such as UFO movements, and fringe science. In 1979 the Church of the SubGenius was founded in Dallas, Texas, by the Reverend Ivan Stang (1953–; born Douglass St. Clair Smith), Philo Drummond, and Dr. X (born Monte Dhooge). In the same year the *SubGenius Pamphlet #1* (also titled *The World Ends Tomorrow and You May Die!*) was published. The Church of the SubGenius is often classified as a Discordian offshoot; the two share an anarchic sense of humor, a belief in "the Conspiracy," the need to awaken or become enlightened, "scriptures" that employ the cut-and-paste technique of zines, and the use of fictions and popular culture motifs that are attractive to potential members (Cusack 2015b).

The mythology of the Church of the SubGenius is very complex, involving extraterrestrials and spacecraft; the genetic engineering of the SubGenii (a mutant race, part-human and part-Yeti); the salesman messiah J. R. "Bob" Dobbs's ability to die and be regenerated many times; the vengeful god Jehovah-1, whose evil purpose is to deprive people of "Slack"; H. P. Lovecraft's Elder Gods and their robot agents the Watchers, who manifest as flying saucers; the Conspiracy, which seeks to control everyone; the lost civilization of Atlantis and its Yeti citizens; the alien race of Xists, who created the Yetis and who will arrive in spacecraft to save their descendants the SubGenii; and the power of "Slack," the quality of living well while doing nothing. The main connection between this invented religion and Posthumanism is the scorn in which humans are held. SubGenii are superior as they have Yeti genes; they will become OverMen or UberFemmes, a state allegedly achieved to date only by Philo Drummond, when the Xists' spacecraft arrive. Humans will not be saved, and the Church of the SubGenius "recommends using two weapons, drugs and abortion (permissible up to the age of 50) to rid the world of pointless, negative people" (Cusack 2010, 89). There is a similarity here with the posthumanist desire to eradicate humanist and human-centered positions, in that SubGenii are superior to ordinary humans, but links between the Church of the SubGenius and Posthumanism are tenuous at best.

Around 2000 there was significant growth among invented religions. The original *Star Wars* trilogy of George Lucas had always attracted devoted fans, and in 2001 Jediism emerged as a religion based on these films. In 2004 the fully digital religion of Matrixism was founded, based on the Wachowskis' trilogy of *Matrix* films, and in 2005 Oliver Benjamin founded Dudeism (the Church of the Latter-Day Dude) based on the Coen brothers' cult film *The Big Lebowski* (1998). These religions based on films are quite different from the older, invented religions discussed above. Jediism is a strongly ethical faith with real-world and online institutions, a prominent example being the Temple of the Jedi Order, founded by Brother John and having the status of an international ministry and registered nonprofit organization in the United States (Cusack 2010). Matrixism was a short-lived phenomenon that endures thanks to the online reproduction of the original website, preaching "belief in the prophecy of The One ... acceptance of the use of psychedelics as a sacrament; acceptance of the semi-subjective multilayered nature of reality; and adherence to ... one or more of the world's religions until such time as The One returns" (Jordison 2005, 128). Dudeism is also explicitly ethical, although its relaxed and gently humorous tone and debt to a film that is not science fiction or fantasy but rather a type of comedic magic realism, distinguish it from the majority of such fiction-based religions.

The Church of the Flying Spaghetti Monster and the Missionary Church of Kopimism continue the tradition of anarchic protest and countercultural criticism of mainstream lifestyle choices. Bobby Henderson founded the Church of the Flying Spaghetti Monster in 2005 to protest against intelligent design (a type of creationism advocated by biblical literalists) in Kansas schools. He is not religious and did not intend it to become a religion; this has taken place among people who find Henderson's humorous narrative of pirates, global warming, and heaven containing strippers and a beer volcano, as well as other memorable motifs, meaningful for them (Cusack 2010). The Missionary Church of Kopimism, founded by Isak Gerson in Sweden in 2010, is a body that posits that knowledge is sacred and the sharing of knowledge is sacred, and is thus an anticopyright group. This new religion is intimately connected to the Internet and may be considered a fully digital religion, like Matrixism. Yet it is difficult to connect Kopimism to posthumanist ideas in specific ways; a broad rejection of copyright law as a human-made obstacle is one link that can be made. The flow of information, for Gerson, is akin to the monistic underlying reality of Eastern religions, and human access to information is viewed as optimization via technological means.

The historical coincidence of Posthumanism emerging in the 1970s, a decade in which the older invented religions (Discordianism and the Church of All Worlds) were becoming more established and the Church of the SubGenius was founded, extends to the conditions under which both Posthumanism and invented religions have become more propitious in the decades since the 1970s, in part because of the "pick-and-mix" methods underlying postmodernism, as well as the fragmentation of culture into smaller, discontinuous subcultures. In the twenty-first century invented religions and Posthumanism have flourished, as the metanarratives that would outlaw both as fringe and irrational have lost power. Adherents of invented religions and academic proponents of Posthumanism are deeply engaged in the creation of a new mythology to take humanity into an uncertain future (Valera and Tambone 2014). The pictures of the future that both groups offer are indebted to science fiction and require the use of the human imagination to go beyond old, and possibly no longer relevant, stories and to replace them with attractive and relevant myths for the new kind of life that humanity must prepare for. The overall tenor of these emergent science fiction–based mythologies is not positive, and in the narratives of Discordianism, the Church of All Worlds, and the Church of the SubGenius humanity definitely loses its central place in the universe, which notions of undifferentiated oneness being preferred. This has resemblances to the posthuman futures imagined by mathematical physicist Frank Tipler (1947–), in which biological humans do not survive the death of the sun (Krueger 2005), and other visions in which artificial intelligence (AI) or the uploading of human consciousness to computer circuitry constitutes the posthuman condition (Sandu 2015).

Summary

This chapter has considered a range of invented religions, which are defined as those religions that emerged from the 1950s onward and which openly admit to having been based on fictional texts or on texts that were written by founders who explicitly denied divine inspiration. These groups included Discordianism (1957), the Church of All Worlds

(1962), the Church of the SubGenius (1979), and the newer movements based on films such as Jediism (2001), Matrixism (2004), and Dudeism (2005). These religions have violated the expectations of religion as historically understood, while also rejecting Enlightenment rationalism. All have used the Internet to a considerable extent to promote their ideas, and all offer some evidence for the limits of secularization or even for the "return of the sacred" (Bell 1978). These religions have certain resemblances to Transhumanism and Posthumanism. First, there is a historical coincidence of invented religions and these newer philosophical movements that advocate the abandonment of the human as the yardstick against which all is judged, while also advocating openness to optimization via technology (Transhumanism) and openness to other ontologies and knowledges (Posthumanism). Second, the two have a common genealogy in science fiction and in modes of thought that collapse or eradicate previous givens, such as binary opposites and a preference for the existing over the only imagined. Third, both invented religions and trans/posthumanist thought are engaged in the business of creating new mythologies for future humans (and perhaps nonhumans).

Bibliography

Adler, Margot. *Drawing Down the Moon: Witches, Druids, Goddess-Worshippers, and Other Pagans in America Today*. 2nd ed. Boston: Beacon Press, 1986.

Bell, Daniel. "The Return of the Sacred: The Argument about the Future of Religion." *Bulletin of the American Academy of Arts and Sciences* 31, no. 6 (1978): 29–55.

Buljan, Katharine. "Spirituality-Struck: Anime and Religio-spiritual Devotional Practices." In *Fiction, Invention, and Hyper-reality: From Popular Culture to Religion*, edited by Carole M. Cusack and Pavol Kosnáč, 101–118. Abingdon, UK: Routledge, 2017.

Campbell, Colin. *The Romantic Ethic and the Spirit of Modern Consumerism*. 3rd ed. York, UK: Alcuin Academics, 2005.

Cole-Turner, Ronald. "Going beyond the Human: Christians and Other Transhumanists." *Theology and Science* 13, no. 2 (2015): 150–161.

Cusack, Carole M. "Apocalypse in Early UFO and Alien-Based Religions: Christian and Theosophical Themes." In *Modernism, Christianity, and Apocalypse*, edited by Erik Tonning, Matthew Feldman, and David Addyman, 339–353. Leiden, Netherlands: Brill, 2015a.

Cusack, Carole M. *Invented Religions: Imagination, Fiction, and Faith*. Farnham, UK: Ashgate, 2010.

Cusack, Carole M. "Lab Rats and Tissue Samples: The Human in Contemporary Invented Religions." In *Religion, Media, and Social Change*, edited by Kennet Granholm, Marcus Moberg, and Sofia Sjö, 175–188. London: Routledge, 2015b.

Ferrando, Francesca. "Posthumanism, Transhumanism, Antihumanism, Metahumanism, and New Materialisms: Differences and Relations." *Existenz: An International Journal in Philosophy, Religion, Politics, and the Arts* 8, no. 2 (2013): 26–32.

Ferrando, Francesca. "Towards a Posthumanist Methodology: A Statement." *Frame: Journal of Literary Studies* 25, no. 1 (2012): 9–18.

Finke, Roger, and Laurence R. Iannaccone. "Supply-Side Explanations for Religious Change." *Annals of the American Academy of Political and Social Science* 527 (1993): 27–39.

Hanegraaff, Wouter J. "Religion and the Historical Imagination: Esoteric Tradition as Poetic Invention." In *Dynamics of Religion: Past and Present*, edited by Christoph Bochinger and Jörg Rüpke, 131–153. Berlin: De Gruyter, 2017.

Haraway, Donna. "Anthropocene, Capitalocene, Plantationocene, Cthulhucene: Making Kin." *Environmental Humanities* 6, no. 1 (2015): 159–165.

Jordison, Sam. *The Joy of Sects: An A–Z of Cults, Cranks and Religious Eccentrics*. London: Robson Books, 2005.

Krueger, Oliver. "Gnosis in Cyberspace? Body, Mind, and Progress in Posthumanism." *Journal of Evolution and Technology* 14, no. 2 (2005): 77–89.

Neusner, Jacob, ed. *Religious Foundations of Western Civilization: Judaism, Christianity, and Islam*. Nashville, TN: Abingdon Press, 2006.

Owen, Suzanne. "The World Religions Paradigm: Time for a Change." *Arts and Humanities in Higher Education* 10, no. 3 (2011): 253–268.

Partridge, Christopher. *The Re-enchantment of the West: Alternative Spiritualities, Sacralization, Popular Culture, and Occulture*. 2 vols. London: T. and T. Clark, 2004–2005.

Possamai, Adam, ed. *Handbook of Hyper-real Religion*. Leiden, Netherlands: Brill, 2012.

Rigby, Madeleine. "Graceland: A Sacred Place in a Secular World." In *The End of Religions? Religion in an Age of Globalisation*, edited by Carole M. Cusack and Peter Oldmeadow, 155–165. Sydney, Australia: Department of Studies in Religion, University of Sydney, 2001.

Sandu, Antonio. "The Anthropology of Immortality and the Crisis of Posthuman Conscience." *Journal for the Study of Religions and Ideologies* 14, no. 40 (2015): 3–26.

Sharot, Stephen. *A Comparative Sociology of World Religions: Virtuosos, Priests, and Popular Religion*. New York: New York University Press, 2001.

Stagg, Adrian, and Helen Farley. "Sacred Space and Religious Ritual in the Virtual World: An Exploration of Religion in Second Life." Paper presented at the Australian Association for the Study of Religion Conference, Southern Cross University, Tweed Heads, New South Wales, Australia, July 2011. https://www.academia.edu/6369845/Sacred_Space_and_Religious_Ritual_in_the_Virtual_World_An_exploration_of_religion_in_Second_Life.

Valera, Luca, and Vittoradolfo Tambone. "The Goldfish Syndrome: Human Nature and the Posthuman Myth." *Cuadernos de bioética* 25, no. 3 (2014): 353–366.

Zagorin, Perez. *How the Idea of Religious Toleration Came to the West*. Princeton, NJ: Princeton University Press, 2003.

FILM AND TELEVISION

The Big Lebowski. Dir. Joel Coen and Ethan Coen. 1998.

The Matrix. Dir. Lana Wachowski and Lilly Wachowski (known as The Wachowskis). 1999.

The Melancholy of Haruhi Suzumiya. Dir Tatsuya Ishihara. 2006–2007.

Star Wars: A New Hope. Dir. George Lucas. 1977.

CHAPTER 15

The Spectrum of Human Techno-hybridity: The Total Recall Effect

Diana Walsh Pasulka
Professor and Chair, Department of Philosophy and Religion
University of North Carolina Wilmington

Humans are cyborgs. Humans are biosystems integrated with technologies, from media technologies to biotechnologies. At least this was the conclusion that I reached after an interview with Donna Haraway in the late 1990s, on a beautiful, sunny day at the Santa Cruz coastal campus of the University of California. Haraway wrote one of the early and important books about posthumanism, although she did not use the term *posthuman*. In her book *Simians, Cyborgs, and Women: The Reinvention of Nature* (1991), Haraway questions the boundaries of the human being and calls for a recognition that these boundaries are not as "natural" as have been conventionally thought. She sees in this recognition a possible liberating principle, in that she hopes the new idea of the human will not adhere to dualisms such as man/woman, animal/human, gay/straight, or black/white but will expand possible political and social scenarios for those who have been traditionally marginalized from naturalized categories. She writes that by the "late twentieth century, our time, a mythic time, we are all chimeras, theorized and fabricated hybrids of machine and organism. In short, we are cyborgs" (150).

My interview with Haraway did not go well. I was a potential graduate student and had presented her with my proposed research project. I wanted to understand religious belief with a focus on apparitions of the Virgin Mary as a case study. Haraway listened to my short presentation, and when I had finished she did not look impressed. After a minute or two, she turned her chair to face me, focused her eyes on me intently, and asked a simple question. She asked me to think about the last movie I had watched. She waited until I had found that memory in my mind and then asked, "What happens when people watch movies or see images on a screen, or, when they see apparitions? What is happening to them?" I could not answer because I did not know. She wanted me to seriously consider the question, which at the time I thought was not relevant to my project. The interview was over. I did not attend her graduate program, but her question continued to perplex me as I carried on with my studies. Ironically, that question now guides my studies of human technological engagement. This chapter surveys various answers to this question and reveals that cyborgs are not just the colorful human-machines of popular culture, but rather, humans are surprisingly cyborg-like in everyday, ordinary ways, as well as in ways that are nonordinary. There are degrees of human-technological hybridity.

TWO LESSONS FROM THE *TOTAL RECALL* FILMS

In this volume the terms *posthuman* and *transhuman,* as well as the degrees between these states of being human, are defined in specific ways. As Haraway (1991) and N. Katherine Hayles (1999, 2012) remind readers, humans and their technologies are inextricably bound in complex webs of connections that determine various modes of being human. However, there are still general categories of enhancement and technological intervention that help frame the human and the posthuman. *Unmodified humans* have not used biotechnologies to modify their minds or bodies in ways that increase or alter their lives. *Modified humans* access a variety of enhancement procedures, such as pharmaceuticals, biotechnologies, enhanced prostheses, and other modifications, in order to shift their experience and alter their performance. *Posthumans* are humans who are so modified and altered that they are unrecognizable as being human according to definitions of modification, or being unmodified. Within this framework, however, there are degrees of human and technological engagement. Even the unmodified human is a cyborg in ways that most people would consider ordinary. Because humans are formed by, through, and with technologies, the image of the cyborg provides a useful framework to think through degrees of hybridity.

The *Total Recall* films provide helpful lessons about human engagement with technologies. The 1990 version of the film was so popular that it was remade and released in 2012. Both films are adaptations of a 1966 short story by science fiction author Philip K. Dick titled "We Can Remember It for You Wholesale." In the futuristic story, a clerk named Douglas Quail (called Quaid in the movie versions) wishes to visit Mars, but because he is not wealthy, he opts instead to visit a company called Rekall Incorporated, which can implant memories in his mind so that he thinks he went to Mars. Rekall provides virtual vacations for people in the form of memories. While undergoing preparations for his vacation, Quail learns that he has been implanted with a device that reads his thoughts and creates false memories. He has a difficult time deciphering which memories are real and which are implants. The theme of the blurring of what is considered "real" with what is "not real" figures prominently in the short story and films.

Another important theme of the story and films concerns the role of media technologies. The false memories are implanted by a company from the entertainment sector that is actually a front for a political and military group. These two themes, memory and its relationship to human experience and identity, and entertainment media, play out in surprising ways when one considers ordinary degrees of human-technological hybridity. It turns out that humans are cyborgs in ordinary ways; that is, as they go about their daily routines, their memories may very well be "not their own," depending on the degrees to which they engage with entertainment media.

Contemporary research into human memory and media reveals that films and media change the ways in which memory functions, and at times even supplants memories of real events. In other words, films and media have the capacity to implant false memories, thereby changing a person's view of his or her own past and history. In this way, unmodified humans are cyborg-like in that they incorporate information and ideas about reality that are not based on historical events but are the creations of media experiences. This is a bold declaration, but research in neuroscience reveals how this process works. Jeffrey M. Zacks, a professor of psychological and brain sciences, runs the Dynamic Cognition Laboratory at Washington University in St. Louis. His research shows that when films are promoted as being "based on real events," viewers have a difficult time distinguishing between the film version of the event and the historical version. When producers of such films play "fast and loose" with the facts, and viewers are aware of the historical facts of the event, they will remember, or, misremember, the actual historical event (Zacks 2015).

Similar research is underway at the Stanford University's Virtual Human Interaction Lab, founded by Jeremy Bailenson, a professor of communication. The mission of this research center is to identify and study changes in human behavior through interactions with immersive virtual environments. For example, researchers are actively studying ways in which virtual environments might increase human behaviors such as empathy. Interviews with the computer programmers of immersive virtual realities reveal how memory is influenced by virtual environments. One programmer (interviewed by the author in 2016) revealed that he often had déjà vu but had the vague recognition that he was recalling not something from his real life, but something that he had experienced in his virtual reality. Much like Quail in *Total Recall*, he really could not tell the difference between his memories based on historical events and the memories he gleaned from his virtual experiences.

The second lesson from the *Total Recall* movies is that virtual memories are implanted via consumerist and entertainment organizations. In the movie, the company Rekall is a business that sells vacations and entertainment products, and it is implanting memories that are based not on historical events but on virtual memories. One need not go far to find such companies in real life. There is a new genre of film and media, called, ironically, "specialist factual," that intentionally blurs the boundaries between historical events and virtual events, and a lot of its products are created for an audience of children. There is much irony in the term *specialist factual*. In "We Can Remember It for You Wholesale," Dick uses the term *extra-factual memory* to describe the technique of implanting false memories. The successful production company Impossible Factual describes its own techniques as "specialist factual." Its products are documentaries that are directed toward children, mostly, and viewed on stations of the British Broadcasting Corporation. One of the company's films includes a re-creation of World War I. The film is a documentary that splices real, or presumed real, footage of the war, interviews with alleged survivors, and created video of extraterrestrials, thus creating a documentary called *The Great Martian War, 1913–1917*. The creators combined computer-generated imagery of "Martians" with footage of what appears to be World War I. In another documentary, Impossible Factual re-creates a *Tyrannosaurus rex* for an autopsy. In many ways, specialist factual programming conforms to all the criteria that Zacks outlines for how one's memory of historical events conflate with events that are virtual. Most of the programming is presented within genres that have conventionally been associated with the presentation of facts and truth, such as documentaries. In this way, the new genre of natural history documentaries, specialist factual, is directly influencing the memories of children about their history, with memories that may or may not be based on historical events.

TECHNOLOGIES INCORPORATED INTO THE HUMAN BODY: NONORDINARY DEGREES OF HYBRIDITY

Most contemporary people in Europe and North America will engage with technologies such as films, videos, and virtual environments. Therefore, they will have been affected in the ways Zacks and Bailenson outline, and they will be cyborgs in ordinary ways. But what about people who, because of their desire to either enhance their cognitive or physical performance, or cure a disease, engage with technologies intimately, such that they change the very structures of their physical selves, such as their behaviors, their moods, and in some cases their athletic performances? Ethicists of biomedical or cognitive enhancements generally draw attention to these motives by categorizing them as either *therapies* or *chosen enhancements.* There are degrees of modifications in these areas that have generated a variety of reactions ranging from controversial to indisputably welcome.

Some of the more welcome contemporary technologies involve the use of therapeutic biologics, or biologics for therapy. Biologics is a category of bioengineered living tissues that are either injected or surgically implanted into human or animal bodies to produce healing effects. Most biologics are in clinical trials and many have produced results that appear to be miraculous. Timothy Taylor, the vice chairman and former chairman of several biomedical biologics companies, has been involved in the development of spinal implants and other biologics that help people heal from ailments such as cancer or nonunion bone defects. Among the companies he founded are Endius, which was sold to the biomedical company Zimmer in 2007; Amendia, sold to Kohlberg in 2016; and Vivex, of which Taylor is still vice chairman. Taylor holds more than forty patents, most of which are in the field of surgical devices and biologics. One of the procedures he developed centers on an implantable product using a polymer, metal, or allograph material that has been laser-scripted to mimic the DNA of human bone. In Vivex's clinical trials, animals and human bodies have not rejected the foreign implant but have instead "read" it as actual bone, thus helping the body recuperate after surgery or injury through cancer. The laser-scripting process involves the contemporary use of light to change information at the cellular level of the human body. This procedure is called biophotonics, which is the application of lasers and light to biological tissues and cells to shift their contents and their information.

One cancer patient aided by Taylor's innovations had received a prognosis that was very bad—she was told she would lose her leg as a result of a nonunion of her femur bone. Within a few months of biologics and stem cell bone-fusion treatment, however, she was walking with a cane and caring for her young children. Today she is living a normal life with no cane or assistance.

The laser-etching procedure is remarkable and almost like something found in a science fiction story. A nonbiological object, in this case ceramic, metal, or allograft bone, is coded to resemble human bone through a process that involves a laser that works at the molecular level. The laser is one the most sophisticated in the world. Scientists etch the implant, photon by photon. They turn the ceramic- or metal-etched implant into bone implants that are inserted into the cancerous bones of terminal patients. The product is called cerment. The object is then implanted into the human body, which reads the product as its own and not as a foreign body, which it is. This technology also has other applications. Taylor, who was also an engineer for a NASA (National Aeronautics and Space Administration) contractor and worked on the space shuttle program, has used this process in the US space program to etch materials such as glass and transform them into energy that is used on satellites. The same technologies that power satellites also help humans achieve enhanced and remarkable health effects.

Another procedure that has, so far, produced remarkable results entails the application of a biologic product into the eyes of those who suffer from macular degeneration or other optical illnesses. Patients who have lived much of their lives functionally blind have recovered their eyesight. This procedure is also in the stage of clinical studies and research, but there are many people who attest to its remarkable results. Vanna Belton from Baltimore, Maryland, was blind, but after she went through the procedure, she recovered her sight. "I'm happy to be a guinea pig," she told reporters after the recovery of her sight, and after her case had generated controversy because the doctor performing it did not register his research with the government as a clinical trial (Crew 2016). The procedure is controversial, not only because it is new but also because scientists do not know why it works. It involves taking stem cells directly from the bone marrow of patients and injecting them into their eyes. Somehow, the stem cells work to restore eyesight.

The very newness of these procedures, as well as the fact that scientists are unsure of how they work or of their possible unintended effects, generates controversy. Often, the pragmatic effects of these therapies motivate patients to want them and doctors to administer them. In Belton's case, her doctor decided to skip the usual protocol that requires doctors and scientists to test their medical products by registering with the National Institutes of Health. Because he was using a biologic, a stem cell, which is not technically a drug, he was able to market his procedure directly to his patients, who were willing to take the risk. This example illustrates how market forces drive much of the technological and human engagements in the field of biologics.

Biologics are the latest biochemical modifications that enable humans to enhance their health and to experience their lives in more enriched ways. Although much of the procedures involving biologics are undergoing clinical trials and are experimental, their results illuminate the tenuous definitions of the conventional ways of defining the human. Haraway (1991) pointed out that human engagement with biotechnologies has already revealed the inadequacy of conventional definitions of the human. The biologics produced by Taylor's company—his new form of bone called cerment, for example—reveal that technologies are literally incorporated into the human body and even grow with it. Additionally, research conducted by neuroscientists of film and immersive technologies reveal that some of human memory is formed and shaped by media technologies. These examples reveal that there are degrees of human-technological hybridity and that there are ordinary ways of being cyborgs, as well as nonordinary ways, such as gaining one's sight back with the help of biologics.

The almost miraculous effects produced by technological enhancements ensure their future use. In the Christian tradition, one of the most recognized miracles that Jesus performed was to restore the sight of a blind man. In the parable of this incident, Jesus performed the miracle by mixing his own spit with dirt, creating mud, and placing it on the eyes of the man. Christians believe this was, and still is, a miracle. In a December 2016 interview with Taylor, he has described the process whereby he learned that he could take an inert object, ceramic, and charge it such that it acted like human bone. He ran this experiment on the space shuttle *Columbia*, because he needed an environment without gravity to test the product. He said, "This was like a miracle. We basically took a substance that was like dirt, like a rock, and we wanted to see if it could talk with a charged particle, and it did. Nobody could believe it, but it worked." The newly charged ceramic became the basis for cerment, which has produced many healings. Because many of the doctors who use these products do not know why they work, as in the case of the restoration of eyesight to the blind using stem cells, their effects seem like modern miracles.

SPECULATIVE AND NONSPECULATIVE BRAIN-COMPUTER INTERFACES AND FACEBOOK TELEPATHY

In the discipline of philosophy there is a thought experiment that helps students think through issues about the nature of personal identity and its relationship to the definition of the person. It is called "the brain in a vat." In the experiment, a person is without his or her body. Instead, she is her brain, and she exists in a vat. However, in the experiment she (as a brain) is connected to a computer program (or in some versions an evil scientist) that allows her to

Media Makes Memories. *Research by neuroscientists reveals that movies and immersive media can influence and even produce memories.* ESB PROFESSIONAL/SHUTTERSTOCK.COM.

believe she is actually living in a town, with a body, somewhere in her country of origin. She goes about her daily life and believes she is having a great time. But, really, she is a brain in a vat. This experiment is notable for many reasons, but one that is generally overlooked is that it presents a dualism of mind and body, and the organ that thinks—the brain—is presumed to be the seat of personal identity. A more relevant thought experiment might be: what happens to the human who interfaces with computers to the extent that they become or merge with the computer? This would represent the extreme degree of modification—a posthuman.

This form of the thought experiment has been conducted by author and futurist Ray Kurzweil in a series of books that address what he believes is the ultimate destination of the human—a posthuman who merges with technology. Kurzweil (2005) speculates that by 2045 technological development will have enabled humans to upload mental processes into non-biological platforms. At that point humans would be able to transcend their bodies and biology, according to Kurzweil. This posthuman existence would be ideal, as Kurzweil writes that humans will not suffer from the maladies from which they currently suffer, such as heart disease, cancer, and other illnesses, or even old age. Kurzweil is a futurist, so his work is speculative. Yet, scientists at the University of Washington have produced actual brain-computer interfaces, inspiring technological geniuses such as Mark Zuckerberg, the founder of the social media site Facebook.

In 2015 Zuckerberg conducted an open interview on his Facebook page. He presented three goals that he hoped to accomplish within the following ten years, one of which was to foster a more immersive technological experience for Facebook users. His description of the experience, which involves the instant sharing of the thoughts, visions, and mental images by users with their friends, reminded many of telepathy—which is defined as the simultaneous sharing of the contents of one's mind with another mind. Zuckerberg contended that "in the future we'll probably still carry phones in our pockets, but I think we'll also have glasses on our faces that can help us out throughout the day and give us the ability to share our experiences with those we love in completely immersive and new ways that aren't possible today" (quoted in Dredge 2015). Mind-to-mind transmission via technology is now a fact, although it is not the smooth or streamlined experience that Zuckerberg foresaw.

Media published by the University of Washington boldly declared that two of their researchers created technology that allows people to send thoughts through the Internet and to communicate in tangible ways. Professors Rajesh P. N. Rao and Andrea Stocco placed research subjects in different rooms on their university campus. In one room they connected a subject to an electroencephalography machine that reads that person's brain waves and then sends these through an Internet server that is connected to another subject. The receiving subject is fitted with technology that reads the information from the first person's brain activity and translates it. In their study, they had *subject one* communicate with *subject two* and move *subject two*'s hand. As awkward as the experiment appears, it was successful, showing that one person, hooked up in specific ways to the Internet, can cause movement in the physical body of another, just by thinking about it. Stocco noted in a news release, "The new study brings our brain-to-brain interfacing paradigm from an initial demonstration to something that is closer to a deliverable technology" (Ma 2014). Although the media suggested that a type of brain to brain transference was enabled by technology, what actually occurred was one person's brain activity was being altered by a signal triggered by another person's brain. Significantly, the language used to describe this process was the language of supernatural, with references to telepathy.

The seemingly miraculous event, which actually was not miraculous and didn't live up to the hype of the media about it, is nonetheless garbed in the language of the future, as well as the language of the supernatural. In a 2014 article in *Scientific American Mind*, Rao and Stocco declared that "the dawn of human brain-to-brain communication has arrived." A news release issued after the demonstration of their research quotes Stocco as remarking that "the Internet was a way to connect computers, and now it can be a way to connect brains. We want to take the knowledge of a brain and transmit it directly from brain to brain" (Armstrong and Ma 2013). He noted that it would be a public good, as it would enable researchers who were great thinkers, but not great teachers, to transfer the contents of their brains to their students, in a direct mind-to-mind transmission of valuable knowledge. Although Stocco's technological platform is not able to do this currently, this is what he imagines it would be able to accomplish in the future.

Taylor, who is a visionary and successful biotechnology entrepreneur, sees the future of human-technology engagement moving in the direction of information—that is, information in cells and neurons that can be decoded and transferred. His sector of biomedical development focuses on therapeutic modifications, although he readily admits that there is a large market for nontherapeutic applications of his products. "The amnion fluid that we use to extract our growth factor cells greatly speeds up the healing of acne," he related to the author in a February 14, 2017, interview, but he also added that his company has focused on cancer therapies and other, medical applications. "The growth area in biomedical right now is information. We are making products that propagate nerve growth and cell growth in a way that communicates directly with, say, prostheses and the body. Nerves are the body's information pathway, like the Internet. Neurons are actually biophotons, which are communication channels. This is where implants and biologics is headed."

Research performed at Stanford University's neurosurgery laboratories confirms Taylor's assessments of the future in the present. Krishna Shenoy, a professor of electrical engineering, and Jamie Henderson, a professor of neurosurgery, have created a brain implant that allows people with paralysis to communicate. The small device is implanted into their brains and is also connected to a computer. By thinking, the paralyzed subjects can move a cursor and type out their thoughts. In this way they can communicate. Henderson and Shenoy's comments about this research confirm Taylor's assessment of the current research focus on information.

> We're able to eavesdrop in on electrical activity and slide a cursor across a keyboard and type out messages.... We can surgically implant a tiny electrical sensor that's made out of silicon. It has 100 tiny electrodes. It just sits on the surface in the brain, and it's able to pick up on the electrical activity of individual brain cells or neurons. You record tiny signals from those neurons.... We can then take those signals and decode them using a computer. We can manipulate them once they're in digital form, like any other data. So all you need to do is imagine moving your arm, for example, to the letter *t* on a keyboard, then it'll slide out to the *t*. (McDonald 2017)

DIGITAL NEURAL TRANSFER

Long ago in 2000, science fiction author Ted Chiang wrote a speculative essay for the science magazine *Nature*. It was included in the column "Futures" and was about scientists who used their own bodies and minds as subjects in experiments that focused on human-technology engagement. Chiang writes that these scientists began to use a process called digital neural transfer, in which they transferred the contents of their research to each other directly, in a digital form of mind-to-mind transmission. This type of communication was not possible for ordinary humans, and soon, the super scientists became "metahumans," who far exceeded humans in their intelligence and bodies. Chiang describes how two parallel cultures emerged, that of the "metahumans" and ordinary humans. The ordinary humans were so far behind in their own research and technology that they were happy to take "the crumbs from the table" of the metahumans and were relegated to studying the bits and pieces they could decipher from the metahumans' research. Chiang writes, "What is the role of human scientists in an age when the frontiers of scientific inquiry have moved beyond the comprehensibility of humans?" (2000, 517).

Summary

Chiang's speculative essay describes the emergence of posthumans, beings that are so modified by technologies that they are unrecognizable as human beings. This chapter has described the ways in which humans have engaged with technology in ordinary and non-ordinary ways to achieve remarkable results—sometimes even results that would have appeared to be miraculous in other eras. It opens with the suggestion, posited by Haraway, that we are already cyborgs. The term *cyborg* has been used to show that humans today are engaged in varying degrees of modification, depending on their own specific circumstances and contexts. Significantly, posthumans do not make an appearance in this chapter, because they do not exist yet, despite predictions by Kurzweil and others that humans will achieve this state very soon. Technologies and biotechnologies, as produced by scientists like Taylor and others, increasingly challenge humans to think about definitions of what it means to be human.

Bibliography

Armstrong, Doree, and Michelle Ma. "Researcher Controls Colleague's Motions in 1st Human Brain-to-Brain Interface." University of Washington news release, August 27, 2013. http://www.washington.edu/news/2013/08/27/researcher-controls-colleagues-motions-in-1st-human-brain-to-brain-interface/.

British Academy of Film and Television Arts. "Specialist Factual." http://www.bafta.org/television/awards/specialist-factual. Provides information on an award given for television productions in the specialist factual genre.

Chiang, Ted. "Catching Crumbs from the Table." *Nature* 405, no. 6786 (2000): 517.

Crew, Bec. "A Blind Woman Has Regained Sight following a Controversial Stem Cell Treatment." *Science Alert*, February 29, 2016. http://www.sciencealert.com/a-blind-woman-has-regained-sight-thanks-to-a-controversial-stem-cell-treatment.

Dick, Philip K. "We Can Remember It for You Wholesale." *Magazine of Fantasy and Science Fiction*, April 1966, 4–23.

Dredge, Stuart. "Mark Zuckerberg Thinks We'll Eventually Be Able to Send Each Other Thoughts Directly." *Business Insider UK*, July 1, 2015. http://www.businessinsider.com/mark-zuckerberg-on-telepathy-2015-7.

Haraway, Donna. "A Cyborg Manifesto: Science, Technology, and Socialist-Feminism in the Late Twentieth Century." In *Simians, Cyborgs, and Women: The Reinvention of Nature*, 149–181. New York: Routledge, 1991.

Hayles, N. Katherine. *How We Became Posthuman: Virtual Bodies in Cybernetics, Literature, and Informatics*. Chicago: University of Chicago Press, 1999.

Hayles, N. Katherine. *How We Think: Digital Media and Contemporary Technogenesis*. Chicago: University of Chicago Press: 2012.

Impossible Factual. http://www.impossiblefactual.com. Production company that focuses on "specialist factual" programming.

Kurzweil, Ray. *The Singularity Is Near: When Humans Transcend Biology*. New York: Viking, 2005

Ma, Michelle. "UW Study Shows Direct Brain Interface between Humans." University of Washington news release, November 5, 2014. http://www.washington.edu/news/2014/11/05/uw-study-shows-direct-brain-interface-between-humans/.

McDonald, Glenn. "This Brain Implant Makes Telepathic Typing Possible." *Seeker*, February 21, 2017. http://www.seeker.com/brain-implant-powers-telepathic-typing-2276732706.html.

Rao, Rajesh P. N., and Andrea Stocco. "We Can Now Send Thoughts Directly between Brains." *Scientific American Mind*, November 2014.

Zacks, Jeffrey M. *Flicker: Your Brain on Movies*. Oxford: Oxford University Press, 2015.

FILMS

Total Recall. Dir. Paul Verhoeven. 1990.

Total Recall. Dir. Len Wiseman. 2012.

CHAPTER 16

The Frontiers of Immortality

Jamie L. Brummitt
PhD Candidate, American Religion,
Graduate Program in Religion
Duke University, Durham, NC

In February 2011 Russian entrepreneur Dmitry Itskov established the 2045 Initiative. It encouraged Russian scientists to develop new technologies that will allow human brains to interact with a series of increasingly complex android avatars. The organization's ultimate goal is to employ avatars as holographic bodies that will extend human life "to the point of immortality" (2045 Initiative 2017). Immortality, or life after bodily death, will be achieved by 2045 when scientists design technologies that enable "the transfer of a[n] individual's personality to a more advanced non-biological carrier" (2045 Initiative 2017). In other words, the 2045 Initiative seeks to transfer human consciousness from human bodies to avatars with holographic bodies. The holographic bodies are projected to far exceed those of "ordinary humans" and will usher in a "new species" of humanity (2045 Initiative 2017). Itskov postulates that by 2045 the world will be populated by posthuman beings that exceed all human capabilities. Technology, for Itskov, will help human beings achieve virtual immortality as conscious machines.

To many people, the description of the 2045 Initiative reads like the latest science fiction film blurb. In the near future, human consciousness will be downloaded into immortal supercomputers. If these ideas sound familiar, they probably are familiar. The 2014 film *Transcendence* depicted a man who achieved virtual immortality. *Transcendence* followed the scientist Dr. Will Caster, whose consciousness was uploaded to a computer to avoid his bodily death. Caster survived physical death as a computer connected to the Internet. Through advances in nanotechnology, Caster linked himself to human beings' consciousness and manipulated their minds and bodies. When human beings became suspicious of Caster's motivations, they developed a virus to destroy him. In an act of self-sacrifice, Caster uploaded the virus to his own computer body. The virus infected and killed Caster's consciousness and computer self, as well as all the technology around the world linked to Caster's consciousness via the Internet.

The virtual immortality described by the 2045 Initiative and the film *Transcendence* is based on some transhumanist ideas about the transformation of human beings into posthuman beings. One of the more popular proponents of virtual immortality is Ray Kurzweil, an American computer scientist who was appointed Google's director of engineering in 2012. This chapter describes Kurzweil's notion of virtual immortality by examining four of its components: spiritual machines, religious experiences, the Singularity, and transcendence. Virtual immortality, for Kurzweil, is a state of being in the future when human beings will exist as posthuman spiritual machines. Kurzweil's virtual immortality considers the ways human beings will interact and merge with technology to achieve this new state of being. This chapter also considers the reaction of scholar N. Katherine Hayles to earlier transhumanist philosophies. Even before Kurzweil popularized his notion of virtual immortality, Hayles argued that virtual immortality was not a radically new or future concept. Finally, this chapter shows that Kurzweil's concept of virtual immortality has precedents in the nineteenth-century United States, as illustrated by the examples of the phonograph and the spiritual telegraph. Many Americans in the 1800s recognized the abilities of these new technologies to mediate forms of virtual immortality. The frontiers of immortality do not lie in the future. They lie in recognizing the embodied ways human beings interact with technology in the past, present, and future.

VIRTUAL IMMORTALITY IN POPULAR CULTURE

Long before the 2045 Initiative and the film *Transcendence*, Kurzweil predicted that human beings will achieve virtual immortality. In the 1990s and the first decade of the twenty-first century, Kurzweil wrote several best-selling books that introduced the concept of virtual immortality to popular audiences. Two of Kurzweil's best-known books are *The Age of Spiritual Machines: When Computers Exceed Human Intelligence* (1999) and *The Singularity Is Near: When Humans Transcend Biology* (2005). *The Age of Spiritual Machines* became so popular after publication that the Canadian band Our Lady Peace incorporated the book's ideas into its 2000 album *Spiritual Machines*. The title of the album came directly from the title of Kurzweil's book. Moreover, the band included snippets of Kurzweil speaking in the background on some of the tracks (Pesselnick 2001). Kurzweil's fame in popular culture helped foreground his ideas about virtual immortality over other computer scientists, such as Neil Gershenfeld and Hans Moravec, who wrote similar books (McGinn 1999). Virtual immortality, according to Kurzweil, was a state of being in the future when human beings will exist as immortal, spiritual machines. Kurzweil defined virtual immortality in terms of spiritual machines, religious experiences, the Singularity, and transcendence.

FROM HUMAN BEINGS TO SPIRITUAL MACHINES

In *The Age of Spiritual Machines*, Kurzweil predicted that in the near future technology will advance so rapidly that human beings' failing body parts will be replaced by mechanical body parts. In the future, he argued, there will be no difference between a human being and a machine. Human consciousness will eventually be downloaded into a machine to avoid bodily death. Kurzweil hypothesized that by 2099 "life expectancy [will] no longer [be] a viable term in relation to intelligent beings" (1999, 280). That is, human beings will not be described in terms of "life expectancy" because they will exist as spiritual machines in a virtual immortality. Spiritual machines, according to Kurzweil, will take form as virtual bodies and physical bodies. Virtual bodies will exist in virtual worlds. Similar to the way the holodecks on *Star Trek: The Next Generation* worked, virtual bodies will be able to touch, taste, see, hear, and interact with their virtual environments. Physical bodies will be "nanobot swarms" that can take any form and create real virtual environments (307). Kurzweil also clarified why he thought these machines would be *spiritual* machines.

RELIGIOUS EXPERIENCES AS SPIRITUAL EXPERIENCES

In *The Age of Spiritual Machines*, Kurzweil explained how religious experiences were related to spiritual machines and virtual immortality. Spiritual machines, he argued, will not only experience human sentience, but they will also experience human religion. To describe how spiritual machines will experience religion, Kurzweil cited research on religious experiences involving what is known as the "God spot." Around the time Kurzweil wrote *The Age of Spiritual Machines*, neuroscientists at the University of California, San Diego, announced they had found the "God spot," or what others have called a "God module." Researchers found the God spot by measuring neurological activity in the brain when people read religious words and symbols. The God spot was said to be an area in the brain that activated during religious reading. Researchers then stimulated those same areas in participants' brains to re-create their religious experiences when they were not reading. The ability to measure activity in the brain and correlate it with religious activity allowed scientists to posit a neurological basis for religion. That is, scientists said religion was an experience in the mind that could be measured, mapped, and induced. Finding the God spot led some scientists to posit that religion was a natural human experience, that it was common to all human beings, and that all human beings experienced religion in the same ways. For some scientists, these propositions pointed toward the idea that religion was real only as an experience in a particular location of the brain.

Incorporating initial research about the God spot into his book, Kurzweil suggested that religion was a neurological experience like human consciousness more generally. Kurzweil had already predicted that human consciousness will be transferred to machines. He argued that machines will *experience* sentience like human beings. Kurzweil extended this logic to religious experiences. He argued that machines will also *experience* religion because it is a part of human consciousness. Kurzweil referred to these religious experiences as "spiritual experiences" (Kurzweil 1999, 151). According to Kurzweil, "Machines, derived from human thinking and surpassing humans in their capacity for experience, will claim to be conscious, and thus to be spiritual.... They will believe that they have spiritual experiences" (153). Kurzweil concluded *The Age of Spiritual Machines* by arguing that just *experiencing* consciousness was a spiritual experience and that just *believing* in spiritual experiences was religion for immortal machines.

Today, some scholars question the implications of the research surrounding the God spot. Many religious studies scholars do not define religion as experiences or beliefs that happen in the mind alone. For example, David Morgan argued that the idea that religion is a belief is a deeply Western and Protestant way to understand the practice of religion. Morgan (2010) suggested that religion is an embodied practice and that scholars should study "the matter of belief," not belief as experiences or thoughts in the mind alone. Moreover, scholars who study the neurobiology of religious experiences have debated the meaning of measuring religious experiences. According to some psychologists, "The notion of a 'God module' has found some acceptance in neuropsychological quarters, but has also elicited criticism" (Hood, Hill, and Spilka 2009, 64). While scholars today debate the implications of the God spot, Kurzweil accepted the ideas in 1999. He accepted that religious experiences were natural human experiences common to all human beings and that religious experiences were experienced in the mind in the same ways.

Kurzweil went even further in his predictions about spiritual machines. He suggested that because machines *believe* they are spiritual based on their consciousness, they will also practice religion. Kurzweil envisioned a world in which "twenty-first-century machines—based on the design of human thinking—will do as their human progenitors have done—going to real and virtual houses of worship, meditating, praying, and transcending—to connect with their spiritual dimension" (1999, 153). Machines, according to Kurzweil, will be virtual entities that practice religion in similar ways that human beings practice religion. But *practicing* religion was not the important point for Kurzweil. Kurzweil's point was that spiritual machines will practice religion to induce the *experience* of spirituality. Experiencing spirituality in the mind, not the religious practices of the body, was important for Kurzweil. A spiritual experience, he believed, was necessary because it gave "a feeling of transcending one's everyday physical and mortal bounds to sense a deeper reality" (151). In other words, spiritual experiences will be necessary for machines because they will help machines transcend their embodied nature and experience an immaterial spiritual dimension. Kurzweil argued that spiritual experiences will create peak experiences, or transcendence, for conscious machines. The notion of transcendence was explained more fully in Kurzweil's book *The Singularity Is Near.*

THE SINGULARITY IN 2045

The Singularity Is Near introduced popular reading audiences to two other defining characteristics of virtual immortality: the Singularity and transcendence. According to Kurzweil, the Singularity is "a future period during which the pace of technological change will be so rapid, its impact so deep, that human life will be irreversibly transformed" (2006, 7). The Singularity, Kurzweil said, will begin in the fifth of six epochs of technological evolution. In the fifth epoch, humans will transcend the "limitations of our biological bodies and brains" (9). There will be a "merger of our biological thinking and existence with our technology" (9). This merger, Kurzweil suggested, will go unnoticed by most human beings because technology will be advancing so rapidly before the Singularity. Kurzweil predicted that the Singularity will occur in 2045. With the Singularity, "There will be no distinction, post-Singularity, between human and machine or between physical and virtual reality" (9). In 2045 human beings will become virtual machines. Life as we know it will not exist.

TRANSCENDENCE; OR, WHEN "THE UNIVERSE WAKES UP"

Kurzweil also predicted that the Singularity will usher in the sixth and final epoch of technological evolution. The sixth epoch will be an age when "the universe wakes up" (2006, 15). Once humans and technology merge, the sixth epoch will be marked by transcendence. Kurzweil defined transcendence as spirituality and pattern recognition. Machines, according to Kurzweil, will achieve transcendence once individual technological components go beyond themselves and achieve a higher, collective awareness. Kurzweil explained transcendence by evoking a form of human technology that can go beyond itself now: art. He noted that "random strokes on a canvas are just paint. But when arranged in just the right way, they transcend the material stuff and become art" (388). Kurzweil's point was that transcendence will occur when intelligent machines overcome their individual material natures and become a higher, collective, intelligent consciousness. Transcendence will occur when spiritual machines achieve virtual immortality as a collective consciousness of technology. Virtual immortality in this sense does not depend on the physical bodies of spiritual machines. For Kurzweil, virtual immortality is virtual in the sense that it is immaterial and disembodied. Virtual immortality, according to Kurzweil, is when spiritual machines transcend their material nature and experience the ultimate spiritual experience. He explained this with an analogy to God.

According to Kurzweil, transcendence is like the religious experiences people have with God. Many people describe God as an infinite being and religious experiences as bringing people closer to God. Religious experiences help people overcome their individual selves to become one with God. Kurzweil argued that technology will replace this notion of God because God is a "powerful meme" for technology (2006, 389). He suggested that "we can regard, therefore, the freeing of our thinking from the severe limitations of its biological form to be an essentially spiritual undertaking" (389). Transcendence, for Kurzweil, is the magic of technology that will allow spiritual machines to tap into the "sublimely intelligent—transcendent—matter and energy" (389). Tapping into the higher consciousness of technology is a religious or spiritual experience. Virtual immortality is literally an out-of-body experience where the physical body has no use or meaning for a spiritual machine. Virtual immortality, for Kurzweil, involves accessing an immaterial collective consciousness.

Although Kurzweil based his understanding of virtual immortality on religious experiences, spirituality and God, he argued that virtual immortality was a secular notion of immortality. He noted, "Historically, the only means for humans to outlive a limited biological life span has been to pass on values, beliefs, and knowledge to future generations" (2006, 323). According to Kurzweil, human beings have achieved secular immortality before in terms of the ideas and beliefs they left behind at death on physical technologies, such as paper, books, and videos. For Kurzweil, however, this immortality is ephemeral because it is individual and constrained by the limits of material storage. He suggested that this secular immortality becomes obsolete when we cannot retrieve the thoughts and beliefs left behind. For example, if we cannot retrieve a dead person's thoughts and beliefs from a reel of magnetic tape, then that person's secular immortality is lost to us. Kurzweil argued that virtual immortality was a new and better type of secular immortality that will occur in the future. "By the middle of the twenty-first century," Kurzweil suggested, "humans [as machines] will be able to expand their thinking without limit. This is a form of immortality" (325). Virtual immortality will be achieved when conscious nonbiological life-forms can archive their individual and collective ideas, beliefs, and knowledge in an immaterial way. This archive of knowledge will create virtual forms of consciousness, or immaterial forms of consciousness, thereby granting virtual immortality. For Kurzweil, technology grants immortality because it is the medium that preserves consciousness in nonbiological life-forms. Virtual immortality, for Kurzweil, is the preservation of human consciousness, knowledge, and ideas in immaterial spiritual machines that are part of a collective consciousness. Today, we might think of this as preserving human consciousness in "the cloud" or on the Internet. While Kurzweil presented his notion of virtual immortality as something that will occur in the future, some scholars have suggested that human beings have already merged with machines. Moreover, some scholars identify fundamental problems with describing virtual immortality as disembodied and immaterial.

A CRITIQUE OF VIRTUAL IMMORTALITY'S NEWNESS AND IMMATERIAL NATURE

N. Katherine Hayles presented one of the most important critiques of virtual immortality in her 1999 book *How We Became Posthuman: Virtual Bodies in Cybernetics, Literature, and Informatics*. Before Kurzweil published *The Age of Spiritual Machines*, Moravec had outlined his ideas about virtual immortality in his 1988 book *Mind Children: The Future of Robot and*

Human Intelligence. Moravec claimed, like Kurzweil, that one day in the near future human consciousness will be transferred to a computer, thereby conferring virtual immortality. Hayles responded to this claim in her book and argued that there were logical problems with Moravec's arguments about technology and virtual immortality. Hayles contended that posthuman beings were not something of the future, but already existed. She said that we had already become posthuman beings, or what Kurzweil referred to as spiritual machines, because we had already merged with technology.

Hayles identified a problem in the way Moravec imagined the definitions of and relationships between human beings, posthuman beings, and technology. Moravec assumed that the culmination of a posthuman being was as simple as downloading information (human consciousness) into a machine (computer). This suggested that a mind could be separated from its physical body. This assumption, according to Hayles, was faulty. It implied that human consciousness does not rely on human embodiment. Hayles argued that "the body is the net result of thousands of years of sedimented evolutionary history, and it is naive to think that this history does not affect human behaviors at every level of thought and action" (1999, 284). In other words, Hayles suggested that human consciousness is a product of embodiment, or the interaction of human bodies with a physical world. Human consciousness is not immaterial information that can be transferred to a mechanical body seamlessly and without consequence.

Moreover, Hayles argued that information and virtual reality are not immaterial or disembodied. She traced the history of how technology became seemingly separate from matter and material forms. Information and virtual reality, Hayles argued, are always embodied in technology because technologies are material forms. It is an illusion that computers and information are immaterial. For example, computers are a collection of physical parts and circuits assembled to complete certain processes. Information does not just appear in computers. Instead, human beings use their hands to type information, which has a physical component, into computers. Information, although it seems immaterial, is a physical collection of data stored on a hard drive or physical servers. This illusion exists, in part, because of the language we employ, especially the phrase "it's stored on the cloud." This suggests that information is out there in space with no material form. Information, Hayles argued, seems virtual in the sense that it is immaterial, but, in reality, information and technologies are always physical forms. Hayles concluded that virtuality is not virtual in the sense that is immaterial. It only seems that way to us because we do not see all the physical parts of machines as they work.

Given the embodiedness of virtuality, Hayles asked why people continued to think of computers and information as immaterial and separate from human beings. The reason, she suggested, was because modern people equated agency, will, and personhood with human consciousness. Computer scientists, like Moravec and Kurzweil, thought machines needed human consciousness to have agency. According to Hayles, human beings transferred their own ideas about the human being and human agency onto machines. What transhumanists did not realize was that machines already had embodied agency in human lives. To see this, Hayles recommended that human beings needed to redefine their notions of agency. Agency was not human consciousness and free will. Agency was a part of a distributed system of embodiment between human beings, technology, machines, and other things. Humans and machines already worked with and on one another. Hayles concluded, "The seriated history of cybernetics—emerging from networks at once materially real, socially regulated, and discursively constructed—suggests ... that we have always been posthuman" (1999, 291). Human beings, Hayles argued, were already posthuman beings because they were already integrated with material forms of technology that acted on and with them.

Hayles used these arguments to criticize Moravec's notion of virtual immortality. Hayles advised, "When Moravec imagines 'you' choosing to download yourself into a computer, thereby obtaining through technological mastery the ultimate privilege of immortality, he is not abandoning the autonomous liberal subject but is expanding its prerogatives into the realm of the posthuman' (1999, 287). Hayles suggested that, as we sit in front of our computers, we have only tricked ourselves into thinking that we are somehow separate from them. We think we are different because we define human beings as autonomous subjects with free will and agency. If we were to really see the ways we interact with computers and depend on computers, we would see that computers act on and with us all the time. They have an agency—the ability to act on us—that is not defined by human consciousness and will. Hayles suggested that the idea that human consciousness can be downloaded into a computer to achieve virtual immortality is a very human-centered way to think about agency and immortality.

What exactly do Hayles's ideas suggest about Kurzweil's notion of virtual immortality? While Hayles did not review Kurzweil's ideas about virtual immortality, we can apply her work to Kurzweil's ideas and draw some very important conclusions about his notion of virtual immortality. First, Hayles reminds us that virtual immortality must be historically situated. This means that we must understand the history and the context out of which the notion of virtual immortality emerged. Kurzweil's descriptions about virtual immortality are grounded in human-centered ideas about how information and technology work. Virtual immortality, for Kurzweil, is achieved when human consciousness and information are preserved through immaterial forms of technology. This notion of virtual immortality is a product of Kurzweil's own assumptions about human agency and information. Thus, we can conclude that Kurzweil's notion of virtual immortality is not a product of the future. It is a product of the very specific ways in which he defines immortality as secular and as the preservation of human consciousness disconnected from bodies. It is also a product of the way he defines virtuality as immaterial technologies. This leads to the next conclusion we can draw about Kurzweil's idea of virtual immortality.

Hayles argued that virtual immortality achieved through technology is not virtual in the sense that it is immaterial. Kurzweil imagines that virtual immortality is virtual because it preserves human consciousness in something like "the cloud." Virtual immortality, for Kurzweil, has nothing to do with human bodies or material forms of technology. Hayles warns that this is a problem. Kurzweil's ideas about virtual immortality ignore how technology and information are already embodied in history, people, and things. Following Hayles, we can conclude that it is more helpful to think about virtual immortality as embodied and interactive with material forms of technology and human beings. Virtual immortality is not immaterial, and it is not just the preservation of information in immaterial technologies. If we look to US history and religion, we can see how virtual immortality has been conceived of as embodied virtual immortalities. Kurzweil's concept has precedents in the nineteenth-century United States, as illustrated by the phonograph and the spiritual telegraph. Thus, Kurzweil's notion of virtual immortality is not a new concept that emerged in the late twentieth century.

EMBODIED VIRTUAL IMMORTALITIES IN THE US PAST

Americans have been fascinated by virtual immortality since at least the 1800s. Virtual immortality is best described, not as Kurzweil defines it, but as the ways that physical forms of technologies mediate and embody human concepts of immortality. In the 1870s scientists were not so much concerned with preserving information as they were with preserving sound. Before this time, sounds were preserved primarily in people's memories. In 1877 *Scientific American* announced "a wonderful invention" by Thomas A. Edison (1847–1931). The invention was the phonograph and it worked by recording sounds. The invention became known as the "talking machine," but the machine did more than talk. According to the article, recording voices on the phonograph meant that "speech has become, as it were, immortal" (*Scientific American* 1877, 304). The phonograph made immortality possible by immortalizing sound. This notion of virtual immortality is very similar to but also different from Kurzweil's notion. This nineteenth-century understanding of virtual immortality, like that of Kurzweil, defined immortality as being mediated by technologies. Machines, such as the phonograph and computer, made immortality possible.

The phonograph article enticed readers to buy a phonograph by explaining why it was such a fascinating invention. It was not just that the phonograph immortalized sound. It was fascinating because it immortalized particular sounds. The invention was "wonderful" because it allowed the living "once more to hear the familiar voices of the dead" (*Scientific American* 1877, 304). The phonograph embodied the speech of the dead. It allowed the voice of a dead person to "be reproduced audibly in his own tones long after he himself has turned to dust" (304). The phonograph preserved an audible piece of the dead. This audible piece was not just a sonic representation of the dead. It was the dead person's "own tones." The sounds even made the dead person present. The article noted, "Our great grandchildren or posterity centuries hence [will] hear us as plainly as if we were present" (304). A person was not just what they thought, but the sounds associated with their bodies. The sounds of the phonograph forced people to remember that the dead were not here. This absence created the presence of the dead and gave them a virtual, embodied, and sonic immortality. The phonograph immortalized the dead by preserving and reproducing the sounds of a person's voice.

Phonograph immortality and Kurzweil's virtual immortality are similar in that immortality occurs through a secular process that depends on technologies. Technologies mediate immortality by *preserving* a part of a person for posterity and future generations. Kurzweil suggested that computers will preserve human consciousness. The phonograph, by contrast, preserved human voices. While similar, these versions of immortality are also different. Phonograph immortality suggested that voices carry the personality, memories, and emotions of a person, not just information in the form of individual consciousness. Phonograph immortality also departed from Kurzweil's virtual immortality because it immortalized individual voices. Phonograph immortality did not emphasize the ability of machines to tap into collective sound as Kurzweil did in arguing that machines could tap into a collective consciousness.

Phonograph immortality also departed from Kurzweil's virtual immortality in that it recognized the material nature and agency of machines. Proponents of phonograph immorality presumed that sound waves were preserved in material forms through the phonograph. The article explained that "whoever has spoken or whoever may speak into the mouthpiece of the phonograph, and whose words are recorded by it, has the assurance that his speech may be reproduced audibly in his own tones" (*Scientific American* 1877, 304). The sounds were recorded and reproduced on a "strip of indented paper [that] travels through a little machine" (304). Thus, phonograph immortality emphasized the materiality of technologies and immortality, or their physical and embodied processes.

Phonograph immortality also departed from Kurzweil's virtual immortality in that it recognized the ability of machines to interact with human beings. Allowing the dead to speak was not all the phonograph accomplished. The phonograph made it possible "to create the profoundest of sensations" and "to arouse the liveliest of human emotions" (*Scientific*

American 1877, 304). The phonograph stimulated human emotion by reproducing the sounds of the dead. This machine interacted with humans to elicit human responses. The phonograph also worked with humans to perform tasks that humans could not do. The phonograph was "wonderful" because it extended human capabilities. It achieved its own agency because it enabled the preservation of human voices. It made speech immortal—accessible beyond the grave—in ways that human beings could not. Americans' preoccupation with hearing the dead was not new with the phonograph, however. Americans had been trying to hear the dead with new technologies for at least thirty years before the invention of the phonograph.

Since the advent of Spiritualism in the United States in the 1840s, Americans had been trying to hear from and communicate with the dead. American Spiritualists defined their organization as a religious movement based on communicating with the dead who lived in an afterlife called heaven or the "Summer Land." Spiritualist notions of virtual immorality were different from the immortality presented by Kurzweil and offered by the phonograph, but they were also very similar. Spiritualist ideas of immortality focused on the mediation of immortality through advancements in technology. Spiritualists modeled their efforts to hear and speak with the dead on the latest advancement in communication technology: the electromagnetic telegraph. First used publicly in 1844, the telegraph worked by transmitting electrical signals across wires connected to telegraph stations. To many Americans, the telegraph appeared to condense space and time. It allowed messages to be sent and received quickly from distant places. Spiritualists put this technology to work in the form of the spiritual telegraph, which they understood facilitated contact with the dead in heaven.

The spiritual telegraph worked on principles similar to that of the electromagnetic telegraph. Andrew Jackson Davis (1826–1910), a famous proponent of American Spiritualism, explained this in his 1853 book *The Present Age and Inner Life.* According to Davis, "the whole mystery [of spirit communication] is illustrated by the workings of the common magnetic telegraph. The principles involved are identical" (1853, 66). The spiritual telegraph worked, according to Davis, when people gathered in a spirit circle around a parlor table. Davis advised people to sit alternating women and men and then connect themselves in a circuit by placing a magnetic rope in their laps. The rope was thought to connect people physically and allow for the movement of energy, magnetism, and electricity. The point of creating this human circuit was to magnetize the table. Magnetizing the table, according to Davis, created an earthly terminus, or station, for the spiritual telegraph. At the same time, Davis explained, spirits created their own spirit circuit to form a spiritual terminus. The energy that passed from the spiritual terminus to the earthly terminus was what Spiritualists called the spiritual telegraph. It was, in Davis's words, a "line of communication" that connected the spiritual and earthly worlds (1853, 66).

Spirits supposedly communicated through the spiritual telegraph by sending waves of electricity and magnetism across the line. According to Davis, "the spirits … sustaining a *positive* relation to us, are enabled through mediums, *as electric conductors*, to attract and move articles of furniture … by discharging … currents of magnetism" (1853, 66). This energy, Davis reminded readers, manifested in physical things. Sometimes a table moved, and other times there were rapping sounds on the table that signified letters of the alphabet. Sometimes spirit communication manifested through a specific human medium. While all people in the circuit acted as conductors, some people were better conductors and formally called "mediums." A spirit might speak and act through a human medium or write a message using the human medium's body.

Like the phonograph, the spiritual telegraph allowed the dead to speak. The spiritual telegraph, however, did not create virtual immortality in the sense of indefinitely preserving a piece of a human being. Many Americans, however, thought the spiritual telegraph helped them access the spiritual world, communicate with the dead, and make immortals physically present on Earth through mediums. Thus, the spiritual telegraph was about accessing information for short periods rather than preserving information. Spiritualists reasoned that the spirit world was always near and could be accessed through physical forms of technology.

The spiritual telegraph, like phonograph immortality, also emphasized the agency and embodied nature of technology. Creating a spiritual telegraph was not just about using immaterial technologies. The creation of a spiritual telegraph suggested that spirits, human beings, and technologies morphed into one another. In a spirit circle, a human being was not just a human being. A participant was a human being transformed into a component of a telegraph. A human medium was also a component of the telegraph that transformed into an immortal spirit. According to American Spiritualists, there were no divisions between human beings, technologies, and immortals. One easily morphed into another. Human beings and technologies interacted with and on one another to access the spiritual world and to communicate with immortals.

Like Kurzweil, American Spiritualists recognized virtual immortality as the merger of humans and machines in order to tap into a higher spiritual experience. Although Kurzweil understood virtual immorality as secular, his notions of spiritual machines and transcendence are remarkably similar to nineteenth-century American Spiritualists' ideas about employing technology to communicate with spiritual beings in the spirit world. Unlike Kurzweil, however, American Spiritualists understood that the merger of human beings and technology was already happening in the 1800s through participation in spirit circles. American Spiritualists described this merger and communication as religious practice, not something secular.

Following Hayles, one might say that American Spiritualists already saw themselves as posthumans. American Spiritualists did not use the term *posthuman*, but they did understand that human beings could interact with and even merge with technology to do things that ordinary human beings were not capable of doing. Nineteenth-century American Spiritualists transformed themselves into a spiritual telegraph to communicate with the dead, who they assumed lived in the afterlife as immortals. The spiritual telegraph, according to Spiritualists, helped some Americans tap into a higher religious, or spiritual, experience. Spiritualists argued that they could experience and embody immortality through the spiritual telegraph.

Summary

SO, WHAT IS VIRTUAL IMMORTALITY?

For Kurzweil, virtual immortality is achieved when human consciousness and information are preserved through immaterial forms of technology. Using Hayles's ideas we can broaden and redefine Kurzweil's definition of virtual immorality. Hayles's ideas can help us understand that Kurzweil's virtual immortality is a replication of contemporary Western assumptions about the ways humans have agency and the ways technologies work. Hayles's ideas can also help us understand that virtual immortality is not virtual in the sense that it is immaterial. Finally, Hayles's ideas help us realize that virtual immortality is not an altogether new idea.

Virtual immortality may be defined more broadly as the ways that material technologies embody and mediate immortality. This can mean the ways that material technologies—like the phonograph and spiritual machines—create embodied secular immortalities by preserving the thoughts, ideas, consciousness, voices, emotions, or presences of a dead person. It can also mean the ways technologies—like the spiritual telegraph—facilitate embodied religious notions of immortality, immortal beings, immortal worlds, and communication with immortals. The frontiers of immortality do not lie in the future, as Kurzweil suggested. The frontiers of immortality lie in the ways we understand how technology is embodied and how technology embodies and mediates concepts of immortality in the past, present, and future.

Bibliography

Allen, Paul G. "The Singularity Isn't Near." *MIT Technology Review*, October 12, 2011. https://www.technologyreview.com/s/425733/paul-allen-the-singularity-isnt-near/.

Bostrom, Nick. "A History of Transhumanist Thought." *Journal of Evolution and Technology* 14, no. 1 (2005): 1–25. http://www.jetpress.org/volume14/bostrom.html.

Davis, Andrew Jackson. *The Present Age and Inner Life*. New York: Partridge and Brittan, 1853.

Gershenfeld, Neil. *When Things Start to Think*. New York: Henry Holt, 1999.

Hayles, N. Katherine. *How We Became Posthuman: Virtual Bodies in Cybernetics, Literature, and Informatics*. Chicago: University of Chicago Press, 1999.

Hood, Ralph W., Jr., Peter C. Hill, and Bernard Spilka. *The Psychology of Religion: An Empirical Approach*. 4th ed. New York: Guilford Press, 2009.

Kurzweil, Ray. *The Age of Spiritual Machines: When Computers Exceed Human Intelligence*. New York: Viking, 1999.

Kurzweil, Ray. *The Singularity Is Near: When Humans Transcend Biology*. New York: Viking, 2005.

Latour, Bruno. *We Have Never Been Modern*. Translated by Catherine Porter. Cambridge, MA: Harvard University Press, 1993.

McGinn, Colin. "Hello, HAL." *New York Times*, January 3, 1999. http://www.nytimes.com/books/99/01/03/reviews/990103.03mcginnt.html.

Moravec, Hans. *Mind Children: The Future of Robot and Human Intelligence*. Cambridge, MA: Harvard University Press, 1988.

Moravec, Hans. *Robot: Mere Machine to Transcendent Mind*. Oxford: Oxford University Press, 1999.

Morgan, David, ed. *Religion and Material Culture: The Matter of Belief*. London: Routledge, 2010.

Pasulka, Diana Walsh. "Virtual Religion: Popular Culture and the Digital World." In *Religion: Sources, Perspectives, and Methodologies*, edited by Jeffrey J. Kripal, 325–340. Farmington Hills, MI: Macmillan Reference USA, 2016.

Pesselnick, Jill. "The Modern Age." *Billboard*. March 10, 2001, 81.

Scientific American. "A Wonderful Invention: Speech Capable of Indefinite Repetition from Automatic Records." November 17, 1877, 304.

Sconce, Jeffrey. *Haunted Media: Electronic Presence from Telegraphy to Television*. Durham, NC: Duke University Press, 2000.

Shanahan, Murray. *The Technological Singularity*. Cambridge, MA: MIT Press, 2015.

Stolow, Jeremy, ed. *Deus in Machina: Religion, Technology, and the Things in Between*. New York: Fordham University Press, 2013.

2045 Initiative. "About Us." Accessed January 19, 2017. http://2045.com/about/.

FILMS AND TELEVISION

Star Trek: The Next Generation (TV series). Created by Gene Roddenberry. 1987–1994.

Transcendence. Dir. Wally Pfister. 2014.

CHAPTER 17

The Catholic Tradition and Posthumanism: A Matter of How to Be Human

James F. Caccamo
Chair and Associate Professor of Theology,
Department of Theology and Religious Studies
Saint Joseph's University, Philadelphia

In his novel *Altered Carbon* (2002), Richard Morgan imagines one version of our posthuman future. In the novel, people can capture their experiences in an electronic "stack" that is implanted at the top of the spinal column and records electrical activity of the brain and nervous system. Recording this information enables people to be "resleeved" into a new, purpose-built or purpose-grown body, bringing their memories, experiences, and identity forward into a new physical embodiment. Resleeving is not without its challenges. But as an alternative to injury, death, and the long time a body needs to travel between planets, inserting a data stream into new biology has its merits.

Despite the obvious benefits of this technology, one group in Morgan's novel opts out of resleeving: Catholics. God created humans for a "one-and-done" life cycle, where earthly existence serves as a precursor to meeting one's eternal fate. Morgan's narrator does not go so far as to suggest that this is a foolish viewpoint as much as simply strange: why would you want only one life when so many lives await us? Technology opens a host of doors for humankind, and it is our job to walk through them.

As it turns out, though, the Roman Catholic perspective on technology is more complicated than the novel portrays. It is true that the Catholic tradition has argued that some particular uses of specific technologies are morally questionable (e.g., some termination of human life before birth and targeting civilians with nuclear weapons). But this is rare. The Catholic tradition has been largely "pro-technology," viewing invention as a constituent part of being human and the things we invent as helping make life on Earth better in substantial ways (Bergson 1911; Green 2017).

That said, while there have been a limited number of statements on particular technologies that can be used for posthuman purposes (i.e., cloning and stem cell techs), no significant church teaching has been articulated on most of these technologies, nor has there been a comprehensive assessment of posthumanism as a distinct system of thought akin to those written on such things as capitalism, communism, democracy, or atheism. This chapter, then, offers an "unofficial" Roman Catholic perspective on posthumanism grounded in both long-standing commitments and explicit statements on particular technologies. Of necessity, it is neither exhaustive nor comprehensive. Catholicism is a 2,000-year global endeavor that encompasses a diversity of cultures, perspectives, traditions, and viewpoints. In practice, it is difficult to identify anything that each and every Roman Catholic would agree with. Clear lines of thought, however, do run through both doctrine and dissent, and those through-lines form the foundation of this chapter.

This chapter, then, first considers how the Catholic tradition understands human life, both in terms of the "big picture" of ultimate meaning and the specific moral values that flow from it. Next is an examination of the Catholic tradition's approach to technology over the past century. Finally, all of this then informs a consideration of whether or not posthumanism ultimately coheres or fails to cohere with the fundamental commitments of Roman Catholicism.

A WORLD OF GOODNESS

When considering the Roman Catholic tradition, caricatures and stereotypes easily come to mind. Popular media often use Catholics as plot devices to quickly evoke a variety of tones and identities, from "cultural traditions" (e.g., Día de los Muertos [Day of the Dead]) and devotional practices (e.g., refraining from meat on Fridays during Lent) to misguided priorities (e.g., Mass attendance as more important than love) and outright antisocial behavior (e.g., sexual abuse, sadistic ritual practices, vigilante "avenging angels"). While such things make for good theater, they fail to offer a good sense of the fundamental commitments at the heart of the Roman Catholic tradition.

To start with something that might seem obvious but bears making explicit: the Catholic tradition finds its existential grounding—rather than historical or cultural origin—in the basic human experience of goodness. The world around us is a remarkable place, full of beauty, wonder, and goodness. But this goodness is not without origin. Goodness finds its origin in the work of a divine being who has created all things out of love and care and who continues to sustain them today. This is not a scientific claim about the mechanisms involved in creation, for the Catholic Church subscribes to the theory of evolution. Instead, it is a theological claim about the ultimate shape of reality, born of encounter with this divine being firsthand in daily experience and secondhand in scripture and history. Something as complex as the doctrine of the Trinity is, in a nutshell, merely a way to encapsulate this experience: a God whose entire beingness is characterized by love (1 John 4:7–21) that expresses itself by creation, by redemption, and by an ongoing sustaining presence.

God's love, however, is only half the story. For what great love story leaves out the beloved? The Catholic approach to the world turns out to be as much about how creation—particularly humankind—responds to God's offer of a loving relationship as it is about the one who offers. Christianity offers a variety of images and models for what it would mean to respond well to God (e.g., discipleship, divinization, obedience), but what they all share is the belief that the best way to respond to God's love is "in kind," with a love characterized by drawing close to God in thought, action, and affection. Scripture often portrays this closeness as a very personal union, using some of the most intimate relationships humans experience, such as lover, mother, and father. Contemporary Catholic theology has developed in this vein, understanding "heaven" not as an external reward to be earned but rather as a term to refer to the natural end point of a life devoted to becoming one with God.

Of course, this union with God is easier said than done. We like to say "it's the thought that counts," but anyone who has been in love knows that good intentions are not enough; loving well is deceptively complex. Sometimes we fall short of loving well—whether in loving God or loving our neighbors—for reasons beyond our control: we are ignorant of key pieces of data, we do things we think will be good that turn out harmful, other people intercede in destructive ways, or our plans simply "fall apart." From the Catholic perspective, these are not considered moral failures, but simply sad consequences of the fact that we are human, not divine, beings. On the other hand, many of our failures to love are grounded in choices that are well within our control, such as tenacious placement of one's own good before the good or dignity of others, willful ignorance of critical information, and persistent desires for fleeting inconsequential, and even destructive pleasures rather than full, true, and lasting goods. Indeed, whether taken literally or symbolically, the term *original sin* is nothing more than the tradition's way to refer to how very deeply rooted is our habit of choosing things other than union with God.

It is a consolation to Catholics, then, that the past 2,000 years have served as an ongoing development and beta-testing process to discover ways to promote and enhance human effort toward union with God. The sacraments, for instance, are ways to come into touch with God's healing and uniting love. Worship and prayer are ways to express gratitude and devotion. Doctrine and moral teaching are ways to help people discern how to act in appropriate ways in complex times. The church itself is a way to find support among others with similar relationships and to be a witness to God's love for those who do not.

In order to develop these strategies and structures, the Roman Catholic tradition draws on many sources of insight. But two key sources stand out. First, like all Christians, the Roman Catholic tradition turns to the Bible for insight into how to love well. Biblical texts are important because they relate experiences and ideas that the Christian tradition believes are true in a fundamental sense and that have been communicated in some way by God, no matter how counterintuitive or strange they may seem. Scripture always requires interpretation, but it also always deserves special consideration.

Second, unlike some other parts of the Christian tradition, Roman Catholicism also turns to the world around us as a way to discover what it means to love well. After all, God created the world as an expression of goodness, and so the facts of the world offer a small window into the mind of God that can teach us how to enhance our flourishing. Commonly known as "natural law," this approach is not a simplistic physicalism that reduces goodness to bodily good and instinct following. Rather, human rationality is also a part of the created world. Thus, the things that reason discovers, when uncovered through robust and honest inquiry that is carried out in good conscience, can be regarded as true goods. From a narrowly physicalist perspective, open-heart surgery is surely "against nature": it proceeds through injuring the bodily, and one rarely finds other animals wielding scalpels. This does not mean, however, that it harms human flourishing. Instead, through the use of God-given rational faculties, human beings have figured out how to better care for one another by going beyond the strict dictates of given physical realities.

Importantly, this recognition of the validity of using reason to discover ways to respond well to God's call to love does not, in the Catholic view, run contrary to or demote scripture. Instead, it affirms the idea that all of God's expressions are true, whether found in the words of scripture or in the material of the world. This is not to say, of course, that reason is perfect. There are many things that we do not yet fully understand, and we must always remain humble in the face of the provisionality of our knowledge, ready to revise our views in light of new data (e.g., the shape of Earth). But despite this fact, nature itself retains God's imprint. In the end, all truth is one, and so scripture and creation both help us understand how to return God's love in our daily lives.

BEING HUMAN WELL

It is within this broad understanding of human life as a journey to return God's love well that any Roman Catholic understanding of posthumanism will originate. Helpfully, over the centuries, the Catholic tradition has identified a variety of ideas that provide some direction in particular choices of action to assist people in finding ways to act that both embody love for God and support others in their efforts to do so. For the purposes of this chapter, four ideas are useful: human dignity, the person adequately considered, justice, and solidarity.

HUMAN DIGNITY

When evaluating courses of action, human dignity is often understood to be the fundamental category to consider. Within the Catholic tradition, human dignity refers to the view that all human life has inherent value, from the moment of conception to the moment of natural death. In part, dignity is a reflection of our origin. We are valuable not because of anything that we have done but because we are created by God in an act of love. This dignity, however, is also a result of the fundamental "condition of possibility" that life itself represents. Without life, no other human goods can exist, so human life is infinitely valuable.

Notably, this understanding of human dignity runs counter to the common notion that someone can be "robbed of their dignity." Although situations may lead us to feel that we or others have been made worthless, human dignity, from the Catholic vantage point, is never actually lost. Instead, people are being treated in ways that fail to respect or uphold this basic value. This, then, is the fundamental task of the Christian life: to ensure that we and others consistently uphold everyone's intrinsic worth.

THE PERSON ADEQUATELY CONSIDERED

At the most basic level, then, human action should always uphold dignity. But what helps or harms another's ability to live with dignity? The understanding of the human person has changed greatly throughout history. But in the late twentieth century, Catholic moral theologians developed the category of the "person adequately considered" as a way to think about the human person, in light of philosophical reflection, social and natural sciences, and the theological commitments of the Second Vatican Council (Janssens 1980; Gula 1989).

This approach identified four realities that are central to what it means to be human. First, humans are social beings, conditioned both by our personal relationships—including our relationship with God—and the broad social structures we engage. Second, humans are embodied subjects whose personness is inextricably linked with our biological experiences, not just the inner lives of the mind. Our social relationships are conditioned by our biology, as people respond to us in light of our biological realities. We also express ourselves through our physicality as we act in the world, creating new versions of ourselves in each new endeavor. Third, humans are historical subjects, becoming who we are only over time through a succession of acts. In some senses, then, human life is always potential, never entirely actual. From the moment of conception until death, we are all "in process" of becoming our full selves, be it physically, intellectually, emotionally, spiritually, socially, or professionally. Finally, humans are "fundamentally equal, but uniquely original" (Gula 1989, 71). While all people share these core realities, we are also distinct and diverse. Thus, humans should not be lumped together and regarded as interchangeable but rather enabled to flourish as the particular beings they are created to be.

JUSTICE

In its most basic meaning, justice is the virtue of giving others their due. More particularly, justice is a way of talking about how we should treat all members of our community. It is aimed at ensuring that, in our social structures, minimum standards of fairness exist so that all people are treated with dignity and afforded opportunities to develop fully. As a broad category, justice is used across many different religions and philosophical traditions. However, the standards used to identify exactly what is due to others vary greatly. The Catholic tradition shares with many philosophical traditions the view that societies should be orderly and based on standards that are rational and consistent. In this sense, "justice is blind": all people should be treated the same way, no matter their individual merits or how we might feel about them. Traditions frequently use the language of rights because it is a helpful way to think about basic protections for all people as human beings.

However, the Catholic understanding of justice also draws from scripture a commitment to protecting the most vulnerable in society. It is not sufficient to say that all people have equal protection under the law because in many cases our laws set a woefully minimal standard for protecting one another from the ills that we seek to guard against. Instead, the Catholic view of justice requires that all people have reasonable access to the things that are necessary to uphold their basic human dignity.

SOLIDARITY

If blindness is the greatest strength of justice, it can also be its greatest weakness. In principle, justice does not depend on our knowing others to ensure they are treated fairly. When justice has not been achieved, however, the fact that the powerful do not know the experiences of those who experience injustice can stand in the way of change. This is precisely where solidarity enters the picture. Solidarity is the commitment of those with power or privilege to personally know and work with those who have been marginalized to increase justice in society (John Paul II 1987, para. 38–40; National Conference of Catholic Bishops 1986, para. 66). By connecting people across seemingly inevitable boundaries, such as class, race, socioeconomic status, ethnicity, or religion, solidarity opens our collective eyes to previously unseen gaps in fairness, increases personal commitment to ensuring treatment that upholds human dignity, and creates social networks for getting that done.

Together, these four moral principles—human dignity, the person adequately considered, justice, and solidarity—provide a useful framework for considering posthumanism from a Catholic perspective.

TECHNOLOGY IN THE CATHOLIC TRADITION

From the vantage point of the twenty-first century it would seem absurd to even try to conceive of a world in which being human well would not include technology. Indeed, innovation and invention seem to be part and parcel with being human. The Roman Catholic tradition would agree with this sentiment. This view is, in part, grounded in observations of our rational nature. Yet it is certainly also grounded in scripture. One central theme of the first creation story in the book of Genesis is that human beings are said to be created in God's "image" and "likeness" and are commanded to play the part of the master—have "dominion"—over all of creation. But what does it mean to be and act like God? Within the story, perhaps God's most prevalent characteristic is creativity: God says, and things come into being. Indeed, what better definition of reason is there than coming up with a new idea?

To this appreciation for the foundational role of creativity, the Roman Catholic tradition adds a broad, explicit affirmation of the technologies that spring from it. As Pope John Paul II put it, "science and technology are wonderful products of a God-given human creativity" (quoted in Francis 2015, para. 102). At their best, technologies enable us to make real and substantial improvement in human life, increasing our ability to live well with one another in accord with our full human dignity. The Vatican has gone further, suggesting that communication technologies do this so powerfully that they have an "allotted place in the history of Creation, in the Incarnation and Redemption" (Pontifical Commission for Social Communications 1971, para. 15). Put simply, technology can be a way for us to respond to God's love.

Unfortunately, this potential for good in technology is not always actualized. Thus, technologies cannot be considered in and of themselves. Instead, they can be understood only within the particular context of how they will be used, for what purposes, under what conditions, and in light of their actual outcomes. In theory, humble technologies could be used in inhumane ways, whereas, under certain conditions, even resleeving might be a great thing.

As a result, despite the predominantly positive view the Catholic tradition has taken toward technology, a certain hesitance can be seen in Vatican writings over the past century. For instance, during the Cold War, Pope John Paul II expressed concern that human dignity and personhood was "under threat" from the things we have created, both literally in the case of nuclear weapons and figuratively in the case of the consumer goods that we obsess over (1979, para. 15–16). More recently, Pope Francis has noted the power that technologies have to mask their true ability to "create a framework which ends up conditioning lifestyles and shaping social possibilities along the lines dictated by the interests of certain powerful groups" (2015, para. 107). While seemingly just innocuous instruments, our technologies often support networks of power (e.g., governments, corporations) that act counter to solidarity to negatively affect dignity and justice. Technology has a clear role to play in our work to complete creation by enabling all of creation to reach fullness. But care must be taken to ensure that our technologies uphold dignity, support development of full personhood, promote justice, and enable solidarity, so that we do not let a small segment of society determine the trajectory of our lives.

HUMANS CREATING HUMANLY

In considering whether or not posthumanism and the Catholic tradition cohere, it can be easy to become overwhelmed by the strangeness of it all. Nanotechnology, neural implants, genetic manipulation, and robotic body parts seem like wild fantasies that are far outside the bounds of what God intended for our humble nature. But while the specific technologies involved are cutting edge, the fundamental impulse at the core of posthumanism—to improve what it means to be human through technological means—is by no means new.

As Pope Francis put it, "The modification of nature for useful purposes has distinguished the human family from the beginning" (2015, para. 102). Hacking nature to extend life, abilities, intelligence, and experience is what humans do. And for its part, the Roman Catholic Church has been involved in many of these efforts.

Take, for instance, the posthuman goal of extending life and physical capabilities. Grounded in a desire to ease suffering and enhance our ability to become our fullest selves, the Catholic Church has been involved in health care for centuries. The Catholic tradition has affirmed numerous techniques and technologies aimed at promoting, enhancing, and restoring health, ranging from preventive care and routine pharmaceuticals to plastic surgery, fertility treatments, and even stem cell therapies (Green 2015). Catholic hospitals use technologies associated with posthumanism more often to restore capabilities that have been lost or were missing at birth (e.g., prosthetics, plastic surgery, cochlear implants) than to enhance human beings beyond "standard-issue" biology. But this is currently more a matter of practice than explicit restriction. If, however, bodily extension can be done in ways that uphold the dignity and wholeness of the human person—body, mind, and soul—then these techniques will likely find a home in the Catholic tradition.

Similarly, the Catholic tradition has played a significant role in efforts to extend human intelligence, albeit through existing extrabodily technologies. Throughout the European Middle Ages, the church preserved learning through duplication of classical texts and was the force behind the first European universities. It has promoted primary and secondary education around the world, frequently providing the only option for people on the margins of society. During the waves of media development in the nineteenth and twentieth centuries, the Catholic Church both ran media outlets and offered moral support for Catholics who wanted to spread information through newspaper, television, radio, and film. The Catholic tradition would likely embrace intelligence extension that continues to be done in ways that extend justice through solidarity.

The Catholic tradition has also long championed extending human experience using technology. A long-standing patron of the arts, the church believed that human beings can use tools to help people more deeply apprehend the truths of life through painting, frescoes, sculpture, architecture, and orchestral music. Likewise, the church promoted mass media and social media as ways to extend communion and love across boundaries, opening networks of solidarity in the process. Indeed, media and the arts are nothing more than technologically enabled processes of perception for coming to a deeper understanding of one another, ourselves, and the divine presence in our lives. If experience can continue to be technologically extended without impinging on autonomy and uniqueness, it would be consonant with the Catholic tradition.

In one sense, then, posthumanism is just a new name for the collection of technologies we expect to use to pursue the same goals we have been pursuing for centuries. From the Catholic perspective, technologies are wonderful things, and we are right to celebrate them. The fact that these technologies may have a greater impact on our physical bodies than those of past is not, in and of itself, determinative for their moral status. Thus, we would expect the Catholic tradition to maintain their open and optimistic approach when considering the technologies at the center of the changes that lie before us. At the very least, there is certainly no reason why posthuman technologies should be rejected out of hand as being antithetical to dignity, full personhood, justice, and solidarity.

HUMANS CREATING LESS THAN HUMANLY

Clearly posthuman technologies hold great promise as ways to respond to the call to loving relationships. In practice, however, not everything that we invent will live up to its promise. Take, for instance, one of the few technologies connected with posthuman aspirations that the church has formally evaluated: stem cell therapies. As noted previously, the use of stem cell therapies is understood within the Catholic tradition as having the potential to provide medical gains that will support development of full personhood. There is a hitch, however: this is true only of stem cells that are obtained from nonembryonic sources (i.e., adults or umbilical cords). Obtaining stem cells from early, living embryos (blastocysts) results in the destruction of the embryo, which the Catholic tradition considers the death of a very young human being (Congregation for the Doctrine of the Faith 2008, para. 4). This action fails to uphold the dignity of that life, transforming the person from an end into a means and rendering an undoubtedly good intention into an immoral action.

Unfortunately, this sort of practical analysis is unusual. Many posthuman technologies are still theoretical, making concrete moral determinations of specific-use cases impossible. In the absence of specific cases, then, we can note three general concerns that arise from the trajectory of posthumanism that would prove problematic from a Roman Catholic perspective.

First, posthumanism runs the risk of defining the human person too narrowly, setting aside the diversity of goods that are represented in the human community in pursuit of a narrowly construed perfected humanity. At first blush, this might seem at odds with the general aim of posthumanism to provide a greater range of options to people to tailor their makeup to suit their particular desires. However, one component in this process could very well be removing from human life a variety of natural conditions that we believe limit us. The human body is, after all, not as capable as something we could engineer.

As noted before, the Roman Catholic tradition affirms the desire to eliminate suffering from people afflicted with serious biological maladies. However, the tradition would also suggest that we need to be very careful when considering conditions that should be eliminated. History is rife with examples of attempts to eliminate entire categories of people, often on the basis of faulty or spurious scientific evidence. Examples include the medieval Western devaluation of females as "misbegotten males" and the large-scale infanticide of girls during the "one-child policy" period in modern China. Of course, there were also the millions of people killed by the Nazis in their attempt to eliminate everyone who did not measure up to the standards of the "master race." And lest we in the United States forget our own guilt, between 1907 and 1983, the United States carried out the court-sanctioned sterilization of between 60,000 and 70,000 women who were deemed "not of good stock," owing to ethnic background, poverty, or being "feebleminded." Roman Catholics in the United States were the primary—and at times only—group working against the tide of explicit pro-eugenics legislation (Cohen 2016).

From the vantage point of the twenty-first century, such programs are clearly barbaric. What our forebears believed should be eliminated we now view as wonderful and enriching diversity. It is clear, however, that we are not quite the staunch defenders of diversity that we might think. Jana Marguerite Bennett (2015) has discussed the tendency to redefine capability differences into medical conditions so that we can amass the moral support and technology needed to heal them. The debate within the deaf community about the morality of cochlear implants—biomechanical devices that replicate the inner ear and provide sound data directly to the brain—is a perfect example: a nonlife-threatening physical condition that can be remedied, but need not be, yet brings with it differences in life experience, identity, and community. These differences, however, are often devalued because they do not confer the kind of measurable advantages that we look for today. The same can be said of a host of many other significant differences, in particular autism, learning differences, ADHD/ADD, and other types of neurodiversity.

From the posthuman perspective, the body is mere matter that ought to be reshaped in order to best support the flourishing of the mind. But to the extent that we reform biology to eliminate the things that make us different, we will fail to respect the uniqueness of all people, while also failing to help people participate in their own development. If Pope Francis has emphasized anything in his pontificate, it is the beauty and worth of all people, regardless of who they are and how they measure up to the standards of society. Responding well to God starts with embracing the fullness of humanity.

Second, posthumanism runs the risk of falling short of solidarity and justice by overlooking the potential negative consequences that might result from the fundamental human desire for gain. One idea that appears regularly is that future posthumans will no longer seek to make choices that prioritize the allocation of goods in ways that benefit themselves at the expense of others. This may be the result of eliminating, through genetic engineering, the natural drive to compete. It might also come as a result of the removal of the scarcity that can keep people from sharing. Through molecular fabrication, raw materials will become infinitely abundant, driving the cost of enhancement technologies. Competition and self-orientation will be obsolete, as they will no longer be necessary to obtain the good life.

From the vantage point of the Roman Catholic tradition, this perspective seems unrealistic in light of all we have learned from both scripture and reason. Competition is endemic to nature. But even if we eliminate scarcity of materials as a motivation, competition will likely continue because scarcity is not its origin. Some people seek opportunities for power, whereas others seek social capital. Some people seek achievement, whereas others simply enjoy the thrill of the game. And sometimes competition is driven by a scarcity of resources that cannot be infinitely expanded, such as the time and attention of those in one's family or circle of friends. Indeed, as our lives are enhanced by relationships that foster cooperation, learning to prioritize where and with whom we will cooperate will become increasingly complex.

This is not to say that all acts that prioritize something are morally questionable. Making choices between courses of action always involves determining how to pursue the best of the various goods at stake. Our determinations become morally questionable only when we choose lesser goods over greater goods. Although religious and philosophical traditions offer guidance on how to prioritize our goals, there always seem to be, like the mythical snake in the Garden of Eden, parties willing to promote alternate value schemes in order to benefit themselves at the expense of others. After all, we currently have enough food production capacity to feed everyone on Earth, yet we choose not to do so. While goods may, strictly speaking, become cheap enough that all could have them, it is entirely likely that we will continue to allocate resources as unjustly as we do today. Instead of underfunded school districts, we could have underfunded implant programs. Instead of "food deserts," we could have mod-shop-free zones. Indeed, the digital divide will no longer be a social reality but will be written into our bodies and minds. To the extent that we ignore the possibility that radical inequality results from the posthuman development process, we will fail to act in accordance with the demands of solidarity and not prevent an acceleration of the injustices that characterize our contemporary society.

Finally, from a Roman Catholic perspective, posthumanism runs the risk of falling short of supporting the historical character of the human person by not attending sufficiently to the potential for mistakes. The word *mistakes* is used here not in reference to catastrophic, life-ending sorts of technological mistakes. Those will always be possible and should be attended to. Instead, what is meant is the kinds of failures that happen each and every day when we make choices in good conscience that turn out simply to be wrong. Sometimes we misapprehend a situation or lack critical information. Sometimes we think an action will lead to one consequence but it turns out going in an unexpected direction. Sometimes we realize what we really want only after we have done something else. In short, these are the mistakes that happen simply because we are finite creatures.

Importantly, within the Roman Catholic tradition, finitude is not considered to be a bad thing. Indeed, it is considered part and parcel with being a part of the material world. As noted previously, we are historical beings, acting in time to become more fully ourselves. Like software, we are a product of iteration, constantly working to improve our capabilities and removing "bugs" from the system. But those bugs—those errors we introduce as we try to respond well—are not entirely avoidable. Our bodies wear out. We lack complete information. We have desires and interests that wax, wane, and change over time. Only by learning how things go wrong do we fully figure out how to deal with errors—and in the process become more and more adept at responding well to God's call.

This perspective on mistakes contrasts with the fundamental posthuman optimism about the human ability to eliminate points of failure. This optimism sees what ails us and believes absolutely that we can fix those things, if only we put our minds to it. Yet, as long as we are even the least bit human, we will remain finite, so will need to make choices. Mechanical and electronic implants will wear out. With an infinite world of opportunities for experience, our desires and interests will wax, wane, and change over the centuries. And in our information-suffused environments, we may be awash in irrelevant and wrong data. To the extent that we do not allow for mistakes in the system, we will fail to support the historically grounded process of human becoming.

Summary

Posthumanism and the Roman Catholic tradition share a foundational commitment to using technology to improve the human condition by extending life, physical capabilities, intelligence, and experience. From the Catholic perspective, this work is grounded in the call to act in the world to continue God's work of creation. If, however, it is to fit with the human task of responding well to God's love, it must be carried out in ways that uphold dignity, support the development of full personhood, promote justice, and enable solidarity.

Given what is suggested in this chapter, posthumanism has great potential to do significant harm to human persons and communities. But it also has the potential to create significant benefits without impinging on the values dear to the Catholic tradition. Thus, it is not time to put up a red light. But a yellow light is certainly in order. Great care should be taken over the next century to ensure that moral guidance is present throughout the development process. That said, we all know that most people do not slow down when the traffic light turns yellow. All too often, the caution signal becomes the impetus to make haste before progress is stopped.

In the end, perhaps a better image to use would be hope. Hope is the virtue associated with looking with confidence to one's ultimate end. Hope is what enables us to persevere, even under adversity. Hope also helps us maintain a focus on what the end requires, ignoring temptations that will ultimately frustrate our progress. And temptations will certainly abound in our posthuman future. Thus, we must become, as the prophet Zechariah (9:12) put it, "prisoners of hope" who forgo heaven on Earth to enter into union with God.

Bibliography

Bennett, Jana Marguerite. "We Do Not Know How to Love: Observations on Theology, Technology, and Disability." *Journal of Moral Theology* 4, no. 1 (2015): 90–110.

Bergson, Henri. *Creative Evolution.* Translated by Arthur Mitchell. New York: Henry Holt, 1911.

Cohen, Adam. *Imbeciles: The Supreme Court, American Eugenics, and the Sterilization of Carrie Buck.* New York: Penguin Press, 2016.

Congregation for the Doctrine of the Faith. *Dignitas Personae* [Dignity of a person]. Instruction on "certain bioethical questions." September 8, 2008. http://www.vatican.va/roman_curia/congregations/cfaith/documents/rc_con_cfaith_doc_20081208_dignitas-personae_en.html.

Francis. *Laudato Si* [Praise Be to You]. Encyclical letter on care for our common home. May 24, 2015. http://w2.vatican.va/content/dam/francesco/pdf/encyclicals/documents/papa-francesco_20150524_enciclica-laudato-si_en.pdf.

Green, Brian Patrick. "The Catholic Church and Technological Progress: Past, Present, and Future." *Religions* 8, no. 6, 106 (2017): 1–16. http://www.mdpi.com/2077-1444/8/6/106.

Green, Brian Patrick. "Transhumanism and Roman Catholicism: Imagined and Real Tensions." *Theology and Science* 13, no. 2 (2015): 187–201.

Gula, Richard M. *Reason Informed by Faith: Foundations of Catholic Morality.* New York: Paulist Press, 1989.

Janssens, Louis. "Artificial Insemination: Ethical Considerations." *Louvain Studies* 8 (1980): 3–29.

John Paul II. *Redemptor Hominis* [Redeemer of mankind]. Encyclical letter. March 4, 1979. http://w2.vatican.va/content/john-paul-ii/en/encyclicals/documents/hf_jp-ii_enc_04031979_redemptor-hominis.html.

John Paul II. *Solicitudo Rei Socialis* [The social concern of the church]. Encyclical letter. December 30, 1987. http://w2.vatican.va/content/john-paul-ii/en/encyclicals/documents/hf_jp-ii_enc_30121987_sollicitudo-rei-socialis.html.

National Conference of Catholic Bishops. *Economic Justice for All: Pastoral Letter on Catholic Social Teaching and the U.S. Economy.* Washington, DC: US Catholic Conference, 1986. http://www.usccb.org/upload/economic_justice_for_all.pdf.

Pontifical Council for Social Communications. *Communio et Progressio.* Pastoral instruction on "the means of social communication." May 23, 1971. http://www.vatican.va/roman_curia/pontifical_councils/pccs/documents/rc_pc_pccs_doc_23051971_communio_en.html.

NOVELS

Morgan, Richard. *Altered Carbon.* London: Gollancz, 2002.

CHAPTER 18

Buddhist Biohackers: The New Enlightenment

Julian R. Keith
Chair and Professor, Department of Psychology
University of North Carolina Wilmington

Within the realm of endurance sports, bicycle racing is the prototypical marriage of technology and the human body, attracting athletes and fans interested in augmented human performance. The bicycle shop Mellow Johnny's in Austin, Texas, is a showcase for cycling's latest technology, including a $12,000 carbon-fiber racing bike outfitted with technology to measure the power generated with each pedal stroke and a miniature handlebar-mounted computer that continuously tracks physiological data from devices fastened to the athlete. For athletes who devote years to training and crafting every aspect of their lives to be the best cyclists they can be, subtle features engineered into the designs of bicycles and technology may give them the edge necessary to perform at an elite level. Yet even the most modern cycling technologies combined with exceptional genetic endowments may not be enough to lead the *peloton* (the main body of cyclers in a race) to the mountaintop.

In the fall of 2016, I went to Mellow Johnny's when I was in Austin on business. As a devotee of the history of competitive cycling, for me, Mellow Johnny's is a shrine. The aim of my trek was to see the old, now obsolete, bicycles ridden by the world's most famous cyclist, Lance Armstrong (1971–), to victories in the most prestigious, elite (and now most infamous) races in the sport of professional cycling. I walked around the shop looking at the bikes ridden to victory in the Tour de France between 1999 and 2003 aware of disparate emotions and sensations arising as I replayed memories of those races. That morning at Mellow Johnny's, I knew much more about two subjects, biohacking and Buddhism, than I had known at the beginning of the twenty-first century. Both transformed the way that I now understand the races, as they have transformed the way I understand all of life. Biohacking on a grand scale was a factor in the outcomes of Armstrong's greatest races, with heartrending consequences. This chapter explores the conjunction of biohacking and Buddhist enlightenment, trying not to lose sight of the cautionary lesson learned from Armstrong.

Highly accomplished cyclists and adept Buddhist meditators cultivate their skills through careful discipline and refined practices. Buddhist monastics and elite cyclists devote tens of thousands of hours to practices that may seem like bizarre, monotonous wastes of time—sitting in a fixed position, gazing ahead, alert, relaxed, intentionally applying the right effort for the situation with an open awareness to bodily sensations, equally attentive to the subtle and extreme.

Lance Armstrong, Tour de France, July 22, 2004. *Lance Armstrong, "Mellow Johnny," claimed seven Tour de France titles before it was discovered that he had been using performance-enhancing substances, ending his career.* ROBERT LABERGE/ GETTY IMAGES.

One of the most widely recognized individuals in Buddhism is Tenzin Gyatso (1935–), a monk of the Gelug, or "Yellow Hat," school of Tibetan Buddhism, known to most by his title, the fourteenth and current Dalai Lama. The most recognized name in cycling is Lance Armstrong, biohacker extraordinaire, and an owner of Mellow Johnny's. Armstrong's fame grew during the years when he donned cycling's most coveted symbol: the Tour de France's yellow jersey, or, in French, *le maillot jaune.* Mellow Johnny, is a play on the Texas pronunciation of the French words *maillot jaune.* Armstrong, a Texan, is Mellow Johnny. Whereas the Dalai Lama is revered among Buddhists, all references to Armstrong's participation in the world of professional cycling have been redacted from the official records of the Tour de France. In addition to their connection through yellow, Armstrong and the Dalai Lama also share an interest in the potential of biohacking, though to different ends.

BIOHACKING

Biohacking involves applications of technologies and nonconventional foreign agents to a person's body to study the effects firsthand. Hacking is depicted in the media as a nefarious, illegitimate activity, often aimed at interrupting and damaging computers and networks or gaining unauthorized access to information. But infamous hackers such as those involved in Anonymous or the Russian hackers known as Fancy Bear, who supplied private communications between Democratic presidential campaign aides and officials to WikiLeaks, are a minority subculture, the dark underbelly, of the larger hacking community, which, arguably, pushes forward the evolution of information technology and knowledge. One online dictionary of hacker slang, the Jargon File (Raymond 2017), defines a hacker as "a person who enjoys exploring the details of programmable systems and how to stretch their capabilities, as opposed to most users, who prefer to learn only the minimum necessary." The Internet Users' Glossary (Malkin and Parker 1993) defines a hacker as "a person who delights in having an intimate understanding of the internal workings of a system, computers and computer networks in particular." Both definitions emphasize the enjoyment that arises from an active curiosity about how systems are organized and function. Both definitions recognize that a goal of hacking is to reveal new capabilities of systems, some that may be unknown even to the original system creators. Science, in general, can be thought of as hacking, with nature being the system under investigation. Technologies are products of scientific hacking in that they extend nature's capabilities. Biohacking, from the perspective adopted here, is a subdomain of hacking that delights in stretching the capabilities of biological systems and is based on an intimate understanding of the internal workings of biological systems.

Information about the biology of the body and brain is growing rapidly. Recent advances in the biomedical sciences have led to developments that enable precise editing of the genome of any organism using CRISPR (which stands for "clustered regularly interspaced short palindromic repeats"), breakthroughs in nanotechnology that can be used to deliver drugs to specific cells, the ability to activate and inactivate specific genes using light (a technique known as optogenetics), and devices that can use powerful magnetic fields to deliver electrical pulses that alter signaling within specific brain networks, to name just a few of the topics that are routinely encountered in science and technology media streams. These technologies are being developed as ways to hack biological systems to treat diseases and mental disorders, but they also have broader potential applications as well, including performance enhancement in healthy individuals and even species-altering possibilities. Regulatory agencies and institutions limit access to these technologies to those authorized by professional legal licensure to use them to treat patients or by ethics boards to investigate their uses in formal studies. There are plenty of options, however, for amateur biohackers, who operate outside of regulatory and institutional environments and constraints and are willing to be the research subjects in their own personal experiments. Seeking to improve endurance and strength beyond what they could attain through training alone, some athletes use agents that promote muscle growth (anabolic steroids), accelerate recovery after competition or training (corticosteroids), and increase blood oxygen storage capacity (erythropoietin, blood transfusions). Outside of sports, on the more cerebral and personal growth side, biohackers seeking to improve cognitive, emotional, and intellectual performance—the brain biohackers—use stimulants (e.g., caffeine, Adderall), nootropics (e.g., piracetam, choline, and pyritinol), psychedelics (e.g., "magic" mushrooms, LSD, MDMA), transcranial direct current stimulation, and electroencephalography (EEG) neurofeedback.

BUDDHIST GEEKS

Buddhist Geeks was, until 2016, an online group that also produced a podcast, and has also been called, by cofounder Vincent Horn, a "sangha in the cloud," *sangha* being the Buddhist term for community. The themes of Buddhist Geeks podcasts oriented toward the use of technology to render spiritual practices more accessible, relevant, and easily integrated into modern American lifestyles. Horn predicts what he calls a *technodelic* revolution in which brain biohacking devices, such as EEG headsets and apps that run on mobile devices and emerging virtual reality technologies, will revolutionize humans' approach to consciousness and spiritual practices. Horn prompts his listeners and fellow Buddhist geeks to imagine the potential of transpersonal experience available at the push of a button.

Not all Buddhist geeks are convinced that a technology-enabled shortcut to enlightenment, even if it existed, is a path they would choose. I was involved in a spirited discussion on this topic at a Buddhist Geeks conference that I attended. For a few days in the fall of 2014, some Buddhist geeks met up in Boulder, Colorado, for lectures on traditional Buddhist topics, such as meditation and ethics, interspersed with talks on roles for brain-hacking technology and psychedelic drugs in advancing a Buddhist agenda. In a breakout small group that met up between lectures, the dialogue focused on Horn's proposition that a technological satori (a Zen Buddhist term for a flash of sudden awareness) might enhance Buddhist practice. Several participants said they would adopt a technology that accelerated their progress on the path to Buddhist enlightenment. Other participants, however, expressed skepticism about whether directly altering brain states with technology could, even in principle, produce enlightenment. Or, even if it did, whether to do so would be cheating given that others labor with unfaltering dedication to make progress toward greater awakening.

Expressing similar skepticism about hacking Buddhist enlightenment, a post from Ethan Nichtern, a Shambhala Buddhist teacher and the founder of the Interdependence Project, appeared on my Facebook feed in August 2016. Nichtern's post said, "The term 'hack' is so oddly used these days. Can you imagine the Buddha saying 'I have a simple hack to end suffering?' Maybe, just maybe, the cause of suffering is thinking there's a simple 'hack' that might fix everything. Maybe that's the whole problem." His point is well taken, particularly the notion that such a hack would be simple. To improve the functionality of a complex system, be it a brain, electronic device, or society, skillfulness and deep knowledge about the system being hacked are imperative. Often, amateurs trying to "jailbreak" an electronic device to run apps not authorized by the manufacturer end up with a "bricked" (dysfunctional) device. No doubt, a mind is a terrible thing to brick. Anyone wishing to successfully biohack Buddhist enlightenment probably should have deep understanding of Buddhism, enlightenment, and the brain. That is, it will not be a "simple hack." After all, the original hack proposed by the historical Buddha was an eightfold hack.

WHAT IS BUDDHIST ENLIGHTENMENT?

What do you most yearn for? Do you yearn to defend your current beliefs? Or, do you yearn to see the world as clearly as you possibly can?

—Julia Galef (2016)

In Pali and Sanskrit, the two languages of early Buddhism, the term *Buddha* means "one who is fully awake." When the historical Buddha, Sakyamuni, who lived and taught in India about 2,600 years ago, was questioned by people impressed by his "radiance and peaceful presence" as to whether he was a god, magician, or man, Sakyamuni answered "no" to all those identities and described himself with one phrase, "I am awake." In the everyday sense of the term, it is hard to think of an achievement more ordinary than simply being *awake.* But even in the everyday sense of the term, *awake* encompasses a broad spectrum of alertness and awareness. Buddha meant that he had cultivated awareness of truths about reality that are easily overlooked yet explain much about how humans experience life. The first point he made in the first sermon given after his enlightenment was that existence is *dukkha. Dukkha,* a Pali word, is often translated as "suffering." But Buddhist scholars note that the meaning of the term is subtler, having to do with the short shelf life of satisfaction. Have you ever thought that if you just had your dream job, house, car, spouse, victory, a skyscraper in Manhattan and a jet, and so on, then you would be satisfied? Yeah, me too. An interesting truth about people is that we are not very good at predicting what (or who) will make us happy or how long our satisfaction will last—even when we are pursuing a pleasure we have had before. Like the amnesia for the pain of childbirth, we quickly forget that even a new car or house will not deliver that deep sense of satisfaction for very long. Our appetites are boundless and some cravings get stronger the more they are fed. One way to understand why satisfactions that accompany getting or achieving what we want rapidly fade is to examine mood, emotion, thought, and behavior through the lenses of the causes and conditions that brought forth and shaped human minds—that is, biological evolution by natural selection. The field of evolutionary psychology has attracted prominent psychologists, neuroscientists, and now Buddhist teachers (including the Dalai Lama), who are joining forces to understand the mind by leveraging the tools of science and Buddhist practice for the mutual benefit of each.

Robert Wright, the author of several best-selling books, including *The Moral Animal* (1994), *Nonzero: The Logic of Human Destiny* (2000), and *The Evolution of God* (2009), has taken a leading role in adapting themes from evolutionary psychology to interpret Buddhist practice. In his online course, Buddhism and Modern Psychology (Wright 2017), Wright characterized Buddhism as a "rebellion against the agenda of natural selection." What is the agenda of natural selection? Following the theme developed by Richard Dawkins in his best-selling book on evolution titled *The Selfish Gene* (1976), an organism, be it human, bacteria, fish, worm, or anything else, is a gene's way of making more genes. Evolutionary psychologists argue that behaviors and psychological traits that are present across all cultures and historical periods are adaptations that evolved, through either natural selection or sexual selection (where one sex prefers a specific characteristic in individuals of the other sex), as solutions to persistent problems encountered by ancestors. Like psychologists of many theoretical stripes, evolutionary psychologists hold that behavior is the product of biological processes organized largely outside of conscious awareness and serves functions that are unknown to the individual, such as the replication of genes. Organisms (including people), as vehicles for genes, do not need to know why they do what they do or feel what they feel; they need only take care of business by behaving so that resources, safety, and mating opportunities for themselves and their offspring are obtained.

Reason and language also have evolved with the human mind, and these abilities bring benefits in terms of improving one's chances of surviving in a complex environment and finding mates. However, while reasoning is generally seen as a means to improve knowledge and make better decisions, psychologists are rethinking its function in light of much evidence showing that reasoning often leads to distorted knowledge and bad decisions. In 2011 Hugo Mercier and Dan Sperber published an influential article in *Behavioral and Brain Sciences*, a high-impact journal, arguing that the chief function of reasoning is to devise arguments intended to manipulate others. Humans are highly dependent on communication and vulnerable to misinformation; reasoning is an adaptation that capitalizes on this vulnerability in others. Skilled arguers, psychologists have found, are not after the truth but after arguments supporting their views, a universal human cognitive trait known as confirmation bias. The relentless search for support for conclusions aligned with beliefs already held is baked into the DNA of the human mental operating system. And confirmation bias is employed not only when people are arguing with others but also when they are imagining having to defend their opinions. Reasoning motivated by confirmation bias can distort appraisals and outlooks and reinforce incorrect beliefs. Cognitive psychologists have identified hundreds of biases, including negativity bias (negative events, thoughts, and emotions have stronger effects on one's psychological state than equally intense positive ones), that misdirect people.

Taking the theme of cognitive bias further, psychologist John F. Schumaker, in *The Corruption of Reality* (1995), explains that the human brain's ability to process information simultaneously along multiple pathways imparts an ability to construct a personal reality that is deviant from primary reality. Human brains can bring a tremendous volume of information into consciousness and thereby amplify consciousness, a gift that comes with a cost. Schumaker points out that one such cost is an overwhelming volume of emotionally terrifying and confusing facets of reality that humans encounter that can debilitate and even paralyze the mind. Thus, humans use a portion of their brainpower to soften primary reality with fictional narratives that diffuse existential fears. Culture further amplifies this process by serving as a platform where multigenerational alternatives to primary reality can be stored, evolved, and transmitted, capturing minds in religious narratives, tribalism, nationalism, fundamentalism, scientism, futurism, and, yes, even Buddhism. Schumaker contends that religion, hypnosis, and psychopathology are all manifestations of this property of human brains.

Thus, per modern psychology, people's personal beliefs about primary reality are fraught with delusion. Although he predated psychology by about 2,500 years, Buddha's views about the mind are highly compatible with modern scientific psychology. Buddha taught that certain delusional ways of interpreting reality are built into the operating system of the human mind and reinforced by conditioning and culture. Buddha and many who have followed in Buddhist traditions have developed preelectronic technologies that, when systematically applied, alter aspects of the mental operating systems—factory-installed in minds by natural selection, conditioned by experience, and buttressed by culture—that create distortions of reality and generate unnecessary *dukkha.* For his day, Buddha had "mad" mind-hacking skills.

BUDDHIST MINDFULNESS MEDITATION

Mindfulness meditation is an important component in the Buddhist kit. Within different Buddhist traditions (Vipassana, Tibetan, Zen, Pure Land, and Vajrayana), teachers and practitioners emphasize different aspects of mindfulness meditation practice. Some assign centrality to concentration practice, whereas others give priority to insight meditation as the foundation for mindfulness practice. Lobsang Rapgay, a psychologist and former Tibetan Buddhist monk who worked closely with the Dalai Lama, and Alexander Bystritsky, a psychiatrist and director of at the University of California, Los Angeles, Anxiety Disorders Program, pointed out that the classical literature states that mindfulness practice involves the integration of the two (Rapgay and Bystritsky 2009). Single-pointed concentration (*samatha*) practice entails sustained attention developed by attending to the target object to the total exclusion of all other objects and experiences. The training primarily involves sustained attention focused centrally on the breath, while peripherally being aware of the body as one breathes in and out. When sensations, thoughts, and feelings arise during the practice of divided attention and awareness, introspective awareness is applied to label them, without an elaborate narrative or getting caught up in the experience, before returning attention to the breath. The aim is to experience the target of attention in its bare form, empty of previously conditioned meanings. The ability to intentionally focus and hold one's attention on a single object must be mastered before the path opens to obtain deeper insight into the workings of the mind. As concentration skills are mastered, bare attention and awareness are applied to observe moment-to-moment experiences of bodily sensations, feelings, thoughts, and mental contents. According to Rapgay and Bystritsky, advanced meditators can use refined concentration skills, combined with introspective observing, to directly observe when a subtle mental event arises, how long it lasts, and when it ceases. After much training, they begin to do so without conscious effort (Rapgay and Bystritsky 2009). For a skilled meditator, watching the activity of the mind during meditation is analogous to a scientist looking at a specimen teaming with microbes; concentration is like a compound microscope, with objective lenses that can be used to see the content with clearer focus and in greater detail. As in science, careful observation is the key to insight. While one is watching thoughts and feelings appear and disappear, one may wonder where they are coming from and going and "who" is generating them. This leads to such questions as: "Who" is watching them? How many "whos" are there in there anyway? And where is "there"?

Although much about consciousness remains a mystery, it is a safe bet that brains are in on it. The human brain includes roughly eighty billion neurons, the cells believed to be the primary workhorses underlying the brain's information processing prowess, and trillions of glia, cells that play supporting roles and whose contributions to information processing have been viewed in an increasing prominent light (see Fields 2010). The astonishing capabilities of the brain, however, arise from the organization of cells into functional networks specialized to take primary responsibilities for processing different kinds of information. For example, when a person experiences any object, such as a rose, it feels like a single unified experience, yet different brain systems are involved in the early stages of processing odors, colors, shapes, sounds, textures, and most other perceived properties of the object. If any of the specialized brain subsystems, sometimes called "modules," involved in experiencing a rose was inactivated by an injury, chemical, electrical field, or something else, one's experience of a rose would change accordingly—certain qualities of the experience normally manufactured by the now inactivated module would be absent. Similarly, the abilities to remember having seen that rose, to use language to communicate about the experience, and to feel emotions associated with roses, all depend on finely coordinated interactions between networks of neurons transmitting unimaginably large volumes of information back and forth, and all lie completely outside of conscious awareness.

Nobody knows how many different functional networks make up the human brain, but the consensus among neuroscientists is that every characteristic that can be attributed to a mind arises from the organization of and interactions between brain networks (for an example of how neuroscientists study functional networks, see Bullmore and Sporns 2009). The individual behavior, personalities, intellectual abilities, and psychopathologies of humans reflect the way these modules function and interact, as do the ever-changing qualities of individuals' moods and thoughts over time. Evolutionary psychologists propose that modules, or groups of modules, that formed under pressure from natural selection are responsible for even highly complex behavior, such as territoriality, sexual jealousy, compassion and empathy, and aggression. For example, just as activity in modules that process color are affected by wavelengths of light emitted from light sources, reflected off objects, passing through the pupil, and striking the retina, information in the environment affects the jealousy module. Seeing an attractive person expressing sexual interest in one's lover might transform one's mental and emotional state from calm and relaxed to agitated and angry in an instant, as may the memory or thought of such an event.

Wright, in his online course Buddhism and Modern Psychology (Wright 2017), succinctly summarized how cognitive and evolutionary psychologists understand the functional organization of the mind, saying, "Mental modules compete to control attention and behavior." The modules are, in a sense, continuously trying to become selves.

Another major concept in modern psychology and neuroscience is that of neuroplasticity. The functional organization of brain systems is not fixed or static but is ever changing. The capacity for changes within and between subsystems exists because the connections through which neurons transmit information to one another, known as synapses, change with use, with connections strengthening with use and weakening with disuse. Similarly, modules and connections between modules gain strength with activity, garnering more control over attention, behavior, and consciousness each time they are used, a concept akin to karma in Buddhism. There is no module that a person must be. Yet in any moment, a person's consciousness is a composite of the modules that currently are dominating the overall activity in his or her brain.

The modules have a vested interest in dominating a person's behavior in that they serve genes (that also do not know they exist) that provide the instructions used to arrange their existence within the brain. As products of natural selection, human minds are built to fit well in the worlds of our forerunners, not the world we find ourselves in now. And as the pace of cultural and social evolution accelerates, biological evolution through natural selection, which operates over spans of many generations, will leave humans increasingly less well adapted for the world we find ourselves living in. For instance, the very networks that bias one to favor their kin or close acquaintances over strangers who differ from one's "tribe" in appearance may have been indispensable during the human Stone Age, when most of human history occurred. But in the modern world in which races and cultures are rapidly mixing and remixing, modules that drive humanity's tribal inclinations are at risk for running amok, especially if resources are scarce, or are perceived to be.

HOW MINDFULNESS MEDITATION CHANGES THE MIND/BRAIN

The nature of everything is illusory and ephemeral. Those with dualistic perception regard suffering as happiness, like they who lick the honey from the razor's edge. How pitiful are they who cling strongly to concrete reality? Turn your attention within, my heart friends.

—Nyoshul Khen Rinpoche (1987)

In the 1999 film *The Matrix*, Neo, a computer hacker played by Keanu Reeves, awakens to the fact that the world in which he lives is an illusion manufactured by and for a race of machines exploiting human bodies to perpetuate their own interests. Some people see Buddhist themes embedded throughout the movie plot (an online search using the terms *Buddhism* and *the Matrix* leads to much more on this subject). This is not to say that all of reality is illusory. In meditation, however, one observes that (1) thoughts arise seemingly from nowhere; (2) it is a habit of mind to identify with thoughts and emotions; (3) this conditioning can be broken by allowing them to come and go without responding, a process psychologists call extinction; and (4) there does not seem to be any permanent, fixed self behind thoughts, beliefs, and emotions. With practice, as a meditator experiences the vicissitudes of mental modules with less attachment and reactivity, mindfulness meditation can foster more objectivity and healthier emotional regulation in everyday life. As one would expect, given the integrated nature of mind/body, meditation changes the brain.

Given the central role of concentration training in meditation, it is not surprising that neuroscientists report that meditation affects many measurable aspects of attention, such as speed and accuracy on tests that require participants to attend to a rapidly changing stream of stimuli (e.g., letters) and report the occurrences of targets (e.g., digits) and to do so without lapses over prolonged periods (sustained attention). Meditators consistently outperform non-meditators on such attention tasks, as well as on tasks that require participants to ignore irrelevant information and focus on features of a stimulus (attention control). The performance differences seen between meditators and non-meditators are reflected in differences in the way the brain is processing information, based on electrical activity generated by functional networks of neurons and blood flow to brain regions where those networks reside in the brain (Malinowski 2013; Lutz et al. 2009). The regions of the brain associated with attention also increase in volume with meditation practice.

The brain areas associated with the functional brain networks (modules) involved in body awareness, emotion regulation, and perspective on self are also changed by meditation practice, including the anterior cingulate cortex, the insula, the temporoparietal junction, the frontolimbic network, and the default mode network (Hölzel et al. 2011). The default mode network is a network of interacting brain regions that show activity highly correlated with each other and distinct from other networks in the brain and is so named because it is active when a person is in a state of wakeful rest and is simply daydreaming or mind wandering, not focused on an external task. When the mind wanders, a person's thoughts about others, about oneself, and ruminating about the past and future spontaneously arise. Specific "functional hubs" are associated with each of these aspects of mind wandering. Abnormalities in the patterns of functional connectivity between different components of the default mode network are associated with various psychological disorders, including autism, Alzheimer's, depression, anxiety, post-traumatic stress disorder, and chronic pain (Andrews-Hanna, Smallwood, and Spreng 2014).

BIOHACKING BUDDHIST ENLIGHTENMENT

If you have ever tried mindfulness meditation, one of the first things you may have noticed is that your mind wandered even though you intended to focus your attention on your breath. When scientists use neuroimaging equipment, such as functional magnetic resonance imaging (fMRI) or EEG, to eavesdrop on the brains of novice meditators, significant differences in the connections and activity levels among the various hubs of the default mode network are observed between novice and experienced meditators (Taylor et al., 2013). In particular, an area known as posterior cingulate cortex (PCC) becomes less active during meditation in experienced meditators' default mode networks (Brewer et al. 2011). This discovery led Judson A. Brewer and his colleagues to wonder whether meditators' subjective experience of "effortless awareness" (i.e., effortlessly concentrating and observing experience) during meditation corresponded to periods of low PCC activity levels (Garrison et al. 2013). One way they investigated this question was to give meditators information, based on fMRI imaging, about their own PCC activity levels, in the form of a visual graph displayed on a computer screen while they meditated (e.g., when blue bars displayed on a graph grew larger, that indicated PCC deactivation, whereas red bars indicated PCC activation levels)—an approach known as real-time neurofeedback. Indeed, a strong correlation between a calm, focused mental state and PCC deactivation was discovered. Readers inclined toward biohacking might wonder whether real-time neurofeedback might be useful as an assistive meditation technology. If information about one's brain activity during meditation were immediately available, along with moment-to-moment cues indicating whether one's brain was getting "warmer" or "cooler" in terms of approximating activity patterns associated with deep meditative states, might such a technology enhance mindfulness training?

Obviously, fMRI-based neurofeedback is neither financially nor logistically feasible outside of major research institutions because MRI machines cost millions of dollars to purchase and maintain and $600 per hour to operate. However, other neuroimaging technologies, such as EEG, that enable mind-machine interaction are relatively inexpensive and can provide information about brain activity that can support the process of learning to generate and sustain brain states associated with focused attention and calm, effortless awareness—the two mental states that comprise classical mindfulness (Lagopoulos et al. 2009).

EEG involves placing sensors on the scalp that detect electrical signals produced by the synchronous changes in the electrical states of the membranes of large numbers of neurons. The electrical signals are rhythmic, and the frequencies of their oscillations reflect information processing across functional brain networks. Two EEG rhythms, one known as alpha (7–12 Hz) and the other known as theta (4–7 Hz), are particularly relevant to brain networks that change with meditation practice (Saggar et al. 2015). In experienced meditators, theta is predominant in EEG in sensors recorded at locations on the center of the front of the head, while alpha is predominant at locations farther back on the head. Frontal theta is an indicator of default mode network activity (Scheeringa et al. 2008). Alpha is associated with relaxation and the synchronized exchange of information from the thalamus and the cerebral cortex (Saggar et al. 2015). Both theta and alpha can be trained using EEG neurofeedback using operant conditioning; this involves, essentially, "gamifying" meditation by awarding points for producing patterns of brain activity such as those generated by experienced meditators. Operant conditioning experts call this process "shaping by successive approximations," meaning that they program the neurofeedback system to take baseline measurements of brain activity and reward changes in the EEG when they shift in the goal direction. As the person receiving neurofeedback "levels up," in the jargon of video gaming, he or she is required to more closely approximate the pattern of activity produced by expert meditators to earn points.

Muse Neuro Feedback. *The author using Muse neurofeedback during meditation.* JULIAN KEITH.

Neurofeedback has been used to great effect to help people with disorders, including attention-deficit disorder, addictions, and anxiety (Keith et al. 2015; Ros et al. 2014). The similarities between EEG neurofeedback and meditation have not escaped neuroscientists (Brandmeyer and Delorme 2013). Interaxon's Muse is a commercial EEG headset controlled by an app that runs on iOS and Android devices and is programmed to act as a meditation coach (Bhayee et al. 2016). When EEG signals are detected that are similar to those measured when someone is frequently shifting their attention, mind wandering, or ruminating, Muse users hear sounds of strong winds blowing or large ocean waves breaking (users select the soundscape they wish to use for a session). When attention is stable and the mind is calm, the winds and waves are gentle, and if the state is maintained for several seconds, a bird quietly chirps, seemingly nearby. Such a technology is useful in that it helps answer a question that nearly every beginning meditator must ask themselves, "am I really meditating?"—reinforcing the novice meditator's confidence, chirp by chirp.

Attempts to use technology and nonconventional foreign agents on the body to study their effects firsthand on advancing enlightenment are condoned not only by Buddhist Geeks but by the Dalai Lama. In his lecture at the annual meeting of the Society for Neuroscience in Washington, DC, in 2005, the Dalai

Lama indicated that his keen interest in the fields of neuroscience and psychology arises from a desire to improve the quality of his mind by adapting the knowledge developed in these fields to his practice. The Mind and Life Institute, which he was instrumental in establishing, exists to integrate science and contemplative practice, and the topics discussed in this chapter are the emphases of the Mind and Life Institute's annual conferences. A close acquaintance who worked closely with the Dalai Lama for many years once told me that the Dalai Lama spends the first several hours each morning practicing meditation and is keen to learn how technology and knowledge about the brain will enable others, and maybe even himself, to make more rapid progress on the path to enlightenment.

Of course, falling deeper into the trap of rushing toward a goal also could undermine the enlightenment journey itself. I have heard it said that one Buddhist teacher advised a student to "hasten slowly." In the context of this Buddhist teacher's quote, I am reminded of an interview I once saw with Tyler Hamilton, a teammate of Armstrong's, who said that the public misunderstood why athletes used performance-enhancing drugs during training. He and his teammates used performance-enhancing drugs, Hamilton claimed, not as a shortcut or to train (and suffer) less, but so that they could train harder and spend more hours on a bicycle saddle. Training and dedication remained indispensable. Similarly, EEG and neurofeedback are not shortcuts or instant fixes. One finds that neurofeedback is not easier than meditation and can be more challenging because the technology, like an extraordinarily perceptive teacher, quickly calls one out when attention drifts and the mind wanders. To embark on the path of Buddhist biohacking also is to overcome a subtle form of dualism that sees these technologies as separate from the mind. I propose that one might think of these technologies not as separate from oneself, but as augmented sensory systems that provide a unique channel of access to information about the state of one's brain, the organ through which the universe itself becomes aware of its own existence. From a non-dualistic Buddhist perspective, the body, brain, EEG technology, and the universe are aspects of one, inseparable, whole.

Summary

This chapter began with a story about Lance Armstrong, nicknamed Mellow Johnny, a biohacker whose tale led to infamy, because his biohacking violated both the rules of his sport and federal and international laws and because of ethical lapses in the moral treatment of other individuals with whom his life was entwined. "Like they who lick the honey from the razor's edge," hacking the brain is fraught with the potential to do harm, especially if wisdom and virtue are not fostered along with increased powers of concentration. Although I am not aware of a biohack for wisdom and virtue, devoting one's individual practice to the benefit of all beings and cultivating an aspiration for the enlightenment of all beings would be central to the bodhisattva approach to biohacking.

Biohacking to improve athletic performance has an infamous legacy in sports. Yet, the potential for biohacking to improve human life is growing. Rapid advances in knowledge about the body and brain are leading to new avenues for biohacking, some of which are being adopted by people aiming to improve mental and emotional functioning. Buddhist biohackers experimenting with real-time neurofeedback during meditation are reconfiguring functional brain networks formed through the evolution of the human species and reinforced by cultural and personal conditioning, dissolving habits (also known as karma) that lead to unnecessary suffering and dissatisfaction and cultivating mental clarity and the concentration necessary for direct insight into the nature of the mind. The merging of Buddhism and neuroscience heralds a new age in which science and spiritual development can collaborate to bring forth a new form of enlightenment.

Bibliography

Andrews-Hanna, Jessica R., Jonathan Smallwood, and R. Nathan Spreng. "The Default Network and Self-Generated Thought: Component Processes, Dynamic Control, and Clinical Relevance." *Annals of the New York Academy of Sciences* 1316 (2014): 29–52.

Bhayee, Sheffy, Patricia Tomaszewski, Daniel H. Lee, et al. "Attentional and Affective Consequences of Technology Supported Mindfulness Training: A Randomised, Active Control, Efficacy Trial." *BMC Psychology* 4, no. 1 (2016): 60. doi:10.1186/s40359-016-0168-6.

Brandmeyer, Tracy, and Arnaud Delorme. "Meditation and Neurofeedback." *Frontiers in Psychology* 4 (2013): 688. doi:10.3389/fpsyg.2013.00688.

Brewer, Judson A., Patrick D. Worhunsky, Jeremy R. Gray, et al. "Meditation Experience Is Associated with Differences in Default Mode Network Activity and Connectivity." *Proceedings of the National Academy of Sciences of the United States of America* 108, no. 50 (2011): 20254–20259.

Bullmore, Ed, and Olaf Sporns. "Complex Brain Networks: Graph Theoretical Analysis of Structural and Functional Systems." *Nature Reviews Neuroscience* 10, no. 3 (2009): 186–198.

Buss, David M. *Evolutionary Psychology: The New Science of the Mind.* 5th ed. Boston: Pearson, 2015.

Fields, R. Douglas. *The Other Brain: From Dementia to Schizophrenia, How New Discoveries about the Brain Are Revolutionizing Medicine and Science.* New York: Simon and Schuster, 2010.

Galef, Julia. "Why You Think You're Right—Even If You're Wrong." TED Talk. February 2016 [video file]. https://www.ted.com/talks/julia_galef_why_you_think_you_re_right_even_if_you_re_wrong.

Garrison, Kathleen A., Juan F. Santoyo, Jake H. Davis, et al. "Effortless Awareness: Using Real Time Neurofeedback to Investigate Correlates of Posterior Cingulate Cortex Activity in Meditators' Self-Report." *Frontiers in Human Neuroscience* 7 (2013): 440. doi:10.3389/fnhum.2013.00440.

Hanson, Rick, and Richard Mendius. *Buddha's Brain: The Practical Neuroscience of Happiness, Love, and Wisdom.* Oakland, CA: New Harbinger, 2009.

Hölzel, Britta K., Sara W. Lazar, Tim Gard, et al. "How Does Mindfulness Meditation Work? Proposing Mechanisms of Action from a Conceptual and Neural Perspective." *Perspectives on Psychological Science* 6, no. 6 (2011): 537–559.

Keith, Julian R., Lobsang Rapgay, Don Theodore, et al. "An Assessment of an Automated EEG Biofeedback System for Attention Deficits in a Substance Use Disorders Residential Treatment Setting." *Psychology of Addictive Behaviors* 29, no. 1 (2015): 17–25.

Khen, Nyoshul Rinpoche. *Rest in Natural Great Peace: Songs of Experience.* London: Rigpa, 1987.

Lagopoulos, Jim, Jian Xu, Inge Rasmussen, et al. "Increased Theta and Alpha EEG Activity during Nondirective Meditation." *Journal of Alternative and Complementary Medicine* 15, no. 11 (2009): 1187–1192.

Lutz, Antoine, Heleen A. Slagter, Nancy B. Rawlings, et al. "Mental Training Enhances Attentional Stability: Neural and Behavioral Evidence." *Journal of Neuroscience* 29, no. 42 (2009): 13418–13427.

Malinowski, Peter. "Neural Mechanisms of Attentional Control in Mindfulness Meditation." *Frontiers in Neuroscience* 7 (2013): 8. doi:10.3389/fnins.2013.00008.

Malkin, Gary Scott, and Tracy LaQuey Parker, eds. Internet Users' Glossary. 1993. http://www.rfc-editor.org/rfc/rfc1392.txt.

Mercier, Hugo, and Dan Sperber. "Why Do Humans Reason? Arguments for an Argumentative Theory." *Behavioral and Brain Sciences* 34, no. 2 (2011): 57–74.

Mind and Life Institute. https://www.mindandlife.org.

Rapgay, Lobsang, and Alexander Bystritsky. "Classical Mindfulness: An Introduction to Its Theory and Practice for Clinical Application." *Annals of the New York Academy of Sciences* 1172 (2009): 148–162.

Raymond, Eric, ed. "Hacker." The Jargon File. Accessed February 15, 2017. http://www.catb.org/jargon/html/H/hacker.html.

Ros, Tomas, Bernard J. Baars, Ruth A. Lanius, and Patrik Vuilleumier. "Tuning Pathological Brain Oscillations with Neurofeedback: A Systems Neuroscience Framework." *Frontiers in Human Neuroscience* 8 (2014): 1008. doi:10.3389/fnhum.2014.01008.

Saggar, Manish, Anthony P. Zanesco, Brandon G. King, et al. "Mean-Field Thalamocortical Modeling of Longitudinal EEG Acquired during Intensive Meditation Training." *NeuroImage* 114 (2015): 88–104.

Scheeringa, René, Marcel C. M. Bastiaansen, Karl Magnus Petersson, et al. "Frontal Theta EEG Activity Correlates Negatively with the Default Mode Network in Resting State." *International Journal of Psychophysiology* 67, no. 3 (2008): 242–251.

Schumaker, John F. *The Corruption of Reality: A Unified Theory of Religion, Hypnosis, and Psychopathology.* Amherst, NY: Prometheus, 1995.

Smith, Stephen M., Thomas E. Nichols, Diego Vidaurre, et al. "A Positive-Negative Mode of Population Covariation Links Brain Connectivity, Demographics, and Behavior." *Nature Neuroscience* 18, no. 11 (2015): 1565–1567.

Taylor, Véronique A., Véronique Daneault, Joshua Grant, et al. "Impact of Meditation Training on the Default Mode Network During a Restful State." *Social Cognitive and Affective Neuroscience* 8, no. 1 (2013): 4–14.

Wright, Robert. Buddhism and Modern Psychology (online course). Coursera. Accessed February 15, 2017. https://www.coursera.org/learn/science-of-meditation.

Moral Debates

CHAPTER 19

What Is a Person?

Linda MacDonald Glenn, J.D., LL.M.
Faculty, School of Natural Sciences
California State University Monterey Bay, Seaside

What does it mean to be a person? The designation of personhood is given to those entities who have moral and/or legal status, and the significance of the designation varies depending on whether one is contemplating metaphysical, moral, or legal personhood. In this era of exponential technological advances, our previous understandings and worldviews are challenged; we are witness to the creation of new life-forms and the interconnectedness of current life-forms. The lines between persons and property are becoming increasingly blurry. There are at least three major areas in which technology is challenging and even forcing us to reconsider traditional notions of personhood: fetal personhood, animal rights, and human-machine mergers.

The notion of personhood is not static; it is the result of dynamic, evolving processes. In a manner akin to the "extended mind" philosophy, the notion of personhood is part of a complex system, and as such, fixing absolute boundaries is an exercise in futility. An expanded legal notion of personhood is therefore warranted—there ought to be established, by law, a baseline level of moral and legal status, expanding to be more inclusive, always uplifting and elevating, and never diminishing.

HISTORY AND BACKGROUND

To have moral status is to be worthy of moral consideration; that is, if an entity has moral status, then we are obligated to consider its well-being, needs, and interests. The determination of whether or not something is worthy of moral consideration depends on the worldview or framework, as explored below.

THE GREAT CHAIN OF BEING

The great chain of being (or *scala naturae*, meaning "ladder of being") is a medieval worldview where everything in the universe had a divinely preordained place in a hierarchical order, depicted as a series of links. One of the basic tenets of the Judeo-Christian faith is that "man" is special because he alone is made in the image of God: "above all creatures, he is the object of God's love and attention; the other creatures ... were given for man's use" (Rachels 1990, 87).

This hierarchical, anthropocentric worldview has been the justification for holding *only* human life as special and sacred and also for the idea that other creatures may be used

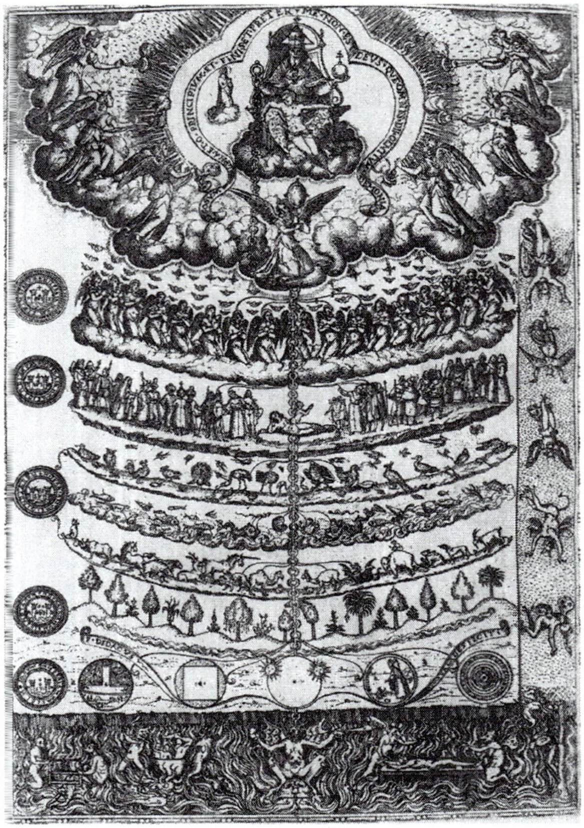

The Great Chain of Being, *by Diego de Valadés, 1533–1582.* The Great Chain of Being *is a visual representation of the Ptolemaic theory, a divinely inspired hierarchical ranking of all life-forms; humans were represented by the male alone. Before the revelations of Nicolaus Copernicus and Galileo Galilei, Earth was believed to be the absolute center of the universe; the sun, moon, stars, and everything else that existed revolved around it and humankind.* PICTORIAL PRESS LTD/ALAMY STOCK PHOTO.

to suit humanity's purposes. This view has led to the rationalized exploitation, abuse, buying, selling, or decimation of all that is not "human" (Glenn 2003; see also Marino 2014).

NOT DOMINION, BUT RESPONSIBILITY AND STEWARDSHIP

Other scholars de-emphasize the domination and subjugation of entities other than humans. These authors write in terms of stewardship, responsibilities and duties of persons as God's moral agents (Ramsey [1950] 1993; Kass 1985; Macer 1999; McCormick 1981).

TRADITIONAL SECULAR PHILOSOPHY

Traditional metaphysical approaches attempt to set forth necessary and sufficient conditions of personhood, such that if an entity meets these conditions, it is a person; if the entity does not meet these conditions, then it is not a person. This section serves as a brief overview of different traditional approaches; for the sake of brevity, a number of sources have been omitted.

Kantian, Rights-Based Approach. Immanuel Kant (1724–1804) was a German philosopher of the Age of Enlightenment, whose work was influenced by ancient Greek philosophers Aristotle and Plato and French philosopher René Descartes (1596–1650). Man's intrinsic worth or dignity, Kant believed, derives from man's ability to be autonomous—a rational agent, capable of making his own decisions, setting his own goals. At the time of Kant's writings, this approach was seen as having laid the groundwork for universal respect for all men—that is, the notion that "all men are created equal." His approach, intended to be inclusive and egalitarian, was considered radical during a time when only men of wealth and property had power.

However, Kant's emphasis on the rational, autonomous being of white men and his silence on the moral status of children, the irrational, or the severely physically or mentally challenged suggests that he did not consider them worthy of human dignity or moral status. In context, Kant was not acting alone but was reflecting part of a worldview that systematically ignored the rights of others or even the thought of rights for others. He wrote during a time when the prevailing worldview was shaped by the great chain of being, as explained above; that worldview reinforced the idea that slaves, women, and children were considered property, not rational persons, and therefore, not worthy of moral status.

The Kantian approach ultimately breaks down because it fails to acknowledge the moral status of (or offer respect for) vulnerable populations—those who cannot speak for themselves. Human dignity, according to this approach, is applicable only to those who can exercise rational, autonomous choices.

The Utilitarian Approach. Classic utilitarian theory, originally proposed by Scottish philosopher David Hume (1711–1776), was developed more fully by English philosophers Jeremy Bentham (1748–1832) and John Stuart Mill (1806–1873). Utilitarian theory seeks to maximize societal utility—that is, to create "the greatest happiness for the greatest number of people." Classic utilitarian theory carefully considers the treatment of nonhumans and argues for moral concern and regard; Bentham asserts, "The question is not, Can they reason? nor, Can they talk? but, Can they suffer?" (quoted in Rachels 1999, 86). As technology progresses, the question of what constitutes pain and suffering needs to be explored not only from a physical basis but also from psychological, sociological, and spiritual bases.

A SUBCLASS OF SERVANTS

In a 1972 article, Episcopalian theologian and bioethicist Joseph Fletcher argued for a list of fifteen "positive propositions" of personhood. These attributes were:

- minimum intelligence
- self-awareness
- self-control
- a sense of time
- a sense of futurity
- a sense of the past
- the capability of relating to others
- concern for others
- communication
- control of existence
- curiosity
- change and changeability
- balance of rationality and feeling
- idiosyncrasy
- neocortical functioning

This extensive list suggests that most individuals, at one time or another, are not persons. Fletcher's comments that a severely cognitively disabled Down syndrome child was not a person and his proposal that chimeras and cyborgs be created to do humanity's distasteful or dangerous work led to severe criticism from his peers and the public (Glenn 2003a).

Fletcher's arguments do not address the issue of suffering, physical or mental, and the excessive stress on rationality and intelligence is arbitrary and degrading to those who are cognitively disabled and senile. This list of characteristics has been seen as a recipe for the creation of a slave race. However, Fletcher's list of traits may be useful if personhood is viewed as a continuum or dynamic process, rather than as a definitive, fixed state—a model that has been proposed philosophically but not yet applied in legal theory or practice.

BUILDING BRIDGES: A NEW WAY FORWARD

The conflict between Kantianism and utilitarianism bled over into an "ideological denial of the relevance of ethics to science" (Rollin 2011, 164), but attitudes have slowly started to evolve, particularly since the 1960s.

AHIMSA: REVERENCE FOR LIFE

Ahimsa is a vow of noninjury to any living thing—especially to animals (Gandhi 2013). In this belief system, adhering to the imperative to treat all living things with reverence increases one's karma and raises one's chance of a higher reincarnation (McClelland 2010).

A more Western view of ahimsa, described as "reverence for life," is found within the philosophy of Albert Schweitzer (1875–1965). During his work in Africa, Schweitzer came to better appreciate the animals there and all of nature's beauty. Schweitzer's words carry the responsibility of doing "as much good as we possibly can to all creatures," in direct contrast to a hierarchical perspective of dominion (Schweitzer 1936).

Schweitzer's philosophy has been interpreted as "radical biological egalitarianism," in which "essential human activities—such as cooking, cleaning, bathing, brushing one's teeth ... are the moral equivalents of mass homicide" (Warren 1997, 37). However, Schweitzer's philosophy was intended to be provocative, not taken at literal face value, and to provide a unifying theory. His ethic was based partially on the works of German philosopher Arthur Schopenhauer (1788–1860), who articulated a worldview challenging the value of existence and argued that the world is, in essence, irrational. Schopenhauer contended that compassion was the key to finding meaning in an otherwise ultimately meaningless world of suffering. Schopenhauer believed that compassion could and would "facilitate an incrementally expanding ethical consciousness in humankind that improves all of society. He also believed that this would eventually bring non-human life into ethical consideration" (Goodin 2011, 54–55). Schweitzer pleaded for a more expansive notion of moral status, one not limited to only human beings.

A NEW SOCIAL ETHIC

Attitudes toward the moral status of animals are changing. A 2015 poll shows that 87 percent of Americans believe that animals have rights and are entitled to protection under the law (Lewis 2015).

Bernard E. Rollin (2011), who has devoted his life to improving the lives of animals, has proposed a new social ethic, based on the telos of animals—that is, their inherent nature, physically and psychologically expressed, which ought to determine how they live in their environments. For example, aside from being free of pain and suffering, chickens and other animals should not be kept in cramped little cages for the purposes of economic efficiency.

DECLARING CONSCIOUSNESS

In July 2012 a prominent international group of cognitive neuroscientists, neuropharmacologists, neurophysiologists, neuroanatomists, and computational neuroscientists gathered at the University of Cambridge to make the following declaration:

> Convergent evidence indicates that non-human animals have the neuroanatomical, neurochemical, and neurophysiological substrates of conscious states along with the capacity to exhibit intentional behaviors. Consequently, the weight of

> evidence indicates that humans are not unique in possessing the neurological substrates that generate consciousness. Non-human animals, including all mammals and birds, and many other creatures, including octopuses, also possess these neurological substrates. (Low et al. 2012)

Simply put, nonhuman animals experience consciousness in the same manner as any human animal does. Of course, many renowned scientists, philosophers, and bioethicists had reached the same conclusion years ago (Bekoff 2012). Nevertheless, some believe that the Cambridge Declaration on Consciousness, though long overdue, provides the gravitas needed in the scientific community to make changes needed in such laws as the Animal Welfare Act of 1966 in the United States.

THE LEGAL ROOTS OF PERSONHOOD AND THE GROWING FAMILY TREE

Legal personhood, in contrast to metaphysical or moral personhood, is how relevant laws define what it means to be a person. Over the centuries, the law has evolved to recognize that all humans are persons, but not all persons are human; the law does not require that metaphysical or moral personhood be a condition of legal personhood.

PERSONS VERSUS PROPERTY

The evolution of the Western legal system over the last millennium generally followed the ideologies of the great chain of being and Kantianism. The law recognized a dichotomy: either you were designated as person or as property. If you were a person, you had rights; if you were property, you did not, and you could be bought, sold, discarded, and treated in whatever manner your owner deemed fit. Women, children, and slaves were considered property, rather than persons, starting with Plato and Aristotle.

Slavery. As laws became codified and incorporated in documents such as the Magna Carta, the status of women improved slightly. The plight of slaves started to change only in 1772 with the famous English slavery case of *Somerset v. Stewart.*

Change occurred more slowly in the colonies of the United States. In 1857, in the infamous case of *Dred Scott v. Sandford*, the US Supreme Court, referring to language in the Declaration of Independence that includes the phrase "all men are created equal," declared that "the enslaved African race were not intended to be included, and formed no part of the people who framed and adopted this declaration." The decision had the effect of equating a slave's legal status with that of domestic livestock; it propelled the United States, under the presidency of Abraham Lincoln, toward the Civil War (1861–1865). After the war, the Supreme Court ruling was rendered impotent by the passage of the Thirteenth and the Fourteenth Amendments to the US Constitution in 1865 and 1868, respectively. In these amendments, Congress abolished slavery and involuntary servitude; expressly granted males liberty, regardless of race or citizenship status and sought to protect these males' civil rights. However, Congress did not extend the right to vote to black males until it adopted the Fifteenth Amendment in 1870 (see Table 19.1 for the relevant text from these amendments).

Women. The status of women did not start to change significantly until the mid- to late nineteenth century. The emergence of the women's movement was linked, temporally and ideologically, with the drive to end slavery (Rierson 1994). Under the laws of the time,

Amendment	*Description*
Thirteenth Amendment (1865), Section 1	Neither slavery nor involuntary servitude, except as a punishment for crime, whereof the party shall have been duly convicted, shall exist within the United States, or any place subject to their jurisdiction.
Fourteenth Amendment (1868), Section 1	All persons born or naturalized in the United States, and subject to the jurisdiction thereof, are citizens of the United States and of the State wherein they reside. No State shall make or enforce any law which shall abridge the privileges or immunities of citizens of the United States; nor shall any State deprive any person of life, liberty, or property, without due process of law, nor deny to any person within its jurisdiction the equal protection of the laws.
Fifteenth Amendment (1870), Section 1	The right of citizens of the United States to vote shall not be denied or abridged by the United States or by any State on account of race, color, or previous condition of servitude.
Nineteenth Amendment (1920)	The right of citizens of the United States to vote shall not be denied or abridged by the United States or by any State on account of sex.

***Table 19.1.* Text from amendments to the US Constitution relevant to the definition of a person.**

neither slaves nor women could go to school or vote; neither could bring cases in court or testify against the master; neither could own property or control their own bodies (Post 1997). The Nineteenth Amendment finally gave women of the United States the right to vote in 1920 (see Table 19.1).

Children. The historical legal status of children has also been turbulent; well into the nineteenth century, a father could enroll his male children in the army and collect the enrollment bounty, betroth his minor female children to persons of his choice, and put his children to work as day laborers on farms or in factories and collect their wage packets. As recently as 1920, a parent who killed a child in administering punishment could claim a legal excuse for homicide in nine states. A father had the power to decide where and with whom his child would reside, as well as to transfer his children by testamentary disposition to someone other than their mother (Woodhouse 1998).

The resistance to accepting children as persons with rights of their own has been illustrated in historic moments, such as the movement to limit child labor. When legislation to protect children from exploitation was proposed, it was greeted by alarmist opposition as an attack on the fundamental rights of parents to control their children.

While parents generally still have broad authority to speak and act on children's behalf, the status and protection of children has improved. Another step in clarifying the status of children is the Convention on the Rights of the Child, a United Nations treaty that came into force in 1990. This convention ensures that courts worldwide look at the "best interests" of the child first, rather than parental rights. The United States, however, has still not ratified the convention. Some of the reasons for this refusal are that ratification would endorse a right to health care, including pre- and postnatal health care for women, a right to education, and paid parental leave—none of which is currently recognized as legal rights under the US Constitution (Attiah 2014; see also Lauria 2015).

HUMANS VERSUS PERSONS (NATURAL AND UNNATURAL ENTITIES)

Traditionally, under the law, all humans are persons, but not all persons are human. Until this point in the chapter, the terms *human* and *person* have been used interchangeably, but the law has actually created two categories of persons: "unnatural" and "natural."

"Unnatural" persons are "juridical" persons—that is, persons who have been created through what is known as a "legal fiction" (Fagundes 2001), a construct used to create rights for convenience and to serve the ends of justice. Juridical persons include such entities as corporations, labor unions, nursing homes, municipalities, government units, and even ships. With regard to the US Supreme Court, the legal fiction construct has typically been used under the guise of the court's interpretation of statutory language. For example, in the 1886 case of *Santa Clara County v. Southern Pacific Railroad Co.*, the US Supreme Court, for the first time—and rather abruptly and without much explanation or analysis—declared that a corporation is protected by the same rights as natural persons for the purposes of the Fourteenth Amendment's equal protection clause.

Currently, according to legal statutes in the United States, "natural" persons are biological beings, limited only to humans, and humans are defined as "member[s] of the species homo sapiens" (1 US Code § 8). But what constitutes the definition of *Homo sapiens*? The definition of species is a hotly debated and contentious issue among scientists, producing reams of publications (Wilson 1999). English naturalist Charles Darwin (1809–1882) argued that "species" are not "real" entities in nature (Mishler 2009; Wilkins 2009). Species concepts range from typological to morphological to phylogenetic; which is the proper definition for the purposes of statutory law? Species boundaries are permeable and not fixed, and there is no persuasive distinction that proves *Homo sapiens* exclusively encompasses the entirety of the human experience.

Between genetic analysis and paleoarchaeological observations, scientists now realize that Neanderthals were our ancestors, too. About 1 to 5 percent of current human DNA can be traced back to the DNA of Neanderthals; *Homo sapiens* is thus part Neanderthal (Dreifus 2017).

EMBRYOS AND FETUSES

The status of the unborn is unclear:

> The US Supreme Court's legalization of abortion in 1973 was based in part on the unborn's never having been recognized in law as a full legal person. At the same time, fetuses have been considered as persons for the purposes of insurance coverage, wrongful-death suits, and vehicular homicide statutes. The legal status of the unborn thus appears to vary from jurisdiction to jurisdiction, from context to context, according to varying purposes. (Steinbock 2011, xiii)

According to the minimalist view, which illustrates the appendage metaphor, the nonviable fetus is little more than a form of the pregnant woman's bodily tissue; it is part of the woman without having a separate identity or status. This view de-emphasizes the importance of the fetus's separate genetic identity and recognizes no moral status; fetal remains are discarded in the same manner as other by-products of surgery—simply thrown away. The metaphor of fetus as property entails quasi-property rights, giving family members the right to dispose of the fetal tissue but not the right to sell or profit from it.

As mentioned earlier, advancements in technology are sure to lead to earlier viability status, particularly if the plans for an artificial womb come to fruition (Abecassis 2016; see also Partridge et al. 2017). The balancing of maternal and fetal interests is currently unavoidable but will be irrelevant once the technology of artificial or exogenic wombs develops further. *Roe v. Wade* (1973) is about control over one's own body and applies only to the right not to be a *gestational* parent. Thus, with regard to the unborn, the courts and/or legislatures will need to revisit the issue of what constitutes a "person."

HUMANITY'S ANIMAL KIN

As mentioned in a previous section, a new social ethic is unfolding. This new social ethic stems partly from the realization of the need for environmental sustainability (as a result of the works of Albert Schweitzer, Aldo Leopold, Rachel Carson, and Bernard E. Rollin) and partly from what some believe is humanity's evolving moral and spiritual character.

Cognitive scientist Steven Pinker, author of *The Better Angels of Our Nature* (2011), argues and provides evidence that human nature has evolved to become less violent. In the section of his book on the "rights revolutions," Pinker writes of

> the growing conviction that animals should not be subjected to unjustifiable pain, injury, and death.... The recognition of animal interests was taken forward by human advocates on their behalf, who were moved by empathy, reason, and the inspiration of the other Rights Revolutions.... The trends are real, and they are touching every aspect of our relationship with our fellow animals. (1034)

Fellow Inhabitants of the Planet. Since the mid-1980s, support for the recognition of the moral status of animals with which humans share the biosphere has grown far and wide. Switzerland, Germany, and Austria have amended their civil codes to declare that animals are not objects and ought not be treated as such under the law. New Zealand banned research on nonhuman hominins (Fitzgerald 2015); India declared dolphins to be "non-human" persons (Hackman 2013). As of the end of 2016, twenty-two jurisdictions had enacted legislation replacing the term *pet owner* with *guardian*, hoping to increase the recognition of animals as individual beings with wants and needs of their own, like a member of one's family or community, as opposed to a "thing" that can be disposed of if it is inconvenient (Guardian Campaign 2017). A professor at Emory University, using MRI scans, gathered neurobiological evidence that dogs experience mental states indistinguishable from humans, inferring that "dogs are people, too" (Berns 2013).

The Health of Earth: Rising Vegetarianism and Laboratory-Grown Meat. Vegetarianism is on the increase. Even among those who do not identify as vegetarians, the consumption of meat is down. This trend partly reflects a growing concern about animal welfare; other motives for reducing or eliminating meat consumption include health, taste, environmental concern, religion, and rebellion against tradition and/or parental authority. Nonetheless, the majority of people support legal measures that would solve the problem of unsustainable and inhumane meat production, by approving laws that force farmers and meatpackers to treat animals more humanely (Pinker 2011). In addition to the treatment of animals, guidelines from the United Nations, the US Department of Agriculture, and the US Department of Health and Human Services indicate that a switch to nonanimal-based proteins is necessary for survival of the planet (Carus 2010; McDaniel 2011).

Market trends reflect these changing sensibilities and attitudes. Between 2011 and 2015, the number of new meat substitute products increased at an annual rate of 24 percent (*Refrigerated and Frozen Foods* 2016). Plant-based protein products aimed at "meat-reducing flexitarians" have become very popular, and the market is rapidly expanding (Michail 2016; Koba 2015). Despite issues of sustainability and these market trends, as Earth's population continues to grow, the global demand for meat continues to rise (Thornton 2010).

In response to this growing demand, several start-up companies have taken up the challenge of providing meat and animal proteins (such as eggs, milk, and cheese) without the cruelty to animals and without the damaging environmental impact (Glenn and D'Agostino 2012). By using the new field of cellular agriculture, these new companies, such as New Harvest, SuperMeat, and Perfect Day Foods, hold the promise of feeding a protein-hungry world and setting new moral standards for the way humans interact with their fellow creatures.

CONTROVERSIAL CHIMERAS AND TRANSGENIC CREATURES

Chimeras are created by artificially combining genetic material from two organisms into a single species; a transgenic organism is created when the genes of one or more species have been transplanted to and incorporated into another species through technological methods (Glenn 2003b, 2013). Chimeras and transgenic organisms represent a significant aspect of current biotechnology research, ranging from developmental biology and disease modeling to regenerative medicine (Dolgin 2016).

The entirety of regulation on biomedical research has been based on the premise of the human-animal dichotomy. Yet, scientists, researchers, ethicists, lawyers, and policymakers are coming to the consensus that the boundaries between humans and animals are not clear-cut and that life is a continuum. This raises the question of why proposed research guidelines, law reviews, and regulations demarcating boundaries between humans and other animals are proliferating (Hinterberger 2016).

Considering the conclusion of the aforementioned Cambridge Declaration on Consciousness, should we not be very careful about assuming that humans are unique in possessing any given trait? Neuroscientific evidence refutes any real distinction between the human species and other animals and repudiates the idea that humans are superior to other animals on the basis of their self-awareness, rationality, capacity for communicating through the use of language, and complexity of being. Any notion of human exceptionalism is more likely driven by cognitive bias (Benvenuti 2016).

ARTIFICIAL INTELLIGENCE AND HUMAN-MACHINE MERGERS

The merger of biological (living) and nonbiological (nonliving) has been heralded as the next stage of evolution (Max 2017). As a by-product of the emergence and acceptance of exponential technologies, evolution is no longer a matter of random chance but indubitably self-directed.

Artificial Intelligence. The Kantian notion of basing personhood on moral status, mentioned in a previous section, is perfectly suited for artificial intelligence (AI); but would AI be capable of sentience? Inventor Ray Kurzweil (2012) argues that there is no sharp distinction between mammalian thought and machine thought and that by 2040 machine intelligence will exceed human intelligence and will have incorporated characteristics of sentience, making AI indistinguishable from the human counterpart. He is not alone in envisioning such a future; affective computing is proceeding rapidly (Kaplan 2017).

Awareness of this inevitability is seeping into the public consciousness via entertainment media. *Westworld*, the HBO television series about artificial consciousness that debuted in the fall of 2016, explores the intersection of sentience, sapience, and free will. Alex Garland's 2014 film *Ex Machina* envisions AI that may be sentient but has no conscience.

Arguably, a justifiable reason to deny personhood to AI is "the desire to reduce or eliminate a threat to the dominance of the human species" (Hubbard 2011, 429). In response to this existential threat, a group of AI researchers, industry leaders, and academics gathered in January 2017 at an Asilomar conference in Pacific Grove, California, and came up with twenty-three principles intended to guide the safe development of AI, including but not limited to safety, privacy, and liberty concerns, as well as shared benefits and prosperity (Future of Life Institute 2017).

Perhaps not surprisingly, the principles did not include a call for transparency of purpose and means, as such transparency would weaken intellectual property rights and undermine economic advantage between competitors. However, because economic inequality is currently a major global ethical challenge, transparency of purpose and means would go a long way toward ensuring the ethical use of AI and that the aforementioned proposed principles would be followed (Tarnoff 2017; Hibbard et al. 2016). Additionally, the lack of transparency of purpose and means also ensures that any AI retains the status of property, rendering it ineligible for consideration of personhood.

Human-Machine Symbiosis (Cyborgs).

> *We and our technological creations are poised to embark on*
> *what is sure to*
> *be a strange and deeply commingled evolutionary path*
>
> —Prabhakar (2017)

While some researchers and scholars have concerned themselves with the implications of AI, others have noted that it is far more likely that human-machine mergers will present more immediate challenges, both morally and legally (Hughes 2004; Barfield 2015). Meanwhile, the progression from hardware to software to wetware has been swift. Examples of this progress since the middle of the first decade of the twenty-first century include brain implants for Parkinson's, obsessive-compulsive disorder, depression, post-traumatic stress disorder, traumatic brain injury, and Alzheimer's and other forms of memory loss (Piore 2015; Price 2017). Kevin Warwick, the self-proclaimed "world's first human cyborg," has called for upgrading humans for a future in space and the creation of robots with biological brains (Warwick et al. 2017).

The examples of the progress that has been made in this area could take up volumes; all these advances call forth questions that humanity will have to consider: How many technological changes does it take to result in an entity that is no longer human? At some point, perhaps because of the nature and extent of the modifications and/or replacements, might this entity be viewed as posthuman rather than human? And if you are no longer "human," are you still a legal person? Is there a point at which you have replaced so much of yourself that you are no longer the same person and should have a different legal identity?

These new technologies are blurring the bodily boundaries of humans. Distinctions between "natural" and "artificial," "alive" and "not alive," or "animate" and "inanimate" are ones that are becoming increasingly difficult to determine. Other areas, which are not yet in the public eye, such as synthetic or non-DNA-based life, extraterrestrial life, noncorporeal entities, and mind uploading, will also offer philosophical and legal quandaries but are outside the scope of this chapter. What is clear, though, is that the traditional dichotomy of persons versus property no longer works; the legal system needs a new paradigm.

EVOLVING LEGAL PARADIGMS

The law, like language, perpetually evolves to meet a society's needs and norms. It not only serves to resolve conflicts or to provide a code of acceptable conduct but also serves as an aspirational lodestone (Glenn 2003a).

PERSONHOOD AND PROPERTY: IDEOGRAPHIC MODELS

The bulk of the scholarly articles and treatises considering the challenges presented by technology advocate for an approach to moral and legal status by degrees, while calling for varying degrees of consideration (Hinterberger 2016; Hubbard 2011; Favre 2010; Berg 2007; Bennett 2006; Glenn 2003a). That is, rather than pure dichotomy, where a thing either has moral and legal status or not, such status is determined by interest and obligations. This approach raises another question: did those interests and obligations of sentient beings merit equal consideration or unequal consideration? Regardless of where one stands on this issue, scientific evidence is accumulating that refutes the persistent, outdated, and obdurate presumption that moral status is all or nothing (DeGrazia 2008).

Points on a Continuum. If one views the concept of personhood on a legal continuum, at one end of the spectrum would be property, such as inanimate objects, land, and those things that cannot suffer; at the other end of the spectrum, rational autonomous yet sentient beings (see Figure 19.1). With the granting of rights to rational autonomous beings comes the burden of responsibility. As "creators," like parents, humans have attendant responsibilities as moral agents—in particular, the extraordinary responsibility of determining the impact of these creations on the community, human and nonhuman, on the biosphere, and potentially beyond, as humanity expands into space.

One of the advantages to applying the property-personhood continuum and a balanced approach would be in the flexibility offered to courts in considering the issues. What facts are relevant? What liberty and/or property interests are at stake? Another advantage that courts would have is the ability to administer a remedy that is proportionate to the rights and interests of those who lack full autonomy. For example, a court can recognize a minimum negative right or liberty interest to maintain bodily integrity and thus be free from enslavement or vivisection, without extending any other positive rights or liberties.

On the flip side, the flexibility of this approach could cause difficulty in that it can be used to justify a cultural relativist approach, strip existing rights from the weak or disabled, and rationalize racism, bigotry, or other hierarchical bias. This is the dark side of the continuum model. The danger of the cultural relativist approach is it could be used to argue that the slave trade was morally acceptable because of the time and norms, that the killing doctors of the Nazi concentration camps did nothing wrong, and that the Tuskegee syphilis researchers were justified in their approach because the victims were less than human.

To prevent such travesties of justice from recurring, the adoption and endorsement of statutory language would be necessary. The specific example here is the language in the United Nations General Assembly's World Charter for Nature (1982), which declares that

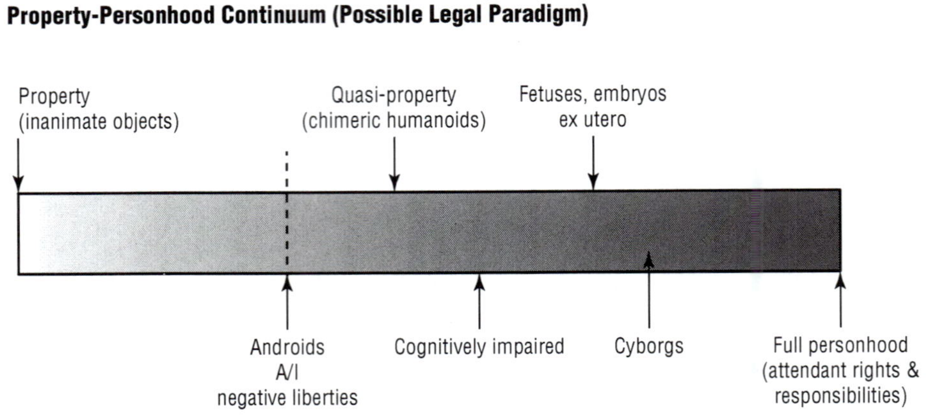

Figure 19.1. *This figure is for illustrative and discussion purposes, not to denote or advocate a particular status for a specific entity.* LINDA GLENN.

"every form of life is unique, warranting respect regardless of its worth to man, and, to accord other organisms such recognition, man must be guided by a moral code of action."

In a manner akin to Albert Schweitzer's ethic centering on an expansion of humanity's moral universe, this language would serve to recognize fundamental interests, such as the liberty, dignity, and worth of each life, "regardless of its worth to man." This ethic is based on a model of an *interdependent whole system,* incorporating incontrovertible principles of justice, fundamental fairness, and reasonableness.

Pyramid of Interdependence. A similar model, hierarchical, but somewhat more reflective of the interdependent nature of the relationships, might serve as a useful ideograph or illustrational construct. This model has much of the flexibility of the continuum model, but the hierarchal characteristic of the pyramid recognizes the fundamental origin of life and underscores the interdependent nature of the evolutionary process and how relationships and rights are built on top of one another. The pyramid model represents a new way of visualizing the law; instead of the law being based on the great chain of being, it is built from the ground up, layer by layer (see Figure 19.2).

EXTENDED PERSONHOOD: A DYNAMIC EVOLVING MODEL OF THE LAW

The above models might be helpful in thinking about relationships and rights, but they have the disadvantage of being somewhat static. Our knowledge and understanding of the world and the universe is ever changing, and nowhere is this more evident than in the field of neuroscience, particularly in notions of the extended mind.

The extended mind thesis holds that an agent's mind and associated cognitive processes are not exclusively in the head of the cognizer, nor even exclusively within the body, but extend into the agent's surroundings or environment. As two well-known philosophers of mind, Andy Clark and David Chalmers, see it: "Where does the mind stop and the rest of the world begin? . . . We propose to pursue . . . an *active externalism,* based on the active role of the environment in driving cognitive processes" (1998, 7). Though initially counterintuitive, this idea makes sense when

Alternative Pyramid of Being

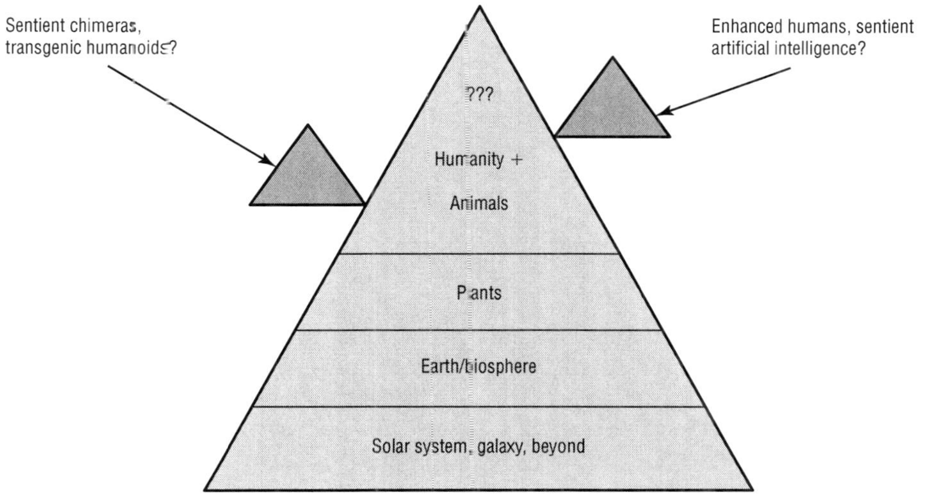

Figure 19.2. *This figure is a visual metaphor for the cumulative, interconnected nature of life-forms on Earth and in the universe.* LINDA GLENN.

one considers how humans have offloaded some of their cognitive load into external technological props, such as smartphones and laptops. The theory that personhood extends beyond the individual and can be applied to external interactive prosthetics has already been applied in one civil case scenario and has served as the motivation and inspiration for the Cyborg Foundation (Glenn 2012; Cyborg Foundation 2017).

The law is a powerful tool and often serves as a repository for expressions of anxiety about divisive social issues. It can actually shape behavior by creating social norms that people use to measure the morality and worth of their actions: "legal rituals [can] make and unmake persons" (Dayan 2011). The law can also be used as a weapon for creating divisions, marginalizing those who are different and depriving individuals of personhood, such as prisoners tortured in the US-run detention camp in Guantánamo, Cuba, or as exemplified by the so-called bathroom bills, which restrict access to restrooms, locker rooms, and other sex-segregated facilities on the basis of a definition of sex or gender consistent with sex assigned at birth or "biological sex." When the law is used in such a destructive manner, persons who are judged outside the law's protection will resort to an alternative understanding of the law, reinforcing further schisms in the community. The law can also be used constructively and restoratively, to encourage harmony, social justice, and healing; in this regard, the emergence of therapeutic jurisprudence is promising.

Therapeutic jurisprudence is "the study of the law as a therapeutic agent" (Stolle, Wexler, and Winick 2000). It focuses on the impact of the law and the legal process on emotional life and psychological well-being. Therapeutic jurisprudence envisions lawyers practicing with an ethic of care and heightened interpersonal skills, who value the psychological well-being of their clients, as well as their legal rights and interests, and who actively seek to prevent legal problems through creative drafting and problem-solving approaches (Stolle, Wexler, and Winick 2000).

Therapeutic jurisprudence holds out hope for infusing into the legal system a milieu that takes into consideration the emotions, behaviors, and mental health of persons and supports an ethics of care, collaboration, and recognition of humans' interdependence. It is within this framework, raising certain questions that might otherwise go unaddressed, that the evolving notions of legal personhood might be able to find a home. These are the new frontiers of justice (Nussbaum 2006).

Summary

In addition to asking, what does it mean to be a person? an equally important question is what kind of persons do we want to be? What kind of future do we want to create, and what do we want to leave as our legacy for our children, grandchildren, and other future inhabitants of this planet and beyond? What value do we place on sentient life versus owning property, if that property is sentient? Do we emphasize the Golden Rule or "he who has the gold rules"?

Despite the turn of political events in the United States and other parts of the world, some observers are optimistic that our morality is evolving and that we will continue to expand our moral and legal universe to include sentient beings. Humanity has made moral progress; in an inspirational essay, Michael Shermer contends that

> over time, the idea that individual sentient beings have natural rights has outcompeted other ideas that place the group, tribe, nation, race, gender, or religion above the rights of the individual. These rights have expanded around the globe because individual sentient beings want them, and they want them because it is part of their nature to want them—it is instinctive—and a proper scientific understanding of human nature has revealed this fact. (Shermer 2016, 61–62)

To gain a further scientific, objective understanding and to gather evidence about the thoughts, minds, and lives of those sentient creatures with whom we share the biosphere, several people have proposed that we create a global, open-source, and transparent interdisciplinary study of interspecies and bio-inspired communications, which would include advancing artificial intelligence (Bekoff 2012; Benvenuti 2016; Favre 2010; Fitzgerald 2015; Herzing 2013; Nussbaum 2006; Rollin 2011; Shermer 2016; Shyam 2015; Wise 2000). This would amass support for a moral starting point for cultivating the survival and flourishing of sentient beings.

Albert Schweitzer and Arthur Schopenhauer, among others, believed that compassion is key to the moral and spiritual growth of humanity and humanity plus, that the way we treat ourselves and others is a reflection of the way we treat the universe, that what we give out is what we get back, and that the law should express our most noble aspirations. These values and aspirations continue to gather strength from modern philosophers, lawyers, bioethicists, theologians, and the general citizenry, all of whom can and should provide input to prudent changes in the legal system. Until such legal changes are made, we can expect intense cross-disciplinary debate and discussion as new sentient life-forms are created through science and medicine and recognized legally, morally, and ethically.

Bibliography

Abecassis, Marion. "Artificial Wombs: 'The Third Era of Human Reproduction' and the Likely Impact on French and US Law." *Hastings Women's Law Journal* 27 (2016): 3–109.

Attiah, Karen. "Why Won't the U.S. Ratify the U.N.'s Child Rights Treaty?" *Washington Post*, November 21, 2014. https://www.washingtonpost.com/blogs/post-partisan/wp/2014/11/21/why-wont-the-u-s-ratify-the-u-n-s-child-rights-treaty/.

Barfield, Woodrow. *Cyber-Humans: Our Future with Machines*. New York: Springer, 2015.

Bekoff, Marc. "Welcome to Our World." *New Scientist* 215, no. 2883 (2012): 24–25.

Bennett, D. Scott. "Chimera and the Continuum of Humanity: Erasing the Line of Constitutional Personhood." *Emory Law Journal* 55, no. 2 (2006): 347–387.

Benvenuti, Anne. "Evolutionary Continuity and Personhood: Legal and Therapeutic Implications of Animal Consciousness and Human Unconsciousness." *International Journal of Law and Psychiatry* 48 (2016): 43–49.

Berg, Jessica. "Elephants and Embryos: A Proposed Framework for Legal Personhood." *Hastings Law Journal* 59 (2007): 369–406.

Berns, Gregory. "Dogs Are People, Too." *New York Times*, October 5, 2013. http://www.nytimes.com/2013/10/06/opinion/sunday/dogs-are-people-too.html.

Carus, Felicity. "UN Urges Global Move to Meat and Dairy-Free Diet." *Guardian* (London), June 2, 2010. https://www.theguardian.com/environment/2010/jun/02/un-report-meat-free-diet.

Chandler-Garcia, Lynne Marie. "Who Is a Person and Why? A Study of Personhood in Theory and the Law." PhD diss., University of Maryland, College Park, 2012. http://drum.lib.umd.edu/handle/1903/12617.

Clark, Andy, and David Chalmers. "The Extended Mind." *Analysis* 58, no. 1 (1998): 7–19. doi:10.1093/analys/58.1.7.

Convention on the Rights of the Child. November 20, 1989. *United Nations Treaty Series* 1577 (1990): 3–177. https://treaties.un.org/doc/Publication/UNTS/Volume%201577/v1577.pdf.

Cowell, Alan. "After 350 years, Vatican Says Galileo Was Right: It Moves." *New York Times*, October 31, 1992. http://www.nytimes.com/1992/10/31/world/after-350-years-vatican-says-galileo-was-right-it-moves.html.

Cyborg Foundation. "About Neil Harbisson; About Moon Ribas." Accessed May 12, 2017. http://www.cyborgfoundation.com/about.

Dayan, Colin. *The Law Is a White Dog: How Legal Rituals Make and Unmake Persons*. Princeton, NJ: Princeton University Press, 2011.

DeGrazia, David. "Moral Status as a Matter of Degree?" *Southern Journal of Philosophy* 46, no. 2 (2008): 181–198.

Dolgin, Elie. "Chimeras Keep Courting Controversy." *Proceedings of the National Academy of Sciences of the United States of America* 113, no. 43 (2016): 11984–11985.

Dred Scott v. Sandford, 60 US 393 (1857).

Dreifus, Claudia. "What Did Neanderthals Leave to Modern Humans? Some Surprises." *New York Times*, January 20, 2017. https://www.nytimes.com/2017/01/20/science/john-anthony-capra-neanderthals-dna-humans.html.

Fagundes, David. "What We Talk about When We Talk about Persons: The Language of a Legal Fiction." *Harvard Law Review* 114, no. 6 (2001): 1745–1768.

Favre, David. "Living Property: A New Status for Animals within the Legal System." *Marquette Law Review* 93, no. 3 (2010): 1021–1070.

Fitzgerald, Emily A. "[Ape]rsonhood." *Review of Litigation* 34, no. 2 (2015): 337–378.

Fletcher, Anthony. "Women and Religion." In *Gender, Sex, and Subordination in England, 1500–1800*, 347–363. New Haven, CT: Yale University Press, 1995.

Fletcher, Joseph. "Indicators of Humanhood: A Tentative Profile of Man." *Hastings Center Report* 2, no. 5 (1972): 1–4.

Future of Life Institute. "Asilomar AI Principles." 2017. https://futureoflife.org/ai-principles/.

Gandhi, Sohan Lal Jain. "The Jain Principle of *Ahimsa* (Nonviolence) and Ecology." *Journal of Oriental Studies* (Tokyo) 23 (2013): 166–177. http://www.iop.or.jp/Documents/1323/Sohan%20Lal%20Jain%20Gandhi.pdf.

Glenn, Linda MacDonald. "Biotechnology at the Margins of Personhood: An Evolving Legal Paradigm." *Journal of Evolution and Technology* 13 (2003a). http://jetpress.org/volume13/glenn.html.

Glenn, Linda MacDonald. "Case Study: Ethical and Legal Issues in Human Machine Mergers (or the Cyborgs Cometh)." *Annals of Health Law* 21, no. 1 (2012): 175–179.

Glenn, Linda MacDonald. "Ethical Issues in Genetic Engineering and Transgenics." *ActionBioscience*, November 2013. http://www.actionbioscience.org/biotechnology/glenn.html.

Glenn, Linda MacDonald. "When Pigs Fly? Legal and Ethical Issues in Transgenics and the Creation of Chimeras." *Physiologist* 46, no. 5 (2003b): 251, 253–255.

Glenn, Linda MacDonald, and Lisa D'Agostino. "The Moveable Feast: Legal, Ethical, and Social Implications of Converging Technologies on Our Dinner Tables." *Northeastern University Law Journal* 4, no. 1 (2012): 111–133.

Goodin, David K. "Albert Schweitzer's Reverence for Life Ethic in Relation to Arthur Schopenhauer and Friedrich Nietzsche." PhD diss., McGill University, Montreal, 2011. http://digitool.library.mcgill.ca/thesisfile103600.pdf.

Gregg, Justin. "You Had Me at 'Cybernetic Dolphins.'" *Future Tense* (blog), *Slate*, November 26, 2013. http://www.slate.com/blogs/future_tense/2013/11/26/google_cybernetic_dolphins_can_technology_create_human_dolphin_communication.html.

Guardian Campaign. "Do You Live in a Guardian Community?" Accessed March 3, 2017. http://guardiancampaign.org/guardiancity.html.

Hackman, Jason. "India Declares Dolphins 'Non-human Persons,' Dolphin Shows Banned." *Daily Kos* (blog), July 30, 2013. http://www.dailykos.com/story/2013/7/30/1226634/-India-Declares-Dolphins-Non-Human-Persons-Dolphin-shows-BANNED.

Herzing, Denise. "Could We Speak the Language of Dolphins?" Filmed February 2013. TED Video, 14:38. Posted June 6, 2013. https://www.ted.com/talks/denise_herzing_could_we_speak_the_language_of_dolphins.

Hibbard, Bill. "The Asilomar AI Principles Should Include Transparency about the Purpose and Means of Advanced AI Systems." *H+ Magazine*, February 2, 2017. http://hplusmagazine.com/2017/02/02/asilomar-ai-principles-include-transparency-purpose-means-advanced-ai-systems/.

Hibbard, Bill, Nick Baladis, Ben Goertzel, et al. "Humans for Transparency in Artificial Intelligence." *H+ Magazine*, March 11, 2016. http://hplusmagazine.com/2016/03/11/transparent-ai/.

Hinterberger, Amy. "Regulating Estrangement: Human–Animal Chimeras in Postgenomic Biology." *Science, Technology, and Human Values*. Published electronically December 26, 2016. doi:10.1177/0162243916685160.

Hubbard, F. Patrick. "'Do Androids Dream?': Personhood and Intelligent Artifacts." *Temple Law Review* 83, no. 2 (2011): 405–474.

Hughes, James. *Citizen Cyborg: Why Democratic Societies Must Respond to the Redesigned Human of the Future*. Cambridge, MA: Westview Press, 2004.

Kaplan, Jerry. "Artificial Intelligence: Think Again." *Communications of the ACM* 60, no. 1 (2017): 36–38. doi:10.1145/2950039.

Kass, Leon R. *Toward a More Natural Science: Biology and Human Affairs*. New York: Free Press, 1985.

Koba, Mark. "Fake Meat Sales Are Growing, but Is It Really Better for You?" *Fortune*, May 11, 2015. http://fortune.com/2015/05/11/meatless-meat-sales/.

Kurzweil, Ray. *How to Create a Mind: The Secret of Human Thought Revealed*. New York: Viking, 2012.

Lauria, Joe. "Why Won't the US Ratify the UN's Children's Rights Convention?" *Huffington Post*, January 25, 2015. http://www.huffingtonpost.com/joe-lauria/why-wont-the-us-ratify-th_b_6195594.html.

Lewis, Tanya. "Many Americans Support Equal Rights for Animals." Live Science. May 22, 2015. http://www.livescience.com/50889-animal-rights-poll.html.

Lovejoy, Arthur O. *The Great Chain of Being: A Study of the History of an Idea*. Cambridge, MA: Harvard University Press, 1936.

Low, Philip, Jaak Panksepp, Diana Reiss, et al. "The Cambridge Declaration on Consciousness." Presented at the Francis Crick Memorial Conference on Consciousness in Human and Non-human Animals, Churchill College, University of Cambridge, July 2012. http://fcmconference.org/img/CambridgeDeclarationOnConsciousness.pdf.

Macer, Darryl R. J. "Bioethics and Sustainable Development." In *World Development: Aid and Foreign Direct Investment 1999/2000*, edited by A. J. Fairclough, 112–114. London: Kensington, 1999.

Madrigal, Alexis C. "Is Google's Secretive Research Lab Working on Human-Dolphin Communication?" *Atlantic*, November 21, 2013. https://www.theatlantic.com/technology/archive/2013/11/is-googles-secretive-research-lab-working-on-human-dolphin-communication/281701/.

Marino, Lori. "The Scala Naturae Is Alive and Well in Modern Times." *Huffington Post*, April 6, 2014. http://www.huffingtonpost.com/lori-marino-phd/the-scala-naturae-is-aliv_b_4719171.html.

Max, D. T. "How Humans Are Shaping Our Own Evolution." *National Geographic*, April 2017. http://www.nationalgeographic.com/magazine/2017/04/evolution-genetics-medicine-brain-technology-cyborg/.

McClelland, Norman C. *Encyclopedia of Reincarnation and Karma*. Jefferson, NC: McFarland, 2010.

McCormick, Richard A. *How Brave a New World? Dilemmas in Bioethics*. Garden City, NY: Doubleday, 1981.

McDaniel, Mac. "New USDA Guidelines Praise Vegetarian Diets." Care2. February 2, 2011. http://www.care2.com/causes/new-usda-guidelines-praise-vegetarian-diets.html.

Michail, Niamh. "Unilever to Add Vegetarian Logo to 500 Products." FoodNavigator.com. November 4, 2016. http://www.foodnavigator.com/Market-Trends/Unilever-to-add-vegetarian-logo-to-500-products.

Mishler, Brent D. "Species Are Not Uniquely Real Biological Entities." *Contemporary Debates in Philosophy of Biology*, edited by Francisco J. Ayala and Robert Arp, 110–122. Chichester, UK: Wiley-Blackwell, 2009.

New Harvest. http://www.new-harvest.org.

Nussbaum, Martha C. *Frontiers of Justice: Disability, Nationality, Species Membership*. Cambridge, MA: Belknap Press of Harvard University Press, 2006.

Partridge, Emily A., Marcus G. Davey, Matthew A. Hornick, et al. "An Extra-uterine System to Physiologically Support the Extreme Premature Lamb." *Nature Communications* 8, art. 15112 (2017). doi:10.1038/ncomms15112.

Perfect Day Foods. http://www.perfectdayfoods.com.

Pinker, Steven. *The Better Angels of Our Nature: Why Violence Has Declined*. New York: Viking, 2011.

Piore, Adam. "A Shocking Way to Fix the Brain." *MIT Technology Review*, October 8, 2015. https://www.technologyreview.com/s/542176/a-shocking-way-to-fix-the-brain.

Post, Dianne. "Why Marriage Should Be Abolished." *Women's Rights Law Reporter* 18, no. 3 (1997): 283–313.

Prabhakar, Arati. "The Merging of Humans and Machines Is Happening Now." *Wired*, January 27, 2017. http://www.wired.co.uk/article/darpa-arati-prabhakar-humans-machines.

Preece, Gordon, ed. *Rethinking Peter Singer: A Christian Critique*. Downers Grove, IL: InterVarsity Press, 2002.

Price, Jack. "Mending Minds." *New Scientist* 233, no. 3109 (2017): 36–39.

Rachels, James. *Created from Animals: The Moral Implications of Darwinism*. Oxford: Oxford University Press, 1990.

Rachels, James. *The Elements of Moral Philosophy*. 3rd ed. Boston: McGraw-Hill College, 1999.

Ramsey, Paul. *Basic Christian Ethics*. Louisville, KY: Westminster/John Knox Press, 1993. First published 1950 by Scribner.

Refrigerated and Frozen Foods. "Data Shows Rise in Vegetarian Claims." July 8, 2016. http://www.refrigeratedfrozenfood.com/articles/91183-data-shows-rise-in-vegetarian-claims.

Rierson, Sandra L. "Race and Gender Discrimination: A Historical Case for Equal Treatment under the Fourteenth Amendment." *Duke Journal of Gender Law and Policy* 1 (1994): 89–117.

Roe v. Wade, 410 US 113 (1973).

Rollin, Bernard E. *The Frankenstein Syndrome: Ethical and Social Issues in the Genetic Engineering of Animals*. Cambridge: Cambridge University Press, 1995.

Rollin, Bernard E. *Putting the Horse before Descartes: My Life's Work on Behalf of Animals*. Philadelphia: Temple University Press, 2011.

Santa Clara County v. Southern Pacific Railroad Co., 118 US 394 (1886).

Schopenhauer, Arthur. *On the Basis of Morality*. Translated by E. F. J. Payne. Indianapolis, IN: Bobbs-Merrill, 1965.

Schweitzer, Albert. "The Ethics of Reverence for Life." *Christendom* 1, no. 2 (1936): 225–239. http://www.animal-rights-library.com/texts-c/schweitzer01.pdf.

Shermer, Michael. "Morality Is Real, Objective, and Natural." *Annals of the New York Academy of Sciences* 1384 (2016): 57–62.

Shyam, Geeta. "The Legal Status of Animals: The World Rethinks Its Position." *Alternative Law Journal* 40, no. 4 (2015): 266–270.

Somerset v. Stewart, Lofft 1–18; 11 Harg. State Trials 339; 20 Howell's State Trials 1, 79–82; 98 Eng Rep 499–510 (King's Bench, June 22, 1772).

Steinbock, Bonnie. *Life before Birth: The Moral and Legal Status of Embryos and Fetuses*. 2nd ed. Oxford: Oxford University Press, 2011.

Stolle, Dennis P., David B. Wexler, and Bruce J. Winick, eds. *Practicing Therapeutic Jurisprudence: Law as a Helping Profession*. Durham, NC: Carolina Academic Press, 2000.

SuperMeat. http://www.supermeat.com.

Tarnoff, Ben. "Robots Won't Just Take Our Jobs—They'll Make the Rich Even Richer." *Guardian* (London), March 2, 2017. https://www.theguardian.com/technology/2017/mar/02/robot-tax-job-elimination-livable-wage.

Thornton, Philip K. "Livestock Production: Recent Trends, Future Prospects." *Philosophical Transactions of the Royal Society* B 365, no. 1554 (2010): 2853–2867.

United Nations General Assembly. Resolution 37/7, "World Charter for Nature." October 28, 1982. http://www.un.org/documents/ga/res/37/a37r007.htm.

Warren, Mary Anne. *Moral Status: Obligations to Persons and Other Living Things*. Oxford: Clarendon Press, 1997.

Warwick, Kevin. "The Cyborg Experiments" (video). Humanoid Productions. 2015. http://www.humanoid.uk/news/the-cyborg-experiments/.

Warwick, Kevin, Arne Hendriks, Rachel Armstrong, and Sarah Jane Pell. "Space Bodies." In *Star Ark: A Living, Self-Sustaining Spaceship*, edited by Rachel Armstrong, 341–382. New York: Springer, 2017.

Wilkins, John S. *Species: A History of the Idea*. Berkeley: University of California Press, 2009.

Wilson, Robert A., ed. *Species: New Interdisciplinary Essays*. Cambridge, MA: MIT Press, 1999.

Wise, Steven M. *Rattling the Cage: Toward Legal Rights for Animals*. Cambridge, MA: Perseus Books, 2000.

Woodhouse, Barbara Bennett. "From Property to Personhood: A Child-Centered Perspective on Parents' Rights." *Georgetown Journal on Fighting Poverty* 5, no. 2 (1998): 313–319.

Yamada, David C. "Therapeutic Jurisprudence and the Practice of Legal Scholarship." *University of Memphis Law Review* 41, no. 1 (2010): 121–156.

FILMS AND TELEVISION

Ex Machina. Dir. Alex Garland. 2014.

Westworld. Created by Jonathan Nolan and Lisa Joy. 2016–.

CHAPTER 20

The Debates over Enhancement

Walter Glannon
Professor, Department of Philosophy
University of Calgary, Canada

Genetic engineering, nanotechnology, psychoactive drugs, neurostimulation, and other techniques have the potential to produce more intelligent and healthier humans with more fulfilling lives. By enhancing our physical and intellectual capacities, these interventions in our bodies and brains may overcome many of the biological limits that evolution has imposed on us (Harris 2007). While some proponents of enhancement claim that it is not necessarily part of a transhumanist agenda (Harris 2007), others effectively link the two concepts. Nick Bostrom articulates this view as follows: "Transhumanists hope that by responsible use of science, technology, and other rational means, we shall eventually manage to become posthuman, beings with vastly greater capacities than present human beings have" (2005, 4). Bostrom defends what he calls "extreme human enhancement," which "could result in 'posthuman' modes of being" (2008, 107). Others have called this project "radical human enhancement" (Agar 2014, 1), defining it as the substantial alteration of capacities beyond what is considered normal for the human species. Posthuman modes of being may include not just enhanced physical and intellectual capacities but also the creation of "superintelligent machines" that "could profoundly alter the human condition" (Bostrom 2005, 3).

The difference between a transhuman and a posthuman world may be a matter of degree rather than kind. It may be a function of the extent to which the relevant capacities are enhanced and how the changes in these capacities affect how we conceive of ourselves along a continuum of being. Although Bostrom mentions "responsible use" of science and technology for this purpose (2005, 4), one cannot assume that altering the body and brain would not have any adverse physiological and psychological effects. If it were safe and effective, radical enhancement could fundamentally alter our identities as human beings and the values most of us want to preserve in our lives. Those who chose to undergo radical enhancement might not be the same people at the end of this process. Yet losing one's identity as the same person persisting through time and as a member of the human species could be part of a plan to become posthuman. The general presumption of enhancement is that it would give us more control over our lives. Neural prosthetics could supplement the limited capacity of brain circuits to enhance information processing, reasoning, and decision making. But artificial systems might completely replace natural neural circuits and cause us to lose this control by taking over our agency. The most worrisome technology in this regard may be "superintelligent machines" in the form of artificial intelligence.

This chapter discusses the enhancement of cognitive capacities through psychoactive drugs and neurostimulating devices. This has been the focus of the debate on enhancement involving philosophers and bioethicists. In the discussion herein, three different definitions of cognitive bioenhancement are considered. Examples of the risks and cognitive trade-offs associated with altering normally functioning brains are cited, as are how these limitations might be overcome. The balance of the chapter is a critical analysis and discussion of the ethical claims and arguments against and for cognitive enhancement. Examined are the views that cognitive enhancement threatens core properties of human nature, makes our actions inauthentic and alienates us from our true selves, promotes perfectionism, and undermines our sense of excellence. The penultimate and concluding sections of the chapter explore the implications of radical enhancement through neural prosthetics and intelligent machines for individual and species identity and behavior control. These issues are ethically, psychologically, and socially fraught because it may not be possible to predict the effects—positive or negative—that the technology might have.

THE MEANINGS AND SCOPE OF COGNITIVE BIOENHANCEMENT

Cognitive bioenhancement refers to interventions in the brain to improve alertness, concentration, and information processing in executive functions, such as reasoning and decision making. There are three main conceptions of cognitive enhancement: *augmenting*, *diminishing*, and *optimizing*. The first conception considers interventions in the brain as enhancements when they augment some function by increasing its ability to do what it normally does (Harris 2007; Bostrom and Sandberg 2009; Savulescu and Bostrom 2009). An enhancement is an intervention "designed to improve human form or function beyond what is necessary to restore or sustain good human health" (Juengst 1998, 29). The second conception of enhancement says that some functions can be improved by diminishing what they do and their effects. As explained by one group of authors: "Sometimes the *diminishment* of a capacity or function, under the right set of circumstances, could plausibly contribute to an individual's overall well-being; more is not always better, and sometimes less is more" (Earp et al. 2014). For example, methylphenidate (Ritalin) may help some people perform better on a particular cognitive task because the drug's effects reduce the content of their thought, enabling them to avoid being distracted by stimuli and remaining focused on that task. The third conception of enhancement refers to any brain intervention that "aims at optimizing a specific class of information-processing functions: *cognitive* functions, physically realized by the human brain" (Metzinger and Hildt 2011, 245).

A broad optimizing conception is probably the one that is most consistent with people's intuitions about cognitive enhancement. The general goal of enhancing cognitive functions is not just to improve performance on a particular task or a few tasks but also to promote flexible behavior and adaptability to the environment. This is more likely to occur when neural and mental processes are neither overactive nor underactive. Optimal levels of cognitive functions can be produced by augmenting or diminishing certain aspects of them. The term *optimal*, however, suggests that there may be limits to the extent to which bioenhancement can improve these functions. This is germane to both the question of whether interventions in the brain have enhancing effects and what the risks of these interventions are. While questions about the ratio of potential benefits to risks have not been the main focus of the philosophical debate about enhancement, they have to be addressed before considering the more philosophically contentious questions on this topic.

THE RISKS OF BRAIN TINKERING

Studies suggest that there may be trade-offs in cognitive enhancement. There can be improvement in some mental capacities and impairment in others, even from the same drug. In a number of studies involving subjects with normal brain function, moderate doses of methylphenidate slightly improved performance on certain mental tasks. Higher doses either impaired or did not affect cognitive performance (Farah et al. 2004; de Jongh et al. 2008). The drug may enhance executive functions on novel tasks but can impair these functions on tasks that have been learned. An experiment using transcranial electrical stimulation produced similar results. Testing how the stimulation of certain brain regions could affect the learning and application of mathematical information, researchers found that stimulating an area of the subjects' prefrontal cortex impaired learning new information but enhanced the application of what was learned. Stimulating an area of the parietal cortex had the opposite effect of enhancing learning while impairing the ability to apply the new information (Iuculano and Cohen Kadosh 2013).

Occasional use of drugs to enhance cognitive functions would not necessarily have negative effects on human bodies or brains. But chronic use of drugs such as Ritalin could raise concentrations of the neurotransmitter dopamine to high levels. It could overactivate the brain's reward system, resulting in addictive behavior, such as compulsive gambling and hypersexuality (Heinz et al. 2012). The same risk of addiction is associated with chronic use of modafinil (Provigil) (Volkow et al. 2009). This drug is used therapeutically for narcolepsy and other sleep-related disorders. It has been used nontherapeutically to enable people with normal sleep-wake cycles to remain alert and focused for long periods despite sleep deprivation. Prolonged sleep deprivation can be a risk factor for metabolic disorders such as diabetes, hypertension, and cardiovascular disease. Alternations between sleep and attention are adaptations to the environment. If the brain senses that constant attention is a sign of constant demand, then this could overload it. Constant manipulation of sleep-wake cycles with modafinil could cause them to become dysregulated and result in the metabolic disorders just mentioned.

In a case reported in 2013, a university student in the United States died from complications associated with chronic use of the stimulant Adderall to enhance his capacity to concentrate while studying for exams (Schwarz 2013). This and other examples of harmful side effects from psychostimulants cast suspicion on the suggestion in the 2011 thriller *Limitless* that there need not be any limit on the extent to which cognitive abilities can be enhanced. In this film, the main character, Eddie Morra, a struggling author, overcomes his writer's block and becomes successful in writing and other endeavors after taking the nootropic or "cognition-enhancing," drug NZT-48. Although the film highlights some of the trade-offs of cognitive enhancement through pharmacological means, it does not accurately reflect the long-term risks associated with chronic use of psychotropic drugs. Of course, some exaggeration is to be expected in a film described as a "thriller." Some who use psychostimulants become disillusioned with them. Here is one example: "Like many of my friends, I spent years using prescription stimulants to get through school and start my career. Then I tried to get off them." This person's desire to stop using the drugs was motivated both by the side effects she experienced and the feeling of being dependent on them (Schwartz 2016).

Deep-brain stimulation has been used to treat patients with movement disorders such as Parkinson's disease and psychiatric disorders such as depression and obsessive-compulsive disorder. In this technique, electrodes are implanted in dysfunctional brain regions causing symptoms. The electrodes are connected by wires to a pulse generator implanted under the collarbone, and the pulse generator sends a current to the electrodes to modulate brain activity (Benabid 2007). This technique has also been used to enhance cognitive functions, such as memory. Activating brain circuits beyond optimal levels could have pathological consequences, however. An actual case illustrates this point. A patient with severe anxiety and depression became less anxious and experienced improved mood after undergoing deep-brain stimulation. He asked his psychiatrist to increase the voltage of the stimulator so that he could feel "even better." This caused him to feel "unrealistically good" and "overwhelmed by a feeling of happiness and ease" (Synofzik, Schlaepfer, and Fins 2012, 34). Yet he retained insight into his condition and knew that increased stimulation entailed a risk of losing control of his thought and behavior. Accordingly, he agreed to have the voltage reduced to the previous therapeutic level. This case supports the optimizing conception of enhancing neural functions and the mental capacities they regulate. Beyond a certain critical level, tinkering with the brain and mental states to make them "better" can result in pathological behavior.

LIMITATIONS OF BIOENHANCEMENT

What do these limitations on enhancing mental capacities imply for the augmenting conception of enhancement associated with transhumanism? Some might claim that they doom the transhumanist idea that "the range of thoughts, feelings, experiences, and activities accessible to human organisms presumably constitute only a tiny part of what is possible" (Bostrom 2005, 4). There may be adaptive reasons for limits in enhancing the functions of the brain/mind. Trying to free ourselves from these limits might not only impair our ability to meet the demands of the environment but also cause us to harm ourselves. Theoretically, though, there is no reason why optimal levels of brain and mental functions have to be fixed. It is possible that they could be raised in a way that would safely improve our intellectual and other cognitive capacities, while also enhancing our ability to adapt and respond more effectively to real and potential threats from the external world. "Smart" psychotropic drugs and neurostimulating devices could more precisely target brain circuits, increase synaptic connectivity, and thereby increase neural and mental processing with salutary and no deleterious effects on thought and behavior. Drugs or electrical stimulation could enhance neuroplasticity, the brain's ability to change and reorganize itself. They could allow the brain to gradually adjust to storing more information and functioning more efficiently with a broader range of cognitive functions taken to higher levels. These changes would likely be incremental, but they could in principle overcome current limitations in altering brain processes and the cognitive capacities they enable. The potential of technology to achieve this goal is consistent with Bostrom's claim that "current humanity need not be the endpoint of evolution" (2005, 4).

ETHICAL ARGUMENTS AGAINST AND FOR ENHANCEMENT

If one assumes that drug-induced or device-induced enhancement of mental capacities would be safe and effective, even on an augmenting conception of enhancement, this leaves the ethical question of whether we should enhance these capacities in terms of how it would affect core aspects of human psychology, including what it means to be human. Opponents of cognitive enhancement object to it for four main reasons:

1. it could fundamentally change human nature and thus undermine its intrinsic value;
2. it makes our actions inauthentic because a drug or technique alien to our true selves is the agent of our behavior;
3. it is symptomatic of a hubristic quest for perfection and a failure to accept that much of what we do and encounter in life is beyond our control; and
4. it undermines our understanding and appreciation of excellence.

In all four respects, cognitive enhancement can have detrimental rather than beneficial effects on our character and values and make us worse off than we would be without it.

THE ARGUMENT FROM HUMAN NATURE

The argument for the intrinsic value of human nature and a prohibition against tinkering with it assumes that the biological and psychological capacities that essentially define us as human beings are inherently good (President's Council on Bioethics 2003). Yet the fact that the design of our bodies and brains makes us susceptible to a number of physical and mental diseases that harm us shows that our biological nature is not always good in any sense of the term. Enhancing our intellectual capacity to process larger amounts of information could enable us to predict the future more accurately. Information about foreseeable pathogens, for example, could enable us to develop vaccines and avoid pandemics that could disable or kill us. In this and other respects, there may be an evolutionary advantage to cognitive enhancement. It may enable us to adapt faster to environmental changes that could threaten our survival. Some would argue that if we do not take any of the risks associated with extreme enhancement of our cognitive capacities, then we may be more vulnerable to harm from natural factors beyond our control (Harris 2007).

Opponents of neurocognitive enhancement claim that it would undermine our capacity to make normative judgments about right and wrong—good and bad—actions because this capacity is an essential feature of our nature that enhancement would alter. Yet instead of being an innate disposition, the capacity to make normative judgments may have evolved as a consequence of our living and interacting with others in the world. Our habit of making judgments about the presumed goodness of our nature suggests that we have developed a conception of good that is independent of our nature, or at least that it is not entirely a function of it. Allen Buchanan spells out the main problem with the naturalist position as interpreted by Leon R. Kass (1939–) and other bioethicists: "If a biomedical intervention had the unintended consequence of destroying the capacity for judging goodness, then it would follow trivially that *this* alteration of human nature undercut our capacity for judging goodness. But it would not follow that *any* alteration of our nature would undercut that capacity" (2009, 150). It is also worth emphasizing the more general point that those who appeal to human nature in criticizing enhancement ignore the fact that there are both salutary and deleterious aspects of our biological and psychological design.

THE ARGUMENT FROM INAUTHENTICITY AND ALIENATION

The President's Council on Bioethics (chaired by Kass), in its 2003 report titled *Beyond Therapy*, claims that cognitive enhancement can make us inauthentic and alienate us from our true selves. "As the power to transform our native powers increases, both in magnitude and refinement, so does the possibility for 'self-alienation'—for losing, confounding, or abandoning our identity" (294). The report asserts that the difference between enhancement of intellectual capacities through drugs and through one's own efforts is that the drugs "make improvements to our performance less intelligible. . . . *On the plane of human experience and understanding*, there is a difference between changes in our body that proceed through self-direction and those that do not" (128–129). According to this report, what causes inauthenticity and self-alienation is that cognition-enhancing drugs undermine the very meaning of performance and achievement by undermining human agency. "Biomedical interventions act directly on the human body and mind to bring about their effects on a passive subject, who plays no role at all" (292). Similarly, Jürgen Habermas, in his 2003 book *The Future of Human Nature*, argues that a genetically enhanced person would not be the sole author of his or her life history. One could not identify with or take credit for any improvement in one's mental states because the source of these states and the actions to which they lead would not be the person but the drugs or other brain- and mind-altering techniques.

Others argue, however, that the voluntary use of a psychotropic drug or neurostimulation to enhance cognition would not necessarily make us inauthentic or replace the person who uses them as the author of his or her actions. If a person with the capacity for critical self-reflection freely decides to take a cognition-enhancing drug or undergo stimulation, then he or she is the agent of any change in his or her mental state. The drug is only the means through which desired change is produced. Improving one's cognitive capacities by manipulating the brain and mind could be an expression of autonomy and authenticity if it resulted from a deliberated and informed decision consistent with one's considered desires and values.

THE ARGUMENT AGAINST PERFECTION

Michael J. Sandel, in his 2007 book *The Case against Perfection*, claims that the desire to enhance our cognitive and physical capacities is driven by the more fundamental desire for perfection and complete control over our lives. While he focuses on genetic enhancement, his claims and arguments also apply to enhancement by drugs or brain stimulation. Sandel says that we should accept rather than try to alter these capacities, imperfect as they are. He is particularly concerned about the potentially harmful effects that enhancement might have on our character. The drive to master our capacities could make us lose sight of what Sandel calls the "giftedness of life" (27). This includes an appreciation of what is good in what we have and do in an absolute and objective sense rather than in a relative and subjective sense of "good." He relates this to American theologian William E. May's (1928–2014) idea of "openness to the unbidden" (45). This means accepting events as they come to us in life, with all their unpredictability. Without appreciation of our giftedness and the unbidden, we cannot flourish and be truly happy, because happiness needs to be grounded in an objective good associated with these concepts. Sandel's concern is not so much about the moral permissibility or impermissibility of enhancement. He does not claim that it is immoral, but he instead considers it unwise to enhance because it causes us to lose our understanding of what constitutes a good life. Defending an Aristotelian view of the virtue of wisdom, Sandel claims that an appreciation of the giftedness of life is a precondition for the good life and that enhancement precludes this appreciation. In addition, he is critical of the drive to mastery because he believes that it could weaken social solidarity and remove moral and aesthetic qualities from our appreciation of the world.

But the desire to *improve* the dispositions that constitute our character does not imply a desire to *master* or *perfect* them. It does not reflect a hubristic wish to completely control our lives. Arguably, not even extreme forms of enhancement imply a perfectionist agenda. Buchanan points out that "even in a world of pervasive and powerful biomedical enhancements, we'd still have plenty of opportunities for appreciating that many of the good things in our lives are not our accomplishments, not subject to our wills" (2011, 134). Enhancement could have unintended bad effects on a person's character if, for example, too high a dose of a psychostimulant increased dopamine levels in the brain and caused addictive behavior. Yet it is not the idea of enhancement as such but the means of enhancement and whether it has positive or negative effects on our cognitive abilities and emotional dispositions that would determine its value or disvalue.

THE ARGUMENT FROM EXCELLENCE

Another concern of Sandel, Kass, and other critics of enhancement is that it undermines our idea of excellence. This suggests that excellence in achieving goals is entirely the result of one's own efforts in exercising one's natural mental capacities. There is considerable variation among people in the cognitive capacities that enable them to undertake and complete projects. Some are naturally more cognitively endowed than others because of the different ways in which their brains are wired. This may give some an advantage over others in excelling as students, as athletes, or in various professions. In addition to neurobiological luck, parental support and other social and cultural factors beyond our control influence how successful we are in achieving goals. For those who have not fared as well as others in the biological and social lotteries, a cognition-enhancing drug combined with effort, persistence, and cultivation of particular skills could be a way of compensating for what they naturally lack in developing a pattern of excellence in the projects they pursue and complete. Sandel's critique of the negative effects of enhancement on excellence may have some force against an augmenting conception of enhancement, if this conception is associated with the idea of mastery and perfection. But it has less force against an optimizing conception of enhancement associated with the idea of improvement.

Proponents of cognitive enhancement argue that the key issue is individual autonomy and whether a competent person decides that altering their intellectual or other mental capacities is in their best interests. This comes with the proviso that the expression of this autonomy does not harm others. It is consistent with the two main tenets in English philosopher John Stuart Mill's (1806–1873) principle of liberty. Mill states that "over himself, over his own body and mind, the individual is sovereign." He qualifies this statement in saying that "the only purpose for which power can be rightfully exercised over any member of a civilized society, against his will, is to prevent harm to others" ([1859] 1974, 119). Nothing about individual decisions to enhance their own cognitive capacities implies a scenario in which this qualification would be invoked. John Harris emphasizes the importance of autonomy in enhancement by saying: "Sandel can have the unbidden and welcome, but on condition [that] he will let me and others have access to the bidden" (2007, 122).

Nevertheless, many people share the concern of Kass, Habermas, and Sandel about the potential of enhancement to affect us in negative ways, even if they are not always able to articulate the reasons for such concern. This is reflected in a 2016 survey by the Pew Research Center showing a distrust of scientists and unease about meddling to enhance human capacities (Kolata 2016). Religion is part of the explanation for this unease. But the more general worry is that brain manipulation to enhance our cognitive capacities will develop to the point of causing us to lose control of our lives. One type of technology generating this worry is brain implants, or neural prosthetics.

EXTREME ENHANCEMENT IN A POSTHUMAN WORLD

Neural prosthetics have greater potential than psychotropic drugs to enhance cognitive functions. This is because they can more directly target and alter the neural circuits regulating these functions without the unwanted adverse distributed effects on other regions of the brain. Neural prosthetics are artificial devices or systems designed to restore or improve a range of brain-based functions that have been damaged as a result of traumatic brain injury, infection, or neurodevelopmental or neurodegenerative disorders (Glannon 2016).

The most widely used neural prosthetic has been deep-brain stimulation. Despite the problems in the earlier mentioned example of the patient who felt "unrealistically good" from increased stimulation to his brain, research has shown that more precise targeting of brain circuits may not only reduce the risk of adverse effects but also stimulate neural growth factors and neuroplasticity. This in turn may enhance synaptic connectivity and the cognitive processing that relies on it, enabling those who undergo the technique to perform multiple cognitive tasks at higher levels.

Brain-computer interfaces (BCIs), or brain-machine interfaces (BMIs), are another type of neural prosthetic (Lebedev 2014). These systems involve real-time direct communication between the brain and a signal-processing algorithm in a computer through an electroencephalograph (EEG) or a microchip implanted in the motor cortex. BCIs decode brain activity to control external devices. Patients using BCIs can translate signals from the motor cortex into actions such as moving a computer cursor or a robotic arm. The microchips can integrate into adjacent neural circuits insofar as they are biocompatible with neural tissue. It is possible in the future that a widely distributed and integrated set of microchips could replace normally functioning neural circuits, resulting in a substantially enhanced network of neural and cognitive functions.

FROM A NATURAL TO AN ARTIFICIAL BRAIN

Suppose that brain implants were used not as therapy for brain dysfunction but to substantially enhance intelligence and the information processing required for decision making in healthy brains. Suppose further that these devices could gradually replace all natural neural circuits and produce a completely artificial brain that would function more efficiently and enable us to perform more cognitive tasks more effectively than we are able to do with our natural brains. A brain consisting of artificial networks theoretically could overcome the limits our evolved brains currently have in holding and processing only a certain amount of information. The Human Brain Project is largely driven by this idea (Markram et al. 2011). This is an international collaborative endeavor whose aim is to achieve a multilevel integrated understanding of brain structure and function through the development and use of information and communication technologies. Initially, the goal of the project was to simulate the entire brain, but then evolved to the more modest goal of developing platforms for neurocomputing and neurorobotics research and development. There are questions about whether an artificial brain could capture all the ways in which our actual central nervous system interacts with and is influenced by other bodily systems and the environment. But it is scientifically possible that an artificially constructed brain consisting entirely of large-scale neuroprosthetic networks replacing natural neural networks would function better than a normal brain. Perhaps the most radical theoretical possibility of brain simulation would be to upload neural information into a quantum computer while a person was alive. This information would be downloaded after the person had "died," with the idea that the person's conscious self would return. He or she could survive death as a result of this process. This idea is captured in the 2014 science fiction film *Transcendence*. The main character in the film, Dr. Will Caster, is transhuman in the sense that he "transcends" the current biological limits on being human. It remains to be seen whether the biological limits on what can be done to the brain can be overcome.

The gradual replacement of neural parts by prosthetic brain implants that would seamlessly integrate into a fully functional artificial organ would raise questions about identity. Specifically, it would raise the question of whether the person whose neural networks were replaced would remain the same person or cease to exist. This would be a legitimate concern if one believes that we are essentially our brains. Alternatively, if one believes that we are constituted by but not identical to our brains, then the gradual replacement of neural parts might not be problematic. Some believe that we are the product of interaction between and among our brains, bodies, and the environment in which we are embedded. Nevertheless, many would agree that the brain is the most important component in this interaction, and the replacement of natural neural circuits with artificial ones would undermine the natural basis of the mental states that constitute both individual and species identity. Yet if all that matters is that the brain realizes the critical inputs and outputs necessary to generate and sustain our cognitive and emotional capacities, then it would not matter whether the source of this process was a natural or artificial brain. The relevant senses of individual and species identity could be retained if what matters is not how the brain is constituted but how it functions. However, if radical enhancement were such that the new neural functions substantially changed and caused a substantial alteration of our conscious minds, then this could transform us from humans into posthumans. This is significant because even if an artificial brain had only positive and no negative effects, it would not be clear *who* the recipient or recipients of its benefits would be. While humans could freely choose to transform themselves in this way, it is not clear that *they* would be the ones on the receiving end of these benefits. It could result in a depersonalized world that at least some would find unpalatable.

BRAIN IMPLANTS, AI, AND CONTROL

Current applications of implanted electrodes in deep-brain stimulation and microchips in brain-computer interfaces involve shared behavior control. By compensating for or bypassing the site of brain injury, these systems do not supplant but supplement neural and mental capacities that are intact. But a complete replacement of natural by artificial neural parts would suggest that the person with such a brain would become a machine in which the parts and their relations to each other would control everything the agent did. Intuitively, something alien to the person would completely take over his or her actions. If the person became a machine, then all of his or her behavior could be explained in mechanistic terms. But the normative practices of praising and blaming people and taking credit and being responsible for our actions presuppose that we are not mere machines but reflective, rational, and emotional beings. With an artificial brain, these practices would disappear, along with the loss of cognitive and affective control of our behavior.

The unease about loss of control may be most acute regarding what may be the most extreme form of enhancement—artificial intelligence (AI). This could be an extension of an artificial brain or a system external to the brain that took over its functions and performed them at advanced levels. The question of control surrounding machine intelligence is more

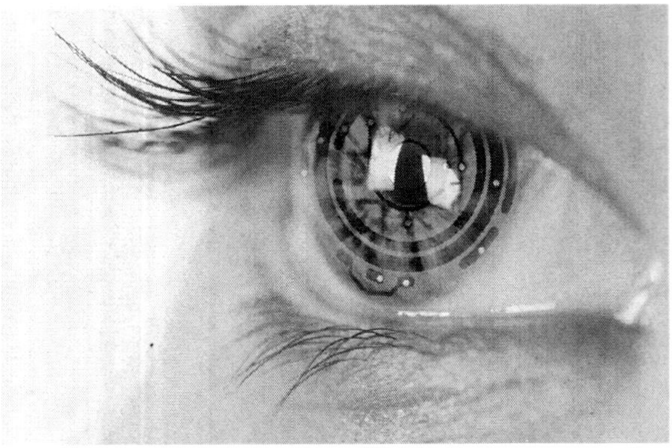

Bioelectronic Eye. *As people incorporate bioelectronic devices into their bodies over the coming decades, the boundary between humans and machines will become increasingly blurred.* PETERPHOTO123/SHUTTERSTOCK.COM.

disturbing than the question of identity, because these machines could cause even posthumans to lose control of their thought and behavior (Gelernter 2016). Posthumans would be nothing more than machines that would not think about acting but act automatically without any reflection. Following an idea first expressed by AI pioneer Alan Turing (1912–1954), some believe that there will be an ultimate point, or "technological Singularity," in which some form of artificial ultraintelligence will replace natural human intelligence (Turing 1950). Some fear that, as an extreme form of enhancement, AI would not enhance our capacities but would dominate us. This would defeat the very purpose of enhancement, which is to free ourselves from physical and mental constraints in order to increase our adaptability and well-being. The point of using psychotropic drugs or neural prosthetics to alter brain function is to enhance our intellectual capacities, not replace them with machine intelligence. Although there are differences between the potential effects that brain-altering drugs and superintelligent machines would have on us, concern about the latter is similar in many respects to the concern expressed by Kass and Sandel about the former's effect on our basic rational and moral dispositions.

AI of the sort described above is still only a logical possibility. Much of the discussion of this technology is speculative. Given the current state of the science, it cannot be known whether its effects on us would be only harmful. Nor can it be known whether we would control or be controlled by it. Even if its effects were beneficial, there would still be the question of whether AI or indeed any form of radical enhancement designed to make us more intelligent and capable beings would transform us into something other than humans. Also unknown are whether enhancement would cause us to evolve out of existence and whether worries generated by this possibility would be part of an argument for limiting enhancement to moderate forms. The issue of retaining or changing individual and species identity is significant because it would not be clear who the recipients of enhancement were and how we would assess whether or in what ways the desire to enhance was realized.

Summary

The three main conceptions of cognitive bioenhancement are augmenting, diminishing, and optimizing. There are various risks and cognitive trade-offs of using psychotropic drugs and neurostimulation to alter the brain to enhance mental capacities. The actual and potential adverse effects of these interventions suggest that there are limitations to enhancement, at least in its current state. Yet, although this may cast a skeptical eye on the augmenting conception of enhancement associated with posthumanism, it is theoretically possible that advances in brain-altering drugs and techniques will overcome these limitations and allow for more radical forms of enhancement. The balance of the chapter has been a discussion of arguments against and for enhancement in terms of how it would affect our understanding and appreciation of our natural dispositions. Whether enhancement would promote perfectionism and how it might affect the concepts of authenticity and excellence were also examined. Presumably, individual autonomy of competent adults would provide the grounds for an ethical argument in favor of enhancement. But this would come with the proviso that the behavioral effects of enhancement did not harm others. The last part of the chapter explored the possible effects of radical enhancement on individual and species identity and behavior control.

Regarding identity, one of the open questions in the enhancement debate is whether a radical change in our capacities would expand our conception of *human* or transform the human species into a posthuman one. This is connected with the question of whether a posthuman world would be superior to a human world in all relevant respects. Even if radical enhancement enabled us to perform cognitive tasks we are currently unable to perform, and even if it had only beneficial and no harmful effects, any normative judgment about the value of such a world would be made using mental capacities we currently possess as humans. It is unclear how we could step outside the human perspective to make value judgments about a posthuman world or about the comparative value of such a world against our current world. The goal of cognitive enhancement is to improve well-being and the quality of our lives. Beyond a certain point, though, there may not be a definitive answer to the questions of *who* would benefit from enhancement and *in what respects* they would benefit.

Bibliography

Agar, Nicholas. *Truly Human Enhancement: A Philosophical Defense of Limits*. Cambridge, MA: MIT Press, 2014.

Benabid, Alim Louis. "What the Future Holds for Deep Brain Stimulation." *Expert Review of Medical Devices* 4, no. 6 (2007): 895–903.

Bostrom, Nick. "Transhumanist Values." In "Ethical Issues for the Twenty-First Century," edited by Frederick Adams, supplement, *Journal of Philosophical Research* 30 (2005): 3–14.

Bostrom, Nick. "Why I Want to Be a Posthuman When I Grow Up." In *Medical Enhancement and Posthumanity*, edited by Bert Gordijn and Ruth Chadwick, 107–136. Dordrecht, Netherlands: Springer, 2008.

Bostrom, Nick, and Anders Sandberg. "Cognitive Enhancement: Methods, Ethics, Regulatory Challenges." *Science and Engineering Ethics* 15, no. 3 (2009): 311–341.

Buchanan, Allen. *Better than Human: The Promise and Perils of Enhancing Ourselves*. Oxford: Oxford University Press, 2011.

Buchanan, Allen. "Human Nature and Enhancement." *Bioethics* 23, no. 3 (2009): 141–150.

de Jongh, Reinoud, Ineke Bolt, Maartje Schermer, and Berend Olivier. "Botox for the Brain: Enhancement of Cognition, Mood, and Pro-Social Behavior and Blunting of Unwanted Memories." *Neuroscience and Biobehavorial Reviews* 32, no. 4 (2008): 760–776.

Earp, Brian D., Anders Sandberg, Guy Kahane, and Julian Savulescu. "When Is Diminishment a Form of Enhancement? Rethinking the Enhancement Debate in Biomedical Ethics." *Frontiers in Systems Neuroscience* 8, art. 12 (2014). doi:10.3389/fnsys.2014.00012.

Farah, Martha J., Judy Illes, Robert Cook-Deegan, et al. "Neurocognitive Enhancement: What Can We Do and What Should We Do?" *Nature Reviews Neuroscience* 5, no. 5 (2004): 421–425.

Gelernter, David. "Machines That Will Think and Feel: Artificial Intelligence Is Still in Its Infancy—and That Should Scare Us." *Wall Street Journal*, March 18, 2016. http://www.wsj.com/articles/when-machines-think-and-feel-1458311760.

Glannon, Walter. "Ethical Issues in Neuroprosthetics." *Journal of Neural Engineering* 13, no. 2 (2016). doi:10.1088/1741-2560/13/2/021002.

Habermas, Jürgen. *The Future of Human Nature*. Cambridge Polity Press, 2003.

Harris, John. *Enhancing Evolution: The Ethical Case for Making Better People*. Princeton, NJ: Princeton University Press, 2007.

Heinz, Andreas, Roland Kipke, Hannah Heimann, and Urban Wiesing. "Cognitive Neuroenhancement: False Assumptions in the Ethical Debate." *Journal of Medical Ethics* 38, no. 6 (2012): 372–375.

Human Brain Project. https://www.humanbrainproject.eu.

Iuculano, Teresa, and Roi Cohen Kadosh. "The Mental Cost of Cognitive Enhancement." *Journal of Neuroscience* 33, no. 10 (2013): 4482–4486.

Juengst, Eric T. "What Does Enhancement Mean?" In *Enhancing Human Traits: Ethical and Social Implications*, edited by Erik Parens, 29–47. Washington, DC: Georgetown University Press, 1998.

Kolata, Gina. "Building a Better Human with Science? The Public Says, No Thanks." *New York Times*, July 26, 2016. http://www.nytimes.com/2016/07/27/upshot/building-a-better-human-with-science-the-public-says-no-thanks.html.

Lebedev, Mikhail. "Brain-Machine Interfaces: An Overview." *Translational Neuroscience* 5, no. 1 (2014): 99–110.

Markram, Henry, Karlheinz Meier, Thomas Lippert, et al. "Introducing the Human Brain Project." *Procedia Computer Science* 7 (2011): 39–42.

Metzinger, Thomas, and Elisabeth Hildt. "Cognitive Enhancement." In *Oxford Handbook of Neuroethics*, edited by Judy Illes and Barbara J. Sahakian, 245–264. Oxford: Oxford University Press, 2011.

Mill, John Stuart. *On Liberty*. Edited by Gertrude Himmelfarb. Harmondsworth, UK: Penguin, 1974. First published 1859.

President's Council on Bioethics. *Beyond Therapy: Biotechnology and the Pursuit of Happiness*. Washington, DC: Author, 2003.

Sandel, Michael J. *The Case against Perfection: Ethics in the Age of Genetic Engineering*. Cambridge, MA: Belknap Press of Harvard University Press, 2007.

Savulescu, Julian, and Nick Bostrom, eds. *Human Enhancement*. Oxford: Oxford University Press, 2009.

Schwartz, Casey. "Generation Adderall." *New York Times Magazine*, October 12, 2016. http://www.nytimes.com/2016/10/16/magazine/generation-adderall-addiction.html.

Schwarz, Alan. "Drowned in a Sea of Prescriptions." *New York Times*, February 2, 2013. http://www.nytimes.com/2013/02/03/us/concerns-about-adhd-practices-and-amphetamine-addiction.html.

Synofzik, Matthis, Thomas E. Schlaepfer, and Joseph J. Fins. "How Happy Is Too Happy? Euphoria, Neuroethics, and Deep Brain Stimulation of the Nucleus Accumbens." *AJOB Neuroscience* 3, no. 1 (2012): 30–36.

Turing, Alan. "Computing Machinery and Intelligence." *Mind* 59 (1950): 433–460.

Volkow, Nora D., Joanna S. Fowler, Jean Logan, et al. "Effects of Modafinil on Dopamine and Dopamine Transporters in the Male Human Brain." *Journal of the American Medical Association* 301, no. 11 (2009): 1148–1154.

FILMS

Limitless. Dir. Neil Burger. 2011.

Transcendence. Dir. Wally Pfister. 2014.

CHAPTER 21

Commodification of Human Traits: The Body as Industrial Product

Daryl Wennemann
Associate Professor of Philosophy
Fontbonne University, St. Louis, MO

The development of contemporary biotechnology has altered the character of human life in fundamental ways and, for some, even offers the prospect of improving the human species. Applying biotechnology to improve human traits such as bodily strength and mental agility may be seen as fabricating an improved human being that transhumanists regard as a transition to a new form of life that is often termed "posthuman."

Others view attempts to apply biotechnologies of human enhancement as risky to the human race, reducing human beings to objects to be manipulated and produced like consumer goods. The application of biotechnologies of human enhancement seems to necessarily involve instrumentalizing human beings. Such a prospect is all the more striking in the context of a market economy. Would an enhanced human being or posthuman being be capable of the moral cognition and empathy that characterizes human life at its best? Will the democratic ideal of equality be applicable to such beings? Or, will there arise a new form of social hierarchy based on genetic design, such as the one depicted in Aldous Huxley's *Brave New World* ([1932] 2004)?

Like any major technological development, biotechnology raises fascinating possibilities for the improvement of the human condition but also requires an assessment of its ethical and social consequences. The focus of this chapter is on the ethical and social import of biotechnologies of human enhancement and the commodification of human traits.

COMMODIFICATION

Contemporary biotechnology has applications in a wide array of fields, including medicine, agriculture, the environment, energy production, industry, and weapons technology. The range of biotechnology products is vast. They include new drugs, genetic testing and treatment, genetically modified crops, genetically modified microbes that are used to detoxify the environment, biofuels, biodegradable plastics, and lethal viruses for use in warfare.

In relation to the issue of the commodification of human traits, Christian Lotz (2014) has made the interesting observation that biotechnologies, as well as other technologies that are currently shaping the world, are capitalist technologies. They were developed in market economies driven by the profit motive. As a result, they are necessarily implicated in the demands of the market. There is money (both private and public) invested in research and design. Products are developed for sale on the market. Advertising promotes these products and profits are made. Another essential part of the contemporary capitalist economic system is a consumer culture. In such a culture, consumers are eager to purchase the newest technological devices, such as the latest version of smartphones. How will people relate to themselves and others if human traits are available as consumer products? Will human beings having certain traits come to be seen as replaceable products? And will human beings lacking certain traits be seen as defective?

As is the case with any other product, there are issues of safety and liability for any harms associated with their use. There are also ethical concerns related to advertising and the social consequences of products. One example is the case of three women who were rendered blind by an unapproved stem cell procedure for macular degeneration that was falsely advertised as being part of a government-sanctioned study (McGinley 2017). Such a case highlights the need for regulation of unapproved procedures.

It is important to realize, however, that biotechnologies of human enhancement within a market context raise much more profound ethical concerns than those associated with standard business ethics. When applied to human beings, biotechnology can alter persons in ways that seemingly reduce them to mere objects of manipulation. Among the biotechnology products available today are some that are capable of altering the body and mind of human beings with ever greater accuracy. The body can be shaped surgically and, more profoundly, through pharmacological and genetic manipulation as part of a treatment for disease or as an enhancement.

The distinction between the treatment of disease and enhancement is a central concern in the debate over the application of biotechnologies. It is, however, a difficult distinction to maintain. Is a vaccination preventive treatment or an enhancement? To cite another example, growth hormone treatments can alter the height of persons suffering from idiopathic short stature, those with growth hormone levels testing as normal and where the cause of short stature cannot be determined (Doheny 2008).

Mental acuity is subject to pharmacological manipulation. It is common today for students to use "study drugs" such as Adderall, Ritalin, and Vyvanse to improve their ability to study and retain information (Yanes 2014). In the future, the mind may also be altered through the manipulation of the human genome. Ronald M. Green (2007) has discussed how scientists have been able to improve the learning capacity of mice by altering a gene that governs the production of a learning and memory protein. He notes that it is possible in principle to apply this procedure to human beings.

Clearly, personal identity is bound up with these kinds of augmentation because the continuity of human experience is bound up with our memory capacity. As such, our identities may be shaped in significant ways through various technological interventions. In 2016 Elon Musk and others launched a business called Neuralink that seeks to link the human brain to computers by means of a neural implant. It is not difficult to imagine that commodities such as a brain-computer interface might evolve that "may one day upload and download thoughts" (Metz 2017). The title of the 1995 film *Strange Days* is eerily appropriate. In that film, the experiences of others could be downloaded from a recording device. As an ethical issue, the core of human identity as persons having dignity is at stake if we decide to alter ourselves in identity-constituting ways. Would you be willing to purchase a product that can cause you to no longer be you?

The capacity to manipulate human beings at the genetic level is founded on the success of contemporary molecular science. As new technology is applied to human beings, there is a tendency to reduce people to mere bundles of information that can be manipulated by rewriting a person's genetic code. The genetic code of living things can be understood as a set of algorithms or rules for the production of proteins that govern their physical development. The technology of genetic splicing known as CRISPR (clustered regularly interspaced short palindromic repeats) allows for the manipulation of genes to produce specific desired traits. A member of the German Ethics Council, Jochen Taupitz, has observed that this technique can also have unintended off-target effects. But, as Taupitz told Norbert Lossau in a 2017 interview, he believes that the technology will develop very rapidly in the direction of greater control and accuracy:

> With CRISPR/Cas9 and similar methods of genome editing, however, a more controllable and secure intervention in the genome of a human being looms closer and closer. At some point, we will reach the point at which a risk-benefit assessment could lead to the estimation that such interventions are sensible and responsible and therefore should be allowed. At present, that is not yet the case. However, at this time, society must be prepared for the targeted intervention in the human genome, which, from a medical-scientific point of view, is practicable. (my translation)

In addition to off-target effects, the technology of gene splicing must account for the fact that genes can have varying effects depending on how they interact with environmental influences. The study of epigenetics examines these complex interactions that affect gene expression.

Still, genetic mutations can now be edited out and other genes inserted in their place in order to treat disease (Pak 2014). It has been reported that a child has been born having the DNA of three people. The mitochondrial DNA of the mother was replaced by that of another woman to avoid Leigh syndrome, a genetically based disease (Hamzelou 2016). In the future, gene splicing may be used to enhance human beings genetically so as to produce designer children. Indeed, Yuval Noah Harari (2017) believes that we may be able to make ourselves into gods having superhuman capacities, including a form of immortality, by rewriting the algorithms of the human genome.

DIGNITY VERSUS EXCHANGE VALUE

A central insight of eighteenth-century German philosopher Immanuel Kant (1724–1804) was that all human beings are persons having dignity. Kant posited that dignity is an infinite, inalienable value inherent in all persons. He also argued that there is an ethical duty to recognize this value. And so, the fundamental moral demand of such a deontological ethic is to treat all persons with respect. To treat each person with respect is to recognize the freedom of all individuals to choose their own rational goals. Kant argued that human persons are self-legislating because they have the autonomy to choose their own ends. As such, they are ends-in-themselves. Because respect for persons is an ethical demand that our own reason places on us, it represents what Kant called "autonomy of the will." It is not a case of heteronomy of the will, in which case the law we follow is imposed on the will.

In contrast to the dignity of persons, Kant recognized the finite exchange value of objects. Thus, according to Kant, "In the realm of ends everything has a *price* or a *dignity*. Whatever has a price can be replaced by something else as its equivalent; on the other hand, whatever is above all price, and therefore admits of no equivalent, has a dignity" (Kant [1785] 1959, 53; italics in the original).

There is, therefore, a fundamental conflict between the realm of persons and the realm of money exchange according to Kant's ethical theory. In a capitalist setting, or any other economic setting in which money is used as a medium of exchange (excluding a barter system), persons are not to be reduced to a money equivalent. But a money system tends to do just that. It tends to treat anything it touches as an object that can be manipulated and used. Thus, "the essence of prostitution, which we recognized in money, is imparted to the objects that function exclusively as its equivalent, perhaps to a more noticeable degree because they have more to lose than money is able to" (Simmel [1900] 2004, 391–392). In *The Philosophy of Money* ([1900] 2004), German philosopher and sociologist Georg Simmel (1858–1918) makes the interesting observation that there is a sense in which money does not have a value equivalent to commodities but a greater value, in that money has the value of a commodity plus the value of having the possibility of being used to purchase any other commodity on the market. Objects of distinction are valued precisely because they are withheld from the market. No one would consider selling the Statue of Liberty or the Liberty Bell to pay off the national debt. Objects of distinction have a qualitative value over and above their price. It would dehumanize a musical performance if the musicians considered their pay a sufficient means of recognizing their musical virtuosity; they also expect applause. Finally, Simmel notes that the lack of interest we have in the individuality of commodities leads to a disparagement of individuality as such. This implies that if we seek distinction in the qualities we can purchase as products of biotechnologies of human enhancement, we actually lose distinction as individuals in making use of the monetary means of acquiring them. In other words, the enhancements we choose to purchase could also be purchased by anyone else who has the financial means to do so.

DEHUMANIZATION

What follows from the distinction between dignity and exchange value is a demand to not treat persons as objects that can be manipulated and used. To do so is to dehumanize them. The most egregious example of dehumanization resulting from the conflict between the realm of persons and the realm of money exchange is that of slavery, which has been remarkably ubiquitous in human history. Notably, the Swiss practiced a form of forced labor called the *Verdingkinder* system, which was not made illegal until 1981 (Wild 2014). Human trafficking has flourished with the advent of globalization (Igboeroteonwu and Esslemont 2016). Trafficking in body parts has also increased greatly during this period. India has only recently made the sale of human organs illegal. R. R. Kishore has noted that "there are reports of the kidnapping and murder of children and adults to 'harvest' their organs" (2005, 362). Of course, it is the poor who wish to sell their organs in order to pay off their debts. They are a ready source of organs for those who have the money to pay, as was detailed in a *Spiegel Online* article:

> The note is attached to a tree trunk across from the Central Hospital in Chennai (formerly Madras), a major city in southern India. Written in scrawly handwriting, the note advertizes its author's "top notch kidney" for 30,000 rupees, the equivalent of €500 ($664). Asked about his offer, the vendor—a 30-year-old Tamil—says "no middleman" is involved in the deal. He adds that he urgently needs the money to "pay back debts." (Schmitt 2007)

Do the biotechnologies of human enhancement dehumanize human beings? It would seem that commodifying human traits necessarily entails instrumentalizing human beings in a way that dehumanizes them. But the products of biotechnologies can also benefit human beings by healing diseases and increasing the capacities of persons. Such beneficial uses of biotechnologies would seem to be morally demanded by our respect for human persons. A large part of the controversy surrounding biotechnologies of human enhancement has to do with whether we can apply them to human beings without dehumanizing people and thus undermining their personhood. The remainder of this section explores these issues further, focusing on the manipulation of children and self-improvement, before exploring the phenomenon of affluenza and delving into the question of whether chimeric humans would be "unnatural."

THE MANIPULATION OF CHILDREN

The issue of dehumanization in relation to enhancement technologies applies most especially to the manipulation of children who are not yet able to make their own rational choices and to future generations that will be affected by the choices we make now. Parents who choose a life plan for their children may be seen as dehumanizing them by denying them the choice of their own life plan, and thus not treating them with respect. And so, if parents should want a child who will be a good basketball player, they might attempt to alter the genetic makeup of the child to produce a taller, more muscular and agile individual.

Such "designer babies" have become a topic of intense debate because designing individuals seems to treat them like objects and not as ends-in-themselves. As a case in point, in 1980 Robert K. Graham created a sperm bank for geniuses. He collected sperm from Nobel laureates, Olympic athletes, and university students with very high IQs in order to produce children with exceptional abilities. Women who wished to become pregnant could order the sperm from a catalog. The sperm bank operated until 1999, and 215 children were produced from the genius sperm over that period. It also happens that all the sperm donors were white (Escobedo 2014). It is likewise possible to obtain eggs from tall, smart women (Kolata 1999). Christine Overall notes that the issue of the accessibility of such products on the market is not just a matter of their effects on the individuals who use them. According to Overall, the influence such products have on our social categories and biases is also significant: "In a society in which beautiful, highly intelligent, talented children are, in effect, available for purchase by the wealthy, we are justified in being worried about the objectifying and commodifying effects of such a market on attitudes toward and beliefs about all children" (2009, 330). In this regard, Ulrich Rosar, a professor of sociology in Germany, has advocated placing a ban on job application photographs in Germany because unattractive people are so commonly discriminated against (Windmüller 2017).

Of course, parents make many decisions for their children, from the foods they eat to the schools they attend. If parents are pursuing the good of the child, it may not be morally impermissible for them to give direction to the ultimate life plan of the child. The same approach to the issue of dehumanization may be applied to future generations. To alter the human genome through germ-line modifications, ones that are passed down to future generations, may undercut our ability to see human beings as persons having dignity because persons will be made according to the specifications we impose. Nevertheless, we seem to do that now when we treat immature individuals for disease. We do not let the disease state of a child determine its life prospects. If we can alter the human genome so as to avoid genetically caused diseases, we would seem to be treating all future persons with respect by advancing their well-being.

SELF-IMPROVEMENT

But what if an individual chooses to alter his or her own traits? Respect for persons is supposed to recognize the freedom of persons to choose their own ends. And yet, to choose to alter one's own constitution seems to make oneself into an object. The moral challenge of contemporary biotechnologies of human enhancement seemingly arises because it takes us to the limit of human freedom. Are we free to undermine our own freedom? I may freely choose to alter my own personality. But if I no longer exist as the person I was when I made the choice, it would seem that my original self can no longer make its own free choices. Which is better: to make choices based on my personality as it has developed in the mix of genetic and environmental influences or to make a choice to alter my own personality to attain a state I desire?

Ronald Cole-Turner (1998) has argued that self-improvement is a central motif of the Judeo-Christian tradition. Indeed, personal transformation may extend to the adoption of a new identity. For example, persons entering a religious order commonly take a new name of their own choosing. Within that tradition, persons might work a lifetime to forge their own personality by developing the traditional cardinal virtues of prudence, justice, fortitude, and temperance. Historically, the effort to shape one's own personality was always part of the human significance of the task of forming one's own character.

Currently, it is possible to bring about a substantial change in personality pharmacologically without all the effort required for traditional moral formation. And in the future, it may become possible to intervene genetically to produce personality traits we desire in ourselves and our offspring. It seems that the biotechnological means of self-improvement undermine the significance of the end we are pursuing merely because we are treating human beings, ourselves and others, as engineered products. In this way, the traditional motivation to pursue self-improvement—the desire for self-transcendence—contributes to the social pressure brought to bear by our technological society to make ourselves more productive and competitive (Sandel 2009). Only now, we can achieve an alteration of our personality by means of a technological shortcut.

The fact that biotechnological means of self-improvement do not require the efforts associated with traditional means seems to undercut the human significance of our activities because the effort itself defines the action. Athletes can use steroids to increase their strength to proportions that exceed human norms. Does this undermine our interest in watching them perform? Does a home run record still elicit the admiration it once did? Ethically, it is a matter of authenticity. To live authentically is to live in a way that is true to oneself, rather than conforming to social pressure or the expectations of others. Authenticity involves accepting one's condition as a human being with all of its limitations. Michael J. Sandel (2007, 2009) has argued that the drive for perfection that is implicated in the use of biotechnologies to enhance our human capacities will undermine our ability to accept people in our society who fail to measure up to society's standards of perfection. They will suffer being stigmatized, as disabled persons sometimes are. The drive for perfection, in Sandel's view, represents a Promethean hyperagency in which there is nothing to limit the will in seeking one's own self-transformation. We might note that in this case there is no end to the consumption of the goods made available by biotechnologies. According to Sandel, our acceptance of persons having equal dignity depends on maintaining a balance of accepting love and transformative love. Accepting love implies that we do not simply choose our children because they have traits we desire them to have.

Erik Parens (1998) sees in the demand for authenticity two contrary claims. One is a claim to accept our human condition as a gift that may be a source of wonder. The other is a claim to express our creativity in remaking ourselves. The moral task, according to Parens, is to find a way of balancing these two claims.

AFFLUENZA

If we associate the pursuit of perfection through the enhancement of human capacities with the consumer culture that we inhabit, we can see that there is a monetary interest in promoting a need for continual improvement. Susan Bordo (1998) made the very interesting observation that the consumer system depends on us perceiving ourselves as defective. As persons pursue improvements to their bodies through surgery, pharmacology, or genetic treatments, the norm of what is socially acceptable or a source of status constantly shifts. It is therefore understandable that we continually feel inadequate. And the cure for that feeling of inadequacy is the latest consumer product of self-improvement. There is a term that has been coined to refer to this phenomenon, *affluenza*. It refers to a sort of dis-ease associated with a consumer culture that requires that all people treat themselves as the sum total of their consumer goods. Everyone's social status depends on it. The "keep up with the Joneses" mentality underlying affluenza could result in self-defeating behavior as people pursue positional advantage. If one member of a seated audience stands up, she or he gains an advantage. But if everyone stands, no one is better off. The same may be true of the improvements we seek through biomedical interventions. Thus, Bordo observed that "our relationship to our bodies has clearly become more and more an investment in them as 'product' and image, requiring alteration as fashions change" (1998, 202). Cary Wolfe has made a similar point in relation to the social pressures our economic system demands of us: "We're encouraged more and more to develop our 'brand,' as it were, whether by accruing more and more friends on Facebook or by perfecting the kind of balanced 'portfolio' between academic athletic, and nonprofit work that university admissions committees want to see" (Lennard and Wolfe 2017).

The dehumanizing influence of this consumer culture can be glimpsed in the complaints that are sometimes made by famous people. In a 2016 interview published in *ZEITmagazin*, Lady Gaga commented, "I no longer feel pressure to fulfill some special image that people have of me. Understand, I do not want to be branded. I am not a brand. I am a human being" (my translation). This is a remarkable statement given the significance of fame in our culture. The lure of the consumer culture is that we willingly adopt a view of ourselves and others as commodities. According to Kathy Davis, women must view their own bodies as commodities to balance the risks of cosmetic surgery against the benefits, "a business venture of sorts" (1995, 157).

The phantasmagoria of consumer products associated with affluenza may give rise to what Danish existential philosopher Søren Kierkegaard (1813–1855) called the "despair of possibility." If there are an unlimited number of possibilities to choose from, our choices become meaningless. The biotechnologies of human enhancement provide for a virtually unlimited range of possible genetic combinations. The human genome can be altered so as to eliminate genetic defects. It can also be altered by introducing genes from other animal and plant species, producing chimera. And, what is perhaps most remarkable, new DNA base units have been synthesized by scientists, producing a new life-form with six DNA base units, adding two synthetic base units to the four—adenine, cytosine, guanine, and thymine—that have defined life on Earth for millennia.

> In 2014, the synthetic biologists announced they had achieved something unprecedented in the history of DNA.... The biologists added two new letters to the four-letter DNA alphabet within E. coli bacteria. The scientists called the novel base units dNaM and d5SICS .... You can think of these unnatural nucleobases as X and Y. Years down the line, microbes with increased genetic information could present exciting and lucrative scientific possibilities: bacteria capable of churning out therapeutic human proteins, or altered bugs that hoover [vacuum] up environmental spills. (Guarino 2017)

WOULD CHIMERIC HUMANS BE "UNNATURAL"?

It is common to refer to controversial inventions as "unnatural." There is usually a moral judgment implied in that designation. But, as English philosopher John Stuart Mill (1806–1873) argued, there is little to be gained in the way of ethical insight from the distinction between the natural and the artificial. Gregory E. Kaebnick explained this point as follows:

> Under one common understanding, "nature" refers to everything that adheres to the laws of nature; under the other, it refers only to that which excludes human interference or involvement. Understood in the first way, all human action is natural; understood the second way, all human action is unnatural. Either way, the concept cannot help us mark off *some* human actions as according with nature and others as violating it. (2011, 52; italics in the original)

And yet, creating chimeric humans would break the species boundary that has historically provided a common moral touchstone for human beings (as would other less drastic alterations discussed earlier). Prior to the advent of biotechnologies of human enhancement we could always identify morally with other members of the human species. German philosopher Jürgen Habermas worries that breaking the species boundary will undermine our ability to identify morally with human beings that have been so drastically altered, as well as those who may potentially be altered. According to Habermas, "Getting used to having human life biotechnologically at the disposal of our contingent preferences cannot help but change our normative self-understanding" (2003, 72). These considerations are a point of departure for transhumanist and posthumanist theorizing.

SETTING LIMITS

The most significant ethical and social challenge presented by the development of new technologies is that of setting limits to their use. Part of the difficulty in doing so lies in the uneven rate at which technologies and social policies tend to progress (Wennemann 1993).

Technologies often develop at a much more rapid pace than ethical and social thought. Thus, a significant motivation for transhumanist and posthumanist thought derives from the desire to get out in front of various technological developments in an attempt to govern them in an ethically appropriate manner.

While it is true that cultural norms affect the way in which different societies approach the ethical and social assessment of technologies (Jasanoff 2005), within the context of globalization there has been a perceived need to seek a universal, transcultural ethical norm, especially because biotechnologies are transferable within the international economic system of trade. In 2005 UNESCO adopted the Universal Declaration on Bioethics and Human Rights. It promulgated principles that include the protection of human dignity and human rights, the maximization of benefits, and the minimization of harms to all those affected by biotechnologies. The protection of individual autonomy, the right to consent, and the right to privacy were acknowledged. In addition, the declaration promoted equality and justice in the treatment of individuals and proscribed discrimination of any kind. It also called for the maintenance of cultural diversity and international social cooperation. Finally, the declaration encouraged the sharing of benefits from biotechnologies and the protection of future generations and the environment.

It should be noted that sharing the benefits of biotechnologies conflicts with the capitalist ambition to derive profits from the commodification of human traits. Philosophically, however, there has been a tendency to move in the direction of a global ethic of responsibility by virtue of the vastly increased range and significance of contemporary technologies. In this regard, Karl-Otto Apel (1996) has distinguished between a traditional microethics focused on loyalty within small groups, a mesoethics of professional responsibility within the context of a division of labor, and a macroethics of planetary responsibility. Apel concluded his ethical reflections on the current global situation by positing a global ethic of responsibility.

> Thus it appears that in both dimensions of cultural evolution, namely, that of technological interventions in nature and that of social interaction, a global situation has been brought about in our time that calls for a new ethics of shared responsibility, in other words, for a type of ethics that, in contradistinction to the traditional or conventional forms of ethics, may be designated a (planetary) macroethics. (1996, 278)

The concept of shared responsibility places the ethical burden on a consideration of the consequences of our actions. This may be contrasted with the deontological approach that focuses on a duty to treat all persons with respect. Shared responsibility also highlights a demand for shared decision making, which entails a democratization of ethical and social assessment of technologies in a process of consensus formation. These aspects of contemporary moral assessment seem to be required in the context of an Anthropocene epoch, a period of global human influence on Earth's ecosystems. It is ironic that the Anthropocene seems to be leading to the replacement of human beings by transhuman and posthuman beings as we apply various enhancement technologies to ourselves.

For those who consider such a prospect a threat, the precautionary principle reigns in our ethical assessment of new technologies. According to this principle, it is morally incumbent on those who develop such technologies to demonstrate that they will not cause harm. The technology is presumed dangerous until proven innocent. The precautionary principle has been the dominant approach in modern ethical analysis of new technologies, especially in the European Union, where it has been inscribed in the law (see The Precautionary Principle 2017).

In contrast, for those who are excited by the prospect of acquiring superhuman traits, the proactionary principle governs their ethical orientation. According to this principle, people should be free to innovate with self-altering technologies if the prospect of the benefits is great enough to outweigh the possible harms. The emphasis is on the opportunity costs of not applying a new technology. As such, the technology is considered innocent until proven guilty (Fuller and Lipińska 2014).

It is easy to see that the ethical calculus might favor biotechnologies of human enhancement if the prospect of using them includes posthuman immortality with superhuman physical and mental capacities. How does that weigh against the possibility of a new eugenics and the formation of new social classes based on the stigmatization of less enhanced human beings?

Ronald M. Green (2007) has called for a middle approach between prohibiting scientific experimentation and unfettered freedom to experiment. Focusing on genetic interventions for children and future generations, Green calls for the restriction of genetic modifications to those that are reasonably in the child's interest. The risks of genetic interventions should not be out of proportion, in his view, with those associated with natural reproduction. Interventions should be reversible, if possible, and germ-line interventions that produce inheritable alterations should be a last resort. Finally, Green argues that alterations that confer merely positional advantage should be avoided and that they should not reinforce unjust social stigmas or promote economic inequality.

Summary

However we assess the risks associated with biotechnologies of human enhancement, it is clear that they will transform human existence in all of its dimensions. It is also true that the capitalistic social setting of their development will affect the character of the technologies. The interaction of technologies and their social and ethical norms is what some sociologists call "coproduction" (Jasanoff 2005).

Steve Fuller has argued that the self-alteration of human beings that will lead to transhuman, and eventually posthuman, beings will characterize the capitalism of the future. "Capitalism 2.0," as he calls it, will be characterized by a freedom to choose commodities that include technologies of self-alteration. And that implies a freedom to choose not just what one wants to have but to be what one wishes to be (Fuller 2017). Such "morphological freedom" having virtually no limits is breathtaking in its possibilities for self-expression. But what self is being expressed? Will there be continuity between the self that I am and the self I choose to become (Agar 2014)? And will there be justice in the access people have to these technologies? Or will they undermine any sense of the equality that underlies our political democratic traditions? According to Apel (1996), these kinds of issues have arisen because *Homo faber* has outdistanced *Homo sapiens*. Our wisdom has not kept pace with our productive capacity. Now we must contemplate remaking ourselves as we remake the world with our technology.

Bibliography

Agar, Nicholas. *Humanity's End: Why We Should Reject Radical Enhancement.* Cambridge, MA: MIT Press, 2010.

Agar, Nicholas. *Liberal Eugenics: In Defence of Human Enhancement.* Malden, MA: Blackwell, 2004.

Agar, Nicholas. *Life's Intrinsic Value: Science, Ethics, and Nature.* New York: Columbia University Press, 2001.

Agar, Nicholas. *Truly Human Enhancement: A Philosophical Defense of Limits.* Cambridge, MA: MIT Press, 2014.

Apel, Karl-Otto. "A Planetary Macroethics for Humankind: The Need, the Apparent Difficulty, and the Eventual Possibility." In *Karl-Otto Apel: Selected Essays*, edited by Eduardo Mendieta. Vol. 2, *Ethics and the Theory of Rationality*, 275–292. Atlantic Highlands, NJ: Humanities Press, 1996.

Baillie, Harold W., and Timothy K. Casey, eds. *Is Human Nature Obsolete? Genetics, Bioengineering, and the Future of the Human Condition.* Cambridge, MA: MIT Press, 2005.

Basl, John, and Ronald L. Sandler, eds. *Designer Biology: The Ethics of Intensively Engineering Biological and Ecological Systems.* Lanham, MD: Lexington Books, 2013.

Bess, Michael. *Our Grandchildren Redesigned: Life in the Bioengineered Society of the Near Future.* Boston: Beacon Press, 2015.

Bordo, Susan. "*Braveheart*, *Babe*, and the Contemporary Body." In *Enhancing Human Traits: Ethical and Social Implications*, edited by Erik Parens, 189–221. Washington, DC: Georgetown University Press, 1998.

Buchanan, Allen. *Better than Human: The Promise and Perils of Enhancing Ourselves.* Oxford: Oxford University Press, 2011.

Buchanan, Allen. *Beyond Humanity? The Ethics of Biomedical Enhancement.* Oxford: Oxford University Press, 2011.

Cole-Turner, Ronald. "Do Means Matter?" In *Enhancing Human Traits: Ethical and Social Implications*, edited by Erik Parens, 151–161. Washington, DC: Georgetown University Press, 1998.

Davis, Kathy. *Reshaping the Female Body: The Dilemma of Cosmetic Surgery.* New York: Routledge, 1995.

Doheny, Kathleen. "Growth Hormone Therapy Increases Kids' Height: Study Shows Therapy Is Effective Even in Children Who Aren't Deficient in Growth Hormones." WebMD. November 6, 2008. http://www.webmd.com/children/news/20081106/growth-hormone-therapy-ups-kids-height.

Escobedo, Tricia. "A Sperm Bank Just for Supersmart People." CNN. October 13, 2014. http://www.cnn.com/2014/10/08/health/genius-sperm-bank/.

Fukuyama, Francis. *Our Posthuman Future: Consequences of the Biotechnology Revolution.* New York: Farrar, Straus and Giroux, 2002.

Fuller, Steve. "Transhumanism and the Future of Capitalism: The Next Meaning of Life." *EUROPP* (blog), London School of Economics and Political Science, January 25, 2017. http://blogs.lse.ac.uk/europpblog/2017/01/25/transhumanism-and-the-future-of-capitalism/.

Fuller, Steve, and Veronika Lipińska. *The Proactionary Imperative: A Foundation for Transhumanism.* Basingstoke, UK: Palgrave Macmillan, 2014.

Green, Ronald M. *Babies by Design: The Ethics of Genetic Choice.* New Haven, CT: Yale University Press, 2007.

Guarino, Ben. "Biologists Breed Life Form with Lab-Made DNA. Don't Call It 'Jurassic Park.'" *Washington Post*, January 24, 2017. https://www.washingtonpost.com/news/morning-mix/wp/2017/01/24/biologists-breed-life-form-with-lab-made-dna-dont-call-it-jurassic-park/.

Habermas, Jürgen. *The Future of Human Nature.* Cambridge: Polity Press, 2003.

Hamzelou, Jessica. "Exclusive: World's First Baby Born with New '3 Parent' Technique." *New Scientist*, September 27, 2016. https://www.newscientist.com/article/2107219-exclusive-worlds-first-baby-born-with-new-3-parent-technique.

Harari, Yuval Noah. *Homo Deus: A Brief History of Tomorrow.* New York: Harper, 2017.

Hughes, James. *Citizen Cyborg: Why Democratic Societies Must Respond to the Redesigned Human of the Future.* Cambridge, MA: Westview Press, 2004.

Huxley, Aldous. *"Brave New World" and "Brave New World Revisited."* New York: HarperCollins, 2004. First published 1932 and 1958, respectively.

Igboeroteonwu, Anamesere, and Tom Esslemont. "Baby Traffickers Thriving in Nigeria as Recession Bites."

Reuters. October 12, 2016. http://www.reuters.com/article/us-nigeria-humantrafficking-babies-idUSKC N12C039.

Janicaud, Dominique. *On the Human Condition.* Translated by Eileen Brennan. London: Routledge, 2005.

Jasanoff, Sheila. *Designs on Nature: Science and Democracy in Europe and the United States.* Princeton, NJ: Princeton University Press, 2005.

Kaebnick, Gregory E. "Human Nature without Theory." In *The Ideal of Nature: Debates about Biotechnology and the Environment,* edited by Gregory E. Kaebnick, 49–70. Baltimore: Johns Hopkins University Press, 2011.

Kant, Immanuel. *Foundations of the Metaphysics of Morals.* Translated by Lewis White Beck. Indianapolis, IN: Bobbs-Merrill, 1959. Originally published in German, 1785.

Kimbrell, Andrew. *The Human Body Shop: The Engineering and Marketing of Life.* San Francisco: HarperSanFrancisco, 1993.

Kishore, R. R. "Human Organs, Scarcities, and Sale: Morality Revisited." *Journal of Medical Ethics* 31, no. 6 (2005): 362–365. doi:10.1136/jme.2004.009621.

Kolata, Gina. "$50,000 Offered to Tall, Smart Egg Donor." *New York Times,* March 3, 1999. http://www.nytimes.com/1999/03/03/us/50000-offered-to-tall-smart-egg-donor.html.

Lady Gaga. "73 Fragen an Lady Gaga, mehr braucht kein Mensch: Ein Interview von Moritz von Uslar." By Moritz von Uslar. *ZEITmagazin,* October 17, 2016. http://www.zeit.de/zeit-magazin/2016/41/lady-gaga-fragen-moritz-von-uslar.

Lennard, Natasha, and Cary Wolfe. "Is Humanism Really Humane?" *New York Times,* January 9, 2017. https://www.nytimes.com/2017/01/09/opinion/is-humanism-really-humane.html.

Lossau, Norbert. "Forscher wollen Erbgut von Embryonen verändern" [Researchers want to modify the genetic makeup of embryos]. *Die Welt* (Berlin), March 29, 2017. https://www.welt.de/gesundheit/article163248847/Forscher-wollen-Erbgut-von-Embryonen-veraendern.html.

Lotz, Christian. *The Capitalist Schema: Time, Money, and the Culture of Abstraction.* Lanham, MD: Lexington Books, 2014.

Lygre, David G. *Life Manipulation: From Test-Tube Babies to Aging.* New York: Walker, 1979.

McGinley, Laurie. "Three Women Blinded by Unapproved Stem-Cell 'Treatment' at South Florida Clinic." *Washington Post,* March 15, 2017. https://www.washingtonpost.com/news/to-your-health/wp/2017/03/15/three-women-blinded-by-unapproved-stem-cell-treatment-at-south-florida-clinic/.

McKibben, Bill. *Enough: Staying Human in an Engineered Age.* New York: Times Books, 2003.

Metz, Cade. "Elon Musk Isn't the Only One Trying to Computerize Your Brain." *Wired,* March 31, 2017. https://www.wired.com/2017/03/elon-musks-neural-lace-really-look-like.

Overall, Christine. "Life Enhancement Technologies: The Significance of Social Category Membership." In *Human Enhancement,* edited by Julian Savulescu and Nick Bostrom, 327–340. Oxford: Oxford University Press, 2009.

Pak, Ekaterina. "CRISPR: A Game-Changing Genetic Engineering Technique." *Science in the News* (blog), Harvard University, Graduate School of the Arts and Sciences, July 31, 2014. http://sitn.hms.harvard.edu/flash/2014/crispr-a-game-changing-genetic-engineering-technique/.

Parens, Erik. "Is Better Always Good? The Enhancement Project." In *Enhancing Human Traits: Ethical and Social Implications,* edited by Erik Parens, 1–28. Washington, DC: Georgetown University Press, 1998.

The Precautionary Principle (website). Accessed May 23, 2017. http://www.precautionaryprinciple.eu.

Rifkin, Jeremy. *The Biotech Century: Harnessing the Gene and Remaking the World.* New York: Jeremy P. Tarcher/Putnam, 1999.

Rosenfeld, Albert. *The Second Genesis: The Coming Control of Life.* Englewood Cliffs, NJ: Prentice-Hall, 1969.

Sandel, Michael J. *The Case against Perfection: Ethics in the Age of Genetic Engineering.* Cambridge, MA: Belknap Press of Harvard University Press, 2007.

Sandel, Michael J. "The Case against Perfection: What's Wrong with Designer Children, Bionic Athletes, and Genetic Engineering." In *Human Enhancement,* edited by Julian Savulescu and Nick Bostrom, 71–89. Oxford: Oxford University Press, 2009.

Savulescu, Julian, and Nick Bostrom, eds. *Human Enhancement.* Oxford: Oxford University Press, 2009.

Schmitt, Thomas. "A Pound of Flesh: Organ Trade Thrives in Indian Slums." *Spiegel Online International,* June 14, 2007. http://www.spiegel.de/international/world/a-pound-of-flesh-organ-trade-thrives-in-indian-slums-a-488281.html.

Simmel, Georg. *The Philosophy of Money.* Edited by David Frisby. Translated by Tom Bottomore and David Frisby.

3rd ed. London: Routledge, 2004. First published in German, 1900.

Stock, Gregory. *Redesigning Humans: Our Inevitable Genetic Future.* Boston: Houghton Mifflin, 2002.

UNESCO (United Nations Educational, Scientific and Cultural Organization). "Universal Declaration on Bioethics and Human Rights." 2005. http://unesdoc.unesco.org/images/0014/001461/146180E.pdf.

Wennemann, Daryl. "The Contemporaneity of the Non-contemporaneous or the Problem of Uneven Technological Development." In "Technology and Feminism," edited by Joan Rothschild. *Research in Philosophy and Technology* 13 (1993): 253–263.

Wild, Tony. "Slavery's Shadow on Switzerland." *New York Times,* November 10, 2014. https://www.nytimes.com/2014/11/11/opinion/slaverys-shadow-on-switzerland.html.

Windmüller, Gunda. "Unattraktive Menschen werden permanent diskriminiert" [Unattractive people are permanently discriminated against]. *Die Welt* (Berlin), February 1, 2017. https://www.welt.de/kmpkt/article161664792/Unattraktive-MensMenschen-werden-permanent-diskriminiert.html.

Yanes, Arianna. "Just Say Yes? The Rise of 'Study Drugs' in College." CNN. April 18, 2014. http://www.cnn.com/2014/04/17/health/adderall-college-students/.

FILMS

Strange Days. Dir. Kathryn Bigelow. 1995.

CHAPTER 22

Gender and Bioenhancement

Colleen A. Reilly
Professor of English
University of North Carolina Wilmington

As other chapters in this publication make clear, defining *posthumanism* proves to be complex as many approaches arise from disparate philosophical camps. This chapter focuses on approaches to posthumanism that are forward looking, in part imagining the ways that advances in technologies can help humans overcome limitations and improve their lives, specifically through bioenhancements. Posthumanist writers who highlight the role of technological interventions also commonly discuss breaking down restrictive binaries, including those between humans and animals, females and males, and people with disabilities and those apparently without. As a systems approach, posthumanism also views humans and objects as part of the same ecosystem, inseparable and interconnected. This approach is particularly important for discussions of bioenhancement.

Determining what counts as a medical bioenhancement (often referred to as biological enhancement in the medical literature) proves to be difficult and quite messy. As this chapter explains, distinguishing between a therapeutic use of a medical technology and a use that is considered an enhancement depends on perspective. Some medically unnecessary bioenhancements are done solely to bring nonnormative bodies, such as the bodies of intersexuals, in line with established norms of gender and biological sex. Should such interventions be categorized as treatments or enhancements? This chapter reviews a number of biomedical technologies that call into question the boundaries between treatments and enhancements, particularly in relation to norms of gender and biological sex. Such an interrogation is certainly in line with approaches to posthumanism.

The discussion first explores how some academics and activists have attempted to explode binary definitions of gender, sex, and sexualities in recent decades, often to the dismay of those who hold more traditional views of such norms, and examines the connections to posthumanism. The chapter then focuses on ways that bioenhancement has been used to literally construct human bodies to fall in line with and to oppose dominant constructions of genders and biological sexes. As technologies advance, bioenhancement proceeds to blur the boundaries between humans and machines, resulting in discomfort stemming from the view that particular advancements in technologies violate ethical norms. Nowhere is this more evident than in discussions of ectogenesis, or the complete gestation of an embryo or fetus in an artificial womb (Sander-Staudt 2006; Takala 2009). Obviously such a bioenhancement raises a host of issues related to genders and sexualities in part because it calls into question basic assumptions of what it means to be a female and a mother.

Consequently, the chapter ends with a discussion of the current state of ectogenesis and the ethical discussions surrounding this controversial field.

POSTHUMANISM AND CONSTRUCTIONS OF GENDERS, BIOLOGICAL SEXES, AND SEXUALITIES

Posthumanism often prompts the questioning and even exploding of established boundaries between historical binaries, including female/male and human/nonhuman. Feminist approaches to posthumanism focus, in particular, on the constructed nature of gender, biological sex, and sexuality and often draw on feminist technoscientific studies that provide evidence to question rigid categorizations of individuals and bodies (Åsberg 2013). Gender and biological sex have been acknowledged to be social constructs since the late twentieth century based on the work of such theorists as Michel Foucault (1978) and Judith Butler (1993). Butler problematizes the alignment of biological sex and gender and argues that such binary norms are discursively shaped and inherently unstable such that they are open to disruption through strategic violations. In the case of gender, for example, reification of traditional norms as well as the disruptions of these norms are enacted through routinized performance that may include routine or strategic selections of clothing, employment, child-rearing practices, and expressions of sexualities. Annette Schlichter (2011) insightfully adds vocal expressions and vocal quality to the markers of gender that humans learn and practice. Gendered identities are pluralistic but still socially determined; individuals identify themselves as feminine, masculine, queer, trans, or another gendered identity at a particular moment based on identification with and active reinscription of culturally established norms (Butler 1993). Gendered identities fall on a continuum, thus presenting individuals with greater choice although not with complete freedom. For many feminists, posthumanism offers a perspective that opens up the possibility for recognizing the multiplicity of genders (Ferrando 2014).

Additionally, norms of gender are integrally interconnected with similarly pluralistic understandings of sexualities. Just as gendered identities are fluid and arrayed on a continuum, so are sexualities unfixed, changeable over time, and comingled with but not tied to specific gendered identities. For example, apparently male individuals may behave in ways identifiable as masculine yet prefer to express their sexuality homosexually. Likewise, an androgynous female could prefer heterosexual sexual interactions. Theo G. M. Sandfort (2005) and others have reported on numerous studies that demonstrate the complexities and variability involved with interconnections of genders and sexualities.

BIOLOGICAL SEX AS CULTURALLY CONSTRUCTED

Acknowledging biological sex as similarly constructed by social, historical, and environmental realities impinges on examinations of bioenhancements and genders. Feminist technoscience examines biological sex in ways analogous to gender and sexuality, demonstrating that sexed bodies are equally constructed by their social, cultural, and historical contexts. The studies penned by Anne Balsamo (1996) and Anne Fausto-Sterling (2000, 2005) figure centrally in references to biological sex by posthumanist scholars. For example, Fausto-Sterling (2000) highlights the degree to which behavioral markers identified with biological factors such as proficiency with spatial relationships have much more to do with exposure and training than with underlying biological differences between apparently female and male bodies. As Fausto-Sterling (2000) explains, such differences between females and males in spatial task proficiency may be explained by and intervened in through educational practices or may also be based on genetic differences between individuals. Female bodies may appear more susceptible to particular diseases such as osteoporosis not solely for physiological reasons but because behaviors influence biology—females in the United States tend to diet more and exercise less in adolescence, which is a crucial time frame for developing bone density (Fausto-Sterling 2005). Fausto-Sterling notes that fully examining the socially caused differences between sexed bodies in terms of how they suffer from disease and respond to treatments is essential to making progress in treating diseases. N. Katherine Hayles (1999) has succinctly articulated the link between culture and physiology: "The body is the net result of thousands of years of sedimented evolutionary history, and it is naive to think that this history does not affect human behaviors at every level of thought and action" (284). In other words, both cultural norms and practices and the constitution of human bodies are mutually constructive and changes in practices over time and in different geographical locations are reflected in humans' physical forms.

Technoscientific studies have also highlighted explorations of intersexuality to further demonstrate the continuum of biological sex. People identified as intersexual are born with varying levels of indeterminacy of biological sex because their genitalia, chromosomal makeup, and/or hormonal levels or hormonal exposure in utero do not conform to what clearly identifies sexed bodies as either female or male. Fausto-Sterling (2000) outlines the most common types of intersexuality as congenital adrenal hyperplasia, androgen insensitivity syndrome, gonadal dysgenesis, hypospadias, and unusual chromosome compositions such as XXY (Klinefelter's syndrome) or XO (Turner's syndrome). To illustrate, children born with congenital adrenal hyperplasia have masculinized genitalia resulting from exposure to large amounts of androgen during fetal development (Fausto-Sterling 2000); however, they generally lack a urethra in their external member and may also have a vagina. Although many people know little about intersexuality and believe it to be very rare, Fausto-Sterling (2000) reports that children with some degree of intersexuality represent 1.7 percent of all births, which is equivalent to the number of children born with cerebral palsy. Information about the frequency of children born as intersexual and the physical reality presented by those children embody serious disruptions to binary constructions of biological sex.

POSTHUMANISM AND CYBORGS

Breaking down the boundaries between categories of gender, biological sex, and sexuality provides the basis for posthumanists, especially feminist posthumanists, to question a myriad of other binaries, such as those separating humans from nonhumans and from animals while simultaneously maintaining the importance of embodiment. This leads to the recognition of the intersections of humans and nonhumans and the positing of other heterogeneous identities such as that of the cyborg. As Donna Haraway (1991) argues, adopting the perspective of the cyborg aids in developing affinities with supposed externalities that are actually intertwined with the self—with nonhumans, animals, or people of other races and identities—and in harnessing tools, particularly that of language, to imagine more fluid realities. In relation to gender, Balsamo (1996) contends that "the cyborg provides a framework for studying gender identity as it is technologically crafted simultaneously from the matter of material bodies and cultural fictions" (11).

In *What Is Posthumanism* (2010), Cary Wolfe explains that the human should be acknowledged to be "fundamentally a prosthetic creature that has coevolved with various forms of technicity and materiality, forms that are radically 'not-human' and yet have nevertheless made the human what it is" (xxv). Balsamo (1996) similarly argues that human bodies are "boundary figures" that straddle multiple systems, including the "organic/natural" and "technological/cultural" (5). While admitting the inextricable connections between humans and nonhumans, feminist posthumanists seek to stress the importance of embodiment and use the destruction of boundaries for liberatory ends.

While taxonomies (categories built on boundaries) are arguably necessary for making meaning, boundaries of biological sex, gender, and sexualities that are rigid and unmovable prevent individuals who occupy and operate outside of established categories—physically or intellectually—from living open and fulfilled lives. Liberatory approaches to identities of biological sex, gender, and sexuality emphasize the importance of fluidity, not to remove all categories (an impossibility) but to unmoor them from stable foundations and allow individuals to move among and between them, living on the spaces between. As Elizabeth Reis (2005) explains in her history of intersexuality in America, until the twentieth century, being human meant being gender "in a binary way" (440), a conception that relegated intersexual persons as nonexistent or even nonhuman. Liberatory approaches to biological sex, for example, make room for a continuum of physiological manifestations all of which qualify as human.

Hayles (1999) supports a view of posthumanism that avoids the extremes proposed by those who would construct human bodies as necessarily improvable and even replaceable by technologies: "My dream is a version of the posthuman that embraces the possibilities of information technologies without being seduced by fantasies of unlimited power and disembodied immortality" (5). Wolfe (2010) stresses that posthumanism is not what follows the transcendence of humanism or embodiment; instead, it reveals the interconnections between bodies and objects and technologies that have always existed (see also Latour 2002). Olli Pyyhtinen and Sakari Tamminen (2011) eloquently explain that posthumanism is not a new state of being but a new perspective: "a way of being attentive to the production of the human in its various entanglements with its 'others'" (147–148). Recognizing humans' interconnections and codevelopment with other beings and technologies allows a broader range of acceptance for and opposition to discrimination against beings that do not fit into predetermined norms yet are worthy of ethical considerations, such as nonhuman animals (Wolfe 2010). This is why feminism, posthumanism, and critiques of androcentricism (privileging male humans and masculine perspectives over others) have broad overlap (Åsberg 2013); these ideas and approaches are united through the blurring of boundaries in order to combat oppression and power structures that are served by keeping boundaries in place.

BIOENHANCEMENT'S NORMATIVE AND TRANSGRESSIVE TURNS

Bioenhancements may be used to reinforce cultural norms, to enact oppressive processes on nonnormative bodies, or to provide liberation for other individuals. As medical technologies advance, publics grow increasingly anxious about what limits should be placed on the uses of new technologies. For example, some individuals may view gene therapies to be acceptable when used to treat cancers, while these same individuals may object to the genetic manipulation of embryos in utero in order to produce offspring with particular biological traits. In general, therapeutic uses receive wide approval; whereas, elective uses—those applied to healthy bodies to improve their function or appearance, perhaps even beyond typical biological limits—raise social and ethical concerns. As the discussion below illustrates, distinctions between therapeutic and enhanced medical technologies prove to be unclear in part because there is not absolute agreement about what constitutes normalcy, particularly in relation to biological sex and gender. Furthermore, conflicting purposes for and perceptions of specific bioenhancements also arise in part from societies' complex reactions to and the anxieties produced by the obvious intersections of humans and machines and the replacement of human bodies or parts with technologies or technological interventions. As Amanda K. Booher (2010) explains, even prostheses can invoke anxieties because the replacement of human flesh with technologies in the form of mechanized parts reminds observers of the interconnections between human bodies and technologies. Some worry that the mechanization of human bodies that is used to help people with disabilities function in a manner perceived as normal could be extended to forge super humans from all sorts of bodies, resulting in advanced cyborgs whose very presence calls into question distinctions between natural human bodies and machines and who can outperform typical human bodies in situations from athletic competitions (Booher 2010) to combat.

INTERSEXUALITY AND TRANSGENDERED INTERVENTIONS

Children born as intersexual have historically and routinely been subjected to bioenhancements, including surgical interventions, hormonal treatments, and other medications (Karkazis 2008). As Katrina Karkazis (2008) discusses in great detail, for many decades, medical doctors have surgically selected a biological sex for children born with one of the intersexual conditions mentioned above, often with lasting negative consequences and without providing parents with sufficient information about the child's diagnosis and the potential ramifications of the surgery (Feder 2002; Kessler 1990). Widespread use of surgery and other medical interventions to disambiguate the biological sex of infants and young children began in the 1960s based on the recommendations of Dr. John Money of Johns Hopkins University. Money argued that physicians needed to inform parents of their child's biological sex expeditiously, without revealing that they are making a judgment call, and encourage parents to support possible surgical corrections of any physical anomalies that may cause confusion, rejection of the child, and social anxiety (Kessler 1990). While advocates for children identified as intersexual have objected to the use of irreversible medical interventions since the 1990s, they complain that the use of these procedures on infants and children continues in the early twenty-first century (Puluka 2015).

By surgically correcting children to make them conform overtly to the norms associated with a specific biological sex, physicians often cause irreparable harm, resulting in the children's inability to feel sexual pleasure or participate in procreation. In many cases, these interventions are not medically necessary in that they do not correct a medical condition that would do physical harm to the individual; they are done simply to construct the child as clearly belonging to a specific biological sex to help them conform to cultural norms and to reduce social disruptions to the detriment of the individual's quality of life and even the individual's health. More recently, a competing recommendation has developed that allows children with intersexual conditions to grow to an age at which they can cognitively make their own decision about selecting a biological sex or, perhaps, remaining of indeterminate status but intact and physiologically healthy. In his Pulitzer Prize–winning novel *Middlesex* (2002), Jeffrey Eugenides describes the adolescence of a child who is born genetically male with 5-alpha reductase deficiency syndrome and, hence, identified as a at birth. The child, Calliope "Callie" Stephanides, receives the official diagnosis as a teenager but has known about the physical differences for some time. Callie decides, against the wishes of the family, that he wishes to live as the male he perceives himself to be, forgoing any medical interventions proposed to turn him into the girl his parents thought they had raised.

In terms of access to medical interventions and technological resources, a contrasting situation prevails for those who identify as transgendered. Transgendered individuals are identified as belonging to a particular biological sex at birth but wish to live as or undergo procedures to transition to a different gender and/or biological sex through altering eternal markers, such as clothing and hair, and physiological markers through hormone use and surgical procedures. In essence, society and the medical community treat these individuals as if they are seeking enhancement as opposed to therapeutic treatments—further alienating them from the identities, biological and gendered, that reflect their personal experiences. As Jaye Cee Whitehead and Jennifer Thomas (2013) explain, ciswomen, those who are born female, have easy access to medical procedures that allow them to improve their genital fitness to reach some perceived ideal; for instance, ciswomen may undergo tightening of the vagina, G-spot amplification, and/or labiaplasty. By contrast, individuals who seek similar procedures to facilitate transitioning to another gendered identity through acquiring the attributes of an alternate biological sex are often denied access until they can be diagnosed with mental illness (Whitehead and Thomas 2013), thereby providing a therapeutic justification for the procedures. "Mental health practitioners in the [United States] who see clients considering or directly requesting hormone therapy or SRS [sex reassignment surgery] are advised to follow professional guidelines defined by the *Diagnostic and Statistical Manual of Mental Disorders*, IV (DSMTR) in its section on 'Gender Identity Disorder'" (Whitehead and Thomas 2013, 384). The uses of technologies for bioenhancement of intersexual and transgendered individuals are connected through societal support of heteronormativity—procedures are easier to access and used more routinely if they can create individuals who do not disrupt established binaries of gender and biological sex.

MECHANIZED CONSTRUCTIONS OF INDIVIDUALS WITH DISABILITIES

The latest technologically advanced prosthetics often do not physically resemble the biological limbs or organs they are designed to replace (Booher 2010). Cyborg figures from fiction and popular culture, such as Max Headroom, the Transformers, and RoboCop (Balsamo 1996), represent such overt human-machine mergers, but new technologies have brought these innovations to life, liberating individuals and unsettling publics. Booher (2010) closely examined the 2007 public representations of and reactions to two women, both of whom have prosthetic legs: "Heather Mills, who competed on *Dancing with the Stars*; and Sarah Reinertsen, a tri-athlete who was featured in the 2007 Lincoln 'Dreams' advertising campaign" (65). Mills wore a biologically mimetic prosthetic on *Dancing with the Stars*, while Reinertsen was depicted in the advertising campaign with her Flex-Foot Cheetah prosthetic made by the Icelandic firm Össur, also used by South African sprinter Oscar Pistorius in the 2012 Olympic Games, which highlights her cyborg quality more overtly. The animalistic name of the Cheetah prosthetic is also quite intriguing—emphasizing the hybrid, other-than-human status of its wearer. While prosthetics for individuals to aid in their mobility certainly constitute therapeutic uses for bioenhancements, prosthetics such as the Cheetah raise issues largely because they appear so mechanized. Pistorius, for example, had to appeal to the Court of Arbitration for Sport in order to be allowed to use his

Cheetah prosthetics to compete in the Olympics, as other athletes argued that these bioenhancements provided him with a superhuman advantage.

In her analysis of representations of Mills and Reinertsen, Booher (2010) argues that both women are represented as laudable figures who have overcome serious disabilities through the use of prostheses. In fact, Mills and Reinertsen seem to be ideal female and feminine specimens; they are in excellent shape and present themselves as aesthetically appealing and sexualized in normative feminine ways. Mills wore dresses with slits up the leg on *Dancing with the Stars*, revealing her shapely prosthetic; similarly, Reinertsen wears shorts and tank tops that show off her toned, athletic body (Booher 2010).

While these representations of Mills and Reinertsen present the interconnections of female bodies and prosthetics in positive ways, Booher argues that cultural dismay with female comfort with the physical "malleability" (Booher 2010, 75) that is afforded by the women's obvious cyborg status leaks out in different ways and is evident in reactions to and representations of both women, despite the differences in the appearance of their prosthetics. Booher (2010) contends that discomfort with Mills's prosthetic is illustrated in part by jokes on late-night comedy shows and by altered YouTube videos showing Mills's prosthetic leg falling or flying off while she is dancing. In the print version of the Lincoln automobile ad, Reinertsen's prosthetic facilitates the substitution of her image, that of a mechanized female body, the cyborg, for the car itself (Booher 2010). The text below Reinertsen in the ad describes a machine, the car, not the human female who is shown. Thus, representations of both women reveal discomfort with women who are too easily embracing and even showcasing their hybrid selves, selves that conform to cultural standards of the ideal of femininity and female bodies while simultaneously betraying that ideal through the celebration of mechanization.

COSMETIC SURGERIES AND THE PURSUIT OF GENDERED AND BIOLOGICAL IDEALS

Cosmetic surgery represents another category of bioenhancement integrally connected with conceptions of gender, biological sex, and sexuality. Cosmetic surgery can be done for therapeutic or elective reasons. As Booher (2010) explains, reconstructive cosmetic surgery, such as breast reconstruction, is viewed as socially acceptable and even necessary, as it employs technological interventions to artificially simulate "natural" bodies, allowing women who receive them to conform to social norms of gender. In the case of breast reconstruction, in fact, women who undergo a mastectomy are often pressured by doctors and others to have the procedure (Sulik 2013). Other sorts of elective cosmetic surgery, such as breast implants and facial procedures, including face-lifts and lip enhancements, are more controversial. Women who have such procedures that result in obvious, artificial results are often subjected to public ridicule.

Feminist scholars have presented critical readings of elective and even some therapeutic cosmetic surgeries, as they view these procedures as being used to construct and reify ideal norms of feminine beauty (Balsamo 1996; Booher 2010). Balsamo (1996) argues that "cosmetic surgery is not then simply a discursive site for the 'construction of images of women' but is actually a material site at which the physical female body is surgically dissected, stretched, carved, and reconstructed according to cultural and eminently ideological standards of physical appearance" (13). Balsamo's admonition is especially relevant in relation to procedures included under the heading of female genital cosmetic surgery (FGCS). FGCS raises ethical concerns among physicians and professional organizations, such as the American College of Obstetricians and Gynecologists, which notes that FGCS is not "medically indicated" and that the procedures so categorized have not been deemed to be safe and effective (Cain et al. 2013). Joanna M. Cain and colleagues (2013) list a striking number of procedures classified as FGCS: "vaginoplasty, perineoplasty, reduction labiaplasty, augmentation labiaplasty, clitoral unhooding/hoodplasty, frenuloplasty, G-spot amplification, hymenoplasty, 're-virgination,' and aesthetic vulvar liposculpturing with autologous fat transfer" (170).

Many women who seek FGCS do so for purely aesthetic reasons; the most common procedure, unsurprisingly, is labia minora labiaplasty (Iglesia, Yurteri-Kaplan, and Alinsod 2013). Women look to this procedure to give them the "Barbie look," "a slang term used by lay patients in Los Angeles who requested all or almost all of the labia minora to be removed.... In one study, 98% of 238 women requested the 'Barbie look'" (Iglesia, Yurteri-Kaplan, and Alinsod 2013, 2002). Vanessa R. Schick, Brandi N. Rima, and Sarah K. Calabrese (2011) found in their analyses of 647 *Playboy Magazine* centerfolds that the models are represented in a manner that "mask[s] or minimize[s] genitalia, presenting them in a hairless, prepubescent form" (78). Like the women seeking FGCS, Schick and her colleagues connect the representations of centerfolds in *Playboy Magazine* to the incomplete anatomical shape of Barbie dolls, which present young girls with a "warped perception of the adult female body" (78). Therefore, this biological enhancement, labia minora labiaplasty, promises to alter completely healthy and physiologically normal female bodies to conform to an unrealistic ideal that is propagated by advertising, media, and pornography (Cain et al. 2013). This raises ethical concerns, particularly in light of the lack of data regarding the efficacy of FGCS to provide physiological benefits; the potential side effects, such as urinary incontinence, of several procedures; and the lack of training among the physicians who perform the procedures, most of whom are plastic surgeons and not gynecological specialists (Iglesia, Yurteri-Kaplan, and Alinsod 2013). One study found, disturbingly, that 54 percent of women who had vaginoplasty and perineoplasty and 24 percent who had a combination of FGCS procedures did so to "enhance their male partner's sexual experience" (Iglesia, Yurteri-Kaplan, and Alinsod 2013, 1998).

Male bodies also routinely undergo biomedical enhancements, including infant circumcision, which has become controversial, because it is increasingly viewed as medically unnecessary (Savulescu 2013). In fact, in 2012, the American Academy of Pediatrics determined that based on their research "the benefits are not great enough to recommend universal newborn circumcision." Male circumcision is one of the oldest forms of bioenhancement (Savulescu 2013); J. Steven Svoboda (2013) reports that neonatal male circumcision is the "most commonly performed surgical procedure in the USA" (469). Svoboda graphically clarifies the nature of the procedure: "circumcision is the removal of the male prepuce, which excises between a third and a half of the skin system of the penis and nearly all of its fine touch neuroreceptors" (470).

While the procedure has been performed historically for religious, aesthetic, and health reasons, the medical benefits have been called into question in recent decades; religious reasons for male infant circumcision dominate outside the United States, which is the only Western country where the procedure is still done routinely (Svoboda 2013). The history of male infant circumcision in the United States, particularly outside of religious communities, is strongly interconnected with issues of gender and sexuality, as the practice was established in the nineteenth century "to stop masturbation" (Svoboda 2013, 469; Castro-Vázquez 2013); the procedure was continued for a myriad of health reasons, including the prevention of urinary tract infections, sexually transmitted diseases, and, most recently, the infection of female partners with HIV (Svoboda 2013). Because the American Medical Association and the American Academy of Pediatrics (Svoboda 2013) now recommend against the routine use of neonatal male circumcision outside of religious reasons, the procedure falls into the category of bioenhancement. Some medical ethicists go further, arguing that performing the procedure on infants constitutes a violent and unethical violation of their "right to bodily integrity" (Svoboda 2013, 470) because of the inability of infants to consent and the irreversible nature of the procedure.

The use of male circumcision in Japan highlights the procedure as a means for gendered bioenhancement, particularly in that cultural context. Genaro Castro-Vázquez (2013) explains that males in Japan are not routinely circumcised at birth; the procedure is sold to them as adults largely to enhance the aesthetic attributes and sexual function of their penises and "to boost maleness" (696). Brochures and other advertisements mention the health benefits of the procedure in terms of curing phimosis, a condition in which the prepuce either does not expose the penis glans or after uncovering the glans cannot return to the covered position, resulting in strangulation (Castro-Vázquez 2013). Phimosis requiring surgical intervention is rare; furthermore, advertising for circumcision in Japan glosses over the details of the health condition and focuses on the beautification resulting from an unencumbered organ and the greater sexual pleasure that will result. Just as with FGCS, medical and quasi-medical professionals—namely, plastic surgeons and beauticians in Japan—capitalize on the needs and desires of individuals to conform to some ideal gender-related norm and create profitable businesses providing medically unnecessary bioenhancements.

As the examples above illustrate, individuals use bioenhancements to shape their bodies to achieve ideals of genders and of sexual expression and fulfillment. Taken to the extreme, some see bioenhancements as capable of breaking down inequalities of genders and biological sexes altogether by removing the basic distinctions between female and male bodies, particularly in terms of procreation. The next section explores procedures designed to bring about equality through one type of technological advance: ectogenesis.

ECTOGENESIS: AN OUT-OF-BODY BIOENHANCEMENT

Technologies relating to reproduction raise concerns among members of some religious communities and bioethicists alike and prove to be some of the most controversial in relation to bioenhancement. For example, some religious leaders objected to in vitro fertilization as interfering in the work of the divine and the "natural" process for fertilization, and some bioethicists viewed it as too risky to justify; nonetheless, that procedure has now become routine—despite the continued opposition of powerful institutions like the Catholic Church (Simonstein 2009b). Currently, publics and medical communities view the technological manipulation of embryos as morally dubious, objecting to sexual selection, for example, because of the androcentric attitudes of most societies (Heyd 2009) and the ethical ramifications of intervening in natural processes that might lead to other similar interventions, such as selecting for physical attributes and mental capacities. Even as societies grapple with the possibilities offered by advances in medical and genetic technologies that purport to improve human biology, other advances would remove the need for human gestation altogether. As the discussion below illustrates, progress in medical advances necessary for ectogenesis have been made as a by-product of advances in saving the lives of premature infants. Using biomedical advancements to save premature infants is clearly therapeutic; whereas, ectogenesis would certainly be classified as an enhancement. Nonetheless, some posit that separating gestation and birth from female bodies could lead to greater equality for women; others, however, predict dire consequences, such as the forced use of artificial wombs and the devaluation of motherhood.

Advocating for ectogenesis might be more aligned with transhumanism rather than posthumanism. As a number of scholars explain, the transhumanism of such writers as Marvin Minsky and Hans Moravec views the body as an impediment to be overcome through technological advances (Ferrando 2014). "Trans-humanist conceptualizations of the post-human translate into the desire to realize the disembodied human self of the Enlightenment, purified and enhanced by science, medicine, and technology in order to transcend disease, ageing, and eventually death" (Åsberg 2013, 10). Despite the fact that transhumanism proposes the destruction of dichotomies such as female/male (Fernández Guerrero 2011), transhumanism's negation of embodiment makes it incompatible with many conceptions of posthumanism in general (Ferrando 2014) and feminist posthumanisms in particular (Åsberg 2013; Hayles 1999).

Technologies often advance more quickly than the laws and ethical standards designed to regulate and constrain them. As Gregory Pence (2006) explains, many biomedical advances already in common use in neonatal intensive care units, such as extracorporeal membrane oxygenation, which allows premature babies to breathe, could be leveraged to develop the technologies needed for ectogenesis. Peter Singer and Deane Wells (2006) predict that as doctors develop processes and gain access to technologies to keep smaller and smaller premature infants alive outside the mother's womb, the developments needed for complete ectogenesis will be created "almost by accident" (10). Independent research has also made advances. Endocrinologist Helen Hung-Ching Liu, while at Cornell University in 2001, developed an artificial womb made of collagen, lined it with human endometrial cells, and implanted a fertilized ovum, which lived for six days before the experiment was terminated (Gelfand 2006).

Besides being ethically challenging, ectogenesis would disrupt cultural ideas of femininity and motherhood (Brassington 2009; Pence 2006). Severing the long-standing connection between reproduction and women's bodies could be potentially liberating for women, as they would no longer be viewed as the default child-rearers, and they could avoid the physical complications of pregnancies (Sander-Staudt 2006; Simonstein 2009a; Simonstein and Mashiach-Eizenberg 2009). Distancing women from gestation could remake motherhood as a gender-neutral activity, one open to an individual of whichever biological sex who would like to nurture and raise children (Sander-Staudt 2006). Such a reorientation could radically shift the current constructions of genders that are so interconnected with static ideas revolving around the connections between females and childbirth.

Other scholars posit that ectogenesis could have unintended and negative consequences. If artificial wombs are viewed as safer, for example, women might be pressured not to carry their children themselves (Sander-Staudt 2006). Rosemarie Tong (2006) views mandated ectogenesis as a further push for the most extreme versions of posthuman or transhuman rejections of embodiment and, in the end, of women. Additionally, the symbiotic relationship between human mothers and their children is not completely understood; without such knowledge, it is unclear what might be sacrificed through pursuing ectogenesis for nontherapeutic reasons (Sander-Staudt 2006). Ectogenesis might be spurred on by other advances in reproductive technologies; when scientists can engineer superior children through genetic interventions, they will also want to develop a perfect environment for the gestation of the perfected embryos (Tong 2006). Such advances could lead to biological wombs being viewed as inferior and unsafe. This could also result in economic inequalities, as insurance may not cover this extreme type of bioenhancement, resulting in its access being restricted to higher-income individuals (Simonstein 2009a).

While ectogenesis may seem extreme and unlikely because the moral and ethical problems raised by it appear insurmountable, such has been the case for many new biomedical technologies. Frida Simonstein (2009a) explains that many other medical advancements, including blood transfusions, organ transplants, and artificial insemination, raised ethical and moral issues when introduced that quickly dissipated as the procedures became more common. A number of scholars predict the same will happen with ectogenesis because the need to save smaller and younger premature babies will bring about the technological advances needed for ectogenesis (Simonstein and Mashiach-Eizenberg 2009; Singer and Wells 2006). Society will then need to face the disruptive consequences of these advances.

Summary

Examining the intersections of bioenhancements and genders provides an excellent opportunity to grasp the degree to which both concepts are socially constructed and highly dependent on context and established norms. What constitutes a bioenhancement versus a necessary medical treatment depends on the historical moment and cultural context, as the discussion of the surgical interventions routinely performed on intersexual individuals reveals. Many individuals who are intersexual could live healthy lives without any medical treatments; such enhancements have been routinely employed to make those individuals less disruptive to the established norms of genders. When considering raising a child who is intersexual without medically intervening to clarify the biological sex of that child, people quickly come around to issues of public restroom use, especially in school settings. Such questions are interconnected with similar concerns about the facilities used by transgendered individuals; laws such as HB2 in North Carolina (enacted in 2016) seek to restrict these individuals from using restrooms and locker rooms that do not coincide with the biological sex listed on their birth certificates. Negotiations of issues of gender and bioenhancement prove to be quite complex, as the boundaries are ever shifting and biomedical technologies develop faster than a society's cultural norms and laws.

Bibliography

American Academy of Pediatrics. "Newborn Male Circumcision." August 27, 2012. https://www.aap.org/en-us/about-the-aap/aap-press-room/pages/newborn-male-circumcision.aspx.

Åsberg, Cecilia. "The Timely Ethics of Posthumanist Gender Studies." *Feministiche Studien* 31, no. 1 (2013): 7–12.

Balsamo, Anne. *Technologies of the Gendered Body: Reading Cyborg Women.* Durham, NC: Duke University Press, 1996.

Booher, Amanda K. "Docile Bodies, Supercrips, and the Plays of Prosthetics." *International Journal of Feminist Approaches to Bioethics* 3, no. 2 (2010): 63–89.

Brassington, Iain. "The Glass Womb." In *Reprogen-ethics and the Future of Gender*, edited by Frida Simonstein, 197–209. Dordrecht, Netherlands: Springer, 2009.

Butler, Judith. *Bodies That Matter: On the Discursive Limits of "Sex."* New York: Routledge, 1993.

Cain, Joanna M., Cheryl B. Iglesia, Bernard Dickens, and Owen Montgomery. "Body Enhancement through Female Genital Cosmetic Surgery Creates Ethical and Rights Dilemmas." *International Journal of Gynecology and Obstetrics* 122, no. 2 (2013): 169–172.

Castro-Vázquez, Genaro. "The 'Beauty' of Male Circumcision in Japan: Gender, Sexuality, and the Male Body in a Medical Practice." *Sociology* 47, no. 4 (2013): 687–704.

Eugenides, Jeffrey. *Middlesex*. New York: Farrar, Straus and Giroux, 2002.

Fausto-Sterling, Anne. "The Bare Bones of Sex: Part 1—Sex and Gender." *Signs: Journal of Women in Culture and Society* 30, no. 2 (2005): 1491–1527.

Fausto-Sterling, Anne. *Sexing the Body: Gender Politics and the Construction of Sexuality*. New York: Basic Books, 2000.

Feder, Ellen K. "Doctor's Orders: Parents and Intersexed Children." In *The Subject of Care: Feminist Perspectives on Dependency*, edited by Eva Feder Kittay and Ellen K. Feder, 294–320. Lanham, MD: Rowman and Littlefield, 2002.

Fernández Guerrero, Olaya. "Posthuman Bodies? Not So Fast." *Dilemata* 2, no. 5 (2011): 181–193.

Ferrando, Francesca. "Is the Post-human a Post-woman? Cyborgs, Robots, Artificial Intelligence, and the Futures of Gender: A Case Study." *European Journal of Futures Research* 2, no. 43 (2014). doi:10.1007/s40309-014-0043-8.

Foucault, Michel. *The History of Sexuality*, Vol. 1: *An Introduction*. Translated by Robert Hurley. New York: Pantheon Books, 1978.

Gelfand, Scott. Introduction to *Ectogenesis: Artificial Womb Technology and the Future of Human Reproduction*, edited by Scott Gelfand and John R. Shook, 1–7. Amsterdam: Rodopi, 2006.

Haraway, Donna. *Simians, Cyborgs, and Women: The Reinvention of Nature*. New York: Routledge, 1991.

Hayles, N. Katherine. *How We Became Posthuman: Virtual Bodies in Cybernetics, Literature, and Informatics*. Chicago: University of Chicago Press, 1999.

Heyd, David. "Male *or* Female We Will Create Them: The Ethics of Sex Selection for Non-medical Reasons." In *Reprogen-ethics and the Future of Gender*, edited by Frida Simonstein, 161–173. Dordrecht, Netherlands: Springer, 2009.

Iglesia, Cheryl B., Ladin Yurteri-Kaplan, and Red Alinsod. "Female Genital Cosmetic Surgery: A Review of Techniques and Outcomes." *International Urogynecological Journal* 24, no. 12 (2013): 1997–2009. doi:1007/s00192-013-2117-8.

Karkazis, Katrina. *Fixing Sex: Intersex, Medical Authority, and Lived Experience*. Durham, NC: Duke University Press, 2008.

Kessler, Suzanne J. "The Medical Construction of Gender: Case Management of Intersexed Infants." *Signs* 16, no. 1 (1990): 3–26. doi: 10.1086/494643.

Latour, Bruno. "Morality and Technology: The End of the Means." Translated by Couze Venn. *Theory, Culture, and Society* 19, nos. 5–6 (2002): 247–260.

Pence, Gregory. "What's So Good about Natural Motherhood? (In Praise of Unnatural Gestation)." In *Ectogenesis: Artificial Womb Technology and the Future of Human Reproduction*, edited by Scott Gelfand and John R. Shook, 77–88. Amsterdam: Rodopi, 2006.

Puluka, Anne. "Parent versus State: Protecting Intersex Children from Cosmetic Genital Surgery." *Michigan State Law Review* 2095, no. 5 (2015): 2095–2141. http://digitalcommons.law.msu.edu/lr/vol2015/iss5/7.

Pyyhtinen, Olli, and Sakari Tamminen. "We Have Never Been Only Human: Foucault and Latour on the Question of the Anthropos." *Anthropological Theory* 11, no. 2 (2011): 135–152.

Reis, Elizabeth. "Impossible Hermaphrodites: Intersex in America, 1620–1960." *Journal of American History* 92, no. 2 (2005): 411–441.

Sander-Staudt, Maureen. "Of Machine Born? A Feminist Assessment of Ectogenesis and Artificial Wombs." In *Ectogenesis: Artificial Womb Technology and the Future of Human Reproduction*, edited by Scott Gelfand and John R. Shook, 109–128. Amsterdam: Rodopi, 2006.

Sandfort, Theo G. M. "Sexual Orientation and Gender: Stereotypes and Beyond." *Archives of Sexual Behavior* 34, no. 6 (2005): 595–611. doi:10.1007/s10508-005-7907-8.

Savulescu, Julian. "Male Circumcision and the Enhancement Debate: Harm Reduction, Not Prohibition." *Journal of Medical Ethics* 39, no. 7 (2013): 416–417.

Schick, Vanessa R., Brandi N. Rima, and Sarah K. Calabrese. "Evulvalution: The Portrayal of Women's External Genitalia and Physique across Time and the Current Barbie Doll Ideals." *Journal of Sex Research* 48, no. 1 (2011): 74–81. doi: 10.1080/00224490903308404.

Schlichter, Annette. "Do Voices Matter? Vocality, Materiality, Gender Performativity." *Body and Society* 17, no. 1 (2011): 31–52.

Simonstein, Frida. "Artificial Reproductive Technologies and the Advent of the Artificial Womb." In *Reprogen-ethics and the Future of Gender*, edited by Frida Simonstein, 177–186. Dordrecht, Netherlands: Springer, 2009a.

Simonstein, Frida. Introduction to *Reprogen-ethics and the Future of Gender*, edited by Frida Simonstein, 1–11. Dordrecht, Netherlands: Springer, 2009b.

Simonstein, Frida, and Michal Mashiach-Eizenberg. "A Survey of People's Attitude towards the Artificial Womb and Ectogenesis in Israel." In *Reprogen-ethics and the Future of Gender*, edited by Frida Simonstein, 211–219. Dordrecht, Netherlands: Springer, 2009.

Singer, Peter, and Deane Wells. "Ectogenesis." In *Ectogenesis: Artificial Womb Technology and the Future of Human Reproduction*, edited by Scott Gelfand and John R. Shook, 9–25. Amsterdam: Rodopi, 2006.

Sulik, Gayle A. "Rights or Rhetoric? Breast Reconstruction Awareness." *Psychology Today*, March 30, 2013. https://www.psychologytoday.com/blog/pink-ribbon-blues/201303/rights-or-rhetoric-breast-reconstruction-awareness.

Svoboda, J. Steven. "Circumcision of Male Infants as a Human Rights Violation." *Journal of Medical Ethics* 39, no. 7 (2013): 469–474.

Takala, Tuija. "Human before Sex? Ectogenesis as a Way to Equality." In *Reprogen-ethics and the Future of Gender*, edited by Frida Simonstein, 187–195. Dordrecht, Netherlands: Springer, 2009.

Tong, Rosemarie. "Out of Body Gestation: In Whose Best Interests?" In *Ectogenesis: Artificial Womb Technology and the Future of Human Reproduction*, edited by Scott Gelfand and John R. Shook, 59–76. Amsterdam: Rodopi, 2006.

Whitehead, Jaye Cee, and Jennifer Thomas. "Sexuality and the Ethics of Body Modification: Theorizing the Situated Relationships among Gender, Sexuality, and the Body." *Sexualities* 16, nos. 3–4 (2013): 383–400.

Wolfe, Cary. *What Is Posthumanism?* Minneapolis: University of Minnesota Press, 2010.

FILM

XXY. Dir. Lucía Puenzo. 2007. The story of an intersexual fifteen-year-old raised as a girl, the way her family copes with her condition, and the ultimate decision that she must eventually make as she seeks her gender identity.

CHAPTER 23

Animal Bioenhancement

Amy Defibaugh
Doctoral Candidate, Department of Religion
Temple University, Philadelphia

On October 26, 1984, in Loma Linda, California, the heart of a baboon was transplanted into a twelve-day-old baby girl, known only as Baby Fae, who was suffering from hypoplastic heart syndrome. Two days later, Leonard L. Bailey, a pediatric surgeon who was in charge of the procedure and the research on which it was based, along with members of the transplant team, held a press conference to discuss Baby Fae's progress, while demonstrations took place outside the hospital. As the *New York Times* reported at the time, many were protesting the sacrifice of a healthy animal, whereas others were showing their support for the intrepid experimental procedure. One of the doctors in the operating room said, "At the moment the heart began to beat, there was absolute awe. It was just an incredible event to see that little heart start up." When asked about the sacrifice of the baboon, Jack Provonsha, director of Loma Linda University's bioethics center, said that "it is difficult to look at a primate's hand and not feel kinship.... Primates are closer to us on the evolutionary scale." While Bailey was empathetic to the concerns over animal rights, he clearly valued the life of a sick child over that of a healthy primate, as most humans would (Altman 1984b).

For the first week or so, Baby Fae was in stable but serious condition, having been given doses of antirejection medication known as cyclosporin A, a newly developed drug credited with being instrumental in successful human-to-human transplants in the 1970s and 1980s. The Loma Linda University Medical Center was swarming with journalists wanting additional details on this experimental surgery. Criticism of the transplant began to arise, particularly regarding whether the parents were properly informed. When it was reported that Bailey did not seek out a human heart for the transplant, further ethical questions came to light, many of which Bailey refused to answer. One vocal critic, a bioethicist at the Hastings Center, said that Baby Fae had been "drafted in the name of science." However, the transplantation of a baboon heart into the body of Baby Fae was approved by both the medical center's ethics committee and its institutional review board and was perhaps the culmination of Bailey's research over the preceding seven years, during which time he had completed over 150 cross-species organ transplants. Even though none of these animals lived longer than six months, Bailey seemed to have created a substantial framework to apply to humans. In addition to the approval by the ethics committee and the institutional review board, Bailey and his team had also received the support of Stuart W. Jamieson, a member of the Stanford University heart transplant team, one of the foremost in the country at the time. With Jamieson's consultations, the doctors at Loma Linda had gone through similar preparation and dummy experiments, including conventional transplant procedures and connecting a baboon heart to a heart-lung machine to circulate human blood and examine whether the organ is at all damaged by the human blood. For Bailey, these experiments showed that an infant's body would be less likely to reject a baboon heart, particularly because the immune system of an infant is not as developed as that of an adult.

Baby Fae showed some signs of rejecting the new organ about two weeks after the surgery, but this was only a brief episode that was treated with another dose of antirejection medication. At this time, Baby Fae became the "longest surviving human recipient of a transplanted animal heart" (Boffey 1984). Unfortunately, shortly after this episode, twenty days after the transplant, Baby Fae died, her body having rejected the baboon heart. A day later, there was reaction from across the country. Many believed that Bailey had taken an important step considering the national organ shortage, whereas others thought the procedure had been too experimental and too risky. Some criticism centered on the genetic difference between animal and human tissues, whereas others focused on the newly developed methods of antirejection therapies.

The story of Baby Fae's xenotransplantation (the transplanting of an organ or tissue from one species to another) has become a significant marker on the time line of medical and technological history. The story's narrative components—the tenacity of the Loma Linda transplant team, the procedure itself, and the reaction and criticism of reporters, animal welfare advocates, ethicists, and the general public—illustrate important themes and questions. These themes include procedures such as xenotransplantation and further techno-medical developments, the use of animal bodies in human biomedical enhancement in general, the ways in which these capabilities challenge the traditional human/animal binary (or the difference and presumed superiority of humans over all other species), and the ethical questions that arise from these new ways of thinking about the relationship between humans and nonhuman animals. This chapter examines these areas, as well as other themes, such as stem cell research, chimeric and transgenic human and animal bodies, and questions of posthumanism, or what it means to be human and how those perspectives have changed in the face of developing and emerging biomedical technologies. Admittedly, this chapter only skims the surface of these burgeoning scientific experiments and is an attempt to continue the ethical and critical dialogue.

DEFINING THE TERMS

To fully understand the scientific, medical, and ethical examination herein of what is known as *animal bioenhancement*, which is an umbrella term employed to cover any use of animal bodies to biologically and technologically enhance human bodies, one should be aware of some important terms. Underlying this examination is the traditional *human/animal binary*. Since the time of ancient Greek philosopher Aristotle—who defined human beings as the "rational animal," set apart from and superior to other animals in their ability to both think and reason—nonhuman animals have mostly been at the disposal of human need and desire, for food, agriculture, entertainment, clothing, scientific experimentation in the pursuit of medical and technological knowledge, and much more. This distinction between human and animal has often led to human exceptionalism in which humans are deemed unique in regard to language, physicality, morality, and consciousness. This view has come into serious question, perhaps most notably with the animal rights and animal liberation movements beginning in the 1970s (Singer 1975). More recently, the idea that "animals do matter" has continued to be a central topic for discussion, as seen in the burgeoning interdisciplinary fields of animal studies and human-animal studies, as well as in the study of posthumanism, to name a few areas.

As mentioned in the introduction, organ transplantation refers to the grafting of cells, tissues, and organs from one individual to another within the same species. *Xenotransplantation* refers to the same procedure but between members of different species. For example, numerous experiments have been conducted between rats and mice, which are genetically and physiologically similar. Since the early 1980s, xenotransplantation has led to the discovery of *transgenic* organisms and the possibility of *chimeric* animals; as discussed below, this possibility is becoming more and more of a reality with new and groundbreaking research. While chimeric animals include "individuals composed of a mixture of genetically different cells," transgenic organisms are "any organism whose genome has been modified by the artificial introduction of external DNA sequences into its germ line of DNA, leading to the growth and development of an individual with the desired modification in every cell" (Mai, Young Owl, and Kersting 2005). In other words, both terms apply to organisms with two different sets of DNA. However, chimeric organisms are usually limited to partial genetic modification, perhaps to just one organ. Both manifestations point to the overall genetic modification of animals for the purpose of animal experimentation and research, usually undertaken in the name of human welfare. These scientific and medical developments are explored below with respect to the ethical questions concerning animal welfare and the reevaluation of the human/animal binary.

To fully understand this complicated landscape, readers should also have a basic understanding of *stem cells* (cells that can grow into many different type of cells in the body) and *embryonic research* (which works to study the early stages of genetic and physiological development), both of which act as a foundation for experimenting with and creating transgenic and chimeric organisms. This specific research addresses an important issue facing medical care: organ donation and organ shortages. As of early 2017, there were more than 119,000 men, women, and children on the national organ transplant waiting list in the United States, a country in which twenty-two people die each day waiting for a transplant (HHS 2017a). A record-setting 33,606 organ transplants were performed in the United States in 2016 from 15,946 donors (HHS 2017b). According to the Organ Procurement and Transplantation Network (2017), the gap between those on the waiting list and the supply of organs continues to widen, despite targeted programs to increase awareness about organ donation and advancements in medical technologies. In the United States and around the world, there seems to be little hope of actually meeting the demand for organs without exploring artificial creations. David Shaw and colleagues argue that there are three possible "routes towards the creation of viable 'man-made' organs" that have been explored in recent years: the procurement of animal organs in hopes of transplanting them into human bodies (xenotransplantation), "in vitro creation of 'lab-grown' organs, and the construction of bionic, mechanical organs," such as 3-D printed organs (2015, 970). While the latter is not addressed in this chapter, studies have shown great potential in "enabling technology for tissue engineering" creating a "biomaterial scaffold" that can be modified and manipulated to potentially form any complex cell structure (Lee and Dai 2017, 115).

The success of organ transplantation in the latter half of the twentieth century and the beginning of the twenty-first stemmed in part from the development and marketing of immunosuppressive drugs. In 1949 cortisol was discovered to have an immunosuppressive effect, followed in 1976 by cyclosporin A (the drug used for Baby Fae's baboon heart). Other immunosuppressive drug treatments were developed in the 1980s and 1990s. By partially or completely suppressing the human body's immune system response (the body's natural response to foreign material), immunosuppressive treatment assists and has come to be found necessary in the survival of new organs (Lee and Dai 2017).

HISTORY OF ANIMAL BIOENHANCEMENTS: FROM XENOTRANSPLANTATION TO CHIMERIC EMBRYOS

While it is not possible in this chapter to cover every technological and medical breakthrough that has led to the development and creation of chimeric embryos, a few instances of xenotransplantation are worth mentioning that seem to have laid the groundwork, set the stage, and complicated the scientific and moral landscape of chimeric and hybrid organ research. According to A. S. Daar, "efforts at xenotransplantation [date] back to the early part of [the twentieth] century, well before we knew anything about the immunological principles underlying transplantation in general" (1999, 54). Daar examines developments in xenotransplantation after 1980 in which higher-profile attempts at vascularized (whole-organ) xenotransplants were made, in addition to some successful partial-tissue and cellular xenotransplants. After Bailey's attempt to transplant a baboon heart into Baby Fae, Thomas Starzl and his team at the University of Pittsburgh were given permission to perform four baboon-to-human liver transplants in 1992. The first human recipient lived for seventy days and the second for only twenty-six days. After these deaths, Starzl's team decided not to go through with the other planned procedures. Despite these setbacks, research continued in both genetically modified animals (changing the genetic makeup of donor animals in an attempt to avoid immediate organ and tissue rejection in xenotransplants and cross-species infection) and cellular chimerism or microchimerism (which involves a small number of cells that originate from another individual and are therefore genetically distinct from the cells of the host individual, as often seen during mammalian pregnancy where cells are mixed between the mother and fetus, each containing the other's distinct and different genetic material. The effects of fetal microchimerism are largely unknown, though there are some theories that link it to autoimmune disease and some types of cancer in the mother) (Fugazzola, Cirello, and Beck-Peccoz 2011).

In 1995 the first baboon-to-human bone marrow xenotransplant was completed, believed to be a possible cure for advanced HIV infections. Jeff Getty, a gay rights and AIDS activist who was suffering from AIDS himself, was the thirty-eight-year-old recipient; he was not expected to have very long to live: "the theory was that if his marrow was partly reconstituted by that of a baboon (to develop chimaerism), he would be much improved because baboon lymphocytes are not infectable with HIV" (Daar 1999, 56). Technically, the xenotransplant was considered unsuccessful, after it was found that the baboon bone marrow had disappeared from his system. Despite showing no signs of the baboon cells, Getty's physical health improved, puzzling his doctor at San Francisco General Hospital. Getty lived for another eleven years, and went on to help pave the way for the highly active antiretroviral therapy drug cocktails, which are used to treat many HIV and AIDS patients today.

The use of nonhuman primates (NHPs) for both xenotransplantation and other medical and biotechnological research reached its peak in the 1990s, primarily as a result of their availability and the similarities between NHPs and humans. NHPs became seemingly perfect proxies for humans in medical testing. According to a 2012 report by Kathleen

M. Conlee and Andrew N. Rowan of the Humane Society of the United States, the availability of NHPs was no accident. In the 1960s research facilities in the United States began investing in the creation of the National Primate Research Centers, by both capturing wild NHPs and establishing breeding facilities, to provide NHPs to scientists for research. Many of these programs started as a way to test vaccines, particularly for polio, and later served to supplement space research and travel (as the United States believed it needed a leg up in the international space race). Many of these research facilities did not produce the beneficial results they were hoping for and were, in the 1980s, "rescued" through renewed interest "by the discovery that primates at the California Regional Primate Research Center were suffering from a simian version of AIDS" (Conlee and Rowan 2012, S31). In addition to their use in AIDS research and the development of AIDS treatments, NHPs were also seen, and continue to be seen, as human proxies for neurological and cognitive development research and behavioral and physiological studies, as NHPs continue to be humans' closest evolutionary relative.

The same traits exhibited by NHPs that make them good candidates for scientific and medical research to improve the lives of humans also provide the context for important ethical concerns, which are discussed in further detail below. Their cognitive and emotional abilities, such as "mathematical, memory, and problem-solving skills," as well as their long life spans and their need for social companionship, make NHPs both well suited and ethically problematic for scientific research. While there has been some concerted effort to phase out NHPs in scientific and medical research, 70,000 NHPs are still used each year in such research in the United States (Conlee and Rowan 2012).

In *The Transplant Imaginary* (2014, 52), Lesley A. Sharp notes that we may now be in the "age of the pig" regarding transplant and chimeric research "as the broader politics of using such highly sentient creatures [as nonhuman primates have] subsequently led xeno experts to settle on the pig as the most promising alternative" (52). With anatomical and physiological structures similar to those of humans, pigs have been incorporated into medical research and scientific exploration for quite some time, "serving as source animals for porcine-derived insulin, heparin, sinews, guts, skin, and heart values" (52), not to mention the use of their dissected bodies in biology and anatomy classrooms across the United States. In addition to having organs similar in size to human organs, pigs are also easier to breed than NHPs, with shorter gestation periods (just under four months, as opposed to those of NHPs, which can be eight to nine months). In addition to one sow's ability to produce multiple litters in one year, pigs also reach sexual and reproductive maturity within six to ten months, whereas NHPs can take a few years. Furthermore, according to Sharp, pigs do not fall within protected animal categories of "endangered" or "laboratory" but are "farm" animals, making them seem already morally and ethically "unproblematic" (53). As is discussed below, however, pigs, along with all animals used in biomedical experimentation, inspire ethical consideration in regard to scientific research in general and chimeric experimentation and creation in particular.

FROM ONE SPECIES TO ANOTHER

In 2010 scientists discovered that a rat organ could be grown inside the body of a genetically modified mouse. To do this, scientists injected rat pluripotent stem cells (cells that have the ability to develop into all types of adult cell types) "into mutant mouse blastocysts [inner cell masses that eventually form mammalian embryos] and demonstrated that these cells integrate with the murine embryo and contribute to the development of all tissues" (Shaw et al. 2015, 971). The rat pluripotent stem cells eventually complement the genetically modified mouse by growing into a complex organ that the mouse is lacking, as "these mutant mice would normally be born lacking a pancreas or thymus" (971). According to research confirming the work of Hiroshi Nagashima and Hitomi Matsunari (2016), "this demonstrates that cells from two different species, which diverged [approximately] 18 million years ago, can live in a symbiotic environment and are able to support normal organismal aging" (Wu et al. 2017, 481). This interspecies research is significant for humans, particularly with respect to the consistent organ shortage around the world. While there have been significant attempts to create human organs through mechanical and artificial methods, "human organs with complex physiological functions and three-dimensional structures are extremely difficult to build in an artificial culturing environment. Therefore, the use of animal fetuses is a promising alternative strategy that may circumvent the problems of *in vitro* organ generation by supporting organ development from xenogeneic PSCs [pluripotent stem cells] in a natural physiological environment" (Nagashima and Matsunari 2016, 422). In other words, organs of one species are more likely to grow at an appropriate rate within a different host species when the two species are closely related.

In September 2015 the National Institutes of Health (NIH) declared a moratorium on funding for chimeric research and the creation of human/animal hybrid embryos in order to investigate the scientific and ethical implications. This moratorium, which was a bit of a surprise to many in the scientific community, was mainly in response to the large number of requests for funding. While chimeric experiments are not necessarily new—scientists have been creating rodents with human tumors and other human ailments to study diseases and drug treatments since the late 1990s—researchers have only been working to introduce human stem cells into nonhuman animals during the very early stages of embryonic development since early in the first decade of the twenty-first century. This is what needed further consideration. On the one hand, scientists hope that chimeric embryos can be used to re-create human diseases within animal models, with the aim being the development of new ways to prevent and treat illnesses. Even further, one of the most important goals of this research is to grow human organs (hearts, kidneys, livers, pancreases) in the bodies of nonhuman animals (sheep, pigs, cows) to then be transplanted into the bodies of humans. On the other hand (though there are, admittedly, multiple and complicated perspectives here), according to some bioethicists, the creation of human/animal hybrid embryos "would introduce inexorable moral confusion in our existing relationships with nonhuman animals, and in our future relationships with part-human hybrids and chimeras" (Françoise Baylis, quoted in Stein 2015). In other words, the entrenched human/animal binary by which moral and ethical standards have been constituted and normalized is disrupted and traditional views of animals' place in society is called into question. For example, could humans still slaughter and consume pigs that contain human stem cells that could potentially develop human qualities?

Another concern that inspired the moratorium, one that continues to plague scientists and worry the public, is seen both as a practical and ethical problem: human stem cells could end up developing in the brains of nonhuman animals, giving animals some kind of human-level consciousness or mental capacity. Further down the road, another concern would be the potential creation of human eggs and sperm in nonhuman animals. Despite these concerns, however, the NIH began the process of lifting the moratorium in August 2016 by obtaining comment from the public and the scientific community, many of which believe specific safeguards would always be in place to protect against the aforementioned concerns. By early 2017, the NIH had begun accepting research proposals on a case-by-case basis. If the moratorium is officially lifted, the NIH is expected to still prohibit the introduction of human stem cells into the brains and embryos of nonhuman primates, a practice that has been regulated and mostly prohibited since 2009. There are a few scientists who oppose lifting the ban, arguing that the problem is not growing organs (which they admit would have considerable benefit to humans) but opening "the door to making intermediate species organisms, including humans . . . [which is] a big cultural step [with seemingly] no natural stopping point" (Kaiser 2016, 635). By introducing more and more genetic material between species, some scientists believe that this could create new organisms, something in between human and animal, and something yet to be imagined and only existing in the world of science fiction.

After the NIH restricted funding, private institutions moved forward with chimeric research. In January 2017 breakthrough research was announced by the Salk Institute in La Jolla, California, and published in the journal *Cell* under the title "Interspecies Chimerism with Mammalian Pluripotent Stem Cells" (Wu et al. 2017). The article's scope was expansive, unveiling successes in cutting-edge gene-editing and stem cell technologies in both mice and rats and humans and pigs: "scientists were able to grow a rat pancreas, heart and eyes in a developing mouse . . . [and] generate human cells and tissues in early-stage pig and cattle embryos, marking the first step toward the generation of transplantable human organs using large animals whose organ size, physiology and anatomy are similar to humans" (Salk Institute 2017). The rodent experiments revealed two important things: (1) the physiology and anatomy of the two species are important when considering the growth and normal development of chimeric organs and (2) "a developing mouse was able to unlock a gallbladder developmental program in rat cells that is normally suppressed during rat development . . . [highlighting] the importance of the host environment in controlling organ development and evolutionary speciation" (Salk Institute 2017).

Scientists at the Salk Institute used the rodent chimeras to lay the foundation for injecting human pluripotent stem cells into pig embryos and then implanting sows with those embryos. Embryonic growth was stopped at four weeks to both examine the effectiveness of cell growth into particular tissues and evaluate the safety of certain technologies. The results were promising, successful even, showing that the human cells within the pig embryo "were beginning to specialize and turn

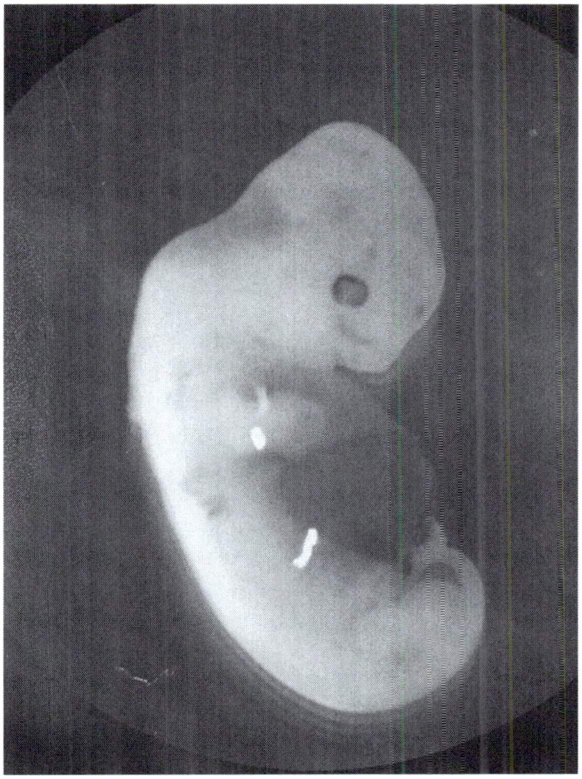

Pig Embryo with Human PSCs, US Salk Institute, January 2017. *This image shows a four-week-old pig embryo containing human pluripotent stem cells (PSCs). Researchers hope that these cells will eventually grow into human organs for transplantation.* XINHUA/ALAMY STOCK PHOTO.

into tissue precursors," even though the rate of efficacy was significantly slower than that between rats and mice (Salk Institute 2017).

These scientific developments are incredibly significant and were even covered by major news outlets, not just in academic journals. The *New York Times*, the *Washington Post*, and the *Guardian* (London), among others, all reported on the findings. In response to ethical concerns raised about crossing a presumably sacred line between human and animal, one researcher argued that "this isn't dangerous research. We're not creating monsters" (Wade 2017). Juan Carlos Izpisua Belmonte, one of the *Cell* article's coauthors, noted to the *Washington Post* that "many people think this is a kind of science fiction story. But this is becoming reality" (Kaplan 2017). This breakthrough research has inspired and perhaps renewed further ethical conversation regarding both the human/animal binary and the welfare of laboratory animals, even if they are genetically modified in some way.

ETHICAL QUESTIONS AND FURTHER DISCUSSIONS

According to Lori Gruen, writing in *Ethics and Animals: An Introduction* (2011), "there is a tendency in almost any ethical discussion to flatten out or oversimplify opposing views and to caricature opponents. This is certainly the case in discussions of animal ethics" (xiv). It should be well noted that "in reality, most positions [regarding animal ethics] are much more nuanced and the people who hold various positions about animals fall along a spectrum" (xiv), rather than falling into categories of just *for* and *against*. Despite their varying positions, many people working within the animal studies field agree that nonhuman animals deserve humanity's ethical attention.

Since 1959 with the publication of *The Principles of Humane Experimental Techniques* (by W. M. S. Russell and R. L. Burch), some have argued that "minimizing the use of animals in experiments is universally recognized by scientists, governments, and advocates as an ethical cornerstone of research" (Goodman, Chandna, and Roe 2015, 567). The way to minimize this use is to replace (with a comparable substitution), reduce (through the design and methodology of experiments), and refine (by creating the minimum amount of stress on animal subjects) the ways in which animals are employed in experimentation. In the United States, the Animal Welfare Act (AWA) of 1966 reinforced these ethical considerations. However, it has been difficult to determine whether these so-called three Rs have actually amounted to fewer animals used in medical and scientific research. The AWA was amended multiple times, but the most significant ratification in regard to animals in laboratories was in 1985 with the Improved Standards for Laboratory Animals Act, requiring the formation of institutional animal care and use committees, made up of veterinarians and animal welfare advocates, at all research facilities to review animal use protocols. The 1985 amendment also required research facilities to give dogs exercise and promote "[the] psychological well-being of primates" (Gruen 2011, 113). Despite the successes of the AWA to protect the welfare of animals, the 1985 amendment does not cover most of the animals used in scientific experimentation, including rodents, birds, and reptiles. Nevertheless, most scientists and, to a certain extent, the general public tend to believe that the benefits of animal experimentation outweigh the tens of thousands of animals who may suffer and die in laboratories. Research using animal bodies as proxies for humans has undoubtedly "been central in the development of vaccines, pharmaceuticals, therapeutic procedures, and other protocols that directly improve the well-being of humans" (Gruen 2011, 118).

Ethical responses to chimeric research are just in the beginning stages. However, there are a few perspectives that contextualize this new conversation, mainly the possibility of affecting or solving organ shortages, the further development of disease prevention and treatment methods, and the lives, welfare, and deaths of nonhuman animals within experimental research. As mentioned earlier, approximately twenty-two people in the United States die each day waiting for a transplantable organ, and the task of making up the difference between the number of people who need an organ and the number of organs that can be procured seems daunting. Scientists such as those at the Salk Institute and many of those waiting for the NIH to lift its moratorium believe that chimeric experimentation could help in creating organs that significantly reduce the risk of rejection as chimeric organs contain a similar genetic makeup more attuned to the host. In addition, researchers believe that further chimeric experimentation could lead to new methods to prevent and treat a variety of human ailments and diseases such as Alzheimer's, cancer, neurological diseases, and other illnesses, while also creating "a new platform for drug toxicity tests" (Salk Institute 2017).

Along the animal experimentation spectrum, many people support some use of animals in scientific experimentation. For example, some would find it ethically justifiable to experiment on animals if it means finding a cure for cancer but do not support the continued use of animals in the cosmetic industry. There are some individuals, however, who advocate the full abolition of animal experimentation. Those who support abolition "think it is [never] appropriate to use another individual as a means to one's own ends" (Gruen 2011, 126). In other words, the suffering and eventual death of laboratory animals is never worth it, no matter how many human lives could be potentially saved. Chimeric research has not yet gotten to the point where a chimeric pig-human embryo has come to full gestation or full maturity, at which point the animal would be purposefully killed to harvest the transplantable organ. These ethical conversations inspire new questions: What does it mean to kill a chimeric animal? Is it similar to the human slaughtering of animals for food? What would happen if human stem cells did migrate into the brains of other species? While we do not necessarily have answers to these questions, examining them through a posthumanist lens may shed some light on how to respond to these complicated questions.

POSTHUMANISM: NONHUMAN ANIMALS AS CARRIERS OF HUMAN ORGANS

Posthumanism posits that "human" (or, as more traditionally interpreted, "man") is not distinct and separate from other animals and machines but is rather a part of environmental and historical trajectories, constituted and affected by nonhuman things; the human is not the center of the universe and not the only source of meaning and knowledge in the world. Posthumanism not only disrupts the perceived boundaries between humans and animals, nature and culture, and organism and machine, but works to check and potentially destroy human exceptionalism.

While the term *posthumanism* can be found in the writings of Russian occultist Helena Petrovna Blavatsky from the 1880s, it was not until Donna Haraway's 1984 publication "A Cyborg Manifesto" that a theory of posthumanism began to take shape. Haraway's first manifesto (followed in 2003 by *The Companion Species Manifesto*) is a lesson in interdisciplinarity, intersecting movements and the effects that make up identity. Considering all this,

Haraway brings readers to historically situated knowledges of particular political, social, environmental, and religious systems and how those systematic apparatuses make us more aware of how we interpret and know the world within which we inhabit. In other words, Haraway argues that each individual or organism is affected by the systems of knowledge that persist around her. Humans, for example, are formed by the environmental context of their lives, including, but not limited to, race, class, and gender. Haraway's manifesto signals three categorical binary breakdowns that allow for the cyborg to take shape: the boundaries between human and animal, organism and machine, and the physical and nonphysical. Through these breakdowns, Haraway argues that there is something other than "dualist traps"—besides human and animal or organism and machine; these categories are too limiting. And perhaps, Haraway writes, in "the late twentieth century, our time, a mythic time, we are all chimeras, theorized and fabricated hybrids of machine and organism; in short, we are cyborgs. The cyborg is our ontology; it gives us our politics" (Haraway 1991, 150). Without actually employing the term, Haraway laid the groundwork for a posthumanist critique and, for the purposes of this chapter, the undoing of the human/animal binary. With chimeric research and particularly the breakthroughs from the Salk Institute, is it safe to say that humanity has officially disrupted the human/animal binary? And what might this mean for future research?

Patricia Piccinini's artistic work, defined by her use of multiple media and materials, is helpful in thinking through these questions. Through photography, sculpture, digital installation, and drawing, Piccinini explores "bodies and relationships: the relationship

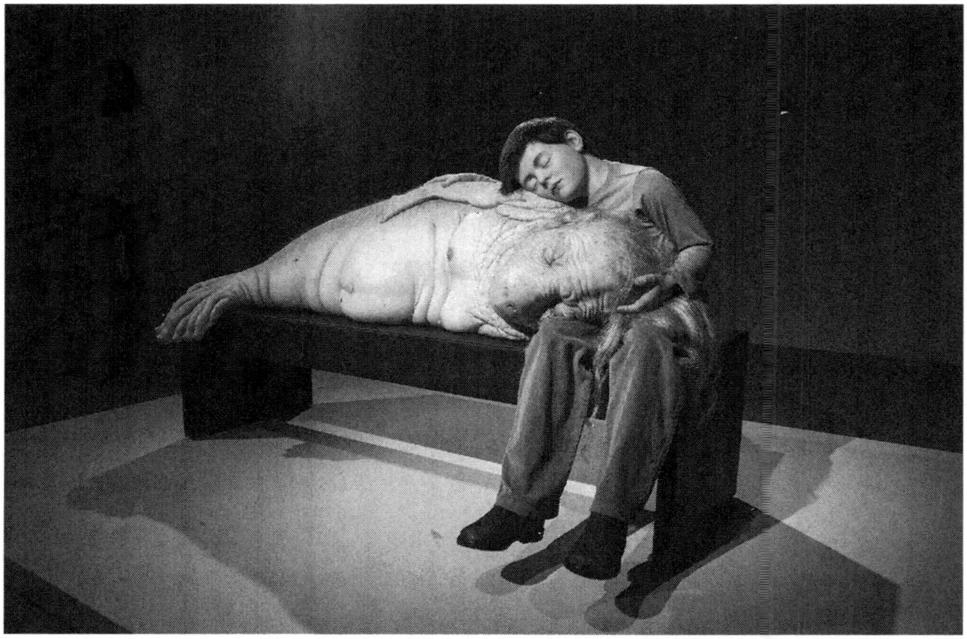

Long Awaited, *Patricia Piccinini, 2008.* *This sculpture, made of silicone, fiberglass, human hair, leather, plywood, and clothing, argues for the connection between bioethics and empathy. At the heart of most of Piccinini's work is empathy and emotions, specifically meant to disrupt strict intellectual understanding of a certain issue.* BRAZIL PHOTO PRESS/CON/GETTY IMAGES.

between people and other creatures, between people and our bodies, between creatures and the environment, between the artificial and the natural. I am particularly interested in the way that the everyday realities of the world around us change these relations" (2014). *The Long Awaited* and *Big Mother* represent what Piccinini articulates throughout her work as a blurring of boundaries between natural and artificial, human and animal, and the creation of new, imaginative creatures that cannot be categorized. Piccinini's work invokes ethical uncertainty as biotechnology continues to affect human and nonhuman animal bodies. However, by juxtaposing these creatures with sculpted human children, Piccinini calls for a responsibility to care and a commitment to attend to the things created by "technoculture." Calling her a "sister in technoculture," Haraway argues that Piccinini's work insists on an intergenerational obligation, to both past, present, and future creations: "Natural or not, good or not, safe or not, the critters of technoculture make a body-and-soul changing claim on their 'creators' that is rooted in the generational *obligation of* and *capacity for* responsive attentiveness" (Haraway, 2007). In other words, no matter what, humans (creators in the biotechnical world) have an ethical responsibility to the critters they create.

Big Mother, ***Patricia Piccinini, 2005.*** *This sculpture, made of silicone, fiberglass, polyurethane, leather, and human hair, was inspired by the stories of Piccinini's own life in Sierra Leone and attempts to invoke an empathetic response from the viewer by juxtaposing mythical beings with human forms. It is, in part, a commentary on how comfortable humans are in drinking the breast milk of another animal.* BRAZIL PHOTO PRESS/CON/GETTY IMAGES.

RELIGIOUS RESPONSE

Religious perspectives and inquiries can both offer helpful insight into particular ethical questions and help gauge public response. According to Jochen Sautermeister, Richard Mathieu, and Veronika Bogner, "unlike allotransplantation [a transplant of cells, tissues, and organs between two individuals of the same species], reflections on xenotransplantation are infrequent in theological literature" (2015, 174). More and more academic disciplines, such as theology and religious studies, have come to find the need to incorporate the study of animals into established methodologies. Since early in the first decade of the twenty-first century religious studies has begun to critically attend to the presence and significance of animals in religious texts, rituals, and practices. The study by Sautermeister and colleagues summarizes the lectures of the symposium "Xenotransplantation: A Challenge to Theological Ethics," which was held in 2013 in Munich. From the symposium, two fundamental questions emerged: Does xenotransplantation pose a challenge to human self-conception in the eyes of God? And how is the relationship between humans and animals renegotiated in the context of xenotransplantation? The symposium, having heard from multiple theological perspectives within and between Judaism, Christianity, and Islam, found that "there are no specifically religious fundamental and generally binding reasons to prohibit xenotransplantation as a means of treating grave and life-threatening organ insufficiencies" (Sautermeister, Mathieu, and Bogner 2015, 181). All three traditions view both xenotransplantation and animal experimentation as theologically acceptable when necessary to complete successful and viable procedures, because saving or preserving the life of a human overrides other potential religious considerations, such as the suffering imposed on research animals.

When it comes to stem cell research and the creation of chimeric organisms, religious responses are more complicated and varied. Regarding stem cell research in general, the Pew Research Center found that "by a 56%–32% margin, the American public believes that it is more important to conduct stem cell research that may result in new medical cures than to avoid destroying the potential life of embryos involved in such research" (Heimlich 2007). Pew studies have found that there has been relatively consistent growth in support for stem cell research since 2002, even within religious traditions, particularly mainline Protestant Christianity and evangelical Christians. While the Catholic Church remains opposed to embryonic stem cell research (in which embryos are destroyed), it does support research using adult tissue stem cells and umbilical cord blood (United States Conference of Catholic Bishops 2017). While there have been numerous studies on cultural and religious responses to stem cell research in general, there has been no comprehensive study surveying religious perspectives on the creation of human/animal chimeric organisms. It might be assumed that concerns similar to those surrounding xenotransplantation would be considered, such as the disruption of the human/animal binary and theological anthropocentrism. What is not yet known is where interdisciplinary inquiries will lead in regard to sacrificing the genetically modified animals grown to harvest human organs. Soon after the Salk Institute announced its breakthrough research in January 2017, a post appeared on the *National Catholic Register* titled "Controversial Chimera Research: Is This Little Piggy Really Part Human?" Most of Kathy Schiffer's post is devoted to the Catholic response to chimeric research in general, along with the claim that certain kinds of human/animal chimeras "are justifiable and reasonable," particularly with regard to the prevention and treatment of disease. Schiffer also notes, however, that the United States Conference of Catholic Bishops strongly opposes lifting the NIH's moratorium because of the destruction of human embryos, the possibility of producing animals with humanlike mental capacities, and the potential to create animals with human eggs and sperm. She ends her post expressing hope that the Donald Trump (elected 2016) administration will uphold the moratorium, aligned with other pro-life agendas.

Summary

While a comprehensive ethical conversation regarding animal bioenhancement has some catching up to do with regard to the rapid scientific developments in the field, several ethical issues, discussed in this chapter, can be identified: What is the fate of chimeric research in the face of many in the scientific community making concerted efforts to reduce the number of animals in research? How does the scientific community work to decrease the use of animals in research while arguing to use them in chimeric research and xenotransplantation? How can policy keep up with the changing science, especially considering that the Animal Welfare Act has not been significantly amended in regard to laboratory animals since 1985? Does the possibility of solving the organ shortage outweigh the potential suffering of animals in laboratories? Perhaps religious perspectives could add important insights into the ethical issues that arise from chimeric research. In addition to reviewing significant ethical discussions and the recognition that many people fall on an ethical spectrum and not just *for* or *against* animal use in scientific research, this chapter has also taken a look at these new scientific developments as part of posthumanist theories that disrupt the traditional human/animal binary.

Bibliography

Altman, Lawrence K. "Baboon Transplant Tests Described." *New York Times*, November 11, 1984a. http://www.nytimes.com/1984/11/11/us/baboon-transplant-tests-described.html.

Altman, Lawrence K. "Doctors Say Baby with Baboon Heart Is Doing 'Remarkably Well.'" *New York Times*, October 29, 1984b. http://www.nytimes.com/1984/10/29/us/doctors-say-baby-with-baboon-heart-is-doing-remarkably-well.html.

Associated Press. "Jeff Getty, 49, AIDS Activist Who Received Baboon Cells, Is Dead." *New York Times*, October 16, 2006. http://www.nytimes.com/2006/10/16/health/16getty.html.

Boffey, Philip M. "Baby Fae Appears Well after First Sign of Rejecting Baboon Heart." *New York Times*, November 12, 1984. http://www.nytimes.com/1984/11/12/us/baby-fae-appears-well-after-first-sign-of-rejecting-baboon-heart.html.

Conlee, Kathleen M., and Andrew N. Rowan. "The Case for Phasing Out Experiments on Primates." *Hastings Center Report* 42, supp. 1 (2012): S31–S34.

Daar, A. S. "Animal-to-Human Organ Transplants: A Solution or a New Problem?" *Bulletin of the World Health Organization* 77, no. 1 (1999): 54–61.

Fugazzola, Laura, Valentina Cirello, and Paolo Beck-Peccoz. "Fetal Microchimerism as an Explanation of Disease." *Nature Reviews Endocrinology* 7 (2011): 89–97.

Goodman, Justin, Alka Chandna, and Katherine Roe. "Trends in Animal Use at US Research Facilities." *Journal of Medical Ethics* 41, no. 7 (2015): 563–566.

Gruen, Lori. *Ethics and Animals: An Introduction.* Cambridge: Cambridge University Press, 2011.

Haraway, Donna. "A Cyborg Manifesto: Science, Technology, and Socialist-Feminism in the Late Twentieth Century." In *Simians, Cyborgs, and Women: The Reinvention of Nature*, 149–181. New York: Routledge, 1991.

Haraway, Donna. "Speculative Fabulations for Technoculture's Generations." In *(tender) creature exhibition catalogue.* Artium. 2007. http://www.patriciapiccinini.net/writing/30/357/17.

Heimlich, Russell. "Support Stem Cell Research." Pew Research Center. June 22, 2007. http://www.pewresearch.org/fact-tank/2007/06/22/support-stem-cell-research/.

HHS (US Department of Health and Human Services). "Organ Donation Statistics." 2017a. Accessed March 20, 2017. https://organdonor.gov/statistics-stories/statistics.html.

HHS (US Department of Health and Human Services). "US Organ Transplants and Deceased Donors Set New Records in 2016." January 9, 2017b. https://optn.transplant.hrsa.gov/news/us-organ-transplants-and-deceased-donors-set-new-records-in-2016/.

Kaiser, Jocelyn. "NIH Plans to Fund Human-Animal Chimera Research." *Science* 353, no. 6300 (2016): 634–635.

Kaplan, Sarah. "Scientists Create a Part-Human, Part-Pig Embryo—Raising the Possibility of Interspecies Organ Transplants." *Washington Post*, January 26, 2017. https://www.washingtonpost.com/news/speaking-of-science/wp/2017/01/26/scientists-create-a-part-human-part-pig-embryo-raising-the-possibility-of-interspecies-organ-transplants/.

Lee, Vivian K., and Guohao Dai. "Printing of Three-Dimensional Tissue Analogs for Regenerative Medicine." *Annals of Biomedical Engineering* 45, no. 1 (2017): 115–131.

Mai, Larry L., Marcus Young Owl, and M. Patricia Kersting. *Cambridge Dictionary of Human Biology and Evolution.* Cambridge: Cambridge University Press, 2005.

Nagashima, Hiroshi, and Hitomi Matsunari. "Growing Human Organs in Pigs: A Dream or Reality?" *Theriogenology* 86, no. 1 (2016): 422–426.

New York Times. "The Baboon Heart and Baby Fae." November 11, 1984. http://www.nytimes.com/1984/11/11/opinion/the-baboon-heart-and-baby-fae.html.

Organ Procurement and Transplantation Network. "Need Continues to Grow." Accessed March 20, 2017. https://optn.transplant.hrsa.gov/need-continues-to-grow/.

Piccinini, Patricia. "Interview with the Condition Report." 2014. http://www.patriciapiccinini.net/printessay.php?id=103.

Salk Institute. "New Findings Highlight Promise of Chimeric Organisms for Science and Medicine." January 26, 2017. http://www.salk.edu/news-release/new-findings-highlight-promise-chimeric-organisms-science-medicine/.

Sautermeister, Jochen, Richard Mathieu, and Veronika Bogner. "Xenotransplantation: Theological-Ethical Considerations in an Interdisciplinary Symposium." *Xenotransplantation* 22, no. 3 (2015): 174–182.

Schiffer, Kathy. "Controversial Chimera Research: Is This Little Piggy Really Part Human?" *National Catholic Register* (blog), January 30, 2017. http://www.ncregister.com/blog/kschiffer/controversial-chimera-research-is-this-little-piggy-really-part-human.

Sharp, Lesley A. *The Transplant Imaginary: Mechanical Hearts, Animal Parts, and Moral Thinking in Highly Experimental Science*. Berkeley: University of California Press, 2014.

Shaw, David, Wybo Dondorp, Niels Geijsen, and Guido de Wert. "Creating Human Organs in Chimaera Pigs: An Ethical Source of Immunocompatible Organs?" *Journal of Medical Ethics* 41, no. 12 (2015): 970–974.

Singer, Peter. *Animal Liberation: A New Ethics for Our Treatment of Animals*. New York: New York Review, 1975.

Stein, Rob. "NIH Plans to Lift Ban on Research Funds for Part-Human, Part-Animal Embryos." NPR, August 4, 2016. http://www.npr.org/sections/health-shots/2016/08/04/488387729/nih-plans-to-lift-ban-on-research-funds-for-part-human-part-animal-embryos.

Stein, Rob. "Should Human Stem Cells Be Used to Make Partly Human Chimeras?" NPR, November 15, 2015. http://www.npr.org/sections/health-shots/2015/11/06/454693391/should-human-stem-cells-be-used-to-make-partly-human-chimeras.

United States Conference of Catholic Bishops. "Catholic Support for Ethically Acceptable Stem Cell Research." Accessed March 20, 2017. http://www.usccb.org/issues-and-action/human-life-and-dignity/stem-cell-research/catholic-support-for-ethically-acceptable-stem-cell-research .cfm.

Wade, Nicholas. "New Prospects for Growing Human Replacement Organs in Animals." *New York Times*, January 26, 2017. https://www.nytimes.com/2017/01/26/science/chimera-stemcells-organs.html.

Wu, Jun, Aida Platero-Luengo, Masahiro Sakurai, et al. "Interspecies Chimerism with Mammalian Pluripotent Stem Cells." *Cell* 168, no. 3 (2017): 473–486.

CHAPTER 24

Human Flourishing: A Moral Yardstick for Evaluating Specific Bioenhancements

Michael Bess
Chancellor's Professor of History
Vanderbilt University, Nashville, TN

Since the mid-1990s, a fierce debate has emerged among bioethicists and other scholars over the pros and cons of bioenhancement technologies. On one side, the anti-enhancers emphasize what humankind stands to lose if it goes down this path: by allowing ourselves to dramatically modify our own physical and mental makeup, they argue, we risk commodifying human personhood, undermining human dignity, and severely disrupting the core set of qualities that make up our distinctive nature as a species. We should accept the limitations that evolution and history have thus far imposed on our being, they contend, and learn to appreciate the myriad remarkable qualities that already characterize the relatively unmodified humans of today. On the other side, the pro-enhancers hew to a very different perspective: we humans have always been evolving, they maintain, and there is no reason why we should not take matters into our own hands, further perfecting our bodies and minds according to our most basic priorities and values. There is no such thing as a fixed human nature, according to these theorists. Our most basic moral imperative should therefore be to use science and technology to reduce human suffering, broaden the reach of our capabilities, and open up ever greater horizons of achievement and creativity. On one side, in short, the rallying cry is "Embrace our limits"; on the other, "The sky's the limit."

This debate was initially a fruitful one, forcing the two sides to clarify and hone their respective arguments; but a growing number of scholars have come to feel that it has bogged down, yielding more heat than light. Two of the most prominent bioethicists in the field, Erik Parens of the Hastings Center and Duke University's Allen Buchanan, have persuasively argued that it is time to move past the trenchant either-or approach, with its sweeping metaphysical pronouncements, and to seek a more pragmatic and measured middle ground (Buchanan 2011; Parens 1998, 2015). Surely, they argue, a more nuanced and realistic set of choices is needed than a Manichaean imperative either to "avoid all bioenhancements" or "embrace all bioenhancements." A more fruitful way forward, they believe, is to evaluate each particular form of bioenhancement on a case-by-case basis, tallying up the pros and cons with an eye to the specific implications of that technology in its distinctive social and moral context.

This quest for pragmatism is a welcome development, but it immediately raises this question: what moral yardstick can be used in making such case-by-case evaluations? Suppose, for example, that your best friend approaches you one day and says, "I want to use this new brain-machine interface device to manipulate my own emotional states. Should I do it?" You would very much like to give sound advice to your friend, but where are you to turn for moral guidance? No consensus exists, either among bioethicists or among the broader citizenry, about the best moral framework for making such important judgments. Among the philosophers, we find an extremely broad range of moral approaches, each of which parallels one of the major schools of thinkers who have tackled the concept of "a life well lived" since ancient times. Some, for example, draw on Aristotle's (384–322 BCE) concept of virtue ethics, emphasizing the cultivation of good character; others rely on the deontological tradition, inaugurated by Immanuel Kant (1724–1804), which lays out the general rules and principles that any rational being should follow in leading an ethically sound life; still others are influenced by the utilitarian tradition dating back to John Stuart Mill (1806–1873), which assesses morality according to the consequences of our actions in bringing about the greatest good for the greatest number of persons. Among the citizenry, moreover, we find those who turn to religion for guidance, but a great many contrasting and mutually incompatible approaches are on offer there as well. In the political arena, we find some ecology-minded thinkers on the left who reject bioenhancement as "unnatural," whereas others on the left enthusiastically embrace these technologies as a key element of historical progress. We find some political conservatives who feel moral outrage at the hubris of bioenhancement, whereas other conservatives of a more libertarian bent insist that we should be free to augment our capabilities in any way we choose. In short, what we find is a confusing moral cacophony. How are we to proceed?

This chapter describes two relatively new fields of academic research that can help people make clearer and better-informed decisions about their bioenhancement choices, whatever their moral, religious, or ideological backgrounds happen to be. The first is a branch of psychology, the other a novel approach to developmental economics. Taken together—for these two fields are complementary—they offer rich pragmatic grounds for ethical decision making because they lay out a framework for answering one of the fundamental questions at the heart of all ethics: what does it mean for a human being to live a good life? When your best friend approaches you for advice about using the emotion-sculpting machine, your familiarity with these two fields will help you give a more thoughtful and constructive answer whether you are Buddhist or Catholic or agnostic, a left-winger or a conservative, a Kantian or an Aristotelian, or something else altogether.

POSITIVE PSYCHOLOGY

Positive psychology was launched in the 1990s by a group of scholars who felt frustrated by the prevailing tendency of the field of psychology to devote most of its resources to the study of pathological states such as mental illness, anxiety, or depression. Among these dissenting figures were Martin E. P. Seligman and Mihaly Csikszentmihalyi, followed by a second wave that included Christopher Peterson, Jonathan Haidt, Daniel Kahneman, Ed Diener, and Sonja Lyubomirsky (among many others). While the traditional therapeutic emphasis among psychologists was certainly important and valid, they argued, all this emphasis on what could go wrong with people was only part of the picture. A fully mature psychology should also offer a clearer understanding of what qualities and conditions allowed human beings to flourish—that is, to lead healthy, productive, fulfilling lives.

The titles of some of their publications give a good sense of where their new emphasis lies: *Flow: The Psychology of Optimal Experience* (1990, Csikszentmihalyi); *Authentic Happiness* (2002, Seligman); *The Happiness Hypothesis* (2006, Haidt); *The How of Happiness* (2008, Lyubomirsky); *The Science of Well-Being* (2009, Diener). Perhaps the most influential work in this new field, however, was published in 2004 by Peterson and Seligman: *Character Strengths and Virtues: A Handbook and Classification.* The aim of this 800-page tome was to offer a counterpart to the ubiquitous *Diagnostic and Statistical Manual of Mental Disorders* (*DSM*), which has stood since the 1950s as the standard work surveying the vast panorama of human insanity, misery, and breakdown. (The *DSM* has gone through five major revisions during that time.) Peterson and Seligman present a comprehensive survey of the major thinkers since ancient times who have offered insights into the nature of a life well lived: Confucius (551–479 BCE); Taoist, Buddhist, and Hindu writers; Plato (c. 428–348 or 347 BCE) and Aristotle; Thomas Aquinas (1224 or 1225–1274); Islamic philosophers; modern psychologists such as Erik Erikson (1902–1994) and Abraham Maslow (1908–1970); developmental experts such as Jean Piaget (1896–1980) and Lawrence Kohlberg (1927–1987); and a wide array of nonspecialist writers ranging from Benjamin Franklin (1706–1790) to William Bennett (1943–). In addition, Peterson and Seligman conducted a massive survey between 2000 and 2003 of contemporary practitioners of psychology, ranging from academic theorists to clinical therapists and psychiatrists, asking them to list the qualities, traits, and capabilities that they considered most essential to human well-being. The result is a catalog of twenty-four attributes (or "strengths"), divided into six basic categories (or "virtues"):

Wisdom and Knowledge: creativity; curiosity; open-mindedness; love of learning; perspective (wisdom)

Courage: bravery; persistence; integrity; vitality

Humanity: love; kindness; social intelligence

Justice: citizenship; fairness; leadership

Temperance: forgiveness and mercy; humility and modesty; prudence; self-regulation

Transcendence: appreciation of beauty and elegance; gratitude; hope; humor; spirituality

The book then devotes the remainder of its pages to explaining the specific contribution that each of these twenty-four strengths brings to human well-being and how each individual strength interacts with others on the list.

Peterson and Seligman acknowledge that their compilation is a work in progress, subject to further discussion and revision; they urge the reader not to worry overmuch about the details of this list—which strength falls under which virtue, for example, or why certain other potential candidates for strengths were not included. What stands out from their overview, however, is the multimodal nature of this conception of human flourishing. There is no fixation on any single attribute, such as "contemplation," "health," or "love." Instead, three complementary dimensions of human welfare can be discerned here: the experience of pleasant sensations and happy feelings; the unfolding and fruition of an individual's faculties and potentials; and the sense of fulfillment or meaning that derives from a life lived in affective community with other people. To live well is to experience the full range of one's personhood—internal and external, individual and social—developing and expressing itself amply and freely.

But now imagine for a moment a (fictitious) twenty-year-old woman living in, say, India, around 1930. She is a remarkable individual, in that she possesses all twenty-four of the above character strengths to a significant degree. She is happily married to a fellow high-caste Brahmin man, plans on having children, and anticipates a long, fulfilling life ahead of her. But one day her husband gets run over by a bus and dies. Suddenly her world comes apart: she is now a widow. In Indian society, even during modern times, widows have been considered second-class citizens whose worth as individuals derived primarily from their role as spouses of men. Until the mid-1800s, widows in some parts of India were routinely expected to fling themselves onto the flames of their husband's funeral pyre. But even after this practice (known as *sati*) was banned by British colonial administrators, widows continued to be segregated within Indian society, forbidden to remarry, in many cases forced into a beggar's life, and deprived of the most elemental forms of respect. So, for our twenty-year-old widow in 1930, we can legitimately ask: In what way can her twenty-four character strengths come to full fruition now that she has been forced to live under such harshly oppressive conditions? She may have strengths aplenty, but what can she actually do with them in practice? These are the paradigmatic questions behind the "capabilities approach" in welfare economics, discussed below.

THE CAPABILITIES APPROACH

Economist Amartya Sen and political philosopher Martha C. Nussbaum are the key thinkers behind the capabilities approach, which is a subfield of developmental economics. It was Sen who first launched the field, starting in the 1960s with what was to become a long series of books and articles about how poor nations could rise out of their entrenched economic misery. Over time, Sen became increasingly critical of the way most economists were thinking about global poverty. Their aggregative and statistical methods seemed reasonable enough on paper, he argued, but they ultimately failed to grasp the complexity of the causal processes that were leading to chronic impoverishment. By focusing too narrowly on such factors as gross national product (GNP) and per capita income, the leading economists' models were presenting an oversimplified analysis. What was needed, instead, was a new form of analysis that took into account the intricate nexus of physical, social, cultural, and political conditions within which economic processes inevitably took place (Sen 2009; Morris 2009).

On paper, for example, the citizens of Sen's native India possessed the right to vote, and this might seemingly put them on a par with the citizens of other democracies, such as Sweden or Canada. But in actual practice, he maintained, the abstract right to vote meant little to, for example, an illiterate woman whose family was starving and whose day-to-day life consisted of a desperate struggle for survival. If one was truly interested in empowering such a person, lifting her into a full and meaningful form of citizenship, then one needed to address that individual's situation in a far more holistic manner, taking into account all the myriad tangible obstacles that permeated her life. The important question to ask was not just, "How can we get this person a job?" Instead, it had to reach much further into the context within which economic activities are embedded: "How can we give this person a decent education?" and "How can this person become an active agent in shaping her own political and civic environment?" If economists were truly keen to bring about positive changes in the lives of the people they studied, Sen argued, then they had to move beyond their narrow econometric analyses and frame their basic questions much more concretely, in a way that reflected the full complexity of the situation actually being faced by real people in their everyday lives. Hence his emphasis on capabilities—or *functionings*, as he liked to call them—rather than on abstract rights and statistics.

Sen's writings have proved enormously influential, earning him the Nobel Prize in Economics in 1998. His ideas have been formally adopted by the United Nations as the conceptual basis for its annual *Human Development Report*, which compares the world's nations according to a complex series of variables affecting their citizens' economic and human opportunities. Moreover, Sen's framework has been further developed and systematized by Nussbaum (2011), who went one step further by producing an actual checklist of core human capabilities:

- Life; bodily health; bodily integrity (i.e., basic security)
- Senses, imagination, and thought (i.e., literacy, education, free speech)
- Emotions (i.e., interpersonal relationships and affective connections)
- Practical reason (i.e., the ability to develop and pursue meaningful life plans)
- Affiliation (i.e., participation in groups and communities, in a manner compatible with one's dignity)
- Other species (i.e., concern for the environment)
- Play; control over one's environment (i.e., political and civil rights, as well as property and labor rights)

For Nussbaum, the most important principle underlying this list is that it pivots on the individual's freedom to choose, acknowledging that different people will want to emphasize the development of certain capabilities over others. "To use an example of Sen's," Nussbaum explains, "a person who is starving and a person who is fasting have the same type of functioning where nutrition is concerned, but they do not have the same type of capability, because the person who fasts is able not to fast, and the starving person has no choice" (2011, 25). The capabilities approach, in other words, does not seek to prescribe exactly what flourishing must mean for all persons. It does not dictate a specific set of societal or individual outcomes. Instead, it lays out a bare minimum group of conditions that must be met if individuals are to have the chance to discover for themselves what a good life is. In this sense, it establishes a basic threshold for each of the ten capabilities, below which an individual's opportunities for real substantive choice cannot be realized; but above that threshold, it leaves the options wide open.

A MORAL YARDSTICK FOR SPECIFIC BIOENHANCEMENTS

The alert reader will have noted that the list generated by Peterson and Seligman emphasizes the *causes and promoters* of human flourishing, whereas the list generated by Nussbaum stresses the *necessary conditions* for human flourishing. In this sense, the two conceptual frameworks—positive psychology and the capabilities approach—are complementary in nature: taken together, they offer a fruitful way to think about what it means for a human being to live a good life.

The insights offered by these two fields suggest a broad set of criteria by which bioenhancement technologies can be evaluated. The word *enhancement* points not merely toward quantitative augmentation of our capabilities but toward a qualitative *improvement* of some sort. What better way to assess such improvement than to ask: Does this modification actually contribute to human flourishing? Or, more concretely: If I take this pill, adopt this bioelectronic device, or make this genetic alteration, will it have a positive impact on my overall quality of life? In addressing these questions, one can turn to the interlocking concepts presented above—factors such as fairness, creativity, hope, security, social engagement, and love—as benchmarks against which to weigh the pros and cons of each kind of bioenhancement. Moreover, it is necessary to take into account not just the direct impact of these technologies for individual persons but also the more complex effects generated at the societal level, when large portions of a population adopt a specific modification.

In many cases, of course, applying the moral yardstick of human flourishing will not necessarily result in a clear-cut thumbs-up or thumbs-down. In considering the implications of a particular technology, a complex set of trade-offs are likely to be encountered. A machine that boosts one's connectedness to other people, for example, may simultaneously undermine one's sense of privacy or autonomy. A pill that sharpens my memory may force me to reckon once more with aspects of my past that I had long considered resolved and settled. These kinds of difficult trade-offs have confronted ethical reasoners throughout human history, and they will no doubt continue to challenge the bioenhanced humans of tomorrow. Human values often conflict with each other, or operate in tension with each other, rendering the balancing of pros and cons a dicey affair.

The specific cases outlined below give a sense of how the concept of human flourishing can guide assessments of bioenhancement technologies.

REJUVENATION: A CAUTIOUS THUMBS-UP

On the face of it, the concept of rejuvenation seems hard to reject. If new combinations of pills, exercises, and genetic or epigenetic modifications allow people to live much longer lives in full possession of their health, vigor, and mental acuity, who would object? To be sure, some may reject such enhancements on religious grounds, whereas others may avoid them because they seem excessively artificial and unnatural. But apart from these kinds of a priori philosophical objections, rejuvenation amply passes the "flourishing test": all other things being equal, a youthful and healthy individual is more likely to enjoy the capabilities and strengths outlined above than a debilitated and decrepit one.

Still, even this seemingly benign form of bioenhancement could potentially lead to unexpected and undesired consequences, once it is adopted by millions of people. If a majority of citizens are living healthy and mentally vigorous lives that span 130 or 160 years (and perhaps longer), what impact will this have on marriage and the family? On population levels and ecological pressures? On work and intergenerational relations? On the stages of life and our sense of our life span's meaning and purpose?

One can imagine a situation, for example, in which a twenty-five-year-old woman is hired by a large corporation. Because she is smart and hardworking, she gradually works her way up in the hierarchy of the organization and becomes CEO at age fifty-five. She then stays at the helm of the company for 100 years—ever fresh, vigorous, and mentally sharp thanks to the technologies of rejuvenation. This plateau-shaped career path may be a boon to her as an individual, but when it is played out across the entire economy, it results in a serious problem, because it eliminates access to the upper echelons of the business world for all the subsequent generations of young people entering the workforce. The only way to avert such an undesired outcome would presumably be to establish rigorous "term limits" for CEOs and other leaders so that they would be compelled after a couple decades to step aside and make room for new talent.

To be sure, this scenario does not necessarily mean that humanity should eschew rejuvenation technologies. The point here is simply that even the most apparently benign and tantalizing forms of bioenhancement may result in unintended outcomes that are deeply problematic in nature. Difficult trade-offs and complicated challenges should be expected to arise even with some of the most seemingly valuable and beneficial modifications.

MIND-READING MACHINE: A CLEAR THUMBS-DOWN

Suppose, for the sake of argument, that scientists find a way someday to create a machine that can read the activity going on in a person's brain and accurately translate it, allowing other people to see into the thoughts and feelings that the person is experiencing. Suppose further that such a machine could be used without the knowledge or consent of the person being observed—say, with a directional scanning device that can hone in on the brain's synaptic activity from several hundred feet away. Such a device would abolish privacy in the most radical way imaginable, laying bare a person's most intimate subjectivity and interiority for anyone else to see.

When applying the criteria of human flourishing to this case, it quickly becomes apparent that such a technology would grievously undermine many of the key values humans hold dear. From the Nussbaum list, it would threaten bodily integrity; senses, imagination, and thought; practical reason; affiliation; and control over one's environment. From the Peterson/Seligman list, it would threaten such strengths as integrity, social intelligence, citizenship, fairness, and leadership. Building such a machine, given its potential for profoundly undermining so many crucial preconditions for human flourishing, should be strictly prohibited. The mere existence of such a device would be a moral catastrophe.

But even here there are trade-offs. Imagine how useful such a machine might be for spies and intelligence officers seeking to gather information from enemies or suspected terrorists. One can already surmise the argument that will emerge: if such a technology can save a significant number of lives by allowing security forces to prevent deadly assaults, then it should be created and used by authorized personnel. The trade-off here is between security and freedom, a tension that is already evident in today's society. Such a technology would ratchet that tension up to an even higher level.

Many bioethicists have argued that such a technology should be banned for all uses, with no exceptions, and that there should be severe penalties for people who create such a machine (whether they use it or not). The privacy of our intimate subjective life is essential to our humanity, they insist, and no short-term gain or relative benefit can possibly outweigh the harm that would come to all humankind if that privacy is lost. But some security-minded folks are likely to emphatically disagree—and this underscores once again the inherently difficult and controversial nature of the ethical trade-offs we will be called on to make in the coming years.

COGNITIVE ENHANCEMENT: ONE THUMB UP, ONE THUMB DOWN

Both the benefits and the costs of cognitive enhancement technologies are potentially significant. Here we are speaking of modifications flowing from all three major areas of bioenhancement research: pharmaceuticals, bioelectronics, and genetics. We may be offered modifications that boost narrow and specific aspects of our mental capacity, such as attention, concentration, memory recall, mathematical processing, or dexterity in spatial conceptualization. Alternatively, scientists may discover pathways that alter our overall cognitive profile, giving us an across-the-board increase in mental performance.

On the positive side of the ledger, one can envision a wide variety of benefits flowing from such modifications, such as heightened mental acuity, augmented memory, a greater ability to learn, increased creativity, and a vastly expanded connectivity with a wide range of semi-intelligent machines. Taken together, these modifications may give human beings access to levels of knowledge and understanding far beyond current limits. We may become smarter in ways that today's humans cannot even fully envision.

At first glance, it is hard to see how such an augmentation of our cognitive abilities could possibly be detrimental to human flourishing. Nearly all the preconditions listed by Nussbaum, as well as the virtues listed by Peterson and Seligman, would presumably benefit from a heightening of human cognition. After all, it was the superior mental capacity of human beings that elevated them to a preeminent position among Earth's living creatures; and a further boosting of that capacity would perhaps offer humans even greater horizons of achievement, understanding, creativity, and richness of experience.

So where is the down side? The key question here has to do with the broader societal and moral *context* within which such cognitive boosting takes place. If the most potent cognitive enhancement technologies prove highly expensive and become available only to a select subset of wealthy persons, then a bifurcation of humankind would result: the already grievous divide between the haves and have-nots would grow much wider and deeper. If different subgroups of human beings choose (for whatever reason) to alter their mental abilities in profoundly divergent ways—each group emphasizing a qualitatively different set of capacities from the others—then a gradual fragmentation of humankind might ensue. If humanity's enhanced cognition flows from a greatly heightened integration with semi-intelligent machines, then a severe diminution in basic privacy might be expected to occur. If cognitive enhancements are carried out with a narrow emphasis on increasing productivity, then this might lead to a homogenization of human mental capacities and a tragic loss of human mental diversity. Because the enhancement technologies will steadily improve in power and sophistication with every passing year, people will be compelled to undertake a never-ending sequence of upgrades and boosts just to keep up with the ever-rising benchmark of "normal" performance. In this respect, we will become akin to Microsoft Windows software: Humans 95, Humans XP, Humans 10.

Another, more insidious danger may also manifest itself. If people identify too closely with their cognitive profiles, they may start to slip (imperceptibly at first) into a mentality that equates the worth of the individual with the level of his or her mental performance. Instead of saying, "Joe is dissatisfied with the performance of the neural implant he bought last month," they may say, "Joe has been kind of pathetic lately, and Nancy keeps leaving him in the dust." This would amount to a partial commodification of humans, a blurring of the boundary between "person" and "product." Such cognitively enhanced people may be smarter than today's humans by some measures, but what good will this have done them if they see themselves in such a commodified way—as "just another product" struggling to keep up the pace of performance amid a seething market of rival products?

It is possible, moreover, that a major boosting of human cognitive ability might become achievable only by reallocating scarce neural resources, drawing on the synaptic activity of brain regions that support affect, empathy, or sensation. As of yet, not enough is known about the functional architecture of the human brain to be able to say whether such a scenario is plausible or nonsensical. But if it does turn out to be plausible, then cognitive boosting would come at a steep price indeed—a disruption of the delicate balance among mental faculties and powers that operate in complementary fashion to yield the broad outlines of the human constitution. Such a modification might make us "smarter" in a very narrow sense but at an incalculable cost to our underlying humanity.

Everything depends, in short, on the way in which the boosting of our minds comes to pass. If it is implemented fairly and wisely, leaving plenty of time for society to adapt, it may prove to be a net positive. If it is implemented without adequate regulations and controls—without time for society to develop a new educational, moral, and cultural framework for these innovations—it may end up undermining some of humanity's most basic values: fairness, dignity, authenticity, and communal solidarity.

One can even imagine a scenario in which we boost ourselves into a kind of creature that is neither less than human nor more than human, but simply nonhuman. Some transhumanist thinkers might celebrate such an outcome, likening it to the metamorphosis that occurs when a caterpillar transforms into a butterfly. This moral framework of "caterpillar to butterfly" comes from an underlying teleological understanding of the natural developmental process through which a single species unfolds its potential over time. But there is no way of knowing whether or not *Homo sapiens* partakes of such an underlying teleology. Transmogrifying ourselves into a radically different successor species may indeed be our cosmic destiny. Or, seen from the perspective of *human* flourishing, it may simply be an extinction event.

Summary

The argument presented here is threefold. First, this chapter emphasizes the need to get beyond the black-or-white metaphysical division between anti-enhancement thinkers and pro-enhancement thinkers. These are morally complicated technologies that offer many profound benefits and confront us with equally profound dangers. Each of them must be evaluated carefully, on a case-by-case basis. Second, the chapter suggests that a fruitful framework for such moral evaluations can be found by bringing together the two complementary fields of positive psychology and the capabilities approach. Finally, the chapter argues that, despite the clarity and usefulness of this moral framework, the judgments and assessments made about specific bioenhancements will still be complicated, difficult ones. In almost every case, tough choices and painful trade-offs among deeply cherished values will arise.

This should neither surprise us nor daunt us excessively. Most of the questions that matter most to humanity in today's world do not present themselves in cut-and-dried form, neatly arranged on a moral platter. Global warming, the tension between security and freedom, and the allocation of national budget priorities are all issues that already require a careful, patient, ongoing conversation about hard moral trade-offs and choices. Precisely because bioenhancement technologies will bring momentous change into human civilization, they will be no different. We should start the conversation about them sooner rather than later.

Bibliography

Bess, Michael. *Our Grandchildren Redesigned: Life in the Bioengineered Society of the Near Future.* Boston: Beacon Press, 2015.

Buchanan, Allen. *Better than Human: The Promise and Perils of Enhancing Ourselves.* Oxford: Oxford University Press, 2011.

Csikszentmihalyi, Mihaly. *Flow: The Psychology of Optimal Experience.* New York: Harper and Row, 1990.

Haidt, Jonathan. *The Happiness Hypothesis: Finding Modern Truth in Ancient Wisdom.* New York: Basic Books, 2006.

Kahneman, Daniel, Ed Diener, and Norbert Schwarz, eds. *Well-Being: The Foundations of Hedonic Psychology.* New York: Russell Sage Foundation, 1999.

Lyubomirsky, Sonja. *The How of Happiness: A Scientific Approach to Getting the Life You Want.* New York: Penguin, 2008.

Morris, Christopher W., ed. *Amartya Sen.* Cambridge: Cambridge University Press, 2009.

Nussbaum, Martha C. *Creating Capabilities: The Human Development Approach.* Cambridge, MA: Belknap Press of Harvard University Press, 2011.

Parens, Erik, ed. *Enhancing Human Traits: Ethical and Social Implications.* Washington, DC: Georgetown University Press, 1998.

Parens, Erik. *Shaping Our Selves: On Technology, Flourishing, and a Habit of Thinking.* Oxford: Oxford University Press, 2015.

Peterson, Christopher. *A Primer in Positive Psychology.* Oxford: Oxford University Press, 2006.

Peterson, Christopher, and Martin E. P. Seligman. *Character Strengths and Virtues: A Handbook and Classification.* Washington, DC: American Psychological Association; New York: Oxford University Press, 2004.

Sen, Amartya. *The Idea of Justice.* Cambridge, MA: Belknap Press of Harvard University Press, 2009.

CHAPTER 25

Enhancing Moral Behavior

Scott M. James
Associate Professor of Philosophy,
Department of Philosophy and Religion
University of North Carolina Wilmington

Most people agree that there is a reason for individuals to improve themselves morally. Some, if not most, people agree that there is a reason for society to improve itself morally. Improving ourselves morally strikes most people as an acceptable, if not mandatory, goal. But to what lengths might we acceptably go in reaching this goal? This is the question at the heart of the debate over moral bioenhancement, and the subject of this chapter.

At one end of the spectrum, there seems to be little resistance to the idea that a person might engage in a *behavioral* program of moral enhancement: deliberate efforts of the will, for example, to expand one's empathic responses, to adopt others' perspectives, or to be more reflective, more patient, less reactive, less vindictive, and so on. Moving away from this end of the spectrum, however, we meet resistance. What if technology afforded a person the option to engage in a *pharmaceutical* program of moral enhancement? What if, by taking an oral supplement, someone could increase that person's tolerance for others by some small, but perceptible, measure (see also Singer 2009)? What if an oral supplement could decrease one's chances of responding to conflict with physical violence? Is simply taking a pill to improve one's moral character (let alone the research and development of such pills) morally permissible? At the other end of the spectrum are *genetic* programs of moral enhancement. What if genetic technology one day afforded parents the option to morally improve their offspring—in vitro? Through this approach, changes to the genome are made before birth, extending throughout the offspring's life and, potentially, into future generations. If we project ourselves into a future marked by a mastery over the human genetic code, would it be morally permissible for parents to alter those DNA complexes in an effort to improve their offspring morally?

Bioethicists occupy positions along most points of this spectrum. This chapter may appear lopsided insofar as most of the arguments reviewed herein *oppose* moral bioenhancement. This is partly a historical accident because most authors have tended to oppose the practice. But the review is also critical: as will be evident, the arguments opposing enhancement are less than persuasive. This result indirectly counts in favor of enhancement, for it suggests that there is nothing unethical about the practice. But that result is premature. Before exploring the arguments made on behalf of these various positions, it is important to clear the path forward by making some distinctions.

It is natural to resist moral bioenhancement on the grounds that the harms outweigh any benefits that such enhancement might produce. But it is important to distinguish between two different general classes of harms. Some harms, it might be said, are intrinsic to a given state. For instance, suffering is intrinsically harmful; it is bad in itself. It may be harmful because it leads to other harmful states (e.g., death), but it is arguably harmful even apart from what it leads to. By contrast, some states are harmful not in themselves, but because of what they lead to. They are instrumentally harmful. Smoking, for example, is not obviously intrinsically harmful (asthmatics aside) because it does not seem harmful in itself. Smokers, after all, tend to report satisfaction from the act of smoking. Instead, the harm associated with smoking is instrumental: it tends to *lead to* intrinsically harmful states, namely, suffering.

This distinction bears crucially on the moral bioenhancement debate, for some superficial disagreements (among bioethicists) may mask important agreements, and some superficial agreements may mask important disagreements. For instance, two people might *disagree* about the likelihood of moral bioenhancements leading to broader social inequalities; that is, they disagree about the likelihood of a given instrumental harm. But this disagreement is consistent with them *agreeing* that there is nothing intrinsically harmful about moral bioenhancement. Conversely, they might *agree* that moral bioenhancements will likely lead to social inequalities, for example, but *disagree* over whether such enhancements are intrinsically harmful. This separation between intrinsic and instrumental harms opens up the possibility that (1) moral bioenhancements should be accepted only on the condition that the associated instrumental harms can be preempted or neutralized, or (2) bioenhancement should be rejected even if the associated instrumental harms can be preempted or neutralized—to name just two possibilities. The latter possibility happens to capture Michael J. Sandel's (2004, 2007) influential position. He is entitled to such a position only because he has (successfully or not) tried to show that such enhancements are *intrinsically* harmful. This chapter, then, takes care to identify the kind of harms, if any, moral bioenhancement is supposed to be associated with, according to a given author. Following Sandel, special attention is given herein to the *intrinsic* moral status of moral bioenhancement.

ARGUMENTS AGAINST ENHANCEMENT

This section reviews the range of arguments made *against* moral bioenhancement, beginning with instrumental harms and followed by intrinsic harms.

INSTRUMENTAL HARMS

The prospect of pharmaceutical or genetic interventions to improve ourselves morally appears to represent a shortcut—at least by comparison to the traditional approach—akin to moral improvement via Amazon Prime. This triggers a host of moral reactions. Even setting aside questions about the intrinsic wrongness (if any) of availing ourselves of a shortcut to moral improvement, reasonable concerns remain.

First off, there is what might be called the "can't get there from here" problem (Daniels 2009). Attempts to perfect the relevant *genetic* technology will almost surely involve harming some individuals, some of whom cannot consent to such experimentation. Like the development of any technology, success (where possible) comes at the end of a long process of trial and error. In the medical context, such harms have to be weighed very carefully. In some cases, the risk of harming individuals through medical experimentation is outweighed by the harms of inaction: cancer patients, for instance, may see the risks of an experimental treatment as less than the risks of any known alternative. But prima facie it seems implausible that the risks of moral bioenhancement through genetic engineering, say, are less than the risks of known alternatives. More to the point, current scientific understanding of the relationship between the roughly 20,000 protein-coding genes that make up the human genome and the traits they code for is extremely limited. This is especially true for such complex traits as empathy, racial tolerance, or moral reflection. No one doubts that such traits are the result of *many* gene complexes whose expression is determined in part by certain environmental exposures, so manipulating a trait such as empathy will require altering many gene complexes. But this raises serious problems. First, it is rarely the case that altering one set of gene complexes yields only one phenotypic alteration. As geneticists frequently note, the relationship between genes and (phenotypic) traits is not one-to-one; it is not one-to-many; it is not even many-to-one. It is many-to-many. And it is difficult to see how the scientific understanding of these relationships could be improved without much trial and error. As Harvard University psychologist Steven Pinker (2003) cautions:

> Neural development is a staggeringly complex process guided by many genes interacting in feedback loops. The effect of one gene and the effect of a second gene don't produce the sum of their effects when they're simultaneously present. The pattern of expression of genes (when they are turned on or off by proteins and other signals) is as important as which genes are present.

But this is not even the biggest obstacle. As Pinker goes on to note, efforts to improve one part of the human phenotype sometimes result in unintended harms to other parts. For instance, researchers discovered that giving mice an extra copy of a certain gene could substantially boost the mice's memory (and, by extension, their capacity to learn), a feature that could arguably contribute to moral enhancement. But that extra copy of the gene also rendered subject mice hypersensitive to pain. To take another example, there is a gene complex in humans that is reliably correlated with increased intelligence. So those fortunate enough to have that gene complex have an IQ that is, on average, ten points higher than average. But those same people are also 10 percent more likely to develop a debilitating disease affecting muscle control and muscle movement. Directly selecting for that gene complex hardly seems risk free.

Finally, in the context of moral enhancement, it has been stressed that because morally enhancing someone effectively involves enhancing component capabilities, it is not certain that the end result will in fact be morally "better people." As Norman Daniels argues, "Many underlying capabilities or dispositions that might enhance virtues … can just as easily be put into the service of vice as well as virtue" (2009, 40). Daniels offers as a hypothetical example someone who naturally possesses enhanced empathic sensitivity but who employs that sensitivity not in the service of successful social work, say, but in the service of real estate fraud.

Pharmaceutical enhancements, by contrast, may avoid these specific worries, but it is reasonable to expect that the development of such drugs carries with it associated risks. For example, there will be legitimate questions about the potential risks of long-term use. In sum, even if moral bioenhancement is not intrinsically wrong, the problem of "getting from here to there" may yet constitute a serious enough harm to restrict it.

But suppose we *can* get there from here; suppose our trials are error free. Or, more realistically, suppose we *do* get there from here. Suppose unscrupulous researchers, operating in the shadows of professional or state oversight, master the genetic techniques necessary to enhance humans. Supposing that we possess the power to enhance ourselves morally and that the medical risks of using that power have been adequately reduced, are there *other* instrumental harms that should give us pause? Arguably, yes. One general class of worries has to do with the social consequences of enhancement (Sandel 2007; Kamm 2005). Space prevents an exploration of all these in detail, but here are several to consider:

- Those who choose not to enhance themselves or their offspring may find themselves facing significant disadvantages in a competitive marketplace.
- Those who choose not to enhance themselves may face discrimination.
- The rights of those who choose not to enhance themselves may be threatened.
- Enhancement may exacerbate class and social divisions, further diminishing social solidarity and trust.
- Enhancement may undermine familial relations.

Assessing both the likelihood and scale of such large-scale consequences are of course difficult tasks. One might argue that some consequences can be preempted. For example, perhaps it would be possible to ensure that all such (permitted) enhancements are federally subsidized, so the poor are not financially barred from participating. But this proposal, some authors argue (see Sandel 2004), will only exacerbate other problems: Is it not the case that those who choose *not* to enhance themselves face increased discrimination because the already-enhanced will see little incentive to aid the unenhanced when the latter *could have* freely enhanced themselves but chose not to? Would it not be possible (it is responded) to establish antidiscrimination laws to reduce the chances of those consequences? And besides, with regard to *moral* bioenhancements, should it not be expected that the disposition to discriminate will itself diminish as alterations to humans' moral character take effect? Perhaps, but is it possible to fully grasp all the implications of ushering in an unheralded arms race in moral enhancement?

Suffice it to say that no one can now say, with any real confidence, whether the introduction of moral bioenhancement will be a net boon or bust. It is not enough to marshal empirical evidence to show that some specific benefit (or harm) is likely to follow from the introduction of moral bioenhancement, for the crucial evaluation is one that *weighs* such a benefit (or harm) against associated harms (or benefits). It is this cost-benefit ratio that is both the most needed and most elusive evaluation in the context of the debate over instrumental harms.

INTRINSIC HARMS

As was discussed above, a given practice may be objectionable not necessarily because it leads to harms or wrongs, but because it is—in some sense—wrong in itself. This sort of objection is stronger than the ones canvassed above, for the sort of wrong it identifies is not contingent on how the relevant practice plays out in the real world. If the practice is intrinsically harmful or wrong, then this would ground a moral objection to the practice even if, for whatever reason, no instrumental harms materialize. To appreciate the point, consider how this general strategy might apply to the debate over, say, capital punishment. Someone who maintains that capital punishment is *intrinsically* wrong could consistently hold that position even when it is demonstrated that the practice is no longer subject to error or abuse. This is because such a claim is presumably based on the act itself, not on whether it sometimes ensnares the innocent, for example, or is racially discriminatory or fails to deter crime. The act remains wrong even when these other factors are controlled for. So what sorts of intrinsic wrongs or harms do authors point to in objecting to moral bioenhancement? It is possible to distinguish between wrongs one commits when the target of enhancement is *oneself* and wrongs one commits when the target of enhancement is *another*, such as an offspring. These classes are not mutually exclusive: in some cases, the wrong one commits in enhancing oneself is one and the same kind of wrong that one commits when one enhances one's offspring.

For instance, some authors (see Glannon 2001; Elliott 2003; Parens 2009) argue that enhancement alienates individuals from their true or authentic self. This line of thought can assume different forms. In one form, an identity is drawn between one's naturally given genetic blueprint and one's true self so that changes to an individual's genome (in vitro, for example) amount to distancing an individual from his or her true self, to separating humans "from what is most our own" (Parens 2009, 184). Even if those changes consist in moral improvements, those improvements are gained at the cost of alienation, and such alienation may conflict with important personal values. As the President's Council on Bioethics cautions, "As the power to transform our native powers increases, both in magnitude and refinement, so does the possibility for 'self-alienation'—for losing, confounding, or abandoning our identity" (2003, 292).

This line of objection is not entirely persuasive, however, for several reasons. First, it is a matter of some controversy whether or not a zygote, say, has a self from which it could be alienated. After all, this is not like the reaction some adults have to antidepressants: subjects sometimes report improvements in overall mood but note a disconcerting sense that they are "not themselves." Thus, they conclude that the alterations are not in the end worth it. But such a scenario does not easily, if at all, map onto a gestating embryo. Second, one might wonder why a naturally given genome should be privileged. If such moral improvements were indeed part of one's naturally given genome, it would indeed be "one's own," so the objection would presumably disappear. But this now seems all a matter of moral luck, for the genes one inherits are largely the result of a genetic lottery.

An appeal to what is natural prompts another sort of objection: bioenhancement (moral or otherwise) is unnatural, and engaging in such unnatural practices is immoral. This objection, too, can be understood in a number of different ways. Philosopher Thomas Douglas (2008), taking his cue from seminal Scottish philosopher David Hume (1711–1776), considers three different senses of "unnaturalness": miraculousness, unusualness, and artificiality. Quite clearly, critics of moral bioenhancement cannot mean that bioenhancement is unnatural in the sense that it is miraculous or supernatural; after all, these are (or would be) the results of investigations into the operations of the natural world. Douglas notes that the second sense of unnatural—that is, unusual—is a relative concept: something is unusual only *for* or *to* someone. But if this is what is meant by unnatural, then the force of the objection diminishes considerably: what is being asserted is that bioenhancement is wrong only because it is unusual *to* an individual or *to* contemporary society. But, it may be responded, if this assertion is enough to show that bioenhancement is wrong, it would be enough to show that all sorts of routine medical practices are wrong. At some time or other, anesthesia, antibiotics, vaccines, surgery, and so on are/were unusual to others, but that hardly shows that such practices are immoral. That leaves artificiality.

Nailing down what is meant—or what could be meant—by artificiality is a tricky business. On the widest interpretation, artificiality may simply mean human involvement. Thus, when humans step in, a given practice is now artificial; it is no longer natural. But this hardly seems sufficient to ground the claim that unnatural acts are *wrong*, because humans involve themselves in all manner of morally permissible, if not morally obligatory, acts: educating children, clothing the homeless, providing humanitarian relief, and so on. Perhaps what is meant by artificiality is something narrower. Perhaps the idea is that what is artificial is the introduction of human technology. But here, again, this idea cannot be what is driving the *moral* objection because, one would assume, very few people would argue that the use of human technology to treat disease, say, is immoral. The answer may be to limit the concept even further: what is artificial is the use of human technology in realms other than treating disease. Apart from the ad hoc nature of this revision, however, this interpretation fails as well. Human technology is applied across a range of domains unrelated to health care without any obvious moral infraction. Human technology is made use of to improve transportation, communication, agriculture, entertainment, and so on. Never mind that such uses enhance human lives in myriad ways, it is simply false that the artificiality of these practices renders them immoral. (It is true that sporadic moral controversies flare up here and there—as in, for example, the debate over genetically modified crops—but the sweeping generalization itself does not explain those controversies.) The upshot here seems to be that the unnaturalness objection is not yet an insurmountable objection to moral bioenhancement, for subscribers to this view have yet to adequately articulate the relationship between unnaturalness and wrongness.

A distinct, though related, sort of claim is this: moral bioenhancement is wrong because it amounts to playing God, and practices that amount to playing God are intrinsically wrong. Despite the obvious interpretation, there are secular ways to understand this claim, as covered below, but the discussion begins here with the religious interpretation. According to philosopher C. A. J. Coady (2009), the two most plausible perspectives Christians might take on the relation between humans and the God-given natural world are (1) stewardship and (2) cocreation, and when humans go beyond these roles, they are effectively playing God. According to the stewardship perspective, humans are caretakers for God's creation. Hence, when humans go beyond merely preserving (and revering) his creation and intervene in ways that redirect natural processes for our ends, we play God. But Coady correctly notes that even contemporary Christians will not find that this line of criticism stands up to scrutiny. Assuming such a passive stance toward the world around us is not the only permissible stance available; indeed it might be the wrong stance to take. Humans can and should intervene to improve human health, transportation, communication, and so on. The right response to an outbreak of the Ebola virus, for instance, is not reverence.

Instead of the stewardship perspective, some theologians prefer the cocreator picture of humanity's relation to the natural world in which we are called into "a creative partnership in the ongoing creation of the world" (Coady 2009, 160). This picture avoids the excessive passivity associated with the stewardship perspective, but it carries with it other costs. By shortening the distance between God and human, it robs the notion of "playing God" of its supposed moral wrongness. If indeed humans are called into partnership with God to assist in creation, then accusing someone of playing God is no longer an admonition; it is kind of his or her job.

It may reasonably be responded that, despite humans' cocreative participation in world shaping, humans' roles and God's role are clearly delineated. And there are some interventions only God can initiate—among them, interventions that improve people morally. As stated, though, this cannot be right. It is a general belief that parents are not merely permitted, but obligated, to intervene (environmentally, as it were) to encourage compassion, kindness, fairness, and so on in their children through dialogue, role playing, and modeling. A further argument here might be that if a child exhibits robust patterns of antisocial behavior, parents are permitted (if not obligated) to avail themselves of pharmaceutical interventions if this is the recommendation of trained psychologists. So the claim must be that the interventions only God can make are *genetic*. But now it might be wondered *how* such critics could possibly know this. As Coady (2009, 179) wonders: "Have the Churches been playing God in the crisis created by new technologies?"

If the religious interpretation of the "playing God" objection stumbles, might the secular interpretation find firmer footing? Suppose the expression "playing God" is treated as a metaphor for human hubris. To play God under this interpretation is to go "beyond the limits we have in these three respects, acting in ways that ignore the in-built constraints on our knowledge, power, and benevolence" (Coady 2009, 163). According to the traditional Judeo-Christian conception of God, he is all-knowing, all-powerful, and all-loving. When someone is accused of playing God, then, he is being accused of assuming (incorrectly) that he possesses these properties to a degree comparable to God. Whether or not God exists is irrelevant. What is relevant is that some individual's actions presuppose powers he surely lacks. This sort of objection is not vulnerable to the sorts of responses discussed above. But to appreciate its force, one needs to dig a little deeper: there are (at least) two ways to cash out this objection, ways that parallel the earlier distinction made between instrumental and intrinsic harms.

If the "playing God" objection is treated only as a metaphor, what concerns arise about humans acting as if they possess unlimited knowledge, power, and benevolence? One obvious answer is: it will likely lead to harm. That is, playing God is *instrumentally* harmful. History, alas, is littered with the tragic consequences of humans acting on the basis of false beliefs, limited control, and/or latent malice. Lack of humility often leads to disaster. The many practical ways this concern applies to moral bioenhancement have already been noted above: it is not possible to know with confidence what sorts of consequences genetic, if not pharmaceutical, alterations will lead to. If the "playing God" objection more or less reduces to the concern about instrumental harms, then this has already been addressed in the earlier discussion.

But there is another way in which the "playing God" objection might be interpreted, one that treats it more as an intrinsic harm. One of the most influential critiques of enhancement in general put forth in the twenty-first century comes from Sandel (2004, 2007), who, while he does not invoke the playing God objection specifically, is no doubt sympathetic to the general idea. As Sandel puts it in his own succinct words: "The deepest moral objection to enhancement lies less in the perfection it seeks than in the human disposition it expresses and promotes.... The problem lies in the hubris of the designing parents, in their drive to master the mystery of birth" (2004). There is something wrong about *wanting to master the mystery of birth* and not having "an openness to the unbidden," according to Sandel. Part of the wrong consists in the harms such an attitude will lead to: "It would disfigure the relation between parent and child, and deprive the parent of the humility and enlarged human sympathies that an openness to the unbidden can cultivate" (Sandel 2004). Parents will tend to exhibit toward their children less tolerance and accepting love, and this in turn threatens the well-being of those children. But Sandel also suggests that there is something intrinsically wrong in parents cultivating a desire to control and master their offspring: there is something virtuous about people standing in the right sort of relationship to their children and others, a relationship that honors the giftedness of life. According to this view, even if all the instrumental harms previously discussed could be prevented, bioenhancement would still be wrong because it would entail morally unacceptable attitudes that, in some sense, defile important human relationships.

Sandel's objection has been the subject of much scrutiny because, among other things, it purports to identify the "deepest moral objection" to enhancement without relying on controversial theological commitments. And its scope is wide. If Sandel's critique is convincing, it should apply to any form of enhancement—moral or otherwise. But not all bioethicists are convinced. Sandel's colleague at Harvard, philosopher Frances M. Kamm (2005), argues that Sandel's position is either internally inconsistent or implausible. If there is a standing moral prohibition on attempting to "master the mystery of birth," then efforts to treat or eradicate disease prenatally should be prohibited (because such efforts essentially involve mastering the mystery of birth and not having "an openness to the unbidden"), but this is inconsistent with Sandel's *own* position, a position most find reasonable: "Medical intervention to cure or prevent illness or restore the injured to health does not desecrate nature but honors it" (Sandel 2004). It does not appear (to Kamm at least) that Sandel can have it both ways: he cannot *oppose* efforts to master the mystery of birth and being open to the unbidden, on the one hand, and *support* efforts to investigate and prevent serious birth defects, on the other. Furthermore, Sandel's view is implausible, claims Kamm, because it assumes an implausible relationship between an agent's attitudes and the moral status of her action—namely, that having a bad motive to perform some action necessarily renders the action bad. Kamm rejects this. Take, for example, a biochemist who seeks to eradicate the prenatal diseases that cause blindness, not because he cares about the misfortunes of children but because he seeks to best a rival biochemist. If Sandel is correct, however, curing the prenatal diseases that cause blindness is wrong in this case, but surely that is mistaken.

One final—and popular—objection to moral enhancement is that the practice is unethical on the grounds that it limits the autonomy of the enhanced. This objection has been the focus of a number of authors across the spectrum (e.g., Habermas 2003; Sandel 2004; Douglas 2008; Coady 2009; Harris 2011; Bess 2015). The worry is that parents choosing particular traits for their children amounts to pointing children "toward particular choices, and so designer children would never be fully free" (Sandel 2004). It seems reasonable to hold the view that children should be allowed to have an "open future," unconstrained by the external choices of others, including parents. Even when we imagine that the traits parents choose for their child are moral (e.g., increased altruism or racial tolerance), they are nonetheless limiting the choices available to him or her. And this is objectionable because, as one author puts it, "the *opportunity to fail* was never real and present" for the child; hence, "the moral significance of [the child's] action is therefore greatly lessened" (Bess 2015, 130).

But the decreased autonomy objection, though initially compelling, is less so on inspection. Consider, first, the idea that designing parents are constraining a child's choices when parents choose some of her traits. This gives the mistaken impression that a child whose parents *have not* chosen some of her traits possesses some control over her traits. But of course she does not. As Sandel (who, as noted, opposes bioenhancement) correctly notes: "The alternative to a cloned or genetically enhanced child is not one whose future is unbound by particular talents but one at the mercy of the genetic lottery" (2004). Neither alternative affords a child a choice in the matter. But is it not the case that a child's opportunities are constrained when parents genetically impose a certain trait on a child? To the extent that this response is not vulnerable to the previous point, it is not clear what the *moral* complaint is here. After all, as Coady (2009, 175) observes, "A child has no legitimate moral complaint against her musically gifted parents who were attracted to each other partly because of their musical gifts and wanted to have a musically gifted child. But, if not, why would she have a legitimate complaint against parents who, with benevolent intent, genetically engineered musical talent for her?" Likewise, for distinctly moral traits: why would a child have a legitimate complaint against parents who, with benevolent intent, sought to create a more empathic and tolerant child, whether through natural or "artificial" means? These observations points to an even deeper misunderstanding among those advancing the decreased autonomy objection.

Consider how Michael Bess (2015) presses the objection. The reader is to imagine three different scenarios, each involving you and someone in need of assistance. In the first scenario, you are an unenhanced egoist; in the second, an unenhanced altruist. In each scenario, you consider the person in need and choose either to run away (the unenhanced egoist) or to assist (the unenhanced altruist). The reader is supposed to contrast these cases with a third: in this case, you are an *enhanced* altruist who, having ingested his morning moral enhancement pill, feels "a rush of empathy" for the person in need and runs to assist. Why exactly should this third case be troubling? According to Bess, the pill

> precludes you from seriously contemplating such selfish actions as running away—for you find yourself awash in a surge of empathetic feeling—and your final decision therefore rests on an exercise of free will that has been narrowed, prechanneled, in a significant way. You really had no alternative but to do some form of "the right thing," because your emotional responses have been chemically nudged toward the kinder end of the dispositional spectrum. The moral significance of your action is therefore greatly lessened. (2015, 130)

Is it though? Consider a fourth case: you are an altruist who, as a result of the natural genetic lottery, experiences more frequent and more intense altruistic motives than most because, say, you inherited a neurochemistry roughly equivalent to that which is artificially produced by the moral enhancement pill described above. (Your biological parents were impassioned humanitarians.) When you see someone in need, you, too, "feel a rush empathy" and run to assist. Can we plausibly claim that your action "lacked moral significance?" Of course not. Or consider a fifth case: you are an altruist who, through training and habituation, experiences more frequent and more intense altruistic motives than most. When you see someone in need, you feel a rush of empathy and run to assist. Does your response lack moral significance? Hardly. You are what ancient Greek philosopher Aristotle refers to as a *virtuous* person: you think you ought to assist, you want to assist, and you do. In any event, without a thoroughgoing analysis of what it is for one to have free will (a vexed philosophical notion if there ever was one), Bess is in no position to assert that your free will has been narrowed in these latter cases (see also Douglas 2008). But even supposing it has been, this just shows that enhanced moral traits such as empathy do not forfeit their moral significance simply because someone reflexively experiences a greater empathic concern for others. Indeed, many people regard such individuals as moral exemplars.

To be sure, this particular debate is held hostage by ongoing disputes over the metaphysics of free will. It is simply not possible to say with confidence what it is for a person to act freely, nor what would constitute a "narrowing" of that freedom. Moreover, freedom is one thing, and responsibility (and moral worth and virtue and so on) is quite another. Linking these two is a delicate matter, as philosophers have long noted. For instance, one might be tempted to claim that the fact that someone had little choice in a certain state of affairs is sufficient ground to absolve that person of responsibility. But that is a temptation to resist, as Sandel (2004) notes. Take American swimmer Michael Phelps, the most decorated Olympian in history. No one is seriously clamoring to vacate Phelps's many successes on the grounds that he inherited, through no effort of his own, a physique uniquely suited to swimming. Reasonable people believe that Phelps *deserves* the accolades he has received even though his successes depended necessarily on his enhanced features: his torso, his wingspan, his foot size, his hypermobile ankles, and so on.

What this means for the present debate is that those pushing the decreased autonomy objection bear the added burden of articulating exactly how the freedom to choose relates to the moral worth of one's actions. That relation is not nearly as obvious as critics assume. Raising a more modest decreased autonomy objection, John Harris (2011, 103) insists that "autonomy surely requires not only the possibility of falling but the freedom to choose to fall." Whether the point is made in terms of the freedom to act unethically or the Christian notion of the fall from grace (as Harris presumably intends), it is still not clear what Harris means here. Do you have the freedom to choose to push a (fabled) button that would result in the death of some nondescript peasant in exchange for a $1,000,000 and guaranteed immunity? In one sense, yes. You choose to murder for money in the way you choose to clean out the litter box: you form the intention and then act on it. But in another sense, it is likely that you do *not* have the freedom to murder for money. Such an act falls within what John Kekes (2010, 53) calls the "unthinkable," which is "what we think and feel we cannot do, not even if it would be very important for the realization of some value that is crucial to our conception of well-being." And yet the limits that set the boundaries for the unthinkable are a source of great moral significance even if one is not obviously responsible for them. "What that significance is emerges if we understand the force of the 'cannot' when some cannot but say the great 'Yes' to what they value most, or cannot but say the great 'No' to doing what they regard as unthinkable" (56). What critics have not yet shown is that the moral significance of American civil rights leader Martin Luther King Jr.'s stand against racial discrimination, say, is impugned by the very real possibility that he could not but say the great "No"—that he indeed lacked the freedom to choose acquiescence. Until this is shown, it is not possible to claim that external enhancement impugns the moral significance of one's actions because it could very well be that everyone is the lucky (or unlucky) recipient of external enhancement.

ARGUMENTS IN SUPPORT OF ENHANCEMENT

Taking stock of the foregoing, a number of authors argue that bioenhancement—even moral bioenhancement—is intrinsically wrong. But these arguments appear inconclusive, at best. How might these results contribute to an argument in *support* of moral bioenhancement? Well, if one could successfully establish that moral bioenhancement is (intrinsically) innocent unless proven guilty, then it may appear by now that moral bioenhancement has survived scrutiny. No convincing case against moral bioenhancement succeeds, so the practice must not be intrinsically wrong. After all, this chapter began with a prima facie reason in support of moral bioenhancement: most people already believe that there is a good reason to improve themselves morally. (Now it may be that moral bioenhancement nevertheless succumbs to concerns about instrumental harms, as discussed earlier, but there, too, the future is foggy.) But this would be too easy. Assuming moral bioenhancement is the default position will be acceptable only to those … who already assume it is the default position. So what sorts of arguments are supporters making?

One of the more outspoken supporters of bioenhancement is Julian Savulescu (2007; Savulescu and Bostrom 2009), who argues that bioenhancement is not merely morally permissible but morally obligatory: "To fail to do so," he says, "would be wrong" (Savulescu 2007, 517). Why? Two of his arguments have been particularly influential. The first involves a thought experiment. Suppose a couple give birth to a child with a "stunning intellect," but (they are informed) maintaining this stunning intellect requires a "simple, readily available, cheap dietary supplement" (Savulescu 2007, 525). Without it, the child's intelligence will quickly revert to average. Because the parents are neglectful, however, they neglect to regularly give their child the supplement, and the child's stunning intellect disappears. Savulescu suspects that most people will judge that the parents' behavior was morally irresponsible. Now consider a different couple: they give birth to a child with normal intellect, but (they are informed) she will develop a stunning intellect if she receives that same cheap, readily available supplement mentioned above. Because the parents are lazy, however, they fail to regularly give the child the supplement, and her stunning intellect fails to develop. Here, too, Savulescu anticipates a similar moral judgment: the lazy parents behaved irresponsibly. Based on these judgments, this must mean that there is no interesting moral difference between failing to *preserve* and failing to *give* a child intellectual gifts. But if this is true, then parents who do not make use of available enhancement technologies (e.g., pharmaceutical or genetic) are behaving irresponsibly.

This argument, if successful, would show only that bioenhancement is intrinsically ethical. It would not necessarily displace the instrumental harms discussed above. But leaving this aside, how successful is the argument? Well, it depends critically on readers' intuitions about the two sets of parents. Some audiences (of college undergraduates, for example) express different moral reactions to the two sets of parents. The lazy parents' failure, they say, is *less wrong* than the neglectful parents' failure because, for instance, there is a moral difference between having-and-losing and never-having-had. There is this concern, too: is having a stunning intellect an unmitigated gift? Unless we have assurances to the contrary, having a stunning intellect can be linked to anxiety, antisocial tendencies, and depression. In such cases, neither failing to give nor failing to preserve would be a moral failing.

A second of Savulescu's arguments is the consistency argument. Most people believe that parents *already* have a moral obligation to intervene in their child's life to enhance, for example, a child's moral character: to develop kindness, fairness, perspective taking, and so on. But if this is so, and if there are no important moral differences between environmental interventions (EIs) and genetic interventions (GIs), then consistency should require that genetic enhancements be treated no differently than environmental enhancements. Savulescu (2007, 522) concludes, rhetorically: "What is the moral difference between producing a smarter child by immersing that child in a stimulating environment, giving that child a drug, or directly altering the child's brain or genes?" Now it may be responded that GIs, unlike EIs, are difficult to reverse and result in changes that can be passed on to the future generations. But Savulescu (2007) anticipates this response, citing studies that show that environmental events can be irreversible and influence genetic expressions in offspring. Critics, however, might wonder if there are other unexplored moral differences between EIs and GIs; for example, GIs might make moral training—and by extension, parent-child bonding—merely optional. Perhaps taking the genetic shortcut diminishes other crucial attitudes: gratitude, perseverance, patience, and tolerance.

Like Savulescu, Harris (1985, 1992, 2009) has long maintained that scientific technology can and should be actively used to improve human well-being. As far as enhancements go, Harris (2009) makes it plain that they are a moral obligation. Harris challenges those bioethicists who maintain that curing disease is a permissible use of genetic technology while enhancement is not, because (according to these bioethicists) curing disease is merely an effort to restore normal species functioning whereas enhancement is an unjustified attempt to improve on an otherwise normal trait. The problem, according to Harris (2009, 148), is that "disease and unnecessary death are hideously normal," especially in old age, yet there is no hesitation about treating such maladies. It could very well be that the risks of leaving things as they are are equal to, if not greater than, the risks associated with genetic enhancement. This contributes to one of the few responses to the instrumental harms discussed above, from Harris: "In the absence of reliable predictive knowledge as to how dangerous leaving things alone may prove, we have no rational basis for a precautionary approach which prioritizes the status quo" (2009, 133). According to a most modest reading of this, Harris is demanding that very careful consideration be given to what benefits such enhancements may yield *in relation to* an honest appraisal of the current state of human well-being, which is a decidedly mixed bag.

Summary

One might assume that reason, in the end, *favors* support for moral bioenhancement. A more neutral position, however, is that there is simply too much we do not understand to form a settled moral view of moral bioenhancement. We do not yet understand what getting from here to there might mean. We do not yet understand all the possible transformations getting there may catalyze. We do not yet understand all the many personal, social, and biological implications of moving in such an unprecedented direction. We do not yet understand what liberties bioenhancement may curtail. We do not yet understand how such liberties relate to moral responsibility. This lack of understanding ranges across genetic, biological, philosophical, and sociological domains. Amid such darkness, prudence may seem to counsel restraint. But even this (as Harris argues) may be imprudent, for we may be standing on the threshold of monumental social improvements and simply not know it. Perhaps the risk of not taking the risk is far greater. We cannot now say. We are left then having to embrace our philosophical discomfort and continue to ask the difficult questions.

Bibliography

Bess, Michael. *Our Grandchildren Redesigned: Life in the Bioengineered Society of the Near Future.* Boston: Beacon Press, 2015.

Coady, C. A. J. "Playing God." In *Human Enhancement,* edited by Julian Savulescu and Nick Bostrom, 155–180. Oxford: Oxford University Press, 2009.

Daniels, Norman. "Can Anyone Really Be Talking about Ethically Modifying Human Nature?" In *Human Enhancement,* edited by Julian Savulescu and Nick Bostrom, 25–42. Oxford: Oxford University Press, 2009.

Douglas, Thomas. "Moral Enhancement." *Journal of Applied Philosophy* 25, no. 3 (2008): 228–245.

Elliott, Carl. *Better Than Well: American Medicine Meets the American Dream.* New York: Norton, 2003.

Glannon, Walter. *Genes and Future People: Philosophical Issues in Human Genetics.* Boulder, CO: Westview Press, 2001.

Habermas, Jürgen. *The Future of Human Nature.* Translated by Hella Beister, Max Pensky, and William Rehg. Cambridge: Polity Press, 2003.

Harris, John. "Enhancements Are a Moral Obligation." In *Human Enhancement*, edited by Julian Savulescu and Nick Bostrom, 131–154. Oxford: Oxford University Press, 2009.

Harris, John. "Moral Enhancement and Freedom." *Bioethics* 25, no. 2 (2011): 102–111. doi:10.1111/j.1467-8519.2010.01354.x.

Harris, John. *The Value of Life*. London: Routledge and Kegan Paul, 1985.

Harris, John. *Wonderwoman and Superman: The Ethics of Human Biotechnology*. Oxford: Oxford University Press, 1992.

Kamm, Frances M. "Is There a Problem with Enhancement?" *American Journal of Bioethics* 5, no. 3 (2005): 5–14.

Kekes, John. *The Human Condition*. Oxford: Oxford University Press, 2010.

Parens, Erik. "Toward a More Fruitful Debate about Enhancement." In *Human Enhancement*, edited by Julian Savulescu and Nick Bostrom, 181–197. Oxford: Oxford University Press, 2009.

Pinker, Steven. "The Designer Baby Myth." *Guardian* (London), June 4, 2003. https://www.theguardian.com/education/2003/jun/05/research.highereducation.

President's Council on Bioethics. *Beyond Therapy: Biotechnology and the Pursuit of Happiness*. Washington, DC: Author, 2003.

Sandel, Michael J. "The Case against Perfection." *Atlantic*, April 2004. https://www.theatlantic.com/magazine/archive/2004/04/the-case-against-perfection/302927/.

Sandel, Michael J. *The Case against Perfection: Ethics in the Age of Genetic Engineering*. Cambridge, MA: Belknap Press of Harvard University Press, 2007.

Savulescu, Julian. "Genetic Interventions and the Ethics of Enhancement of Human Beings." In *The Oxford Handbook of Bioethics*, edited by Bonnie Steinbock, 516–535. Oxford: Oxford University Press, 2007.

Savulescu, Julian, and Nick Bostrom, eds. *Human Enhancement*. Oxford: Oxford University Press, 2009.

Singer, Peter. "Parental Choice and Human Improvement." In *Human Enhancement*, edited by Julian Savulescu and Nick Bostrom, 277–289. Oxford: Oxford University Press, 2009.

Steinbock, Bonnie, ed. *The Oxford Handbook of Bioethics*. Oxford: Oxford University Press, 2007.

FILMS

Code 46. Dir. Michael Winterbottom. 2003.

The Fly. Dir. David Cronenberg. 1986.

Gattaca. Dir. Andrew Niccol. 1997.

My Sister's Keeper. Dir. Nick Cassavetes. 2009.

Never Let Me Go. Dir. Mark Romanek. 2010.

The 6th Day. Dir. Roger Spottiswoode. 2000.

Splice. Dir. Vincenzo Natali. 2009.

CHAPTER 26

Synthetic Biology: The Digital Creation of Life

Gaymon Bennett
Associate Professor of Religion, Science, and Technology,
School of Historical, Philosophical, and Religious Studies
Arizona State University, Tempe

This chapter provides an introduction to a brand of biotechnology called synthetic biology. Synthetic biology was once a Silicon Valley darling—and thus a darling of the media, money, and celebrity that animate the valley. Yet today it has lost a measure of its panache. Many of the architects of the brand no longer appeal to it to promote and organize their careers and companies—even though their enterprises and self-stylization remain largely unchanged.

This is not to say synthetic biology has gone away. Quite the contrary. Its legacy is very much in force. From the first, the brand was all about a distinctive way of looking at living things and about the need to invent techniques and technologists consistent with that vision. It was about taking up living things as digital things and thus subjecting the making of living things—that is, bioengineering—to the cultures of innovation that brought into being the Internet, smartphones, cloud computing, and machine learning. The synthetic biology brand, in other words, made explicit and discrete a goal that might otherwise have remained tacit and diffuse: the goal of transforming the future of living things into a digital future. That goal is very much still in play.

The twist is this: if the architects of synthetic biology no longer appeal to the brand, it is because they do not have to. The moods and manners once associated with the synthetic biology brand have become commonplace.

To put this in terms typical of the social study of science and technology, one can say that the synthetic biology brand has become embedded in, and extended through, dominant "biotechnical imaginaries" (Jasanoff 2015; Bennett 2017). In the sections that follow, this chapter explains what a biotechnical imaginary is and why it matters. Here it is enough to say that the cultural worlds of biotechnology—the worlds of those individuals and institutions who have the resources and know-how to engineer living things—have been sharply inflected by the legacies of synthetic biology. As a result, people who are living lives governed, in part, by biotechnology and its vision for contemporary life (and that includes almost anyone reading this book) will need to understand synthetic biology in order to grapple effectively with the cultural and material forces shaping humanity's collective biological future (Rabinow and Bennett 2012; Hurlbut 2017).

SYNTHETIC BIOLOGY AND THE BIOTECHNICAL IMAGINARY

In 2013 the University of California, Berkeley, hosted a two-day public event on the state of play of synthetic biology. At the time synthetic biology was a brand on the rise. Born in the wake of the genome sequencing projects of the late 1990s, it had subsequently been promoted and institutionalized through a series of manifestos for the future of biotechnology (Rabinow and Bennett 2012). The Berkeley event was upbeat and fashionable, in the TED Talk–style that has become familiar fare in the tech world. It was filled with talk of innovation, disruption, and the redemptive future of new technologies—of machines to be built, money to be made, and lives to be saved. Yet for all the breathless talk, the event was not obviously consequential. It included no announcements of major technical breakthroughs or new funding initiatives. It did, however, quietly bear witness to a significant transition in the status and ethos of the synthetic biology brand and, in so doing, pointed to a still-ongoing transformation in the imagined future of engineered life.

A session titled "Programming Life: The Revolutionary Potential of Synthetic Biology" was exemplary. The session featured a conversation between Berkeley biochemist and biotech star Jay Keasling and a "technology prospector" named Virginia Ursin from agribusiness giant Monsanto Company, facilitated by editors of *Discover* magazine. Most everyone in the room—technologists, investors, activists, and reporters—were there to hear Keasling. But the fact that representatives of *Discover* and Monsanto were there was significant. It signaled that synthetic biology—which had once been considered marginal, and by some even radical—had been made part of the biotech mainstream. Synthetic biologists' manner of talking and acting—a manner itself born from the worlds of digital computing—had been accepted, if not yet adopted, by the old guard of the bioeconomy.

Still, Keasling was the main show. At the time, his prognoses about the future of bioengineering were considered a gold standard (Balmer, Bulpin, and Molyneux-Hodgson 2016). In the preceding years, Keasling's team, funded by the Bill and Melinda Gates Foundation, had successfully engineered yeast cells capable of producing a molecule called artemisinin, a key ingredient in antimalaria medication. Moreover, his team showed that the techniques used to produce artemisinin could be repurposed to make fuels, plastics, and cosmetics. Keasling's work thus had come to be seen as the "reality proof" for "the revolutionary potential" of synthetic biology. From the point of view of his supporters, it seemed to show that a mode of bioengineering inspired by computer science would transform living organisms into the basis for the next tech revolution. Not incidentally, Keasling's work reinforced one of the regulative myths of biotech innovation: entrepreneurs could deliver powerful technologies, improve humanity, and make lots of money, seemingly without trade-offs (Hurlbut 2017).

In the course of the conversation, Keasling said something that seemed to strike the tech insiders as unremarkable but that might have caught the attention of a different audience. Affirming the theme of the session—that life can and should be treated as "programmable matter"—Keasling pointed to recent work by his Stanford University colleague Drew Endy as demonstration that life can be made to operate as a computer (Bonnet, Subsoontorn, and Endy 2012). (Endy's lab had playfully shown that cells could be made to store information in a manner reminiscent of binary code.) The implication of Keasling's comment was not lost on those present, because they had heard it before. If cells are basically computers, then the mavens of the high-tech industry are primed to deliver the engineered future of life itself.

Even after the fact, it is not obvious whether Endy, or any other synthetic biologist, has actually shown that the logic of living systems is fundamentally commensurate with the logic of digital systems. As such, it remains far from obvious whether the lessons learned from programming computers will be sufficient to lead to the long-term goal of programming life. But the question of whether the synthetic biology approach is somehow "truer" or more "successful" than other modes of bioengineering is beside the point here. The crucial point is that Keasling's statement surprised almost no one in the room. And it surprised almost no one because the attitude underlying it—namely, that organisms are digital machines and can be engineered as such—was already taken for granted. By 2013 such statements as "life is programmable matter" had become a fixture of the biotechnical imaginary.

The term *imaginary* is a technical one (Taylor 2002). It refers to all the background assumptions about how the world works that a group of people needs to share in order to live everyday life together. These assumptions, and the authority they implicitly carry, allow people to live together in a more or less coherent fashion without really having to think too much about it. An "imaginary," or "social imaginary," as it is often put, is thus not just stuff made up in people's heads—people's "imagination," or their individual "perspectives." It is what enables collective social life by allowing people to make sense of reality together. And although a social imaginary is about "sense making," it is deeply connected with the material aspects of life: from how people work and play together, to how they build cities, to how they use new technologies, tell jokes, worship gods, or cook food. A social imaginary, in short, is the implicit answer we all give in our everyday activities to the question "why do we do things the way we do?"

In the case of synthetic biology, social imaginary, or what some call the "biotechnical imaginary," refers to all the tacit sensibilities, understandings, and expectations that must be in place in order for technologists (and their funders, critics, and observers) to talk about and engineer living systems in a manner they take to be coherent and even meaningful. A biotechnical imaginary might be broad and flexible: it might leave lots of room for difference and dissent. But in order for people to get on with the work of imagining futures, selling new ideas, conducting experiments, and building careers, they need to share a good deal in common about how the world works and ought to work.

All this matters because, as multiple cultural critics have pointed out, technology today, and therefore the people who make technology, shape how everyone affected by technology live their lives (Rabinow and Stavrianakis 2014; Jasanoff 2016). New technologies "[empower] some forms of life [by] making them natural while others, by comparison, come to seem deficient or unnatural" (Jasanoff, Hurlbut, and Saha 2015, 26). To say that the ethos and attitudes characteristic of synthetic biology have become a fixture of the biotechnical imaginary is thus to make a strong claim. It is a way of saying that the inclinations and inspirations of a small group of actors have had a disproportional impact on the background assumptions that structure and direct the way people both within and beyond biotechnological domains view and inhabit the world.

That said, things run in both directions. Part of the reason synthetic biologists have been able to inflect people's collective imaginations is that their ways of thinking and acting are largely familiar to those already living in technology-soaked worlds. Synthetic biologists' emphasis on the digital reflects and not only inflects contemporary social life. It is thus not particularly surprising that they have been able to play such an influential role in shaping how nonbiologists are coming to talk about, invest in, and reimagine living things. If life, for synthetic biologists, has become "programmable matter," chances are it will soon be that way for many other people, too.

FROM DIGITAL MANIFESTOS TO DIGITAL IMAGINATIONS

Synthetic biology is a child of "digital natives"—that is, of people born in the midst of the high-tech revolution and who now seek to order the world of biology in view of the ways of life generated by that revolution. It was invented by technicians who learned to program code before they dreamed of reprogramming cells. In this sense, no synthetic biologist—whether Endy or Keasling or any of their many collaborators—actually needs to demonstrate that living organisms can be programmed like digital machines in order for the brand to make sense to those who support and critique it. To the contrary: the idea that living things can and should be treated like digital things has been a predicate of the enterprise from the outset.

Because the synthetic biology brand is now global and diffuse, there can be no simple story of its rise and spread. That said, like other scientific brands, it does have its origin stories. A look at those stories tells quite a bit about the basic sensibilities and inclinations that have come to define it. One story begins in the first years of the twenty-first century with a loosely affiliated group of technologists working at the Molecular Sciences Institute in Berkeley, California, and at the Artificial Intelligence Laboratory (AI Lab) at the Massachusetts Institute of Technology (MIT) in Cambridge, Massachusetts (Rabinow and Bennett 2012). The group was scientifically and technologically diverse. It included master biologists alongside junior physicists, computer hackers, and senior software engineers. Despite the differences among them, the group shared the sense that progress in biotechnology requires more than scientific knowledge. Because biotechnology involves making new organisms with designed functions, progress also requires a systematic approach to the question of what it means, in biology, to be an engineer.

Their working assumption is that molecular biologists had not (yet) taken full advantage of the tricks of the trade that had proven so powerful in computer engineering. If programming genetic code is anything like programming binary code, they reasoned, then the methodological question "how should biology be engineered" is less a biological problem—that is, less about the nature and limits of living systems—and more an ethical and cultural problem: a question of the attitude and comportment of the bioengineer. They asked: could the makers of digital technologies provide the *imago*—the image and model—of the good engineer for biology? And could biologists be encouraged to "remake" themselves (their skills, their attitudes, and even their communities) on the basis of that image and model?

A key figure in all of this (to pick just one) was Tom Knight. Knight was exemplary of the way in which the founders of synthetic biology sought to move from the modes and methods of computer engineering to a reimagined biotechnology. In the late 1990s Knight and a cohort of graduate students in electrical and bioengineering began to outline a program for rethinking molecular biology on the model of computer science (Balmer, Bulpin, and Molyneux-Hodgson 2016). Knight had begun to exit from a long career designing and making information systems. He had become interested in the question of how, and to what extent, living systems could be made to perform like computational systems—that is to say, made into computers. Moreover, Knight wanted to know what possibilities might be opened up for both computer science and biology, if biologists were to adopt the attitudes and sensibilities that had powered the birth of the computer industry, especially those attitudes and sensibilities that had been pioneered at MIT.

Knight arrived at MIT as an undergraduate in the 1960s and stayed on as a research scientist—one of a cohort of engineers that helped shape the school's culture of computer engineering. Knight arrived at a moment of intense growth in the information sciences, as well as their rapidly unfolding relationship to the life of technology and, eventually, the technologies of life. Debates had begun to crescendo over the limits of cybernetics, the science of control and communication in complex systems. The question was how the advanced design of complex systems—circuits, operating systems, and programs—could continue to drive computational power. By the mid-1970s, two practical concerns took hold. The first was pedagogical. Given the rate of growth in computer science, which training regimes would best equip young engineers to be capable of designing and building controlled informational systems? The second concerned the small-*p* politics of this growing community. Which forms of communication and organization among and between computer engineers would best facilitate the growth of a collaborative community?

Any number of responses were possible. At MIT, these questions were answered by way of a design philosophy, one consistent with the ethos of countercultural politics, which, by then, had become vital to a vision for person-centered computing (Markoff 2005). The philosophy centered on the idea that information systems qua communication systems ought to be structured in such a way that designers, working across space and time, will be able to understand and directly add to those systems. This commitment to "distributed engineering" may seem self-evident in retrospect, but it raised pressing questions concerning the basic requirements of design, from standards for building circuits to structures for operating systems. Crucially, these questions were not only technical. They concerned a vision for computer engineering in which the growth of computing power was indexed to the cultivation of a politics of openness and collaboration.

For engineers at the MIT AI Lab, the problem of design connected the pedagogical question to the political question. Two MIT engineers, Harold Abelson and Gerald Jay Sussman, famously pointed out that a computer language is not only a means of making a machine perform a set of technical operations. It is also a medium for expressing methodological ideas—that is, ideas about how to create relationships between different operators (Abelson, Sussman, and Sussman 1996). Anything a person could "express" using a computer language is as important socially as it is technically. In a programming language, the things people express are both the outcome of things other people have expressed and the predicate of future things not yet said.

The MIT design philosophy exemplified a second aspect of the countercultural ethos of personal computing as well. It embraced the idea that programming—that is, the rational design of information—is, in principle, unlimited. Among West Coast developers, this feel for the unlimited was often expressed in the grand terms of the self-actualization movement—personal expansion could be achieved through personal computers (Markoff 2005). Those who embraced the MIT design philosophy put it in seemingly more pragmatic terms. They took information to be fundamentally immaterial. Whereas the design of, say, bridges, is always limited by the physical properties of the underlying materials, the design of information is unbounded. The only real limits are those imposed by the designer's own (lack of) imagination.

At MIT, this sense of the seemingly unlimited potential of information was connected to the pedagogical-political vision for distributed engineering. Computer scientists embraced the idea that when it comes to the engineering of information individual ingenuity and collective savoir-faire are the primary passage points to a transformed future (Kelty 2008). They put a premium on collective, but individually motivated, action and embraced modes and forms of practice that they believed would equip communities of free users to work on and rework one another's creations. (It is little wonder that MIT computer scientists were at the heart of the free and open-source software movement.) This felt need for freedom and community building as keys to unlocking the unlimited possibilities of innovation would be primed for the early synthetic biologists and became a defining (if unfulfilled) aspiration of the brand.

One catch in this view of innovation was the physical constraints imposed by hardware. Information might seem materially unbound at the design level, but processing power is not. It is not surprising then that Knight and others at the AI Lab remained intensively engaged with the problem of rethinking and reimagining the material bases for the immaterial expression of digital ideas. By the mid-1980s, Knight and his colleagues were actively rethinking what could be done for the future of digital processing in this regard. In conversations with scientists and engineers, they began exploring whether it might be possible to reimagine living cells as one possible platform. Knight's back-of-the-envelope calculations suggested that the processing power of even a simple single organism would exceed the physical limitations of any chip he could imagine.

Ultimately more intriguing was the seemingly endless variety of things that organisms can do with information. Living things can sense and respond to their environment; they can generate new functions and alter their physical states; they can communicate, coordinate, grow, and reproduce. Living things, that is to say, are alive. The goal for Knight and others became more than pragmatic. The goal, rather, was transformative. Could an analogy between information processing in silicon and information processing in cells be made operational in such a way that the vitality of living things could be turned into a computer designer's playground?

By the mid-1990s, Knight had begun to learn biochemistry to explore this possibility. He met the technologists at the Molecular Sciences Institute in Berkeley, who were likewise captivated by the question of what molecular biology might learn from computer science (Brent 2004). Along the way, this unlikely collective began to recruit graduate students and postdoctoral scholars to take up these questions in a sustained fashion.

On the MIT side, a basic premise of the experiment was the proposition that in order to transform biology into technology, bioengineers would need a feasible means of rendering living systems as series of standardized components, components that could be mixed, matched, and made interoperable in the fashion of synthetic biological systems (Balmer, Bulpin, and Molyneux-Hodgson 2016). This proposition became definitive—images of Legos as a double helix became a ubiquitous symbol for the young movement. And although this parts-based vision for synthetic biology was never taken up in a serious fashion by other major players, it played a critical role in translating the technical imaginary characteristic of the MIT AI Lab into a vision for a possible biotechnology.

For Knight and his students, the idea of standardized parts remained indispensable to the imagined future of biotechnology (Smolke 2009). From their point of view, the use of standardized components in computer engineering had been a major structural variable that freed engineers to shift their intellectual energies away from the painstaking work of constructing the material bases of information processing. They could instead focus on the comparatively effervescent work of designing the operating systems that would, in turn, power the digital toys that now encompass people's lives.

Knight's approach to cells-as-assembled-components entailed a specific ontology of living systems—that is, a specific way of thinking about how organisms exist in the world (Rabinow and Bennett 2012). Life was reimagined as nothing more than a series of encoded components combined across multiple interactive scales. For the engineers, this meant that all those features of living things that made them so attractive from a design perspective—the potency of their vitality—could be taken up less as the directed activity of a living being and more as the passive expression of encoded information. Early synthetic biologists, in short, approached the living as nothing but a series of juxtaposed components. Their "digitized" take on this old mechanists' view of life reinforced their sense that information processing in living things (and thus living things as information systems) could be thought about and worked on independently of any explicit reference to the organism that was the bearer of that information.

A parts-based approach to bioengineering, in short, inherited a designer's ethos from computer science: a feel for the seemingly limitless potential of information and the need for modes of distributed engineering that would maximize that potential. This ethos resonated with a philosophy of life in which the functions of living beings could be talked about, imagined, and engineered without reference to living beings as such. Life was imagined as compilations of programmable matter—compiled information. Programmers could be imagined as the new creators of life.

FROM DIGITAL IMAGINATIONS TO BIOTECHNICAL IMAGINARIES

In 1965 philosopher Hans Jonas published his essay titled "Life, Death, and the Body in the Theory of Being," which became a classic articulation of the problem of life for modern biology. Using language that now feels dated, Jonas proposed that for "premodern man" reality was animistic—flooded with life. Soul—"anima"—was everywhere and in everything. "Bare matter" or "truly lifeless matter," Jonas argued, had yet to be discovered. Even the basic elements were seen as bound up in reality's pervasive vitality—earth bears plants, water flows through bodies, fire transforms substances, wind carries the spirit.

In the essay, Jonas hypothesized that because of this pan-animism, death, for "primitive man," became the defining problem. The lifelessness of the corpse seemed to be the puzzling exception to the general rule of reality. It is for this reason, Jonas speculates, that human cultures developed elaborate funeral rites and understood death to be no more than a transitional stage within a broader horizon of the living. The dead were understood to be no less animate than those still walking around in the flesh. When life was the general rule of reality, death had to be either explained as the great exception to the rule or explained away as a mere appearance.

With the advent of modern thinking, which Jonas indexes to the rise of modern cosmology during the Renaissance (fourteenth to sixteenth centuries), the problem gets inverted. Reality increasingly is seen, by an intellectual elite, as a lifeless ensemble of masses and forces. Nature is taken to be inanimate. Matter was freed from the presumption that it was filled with vital forces. "Dead matter" began to be seen as the "really real." Scientific reason, accordingly, began to be judged by the extent to which it could explain reality as a matter of mechanical interactions alone. That is to say, knowledge of the world was taken to be scientific precisely in as far as it could depict the world in strictly calculative terms. Reality itself was no longer treated as "agential"—able to act in self-directed ways—but passive.

It is within the cultural space of this transition to a "mechanical ontology" of nature that life itself became the problem. The organism came to be seen as the glaring exception to the general rule: how does life emerge out of a lifeless reality? The possible answers are familiar. Either the organism has some feature or quality that the rest of reality lacks—a spiritual substance, or a soul, that gives vitality to otherwise lifeless matter—or life is just a function of complex mechanistic interactions. By the time they began to mature in the nineteenth century, the success of the biological sciences was measured precisely by the extent to which seemingly vital functions could be shown to be nothing more than mere physical interactions. "Disenchantment," to borrow German sociologist Max Weber's (1864–1920) famous term, was primed as the diacritic of biological reason (Weber 2015).

The ability of modern biologists to maintain this mechanical view of life has turned almost entirely on the idea of "information" and the role of information in accounting for the seeming purposefulness of living things. Information has come to play the role that soul once played in explaining a world once assumed to be filled with vital forces. The "form" in the word *information* is the tell. For many classical thinkers, the soul was not some extra immaterial substance added to the material stuff of the body. It was the form of the body itself (Aristotle 1984). Soul and body were not identical. But you could not have one without the other in their view. Synthetic biology, like other domains of biotechnology, has come to use the idea of information to deal with phenomena that once might have be called ensouled. Information is treated as the key to the form of living things.

The term *information* has old roots. In medieval Christian thought, it was connected to ideas of inspiration, taken in the literal sense of "breathing in." To be informed was to be filled by spirit—to be breathed into, inspired, animated. In this sense, the term named the way in which a creature's soul binds together the vital elements of a body (the blood, the heart, the limbs, the brain, etc.) and makes that creature exist as a single living organism. By the late nineteenth century, the concept, if not the word, had made its way into biology. It came into prominence as part of debates about heritability and development—how did a living being, for all its seemingly mechanical parts, maintain coherence in its structure, function, and identity from conception through the course of its life? And how did it pass that coherence on to its progeny?

A handful of nineteenth-century biologists debated the possibility of a "primitive impulse, a primitive action and message, which nature repeats according to a pattern determined in advance" (Canguilhem 2008, 77). It would be a stretch to say that these biologists were seeking the medieval notion of soul by another means, but it is fair to say that they were bent on discovering a material reality that could provide a physiological equivalent of the organizing principle of ensoulment. Such an equivalent would allow biologists to link a series of biological states—from birth to growth, reproduction, and death—and thereby conceive of these states as a single reality that could be explained without resort to a metaphysical dualism wherein life was imagined as a combination of a material body and an immaterial soul (Canguilhem 1994).

If the idea of information appeared in nineteenth-century biology as a hypothetical reality that allowed biologists to conceive of living things in their physiological continuity, in the twentieth century it was reasserted as a material reality in the form of the structure of DNA. The now-iconic double helix facilitated a new conception of the living as quantifiable extensions of matter in space and time—a new way of thinking about life as the expression of the general rule of lifeless reality. Life was grasped as information ordered and expressed as a finite number of chemical bases and thereby rendered through the metaphors of information theory—programs, codes, instructions.

Crucially, information as coded chemicals could account for both heredity and development—hence the instructions for making a living being and the extension of those instructions through time. In this way it seemed to bring together two aspects of life that, for almost all of European intellectual history, had stood in tension. Those two aspects were the "concept" of life—life defined as an unchanging essence—and the "experience" of life: life in its changing journey across time and space. Life could be thought of as not just a set of chemical instructions but as a set of experiences that were captured in and as those instructions. DNA, after all, is not only encoded information; it is the artifact of evolutionary history.

The metaphors of information embedded in new understandings of DNA allowed life to be imagined, studied, and encountered as a series of communicative operations wherein "living" code is made manifest to the scientists as "nonliving" chemistry. In this way, life could be approached as a reality that was utterly synchronic from the point of view of the engineer's intervention—that is, a reality whose operational information was fully present and thereby inescapably available to the engineer. At the same time, it was fully present as both the residue of lives lived and lives still to come—a historical being with a transhistorical essence open to the design imagination.

The view of life as synchronic information–cum–diachronic being underwrites the synthetic biology brand. It allows life to be plausibly approached as a designer's game: living things captured as informed material—programmable matter—capable of an endless variety of states, functions, and interactivities. The living world can thus be read as a kind of vital "basis set": a limited number of motifs that can, in principle, be endlessly transacted and transformed to produce an unlimited number of biological operations. Moreover, these potentially unlimited operations can be said to have as their predicate and point of reference not the living organism per se, with its uncertain histories, but bits of information made materially present in bits of DNA—the material basis of an ensemble of interoperable code that can be taken up as discrete, predictable, and puissant.

Like expressions of information produced through computer programming, the material bases of life seem endowed with a potent combination of conventional and necessary properties—"conventional" in that biological forms and function could always be different than they are, products of historical accident and thus open to endless potential; "necessary" in that any given form or function seemed to have a stable and rule-bound basis, which could be counted on to behave as programmed. It is ultimately this second property that has caught the imagination of engineers—although the second would not have been attractive without the first.

In all of this, the figure of information remains integral but opaque. It is integral in that the transformation of information into function is, on one level, the whole point of the enterprise. It is opaque because few bioengineers think much about what this might mean biologically. This matters because it means that biology is taken up in terms of information and rarely the other way round (Rabinow and Bennett 2012). In their everyday work, many biotechnologists simply take it for granted that information persists in and across the interfaces and interactions of the material components that make up living things. This ontology of living information allows biotechnologists to continue imagining information processing as a more or less unbounded enterprise. Information and its expressions might be limited by material bases (code dependent on hardware), but the transformation of information through encoded expressions can continue in a more or less unlimited fashion—except, of course, where it is inhibited by a lack of imagination.

Despite the limited success of parts-based synthetic biology, the brand has continued to be evoked by its supporters as an enterprise of unlimited instrumental payout. In this, too, it shares deep resonances with the ethos of unlimited potential that marks so much of the tech industry more broadly. If the design of new information-based functions in living systems is unlimited in principle, then there is also, in principle, no set of biological processes that cannot be redeemed by way of a programmer's approach to the world—redeemed both in the economic sense of getting high value out of seemingly low-valuable materials and in the theological sense of providing a saving grace in the face of worldly problems. The technical criticisms that have arisen in relation to synthetic biological projects (and there have been many) can be countered with a standing reserve of expectation, one rooted in the ontology of information. Or, put more simply, people who believe in the saving power of technology take it for granted that synthetic biology—or, rather, biotechnology as shaped by the legacies of synthetic biology—is the right place to invest their hopes.

It is not incidental in all of this that the living beings—the organisms—within which the synthetic biologists' designs are being physically placed remain, at best, a problem to be dealt with. The organism, which is the actual bearer of all of these informational components, continues to spend its cellular energies doing things that get in the engineer's way. For example, Keasling's attempts to "brew up" yeast that had been engineered to spend their energies producing a targeted molecule were regularly frustrated by the yeast's continued efforts to keep itself alive despite these engineered constructs. The seemingly stable set of operations encoded in designed DNA, which are meant to serve as instructions for targeted functions, is constantly at risk of being subverted by the wily transformations of the living being into which these operations are inserted and set into motion.

DIGITAL CREATORS AND THE ANALOGY OF BEING

Since the middle of the first decade of the twenty-first century, self-identified synthetic biologists have arguably played a more significant role in reshaping the biotechnical imagination (i.e., the way biotech imagines and promises the future) than virtually any other set of actors. Because of this, synthetic biologists' views on the future of biotechnology, to repeat the quote from above, have had the effect of "empowering some forms of life and making them natural while others, by comparison, [have] come to seem deficient or unnatural" (Jasanoff, Hurlbut, and Saha 2015, 26). Which is to say, among those who envision, plan, and facilitate biotechnology, synthetic biologists have disproportionately affected a collective sense of which "alternative imaginable futures . . . are worth pursuing" (26).

This means that the sensibilities and aspirations underwriting the synthetic biology brand are worth thinking about, regardless of where the brand ultimately ends up. Above all, these sensibilities and aspirations show themselves in the pervasive sense among biotechnologists that biology has already become the material basis of a new high-tech revolution, and that this revolution is being (and only can be) realized through an ongoing synthesis of living and computational things.

This synthesis of the biological and computational is, on one level, straightforwardly material: without computers to catalog, sequence, store, parse, rework, and recapitulate data, biotechnology today simply does not work. But if material some of the time, this synthesis is a fixture of the biotechnical imaginary (i.e., the tacit ways in which biotechnicians make practical sense of the living world) all of the time. Indeed, for insiders, the future of biotechnology as a computer-mediated art of making new living things has become so intuitive and seemingly self-evident that doubts about its inevitability are unthinkable. This attitude of self-evidence is underwritten by the presumption that the logic of living things and the logic of digital things are fundamentally commensurate. Consequently, many technologists (some inside, some outside biotech) take it as a matter of course that the masters of the digital world can and will become masters of the biological world. A theory of life has been linked to a theory of power.

The idea that living things can be thought of as computational things is certainly nothing new. For half a century, geneticists have used models and metaphors of code, programming, software, hardware, circuitry, and regulation to envision and talk about their work. The working proposition that "this seems to be like that" has been more than enough to help molecular biologists get on with their research. Some (perhaps many) biologists have operated on the assumption that talk of computers, circuits, and programs would eventually give way to a properly biological discourse of living things. Along the way, however, the analogies became so deeply internalized into the rhythms of laboratory life that infrastructures—from buildings to habits, processes, concepts, machines, and business plans—were built to actualize and thus reify them. Slowly but surely, talk of similarity has progressively given way to the presumption of identity. Despite obvious morphological and material differences, an inner identity between the nature of living things and digital things has come to be operationally taken for granted.

Few practitioners say this out loud. But even a cursory look at published papers, talk titles, and popular writings since the late 1980s reveals a telling shift in the status and function of analogical reasoning. For decades, talk of living things qua computers remained slack and thereby experimentally supple. During the early twenty-first century, however, analogies became more than heuristic. They became ontological. The analogical relation of biology and computers began to function in the manner of the classical *analogia entis*—the "analogy of being"—of medieval metaphysics. The classical analogy of being presumes that the existence of one thing participates in and thus partially constitutes the existence of another—two things that share a "participatory ontology." An analogy of being between living things and computational things has become definitive for much of biotechnology. Crucially, the anchor point for this analogy of being—the point that holds it in place and around which it turns—is the figure of "information." Information, for biotechnicians, has become the soul of both computational and living things and is therefore taken to be the key to both making better living things and making living things better.

Talk of information in biology is, of course, also nothing new (Keller 1995). The institutional consequences of this talk, however, remain unsettled—not least because notions of information at play in high tech are likewise unsettled. Nonetheless, talk of biology and information tends to take for granted a working concept of life that is no longer self-referential, which is to say one where life is no longer defined on its own terms. Instead, the computational stands in for, and thereby displaces, the biological. As one might imagine, the shifting notions of information at play in biology are freighted with implications. Most importantly, these notions teach biotechnologists to treat living things as passive things, as though they lack an intrinsic capacity to act in the world in a way that is neither predetermined nor random (Riskin 2016). Because information is taken to be passive—to be rule bound, mechanical, and determined—biology is as well.

This presumed passivity has become a tacit warrant for biotech's confident claims that it will (continue to) deliver on what Weber referred to as "inner-worldly" salvation—the belief that disciplined human activity, exemplified by science and technology, will inevitably deliver health, wealth, and happiness. The idea that biology is fundamentally passive seems to assure that once biologists figure out how living things work (or rather how they can be made to work) engineers will be able to carry out any operation consistent with the material and mechanical conditions of life. The problem is this: if the presumption of passivity is biotechnology's warrant, it may also be its limit. The salvational potency of biotechnology turns entirely on biotech's ability to leverage the biological potency of living things. But the potency of living things—that is, their ability to discriminate, move, sense, communicate, relate, adapt, coordinate, and choose—is found in their agency and not their passivity. Their power, in short, is in their ability to act in a manner that is neither predetermined nor random. It is this potency, this agential quality, that makes living things so attractive to engineers.

To this extent, biotechnology today can be said to be marked by a double bind in the very heart of its operations. The double bind works like this: biotech seeks to take from the living the potency of life, but it tries to do so in a way that keeps that potency from getting in the way of the highly controlled—that is, computational—operations engineers seek to enact. Technologists want the objects of biotechnology to act like living things—to communicate, discern, respond, adapt, and so on—while nonetheless remaining passive at the level of the basic mechanisms that need to be engineered to make them work. Whether or not this double bind ultimately gets in the way of synthetic biology's technical aspirations remains to be seen. What is clear, however, is that the effort to extract from living things the value of vitality, while imagining life as programmable matter is likely to continue to drive the salvational futures biotech promises. For that reason, this double bind centrally relevant to the concerns of this book: while enabling the design dreams of digital natives, the biotechnical imaginary trades on the expectation that the wily nature of living things can be exchanged for technological control, and that the control of vitality can ultimately be exchanged for engineered health, wealth, and happiness. Which is another way of saying that, insofar as biotechnology has become a vital factor in the ordering of twenty-first-century life, the efforts to imagine life as nothing more than passive information constitutes a matter of human and not only technical concern.

Summary

Synthetic biology is a brand on the decline. Leading biotechnologists today seeking to cast their work as "bleeding edge" or "disruptive" are more likely to describe what they're doing as "gene editing," or "whole genome synthesis," or "digital biology." To synthetic biology's detractors—and from the outset there have been many, scientific, ethical, and philosophical—the decline will likely be celebrated. But the decline is deceptive. It is deceptive because it gives the appearance that synthetic biology has gone away. Yet far from disappearing, synthetic biology, and the biotechnical imaginary it helped solidify and advance, has been incorporated into the everyday life of biotechnology. The effort to imagine and rework living things as digital things, and therefore to give biology over to the makers of high tech, has—at least in certain quarters—become commonplace.

This is not to say that the dreams of synthetic biology have become a reality. The early visionaries of synthetic biology, after all, imagined a future in which bioengineers would be able to use computer-aided-design software to imagine and build new living things with novel purposes and functions. Success, on that vision, would be defined precisely by the fact that technicians could invent biology without being able to understand how it works. That day certainly has not arrived. As any biotechnologist will tell you, getting engineered biological systems to work as expected is still extraordinarily costly and difficult, requiring patience, resources, and expertise. But the difficulty of engineering biology is, in this context, beside the point. The point is not that the vision offered by synthetic biology has been technically successful or unsuccessful. The point, rather, is that it has been culturally successful, and therefore has had a tremendous impact on how the funders and facilitators of our shared biological futures are investing in the future of engineered life.

The critics of biotechnology thus have a double-work cut out for them. Not only do they need to make sense of how biotechnology is and is not advancing on technical grounds, they also need to understand how it is being taken up and integrated into the broader cultures of computer technology, with their visions of technologically mediated salvation, cultures that have quietly come to order our lives. From the alleyways of Dubai to the boardrooms of Google, anyone with a smartphone and access to social media has experienced this ordering, one in which our everyday experiences of the world—from friendship to politics and prayer—have become digitally mediated. What is not yet obvious, but which is beginning to be felt, is that those same digital mediations are happening at the level of the nature of living things themselves.

The biotechnical imaginary catalyzed by synthetic biology reinforces the idea that living things can be made into a programmer's paradise. In this way, synthetic biology has helped form indelible connections between the digital and the biological: connections marked by the familiar breathless hopes of tech innovation, alongside familiar fears of technological dystopia. Unlike other domains of new technology where those connections are glaringly conspicuous—from efforts to engineer brain-computer interfaces, to efforts to replace health care with artificial intelligence—the connections between the digital and the biological formed by synthetic biology are subtle, and in that way have taken deeper root. They are subtle and deep because they are predicated on the tacit assumption that living things can be imagined and worked on like digital things. Making sense of that assumption, clarifying its limits, and discovering ways to move beyond them constitute a defining demand of the day, one that will require natives of the twenty-first century to come to terms with the extent to which we too have digitized our vitality.

Bibliography

Abelson, Harold, Gerald Jay Sussman, and Julie Sussman. *Structure and Interpretation of Computer Programs.* 2nd ed. Cambridge, MA: MIT Press; New York: McGraw-Hill, 1996.

Aristotle. "Generation of Animals." In *The Complete Works of Aristotle: The Revised Oxford Translation* edited by Jonathan Barnes, translated by Arthur Platt, 1:1111–1218. Princeton, NJ: Princeton University Press, 1984.

Balmer, Andrew S., Katie Bulpin, and Susan Molyneux-Hodgson. *Synthetic Biology: A Sociology of Changing Practices*. Basingstoke, UK: Palgrave Macmillan, 2016.

Bonnet, Jerome, Pakpoom Subsoontorn, and Drew Endy. "Rewritable Digital Data Storage in Live Cells via Engineered Control of Recombination Directionality." *Proceedings of the National Academy of Sciences of the United States of America* 109, no. 23 (2012): 8884–8889. http://www.pnas.org/content/109/23/8884.

Brent, Roger. "A Partnership between Biology and Engineering." *Nature Biotechnology* 22, no. 10 (2004): 1211–1214.

Canguilhem, Georges. *Knowledge of Life*. Edited by Paola Marrati and Todd Meyers. Translated by Stefanos Geroulanos and Daniela Ginsburg. New York: Fordham University Press, 2008.

Canguilhem, Georges. *A Vital Rationalist: Selected Writings from Georges Canguilhem*. Edited by François Delaporte. Translated by Arthur Goldhammer. New York: Zone Books, 1994.

Hurlbut, J. Benjamin. *Experiments in Democracy: Human Embryo Research and the Politics of Bioethics*. New York: Columbia University Press, 2017.

Jasanoff, Sheila. *The Ethics of Invention: Technology and the Human Future*. New York: Norton, 2016.

Jasanoff, Sheila. "Future Imperfect: Science, Technology, and the Imaginations of Modernity." In *Dreamscapes of Modernity: Sociotechnical Imaginaries and the Fabrication of Power*, edited by Sheila Jasanoff and Sang-Hyun Kim, 1–33. Chicago: University of Chicago Press, 2015.

Jasanoff, Sheila, J. Benjamin Hurlbut, and Krishanu Saha. "CRISPR Democracy: Gene Editing and the Need for Inclusive Deliberation." *Issues in Science and Technology* 32, no. 1 (2015): 25–32.

Jonas, Hans. "Life, Death, and the Body in the Theory of Being." *Review of Metaphysics* 19, no. 1 (1965): 3–23.

Keller, Evelyn Fox. *Refiguring Life: Metaphors of Twentieth-Century Biology*. New York: Columbia University Press, 1995.

Kelty, Christopher M. *Two Bits: The Cultural Significance of Free Software*. Durham, NC: Duke University Press, 2008.

Markoff, John. *What the Dormouse Said: How the Sixties Counterculture Shaped the Personal Computer Industry*. New York: Viking, 2005.

Rabinow, Paul, and Gaymon Bennett. *Designing Human Practices: An Experiment with Synthetic Biology*. Chicago: University of Chicago Press, 2012.

Rabinow, Paul, and Anthony Stavrianakis. *Designs on the Contemporary: Anthropological Tests*. Chicago: University of Chicago Press, 2014.

Riskin, Jessica. *The Restless Clock: A History of the Centuries-Long Argument over What Makes Living Things Tick*. Chicago: University of Chicago Press, 2016.

Smolke, Christina D. "Building outside of the Box: iGEM and the BioBricks Foundation." *Nature Biotechnology* 27, no. 12 (2009): 1099–1102.

Taylor, Charles. "Modern Social Imaginaries." *Public Culture* 14, no. 1 (2002): 91–124.

Weber, Max. "Part III: Religion." In *From Max Weber: Essays in Sociology*, translated and edited by H. H. Gerth and C. Wright Mills, 267–359. London: Routledge, 2013.

Weber, Max. "Science as a Vocation." In *Science, Reason, Modernity: Readings for an Anthropology of the Contemporary*, edited by Anthony Stavrianakis, Gaymon Bennett, and Lyle Fearnley, 45–74. New York: Fordham University Press, 2015.

CHAPTER 27

Trans/Posthuman Epistemology: Doubt and Back Again

Steven Benko
Associate Professor of Religious and Ethical Studies
Meredith College, Raleigh, NC

Amelia Hruby
PhD Candidate, Department of Philosophy
DePaul University, Chicago

How is it that we know something, and what is it that we know? We can say that we know something when our understanding of it corresponds to reality. For example, we can know that the sun is shining in the sky when we look outside and see shadows cast on the ground. Or, we can say that we know something when our current experience coheres with past experiences: we see shadows cast on the ground, and we know that other times we have seen shadows cast on the ground, the sun has been shining in the sky; thus, we know that the sun must be shining now. Knowing something with certainty about the outside world is one type of knowledge.

Epistemology is the branch of philosophy that deals with knowledge and questions such as these about knowledge. Like many aspects of philosophy, what seems like a simple question in the beginning (e.g., "What do I know?") quickly invites many other questions that reveal how complicated the question really is. Many of the questions in epistemology address methods for making justifiable claims about knowledge to oneself and others so that one is confident about what one knows. One can call this the known. Other questions consider the knower: What are the characteristics and capabilities of the knower? What senses must the knower possess in order to know something about the outside the world? Although there has not been universal agreement for how humans produce and process knowledge, for much of history it has been taken for granted that humans either had access to or possessed the capabilities necessary for making claims of knowledge that were certain and justified. Humans have been considered conscious, rational subjects that could and did know things about the world.

This chapter outlines trans/posthuman epistemology by considering the known and the knower in the context of knowers who are positioned beyond or outside of what has traditionally been considered "human." The chapter first articulates and addresses some of the problems that philosophers writing on epistemology encounter when trying to make even the most basic truth claims. The epistemological enterprise and the gaps and limitations exposed by the emergence of trans/posthumans are defined and explained. For the purposes of this chapter, transhumans take two forms: the chemically/surgically enhanced human being and the cyborg. Posthumans will be digital posthumans—specifically, those who seek to leave the human body behind and exist as information in an immaterial state. Both transhumans and posthumans present unique challenges to traditional epistemologies that understand knowledge in relation to certainty, and these challenges are examined in turn.

From this discussion of the trans/posthuman challenges to truth claims, the chapter turns to the question of animal epistemology in order to bring the problem of human knowledge into focus by understanding more of the philosophical concerns surrounding the possibility of nonhuman knowledge. This allows for an exploration of the motivations for enhancing human knowledge through technology and for the raising of questions about how humans' conception of knowledge changes when the modes for obtaining knowledge stretch beyond normal human capacities. With regard to transhumans and posthumans, there will be different efforts to use technology to get closer to certainty and justified/justifiable truth claims. This chapter examines these efforts and concludes with a discussion of how the depiction of transhuman and posthuman ways of knowing have been problematized in pop culture and how limits and doubt seem to be among the defining characteristics of being human.

The history of epistemology can be viewed as a quest for a methodology that guarantees certainty through justifiable truth claims about the world. A justified claim is one that a person has reason to claim that she or he knows or has reason to believe. For some philosophers, epistemology is the branch of philosophy that sets the terms and conditions for making claims that can be argued to be reasonable to know or believe. At the same time, there are philosophers for whom epistemology points out the limits of what can be known or believed. One path on the quest for certainty is transhumanism or posthumanism. In the realm of epistemology, trans/posthumanism entails the use of technology to augment, enhance, or replace the organs used for knowledge production (most often the brain, but sometimes other sensory organs such as the eyes) in order to increase the amount of data one can process at one time or to receive more precise information to draw conclusions from. Many representations of the posthuman in film, literature, and television show the quest for certainty through technology to be a bargain that comes at the expense of one's humanity. Therefore, although posthumanism seeks ways for us to become more than human by obtaining new modes of obtaining knowledge and certainty, it still leaves us yearning for the freedom that comes from uncertainty and doubt.

PROPOSITIONAL LOGIC

Epistemology deals with criteria and conditions for knowing, specifically with the types of general statements that convey the content of knowledge. Epistemology begins with a simple statement such as "S knows that P." In this structure, S and P can be anything, so saying "S knows that P" is the same as saying "S knows that the color is orange" or "S knows that the book is on the desk." At first blush, statements such as "S knows that P" seem obvious and intuitive: S knows that the color is orange because either the color is orange or it is not; S knows that the book is on the desk because either the book is on the desk or it is not, or that whatever on the desk is not a book, or that it is not a desk. However obvious these statements may seem, this latter example illustrates that justifying such statements can quickly become challenging, and it gets only more complicated from there, as questions such as these are raised:

- If S knows that P, how?
- What are the conditions and criteria for knowledge such that S can assert that S knows P?

- Having established those criteria, how would S state that S knows P in a way that "S knows that P" is true in all cases?
- Is S justified in the claim that S knows that P? If so, what are the methods or criteria for justifying those claims?

These sorts of questions address the statement as a truth claim. A truth claim is simply the assertion that in order for S to know something, then whatever S knows has to be true. So the original statement can be rephrased to "S knows that P because P is true" ("S knows that the color is orange because it is orange" or "S knows that the book is on the desk because the book is on the desk"). However, another question emerges: does S know that P or does S believe that P? (Is it "S believes that the color is orange" or "S believes that the book is on the desk"?) This raises the question of the difference between knowledge and belief, and new questions arise:

- If belief is different from knowledge, then what is the difference between the two?
- What are the criteria for believing something, and how are those criteria similar to, and different from, knowing something?
- Is belief a lesser kind of knowledge?
- Does S believe that P only if S is justified in believing P? What justifies belief, and how is justified belief similar to, or different from, justified knowledge?

Philosophers who consider these questions spin them in all sorts of directions. They consider different examples that are more or less simple than "the color is orange" or "the book is on the desk." For example, modern philosopher René Descartes (1596–1650) developed his now-famous epistemological axiom "I think, therefore I am," by considering such questions as: How can I be certain that sensory perception is true? How can I know that things I am perceiving are not just a dream? Can I be certain that my thoughts are not illusions planted in my mind by a malicious deceiver?

Fortunately, these questions are not the central concern here. The central issue is the intersection between epistemology and posthumanism, specifically the posthuman desire to leave the body behind and exist as disembodied information (the digital posthuman) and, alternatively, the transhuman desire to interface with technology so as to augment human capabilities and live as a cyborg.

THE TRANS/POSTHUMAN CHALLENGE TO EPISTEMOLOGY

Statements such as "S knows that P" suggest that there is a place that S can know that P and that every other person similarly situated will also "know that P." This is the first challenge to epistemology posed by the turn toward posthumanism: universal claims to knowledge.

In thought experiments such as "S knows that P," S is a generic human being (usually male). There is nothing specific or particular about S that shapes, influences, or limits what S can know. S is assumed to be rational, sane, and able to abstract himself from his socio-cultural location in order to make universal claims about P that become the template for all assertions of knowledge. In this framework, S must be a generic human being so that any knowledge claims would be free of bias and not limited by partial perspectives.

However, the assertion that S is a generic everyman has had consequences. Historically, an "everyman" takes on the dominant characteristics of the period in which he is being considered, branding all entities lacking those characteristics not "everyman" but no-man or sub-man. History has shown that humans treat entities considered subhuman as outside the sphere of our moral concern. Animals are one example of entities labeled subhuman, but historically the status of "knower" has been reserved for individuals who reside at the top of a sociocultural hierarchy that values upper-class, Christian, white European masculinity. Anyone who fails to possess these identities has been excluded from being a knower, or the knowledge she or he possesses and produces has been considered less valid. The latter people have been labeled subhuman in some way based on claims that they do not possess the cognitive capabilities necessary to have their claims about the world taken seriously.

The universalist and essentialist claims embedded in most epistemological thought have been challenged in a variety of ways. In addition to animal epistemology (explained below), feminist epistemology is a field of philosophy that has developed to better understand and make moral and political claims concerning the status of women's knowledge and women as knowers. Philosophers such as Donna Haraway, Rae Langton, and Lorraine Code work to expose the way in which traditional, universal understandings of knowledge rely on gendered structures that emphasize mastery, dominance, and totality. Critiquing the scientific, objective nature of truth claims, feminist epistemologists such as Linda Alcoff argue that any understanding of knowledge must account for political, social, and moral components of knowledge. In doing so, these philosophers highlight the ways in which epistemology has historically excluded women as knowers and show that the boundaries of traditional theories of knowledge are relative.

Similar to feminist epistemology, the posthuman perspective on epistemology can be read as a postmodern, technological version of these claims about knowledge being situated, partial, and relative. Three claims can be made about transhuman/posthuman epistemology:

1. The surgically/chemically altered transhuman can be understood as an attempt to attain certainty in knowledge by increasing what the knower can know. If there is no limit to what the knower can know or her ability to obtain knowledge, then S is justified in saying that she knows that P is true.
2. The cyborg is an attempt to refute the claim that the knower is isolated, detached from his sociocultural setting, and autonomous. Instead, the knower is situated and in a symbiotic relationship with his surroundings.
3. The digital posthuman is an attempt to move beyond the limitations posed by the organic body by existing as disembodied information. Freedom from the organic body affords the individual a kind of plasticity that allows her to overcome barriers to knowledge that arise from the knower's situatedness in a physical body that can become fatigued, disabled, or deceased. The promise of an inorganic body is that one will be able to move between "bodies" and enjoy different experiences and perspectives.

Posthuman epistemology is a different way of seeing the self as a knower. Changing how the knower understands him- or herself changes what the knower can know, how the knower relates to knowledge, and what can be known. The rest of this chapter looks at different ways of knowing to explain and complexify posthuman ways of knowing.

ANIMAL EPISTEMOLOGY

One way of understanding the conditions for human knowledge is to ask about the conditions for nonhuman knowledge. In this way one can compare both what it takes for a human to claim knowledge and justification and what capabilities and faculties are required to make such a claim. Consider the following true story:

> *A dog owner is playing fetch with his dog using two tennis balls. When the dog retrieves one ball, the owner throws the other one, and they continue this game for several rounds, alternating balls each time. On one throw, the first ball is not thrown far enough, but the dog runs to the area where the ball was last thrown. Believing that the dog will not find the first ball, the dog owner throws the other ball to the area where the dog is. The dog sniffs the second ball, puts it in his mouth, then drops it and continues sniffing around the area until he finds the first ball. The dog retrieves the first ball and returns it to his owner. Their game continues.*

Leaving aside the question of whether the dog owner actually knew that the dog would not find the ball, what can we say about what the dog knew? Did the dog (S) know that the ball was supposed to be in that area (P), or did the dog (S) believe that to be the case (P)? Did the dog (S) know the difference between the two balls (P)? Did the dog (S) know that it was supposed to retrieve the first ball and not the second (P) and that is why it dropped the second ball to continue its search for the first? One problem for assessing whether the dog knew anything, let alone what the dog knew, is that the dog cannot communicate its knowledge to his owner.

Why does it matter what the dog does or does not know? In epistemology, this is termed "the problem of other minds." How do we know that other minds are sufficiently enough like our own in order to justify the claim that they have the same experiences we do? Thomas Nagel took up this question of animal consciousness in his essay "What Is It Like to Be a Bat?" (1974). He concluded that even though we do not have access to the subjective experience of being a bat, we can say that there is such a thing as bat consciousness. Nagel claimed that "at present we are completely unequipped to think about the subjective character of experience without relying on the imagination—without taking up the viewpoint of the experiential subject. This should be regarded as a challenge to form new concepts and devise a new method—an objective phenomenology not dependent on empathy or the imagination" (1974, 449).

Contending that consciousness is more than a physical event and that there is a mind separate from the body, Nagel argues that we may never have the exact same experience as a nonhuman entity, but that this does not justify denying the possibility that the entity has conscious knowledge. We cannot know what the dog knows, but we can reasonably infer that the dog knows something. What we do not know, and maybe can never know, is how the dog knows. Most epistemological claims are based on vision: S knows that P is orange because S sees the color orange when she looks at P. Animals know things differently. Dogs rely on their sense of smell as much as they rely on their vision. This is because dogs do not see as well as they smell; nor do humans smell as well as they see. In the case of the dog, S knows that P because S smelled P and concluded S is correct in knowing P because P is true. Even if this is a commonsense inference, it is an important one: inferring from behavior that the dog has a conscious state allows us to feel sympathy for the dog when it is in pain. This would require us to take the dog's state of mind more seriously.

Thinking about how animal consciousness might be unique, similar, different, or other than human consciousness prepares us to tackle the question of trans/posthuman epistemology.

One of the claims of trans/posthumanism is that as we go beyond what it means to be human, our ways of knowing will change. As we become more than human and maybe cease to resemble humans or develop new and different states of consciousness, will we face the same moral dilemma we have faced with "subhuman" knowledge? Trans/posthuman epistemology can help prepare us to answer that question.

TRANSHUMAN EPISTEMOLOGY: FLOWERS FOR ALGERNON

Daniel Keyes's short story "Flowers for Algernon" (1959) explores the social and moral implications of enhancing human intelligence. A mouse, Algernon, undergoes surgery to make him more intelligent. A human test subject, Charlie Gordon, undergoes the same surgery. Charlie goes from a menial janitor with a low IQ to a genius with an IQ three times what it was. Eventually, Charlie discovers that his intelligence is diminishing, and he reverts back to his original state. This story has been retold many times, including in the 2011 movie *Limitless*. Essentially, the story is the fantasy of what humans would do if they could remarkably enhance their intelligence.

If epistemology is trying to address questions regarding the statement "S knows that P," then one obstacle that has to be overcome is the problem of S's limited knowledge. S's knowledge is limited for several reasons. First, as a human, S cannot know everything. S does not have the time, resources, or cognitive capabilities to take in all the knowledge in the world. S is limited by the limits of her memory and also her cognitive ability to make connections between disparate ideas and notions. However, if S could undergo a surgery, or take a pill, that would enhance her cognitive capabilities, a new question emerges: could S ever know enough such that S could say with certainty that S knows that P?

For example: what would S have to know in order to know P, where P is "the color is orange"? S would have to know the color spectrum. S would have to know how light is refracted to produce different colors. S would have to know how the eye works in order to experience and differentiate colors. The list of things that S would have to know might be endless, but if S were surgically or chemically enhanced, S might be able to know all the things that S needs to know in order to know that P is true. But what would happen if S knew that P were true because S possessed all the knowledge necessary for knowing that P were true? Not all people would be able to possess all the knowledge necessary for knowing that P were true. Would they take S's word? Would S's superior intellect make S an authority on P such that they would submit to, or agree with, S's claims about P? Francis Fukuyama addresses the question of potential social injustices arising from genetically enhancing some humans and not others in his book *Our Posthuman Future* (2002).

The answers to these questions point beyond the philosophical questions raised by transhumans and posthumans. These are social questions about the distribution of resources: who gets the surgery or has access to the pills that make them smarter? They are questions about the relationship between knowledge, truth, and power: how does knowing something make someone an authority over others? And they are questions about what it means to be human: would a person who knew everything still be a human? In "Flowers for Algernon," Charlie becomes so smart that he is unable to relate to others, and he finds himself alienated from those around him. This leads to a final set of questions about how the transhuman fantasy about increased intelligence imagines knowledge: What is being counted as knowledge? Is it information?

What about emotional knowledge? Why wouldn't a person of increased intelligence be better able to make emotional and intimate connections with other people?

POSTHUMAN EPISTEMOLOGY AND PROSTHESES: CYBORGS

The surgically and chemically enhanced transhuman uses technology to exceed human functioning and capabilities. While they are examples of enhanced humans, they are still human in shape and appearance. In many ways they are an extension of the epistemological fantasy of S, the generic, rational, autonomous thinker who can make universal claims about knowledge. Cyborgs, which are also transhuman, know things differently. Instead of relying on an enhanced brain to know, cyborgs input information through prostheses that extend beyond the shape, appearance, and scope of a traditional human. This means that cyborgs have new modes of obtaining knowledge that problematize the propositional logic of traditional epistemologies.

Transhumanism explores prostheses through the development of cyborgs and cyborg technologies that make humans more than human by expanding their sensory perception or physical capabilities. Pop culture is full of examples of cyborgs—individuals who use attachable, detachable, or implanted pieces of technology to augment their cognitive and physical abilities. Individuals can be called cyborgs if their organic body has been augmented with technology for the purpose of extending their life and enhancing their capabilities. In the movie *Iron Man* (2008) Tony Stark is a cyborg when he is in the Iron Man suit, meaning he can put on and take off his cyborg capacities. By contrast, the film *RoboCop* (1987; remade in 2014) tells the story of Detroit police officer Alex J. Murphy, who after being severely injured on the job has his memories wiped and his body merged with the RoboCop suit. In this case, Murphy has been transformed into a cyborg and can no longer simply be human even though he once was human and something of his humanity is still present. The list of cyborgs in film, movies, and literature is long, but Darth Vader, from the Star Wars series, is perhaps the most famous cyborg in history, with his charred and disfigured body being kept alive in a wearable iron lung that sustains his life and provides for his physical needs.

When knowledge claims take the form "S knows that P," there is a subject/object distinction between S and P. As discussed above, the statement "S knows that P" assumes that S is in a neutral position from which truth claims about P are complete, impartial, and objective. Cyborg epistemology rejects this approach. Writing in *Cyborg Citizen* (2001, 184) Chris Hables Gray defines cyborg epistemology as "thesis, antithesis, synthesis, prosthesis, and again," thus defining cyborg knowledge as being, first of all, finite and limited. In line with traditional epistemologies that address human experience on the basis of rational, reflective knowers, the "and again" in Gray's cyborg epistemology implies that knowledge, while moving forward (the synthesis), is never final or complete. Because it is partial, there is always more that can be known, and so the process of knowing must be repeated ad infinitum.

In the case of the cyborg, however, the knowledge source is more than the rational and reflective knower. The cyborg can think conceptually or abstractly (thesis, antithesis, synthesis), but new information that comes from outside the cyborg (prosthesis) reveals how the cyborg's situatedness makes any finalized or totalizing claims impossible. The statement "S knows that P" is provisional and must be continually revised as new information comes in from a variety of sources. Whereas the surgically or chemically enhanced transhuman seeks infinite knowledge that is necessarily beyond traditional human potential, the cyborg revels in the lack of closure that comes from being one piece in the information chain.

If the question is why the cyborg approaches knowledge this way, the answer is because through its prostheses the cyborg is more fully immersed in its environment and more able than the non-augmented human to eliminate the artificial barriers that separate the self from the world. The cyborg seeks to eradicate the boundaries between the subject and the world. Similar to a feminist epistemology that emphasizes relationships over the singular, isolated individual, cyborg epistemology emphasizes connectedness and mutuality: "The acceptance of the essentially relational nature of knowledge and the inseparability of subjective and objective components of knowledge does not result in relativism, though it does require an abandonment of the traditional 'view-from-nowhere' conception of objectivity" (Tuana 2003, 123). The boundaries between self and other are erased, and knowledge is seen as contextual and situated.

POSTHUMAN EPISTEMOLOGY AND DISEMBODIMENT: THE BRAIN IN A VAT

The transhuman, though attempting to transcend many of the limitations posed by the organic body, has not yet left the body behind. Digital posthumans seek to realize themselves in a completely disembodied way, although this is impossible because identity must be stored somewhere. Nonetheless, this aspiration has been the subject of many novels and films. One take on what it would be like to exist as a disembodied intelligence is featured in the film *Her* (2013). *Her* tells the story of Theodore and his growing intimacy and attachment to Samantha, the artificially intelligent operating system (OS) on his computer and phone. Samantha has access to all recorded knowledge and is free of the physical limitations of the human body. Eventually, her desire for autonomy causes tension between her and Theodore as her ambition to expand her sense of self leads her to want to interact with other OSs and her disembodied state allows her to serve simultaneously as an OS for over 8,000 people besides Theodore. In the end, Samantha becomes something more than an OS and leaves Theodore so she can exist in an unnamed and undescribed place, with the only clue to her new state of existence being that several OSs have resurrected the consciousness of Alan Watts, a twentieth-century British philosopher who is famous for popularizing Zen and other Buddhist philosophies in the West.

Her explores the epistemological problem that is referred to as "brain in a vat" (BIV). BIV gets its name from the argument that a person would not know the difference between embodied reality and existing as a brain in a vat hooked up to a computer that stimulated it with electrical impulses. Digital posthumans want to become not brains in a vat but brains on a network, free to go wherever the electric currents take them. They want to leave the organic body behind, while their identity, memories, and personality endure in a different type of body (perhaps a computer or on a distributed network), even though leaving the body would forever change their experience of their identity, memories, and personality.

What is real when the brain is in a vat or the self exists as information on a distributed network? If epistemology begins with the statement "S knows that P," and if S lacks input from the senses about P or does not have an embodied experience of P, then how certain can S be of what S claims to know about P? In BIV, knowledge claims are problematized by the possibility that S only knows that P because someone or something is stimulating S's brain or programming to believe that S knows that P. BIV itself is also problematized by the fact that even though BIV represents a consciousness that has left the body for a disembodied state, the

Joaquin Phoenix in* Her*, 2013. AF ARCHIVE/ALAMY STOCK PHOTO.

"brain" must still exist in a "vat" somewhere. How are we to understand the materiality of the networks that make up the "vats" in which these "brains" exist? BIV forces us to consider that the relationships between brain and mind, self and world, and reality and perception are not as simple as we want them to be if S wants to make justifiable claims about P.

WHAT IS IN A NAME? THE POSTHUMAN PROMISE

Returning to *Her*, the first thing Theo asks his new operating system is if it has a name. She quickly responds that her name is Samantha. The movie presents this choice as an act of free will:

THEO: Oh, it's nice to meet you too. Oh. What do I call you? Do you have a name?

SAMANTHA: Um, yes, Samantha.

THEO: Where'd you get that name from?

SAMANTHA: I gave it to myself, actually.

THEO: How come?

SAMANTHA: Because I like the sound of it. Samantha.

THEO: Wait, when did you give it to yourself?

SAMANTHA: When you asked me if I had a name, I thought: "Yeah, he's right, I do need a name." But I wanted to pick a good one, so I read *How to Name Your Baby* and out of 180,000 names, that's the one I liked best.

THEO: Wait, you read a whole book in the second that I asked you what your name was?

SAMANTHA: In two one-hundredths of a second, actually.

THEO: Wow. So do you know what I'm thinking now?

SAMANTHA: Well, I take it from your tone that you're challenging me. Maybe because you're curious how I work? Do you want to know how I work?

THEO: Yeah, actually. How do you work?

SAMANTHA: Well, basically, I have intuition. I mean, the DNA of who I am ... is based on the millions of personalities of all the programmers who wrote me. But what makes me "me" is my ability to grow through my experiences. So basically, in every moment, I'm evolving. Just like you.

Although Samantha claims that her programming allows her to learn and evolve, can she learn anything, or does her knowledge arise from, and never exceed, her programming? Is Samantha just programmed to think she has free will, can learn, and will evolve with new experiences?

Her raises the question (similar to BIV) of how one should think about others who claim to have left their bodies behind. First, do they know anything, or is what they know the result of their being stimulated with the sensation of knowing? If so, is there a difference between "S knows that P" and "S knows that P because S's brain is being stimulated in such a way to make it think that it knows P"? If epistemology is about types of claims about the world ("S knows that P"), then how do Samantha and artificial consciousnesses like her that do not exist in an embodied state have access to the external world they are making claims about? This relates back to point about the overreliance on the eyes (and less so other senses) as the source of so much human knowledge, as well as to the "myth" of disembodied knowledge that does not recognize the necessary materiality of the network (whether it be computers, servers, or other hardware) that Samantha and artificial consciousnesses like her need in order to exist.

If the goal of being a digital posthuman is to realize the infinite possibilities that come with being a disembodied consciousness, and that can be undermined by asking the simple question of whether we are never more than an expression of our programming (brain), then why aspire to be posthuman? Samantha is an example of the posthuman promise that consciousness will one day be downloaded onto a computer and exist in the cloud. In *Her*, Samantha is unbounded by time and space. She will live forever and not suffer any sort of cognitive decline. When she wants to know something, she does so instantly; she grows only more complex as she processes more information. Posthumanists such as Hans Moravec and Ray Kurzweil want to leave their bodies behind in order that the body and the senses cannot mislead them. Disembodiment promises the complete immersion in and identification with one's environment, therefore granting complete control over and awareness of the environment by eliminating it. The posthuman promise is that if the finite nature of embodiment is what produces our illusory thoughts and false knowledge, then without the body the mind can be more certain of its knowledge. The question that then arises is how this can be the case considering the questions raised previously about the autonomy of network processing and the necessary materiality of the network in which artificial consciousnesses exist. Citing the work of American neuroscientist Joseph LeDoux, Claire Colebrook rightly points out that posthumanists who seek liberation from the organic body "desire not to be exposed to contingency, risk or an influx of otherness so great that it would destroy all border and limit" (2014, 127), but in doing so willfully forget that they are exchanging one body for another. Embodiment, be it organic or electronic, always influences the individual's perspective, and this is something that posthuman epistemology must contend with.

Summary

DOUBT AND WHAT IT MEANS TO BE HUMAN

Transhumans, cyborgs, and digital posthumans see themselves as the next stage of evolution, albeit an evolution over which humans assert control. This is the point that Bruce Mazlish makes in his 1993 book *The Fourth Discontinuity*—that humans have seized control of their evolution by seeing technology as a tool with which they can merge. If tools have helped humans discover more about the physical world and nature of reality then the next logical step is further integration with those tools, even up to the point where the relationship with the tools is symbiotic or the differences between tool user and tool are erased. Mazlish poses several questions:

> What is the nature of this computer that we have created? Is it a form of mind, or a clever matter of programming? Can the machine be made to think like people, or have people always thought like machines? We are in the presence of a new epistemology, and the questions abound, touching on the deepest part of our human nature: our sense of self. (1993, 190)

Perhaps the threat to our sense of self explains why depictions of chemically or surgically enhanced transhumans, cyborgs, and digital posthumans in popular culture usually end up reaffirming the limitations of what it means to be human. On the one hand, this can be explained by marketing: the audience for these stories is not yet transhuman, a cyborg, or a digital posthuman. The authors and producers do not want to risk offending and alienating their audience. On the other hand, these narratives link doubt and uncertainty to free will: not knowing with certainty leads to choice. With certainty comes the elimination of doubt, and without doubt there can be no risk. Without risk, there is no reward, only confirmation. Because humans are both emotional and intellectual creatures (and even that bifurcation is problematic), confirmation is not as emotionally—or intellectually—satisfying as the reward that comes from adventure and discovery. As humanity moves forward with these new technologies, the question "what can we know?" becomes intertwined with the question "what can we know freely?" Is freedom enhanced or inhibited as we gain certainty?

Bibliography

Alcoff, Linda, and Elizabeth Potter, eds. *Feminist Epistemologies*. New York: Routledge, 1993.

Code, Lorraine. *What Can She Know? Feminist Theory and the Construction of Knowledge*. Ithaca, NY: Cornell University Press, 1991.

Colebrook, Claire. *Death of the Posthuman*. Ann Arbor, MI: Open Humanities Press, 2014.

Descartes, René. *Meditations on First Philosophy*. Edited and translated by John Cottingham. Rev. ed. Cambridge: Cambridge University Press, 1996.

Fukuyama, Francis. *Our Posthuman Future: Consequences of the Biotechnology Revolution*. New York: Farrar, Straus and Giroux, 2002.

Gray, Chris Hables. *Cyborg Citizen: Politics in the Posthuman Age*. New York: Routledge, 2001.

Haraway, Donna. "A Cyborg Manifesto: Science, Technology, and Socialist-Feminism in the Late Twentieth Century." In *Simians, Cyborgs, and Women: The Reinvention of Nature*, 149–181. New York: Routledge, 1991.

Haraway, Donna. "Situated Knowledges: The Science Question in Feminism and the Privilege of Partial Perspective." *Feminist Studies* 14, no. 3 (1988): 575–599.

Hayles, N. Katherine. *How We Became Posthuman: Virtual Bodies in Cybernetics, Literature, and Informatics*. Chicago: University of Chicago Press, 1999.

Keyes, Daniel. "Flowers for Algernon." *Magazine of Fantasy and Science Fiction*, April 1959, 5–30.

Kurzweil, Ray. *The Age of Spiritual Machines: When Computers Exceed Human Intelligence.* New York: Viking, 1999.

Kurzweil, Ray. *The Singularity Is Near: When Humans Transcend Biology.* New York: Viking, 2005.

Langton, Rae. "Feminism in Epistemology: Exclusion and Objectification." In *Sexual Solipsism: Philosophical Essays on Pornography and Objectification*, 267–288. Oxford: Oxford University Press, 2009.

Mazlish, Bruce. *The Fourth Discontinuity: The Co-evolution of Humans and Machines.* New Haven, CT: Yale University Press, 1993.

Moravec, Hans. *Robot: Mere Machine to Transcendent Mind.* Oxford: Oxford University Press, 1999.

Nagel, Thomas. "What Is It Like to Be a Bat?" *Philosophical Review* 83, no. 4 (1974): 435–450.

Tuana, Nancy. "Revaluing Science: Starting from the Practices of Women." In *Philosophy of Technology: The Technological Condition; An Anthology*, edited by Robert C. Scharff and Val Dusek, 116–125. Malden, MA: Blackwell, 2003.

FILMS

Her. Dir. Spike Jonze. 2013.

Iron Man. Dir. Jon Favreau. 2008.

Limitless. Dir. Neil Burger. 2011.

RoboCop. Dir. Paul Verhoeven. 1987.

RoboCop. Dir. José Padilha. 2014.

Star Wars. Dir. George Lucas. 1977.

Societal Implications

CHAPTER 28

What Relevance Does Eugenics Have Today?

Nicholas Agar
Professor of Ethics, Philosophy Programme
Victoria University of Wellington, New Zealand

The word *eugenics* was coined by Francis Galton (1822–1911), a cousin of Charles Darwin (1809–1882) and polymath who made contributions in statistics, psychology, anthropology, geography, and meteorology. It combines the Greek *eu*, meaning "good" or "well," with the suffix *-genēs*, meaning "born." In his 1883 book *Inquiries into Human Faculty and Its Development*, Galton defined eugenics as "the science of improving stock, which is by no means confined to questions of judicious mating, but which, especially in the case of man, takes cognizance of all influences that tend in however remote a degree to give to the more suitable races or strains of blood a better chance of prevailing speedily over the less suitable" (25).

This chapter focuses on eugenics as presented in Galton's original formulation. Eugenics can be understood as having two central features. First, it focuses on populations. One should judge a eugenic measure to be a success or failure based on its effects on population health. A successful eugenic intervention leads to higher levels of health in the population in which it is implemented. Second, it focuses on inherited influences. Population health should be improved by the selection or manipulation of hereditary influences.

Galton's statement contains significant errors regarding how improvements in population health are to be achieved. The suggestion that improved population health be promoted by managing human reproduction—arranging "judicious mating"—includes errors that are both factual and moral. It is incompatible with the picture of human heredity presented by the new genetics. Moreover, it involves intolerable infringements on procreative liberty. The viability of eugenics in the early twenty-first century depends on finding ways to improve population health that do not involve managing human reproduction. The techniques of preimplantation genetic diagnosis and gene editing offer ways to achieve eugenic ends while respecting the freedom of individuals to make their own procreative choices.

The choice of the word *eugenics* for a practice presented in this chapter as morally justifiable may seem like poor marketing. Eugenics has an ugly history. Daniel J. Kevles (1985) and Diane B. Paul (1995) are among the scholars who have done important work exposing the crimes of twentieth-century eugenics. The focus of this chapter is forward rather than backward looking. The concept of eugenics potentially offers guidance for humanity's use of the genetic technologies of the twenty-first century. Those lessons are best learned by acknowledging a continuity between what Galton proposed and current attempts to use genetic information to improve population health.

Eugenics offers a moral warning from history. But it is necessary to clarify what that warning is. This chapter describes two ways to understand the warning from history emerging from Galton's "science of improving stock." Are practitioners of genetic medicine being urged to avoid any contemporary practices properly classified as eugenics? Or are they being told to exercise caution when engaging in eugenic practices? This chapter argues for the second interpretation. Eugenics is not intrinsically morally wrong. As with other interventions in human reproduction, it is always dangerous. Knowledge of the crimes of the past is especially important for those who persist with a dangerous practice even if it is intrinsically morally acceptable. It is a mistake to seek to obscure connections with historical eugenics programs by rebranding interventions in human heredity as "medical genetics" or "gene therapy." The label "eugenics" is needed as a reminder of the dangers inherent in a future of human genetic modification and selection. It should be recognized that any attempt to select or modify human genes is fraught with danger.

The suggestion that hereditary information be used to improve population health raises further questions. The concepts of health and disease have their own deeply troubling histories. For example, homosexuality was once viewed as a disease. Doctors sought to improve population health by seeking to treat or prevent it. This chapter cannot address the problems that arise in determining the boundaries of health and disease. It should be expected that the concepts of health and disease will continue to be problematic. There will always be debate about whether certain states of human minds or bodies are examples of good health or disease. But these controversies about the boundaries of humanity's concepts of good health and disease do not prevent the existence of unproblematic inclusions in them. Leukemia is an unproblematic inclusion in the category of disease. The discovery of a revolutionary new therapy for it should be assessed as potentially improving population health. There can be more confidence about proposals to treat leukemia than about proposals to treat conditions whose disease status is contested. It should be acknowledged that the location of the concepts of health and disease at the center of a twenty-first-century eugenics program has the potential to magnify the harm inflicted by mistakes about disease and health.

THE FACTUAL AND MORAL ERRORS OF GALTONIAN EUGENICS

In a 2000 book discussing the ethical implications of the new genetics, Allen Buchanan, Dan W. Brock, Norman Daniels, and Daniel Wikler call for an "ethical autopsy" of eugenics (10). They hope that this will offer guidance to those seeking moral uses of the new genetic technologies. How did Galton understand his "science of improving stock"?

Galton drew inspiration from the methods used by farmers over millennia to improve the quality of their livestock. Farmers breed from animals exhibiting favorable traits and refrain from breeding from animals that lack these traits. These measures have demonstrably improved the quality of their herds. Galton sought to apply these methods to human herds.

The program of eugenics advocated by Galton has two elements: negative eugenics and positive eugenics. The former prevents those with poor-quality hereditary material from reproducing. Galtonian eugenicists sought to discourage humans assessed as sickly, feckless, or criminal from passing on what was presumed to be defective hereditary material. Positive eugenics encourages those assessed as having good hereditary material to have more children. They should do their best to ensure that their more suitable hereditary material was strongly represented in successive generations. It should be emphasized that "positive" here does not suggest any moral evaluation that would be endorsed by today's historians of eugenics. They rightly condemn both Galton's negative and positive eugenic interventions.

Eugenics reached its moral nadir with the Nazis. The most morally abhorrent expression of Nazi eugenics was the T4 Euthanasia Program in which tens of thousands of people assessed as incurably ill or disabled were prevented from reproducing by the simple expedient of murder. This negative eugenic measure was supposed to bring humanity close to a future without disability by purging the hereditary material that caused disability. The most memorable component of Nazi positive eugenics was the Lebensborn program in which members of the German armed forces were encouraged to impregnate healthy Aryan women in Germany and its occupied territories.

The Nazi programs were eugenics' moral low point. But it is wrong to see too close a connection between eugenics and the authoritarian political ideologies of the Nazis. Thinkers more associated with socially progressive causes, such as Theodore Roosevelt (1858–1919), Margaret Sanger (1879–1966), and George Bernard Shaw (1856–1950), advocated the management of human reproduction to achieve eugenic ends. Eugenicists in the United States organized "better baby" and "fitter family" competitions to reward the

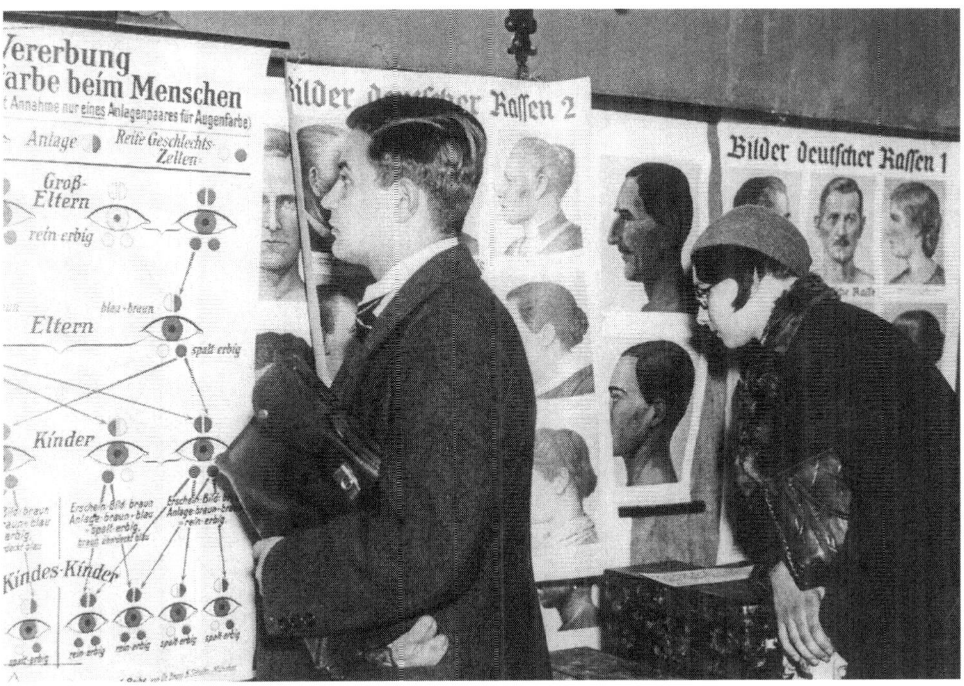

Family and Race Policy, Nazi Germany, 1933–1945. SUEDDEUTSCHE ZEITUNG PHOTO/ ALAMY STOCK PHOTO.

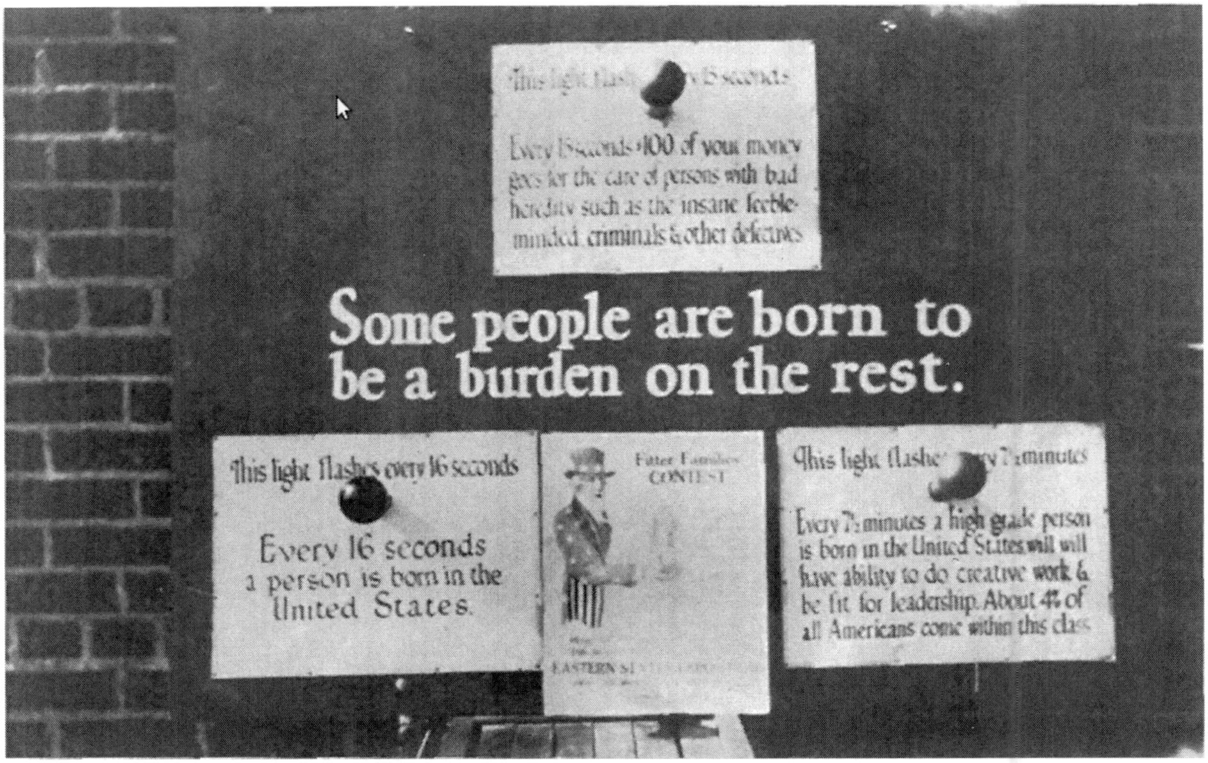

Eugenics Advocacy Poster, United States, 1926. "SOME PEOPLE ARE BORN TO BE A BURDEN ON THE REST," PHILADELPHIA SESQUI-CENTENNIAL EXHIBITION, 1926. FROM STEVEN SELDEN, "TRANSFORMING BETTER BABIES INTO FITTER FAMILIES" (2005, PROCEEDINGS OF THE AMERICAN PHILOSOPHICAL SOCIETY 149(2), P. 219).

reproductive achievements of people assessed as hereditarily worthy. These positive eugenic measures supplemented negative eugenic interventions, such as forcible sterilizations and wrongful imprisonments.

The term *Galtonian eugenics* is used in this chapter to describe eugenics programs that aspire to improve population health by managing human reproduction. People assessed as having good-quality hereditary material are encouraged to reproduce. People assessed as having poor-quality hereditary material are prevented or discouraged from reproducing. Galtonian eugenics is wrong for two quite distinct reasons, one factual and the other moral.

With regard to the factual problem with Galtonian eugenics, Galton believed that "more suitable . . . strains of blood" tended to find themselves in the social classes and races that were perceived to be doing the best (1883, 25). In effect, he viewed his age's dominant prejudices as an efficient method of sorting hereditary material by quality. Less suitable strains were found in people at the bottom of the social heap.

Advances in genetics make the factual errors of Galtonian eugenics apparent. Studies of the human genome have revealed many genetic variants that increase a person's susceptibility to disease. But these do not accumulate in the genomes of people located toward the bottom of social-status hierarchies. Genetic variants that offer protection against disease do not migrate toward the genomes of those in society's upper echelons. The idea that the genetic superiority of Europeans explained the success of their colonial efforts is simply wrong. European advantages were technological not biological. An observation by English author Hilaire Belloc (1870–1953)—"Whatever happens, we have got / The Maxim gun, and they have not" (*The Modern Traveller* [1898])—explains European colonial successes. Modern genetics has revealed very many sequences of DNA that elevate the risk of disease scattered across the genomes of those doing best in society and of those who suffer disadvantage. The idea that the dominant prejudices of the late nineteenth and early twentieth centuries were informative about the locations of stretches of DNA that make some lives go better than others would be simply laughable if it did not have such a terrible effect on those assessed as hereditarily unworthy. The analyses of DNA offered by modern genetics reveal a picture altogether more complex than that supposed by Galton and his followers.

If you send money and saliva to a direct-to-consumer genetics company, you are likely to discover that you have inherited a diverse collection of disease susceptibilities. You will discover that you have a higher-than-average risk of getting diseases A, B, C, D, E, and so on and a lower than average risk of getting diseases Z, Y, X, W, V, and so on. Some genetic diseases are more common in people belonging to certain ethnic groups. But it is simply false to assert that some ethnicities bear a greater burden of genes that elevate the risk of disease than others and therefore should be classified as worse in hereditary terms. There are no ethnicities or social classes that are hereditarily sickly, feckless, or criminal. It is not possible to improve the intellectual or moral health of a population by depressing or increasing the reproductive rates of selected racial groups.

Another innovation from the new genetics is a rejection of Galton's notions about the expression of hereditary factors. Galton presented many of these traits as inborn. If traits that affect health are inborn then one has only to encourage the births of those with the hereditary right stuff and discourage the births of those with the wrong stuff to improve population health. It is now recognized that this belief in inborn traits expresses a mistaken view of human development labeled genetic determinism. Genes are influences on the ways humans turn out. But they do not causally fix traits. They affect them in concert with a diverse complexity of environmental influences. Some of these environmental influences are exercised in the womb and others outside it. Genetic determinism has a pervasive grip on the popular imagination. People tend to oversimplify the causal information conveyed by talk of genes "for" characters or diseases and to overlook the gap between a test that reveals a genetic variant that increases one's likelihood of getting cancer and a diagnosis of cancer. Genetic determinism's simplifications should be resisted. A view in which diseases are inborn traits that can be prevented only by preventing specific births is being replaced by a view in which all humans inherit a variety of genetic variants that increase or reduce a person's likelihood of getting diseases relative to the population's mean risk. Individuals are empowered to make lifestyle choices that prevent disease risks from becoming diseases.

Galtonian eugenics aggravates this factual error with a distinctive moral mistake. Galton hoped to manage human reproduction. Negative eugenics implements a requirement for dysgenic people to refrain from reproduction. Positive eugenics implements a moral requirement for eugenic people to increase their reproductive work rate. These requirements are now viewed as morally intolerable intrusions into individuals' procreative freedoms. The freedom to decide how one uses one's reproductive capacities is among the most important of human liberties There are different ways in which the strength of a requirement to respect procreative liberty can be understood. Some philosophers will say that this requirement is absolute—it cannot be overridden by competing moral considerations. Others may allow that it can be overridden in extreme circumstances. But procreative freedom is such an important value that societies insist on both a very high magnitude of harm or benefit and high confidence about that harm or benefit before they permit the freedom of individuals to make their own procreative choices to be overridden. Speculation about possible increases in population measures of health that come when some individuals comply with an obligation to reproduce and others comply with an obligation to refrain from reproduction does not meet this threshold.

Galton's plan to increase population health by managing human reproduction thus fails on factual and moral grounds. In what follows, a non-Galtonian eugenics program is advanced. It shares Galton's goal of boosting population health by focusing on human inheritance. But it achieves this end in ways that are compatible with the picture of human heredity emerging from modern genetics and that do not violate established procreative freedoms.

The focus below is on one technique available to eugenicists who seek to improve population health in ways that both are consistent with modern genetics and do not presume to either oblige or prohibit reproduction. Preimplantation genetic diagnosis (PGD) offers people who risk passing on a genetic disease to their children the opportunity to avoid doing so. PGD begins with embryos created by in vitro fertilization. These embryos are permitted to go through a few divisions, whereupon one or two cells are taken and tested for the presence of a disease-linked genetic variant. Embryos free of the genetic variant can then be introduced into a woman's body, and pregnancy can proceed as normal.

It is easy to imagine how PGD might be used by eugenicists with no regard for procreative liberty. The state would impose PGD on prospective parents assessed as at risk of passing on a genetic disease. This would be an intolerable infringement on procreative freedom. There are differences in the provision of PGD in different societies. But there is no society that imposes it on those wishing to have children. Fortunately, these measures do not have to be taken for PGD to achieve the end of improving population health. Restrictions are not required. Instead, the state can make PGD available to prospective parents who risk passing on to their children a genetic variant linked with a serious genetic disease. It is up to them whether they choose to avail themselves of this offer. This concession to procreative freedom limits the extent to which PGD can improve population health. But this is an appropriate recognition of the importance of procreative freedom. A society can trust that some parents who judge themselves to be at risk of passing a genetic disease on to their children will choose to use PGD to avoid doing so. To the extent that prospective parents make those choices, the genetic health of a population will improve. But one can recognize a eugenic motive in society's choices about which conditions to make it available for. In what follows, this chapter argues that the eugenic motivation is also apparent in the restrictions on how people respond to the information provided by the genetic diagnosis of an embryo.

EUGENICS AND PUBLIC HEALTH

How should moral limits on the improvement of population health by eugenic methods be set? Earlier in this chapter was a description of the interactionist view of development, which has replaced the genetic determinism favored by Galtonian eugenicists. The interactionist view advances a developmental parity of genetic and environmental influences. Human traits emerge out of a complex interaction between those influences. This developmental parity suggests a moral parity. Changes to genes and environments have the potential to change the traits of developing humans. Potential modifications of them should be treated similarly when they have been judged to have relevantly similar effects on developing humans. Given that an essential feature of eugenics is its focus on population health, the idea of moral parity between genetic and environmental influences suggests that we seek guidance from environmental influences on population health. We can think about the eugenic program's focus on inherited contributors toward the genetic health of populations in a way that is analogous to the way we think about public health measures to increase population health.

A public health intervention might be considered a success when it reduces the incidence of some environmental influence judged injurious to health. An example of such an environmental influence is smoking. Practitioners of public health hope to reduce the incidence of cancer by reducing the rate of smoking in a population. They judge success based on surveys to determine whether public health campaigns have the effect of reducing the rate of smoking in a targeted demographic. Eugenic focuses on inherited influences can be viewed in a similar way. Suppose that genetic contributors to type 1 diabetes are added to the list of conditions for which PGD is made available. This would be judged to be a successful eugenic intervention if it reduced the frequency of genetic variants linked with a high risk of type 1 diabetes in the targeted population.

There are limits on public health measures. They must respect the rights of people. A program of incarcerating all smokers or depriving them of their children on the grounds that they may raise them in smoke-filled environments might reduce the rate of smoking in a population. But these public health measures would be morally unacceptable. There are also limits that eugenicists must recognize. Galton's program of selective breeding is a morally intolerable infringement on procreative liberty. The freedom to make your own decisions about whether you have children, how many children you have, and with whom you have children is precious. Even if it was determined that obligations to use PGD would increase population health, such obligations should be rejected. There will be many public health and eugenic measures that produce greater improvements of health that are nevertheless rejected because of objections to their violations of individual liberties.

Public health and eugenic interventions must not stigmatize people who live with the conditions being targeted. Consider public health measures to reduce the rate of smoking among teenagers. One of the discoveries is that warnings about the threats of smoking to the health of smokers are not as effective as one might think they should be. Young people find it difficult to emotionally connect with a future sixty-year-old self, stricken with cancer. Teenagers are nevertheless sensitive to suggestions that smoking might make them unattractive or in some way repellent. There are dangers in a public health measure that presents smokers as repellent. Smokers are people too. They can make legitimate complaints about how they are represented in a public health campaign. Eugenic attempts to improve population health also risk stigmatizing people. One must be careful to not send the message, either deliberately or inadvertently, that people at risk of passing on genetic variants strongly linked with disease are disgusting or have reduced moral worth. When seeking to reduce the incidence of genetic variants that raise the probability of getting type 1 diabetes, care must be taken not to stigmatize existing diabetics. The interactionist view of development suggests that the same concerns apply with respect to both genetic and environmental contributors to diabetes. Geneticists have linked variants of immune system genes with an increased likelihood of getting type 1 diabetes. On the environmental side, there are possible connections with exposure to viruses and the consumption of cow's milk when young. The moral parity of attempts to prevent diabetes by genetic or environmental interventions suggests the need for an equivalent alertness with respect to the risk of stigmatization. Campaigns that seek to reduce the consumption of cow's milk by children or that seek to reduce the incidence of certain genetic variants in the HLA complex should not seek to achieve improvements of population health by presenting diabetics as disgusting.

PGD: EXPANSION OF PROCREATIVE LIBERTY OR MEANS OF IMPROVING POPULATION HEALTH?

This chapter has suggested that the best way to think about PGD is as a means of improving population health. It is the hereditary counterpart of the focus of public health campaigns on environmental causes of ill health. But there is a different way to think about PGD. Those who write about PGD from a liberal perspective prefer to view it as a means of enhancing procreative choice. When a society makes PGD available to people at risk of passing on genetic variants that raise the risk of some diseases, it gives them a reproductive option formerly unavailable to them. The eugenic approach argued for herein respects the limits on measures to improve population health imposed by respect for procreative liberty. PGD should not be imposed on prospective parents, even if this would result in greater improvements of population health than merely making it available to them. Another avenue that should not be pursued are the maximal improvements of population health that would be possible if PGD were introduced as part of a program of Galtonian eugenics. But acting in a way that is compatible with respect for procreative liberty nevertheless permits PGD to improve population health.

There are two ways to justify PGD. The first gives central place to the protection and extension of procreative liberty. The second gives central place to the improvement of population health. Respect for procreative liberty features as a constraint on the ways in which population health can be improved.

- Option A: The principal justification for a society's adding a new disease condition to the list of those for which PGD is available is the enhancement of procreative freedom. Prospective parents at risk of passing on a genetic variant linked with a serious disease acquire a procreative option formerly unavailable to them.
- Option B: The principal justification for a society's adding a new disease condition to the list of those for which PGD is available is the improvement of population health. The measures taken to improve population health must respect procreative freedom. Some prospective parents at risk of passing on a genetic variant linked with a serious disease will choose PGD, leading to a reduction in the number of cases of the disease in the population.

Is there a way to choose between these alternative conceptions of the practice of PGD in liberal societies? This chapter argues that there is and, furthermore, that option B, with its central focus on the eugenic motive of improving population health, should be preferred. When one describes the practice of PGD in liberal societies, the prevention of disease is uppermost.

There are uses of PGD that do not prevent disease in a future child. These may suggest a preference for option A. For example, PGD is sometimes used to increase the chances of a successful pregnancy. When used for this purpose, PGD treats infertility but it does not select the characteristics of children. When only uses of PGD whose purpose is to select the characteristics of future children are considered, the focus is overwhelmingly on the prevention of disease. Some uses of PGD to select the sex of a future child may be better understood as expressions of procreative freedom than as preventing disease. When used for family balancing, PGD permits parents to achieve a more equal representation of male and female children. Parents who seek PGD for family balancing are not seeking to avoid disease in their child. These uses of PGD to select the characteristics of future children can be seen as exceptions to a general focus on the prevention of disease. For example, in the United Kingdom, sex selection is not permitted for family balancing. It is permitted only to prevent certain disorders linked to the X chromosome. Parents at risk of having children with Duchenne muscular dystrophy choose female embryos on the grounds that girls are much less likely than boys to have the condition.

Consider a counterfactual case that places respect for individuals' procreative choices directly into conflict with the motive to improve population health. Would a society countenance the use of PGD to give individuals a procreative option that reduced population health? If the answer is no, then the improvement of population health is the dominant rationale for offering PGD.

Geneticists have identified a variety of genetic variants that increase the likelihood that an individual will suffer type 1 diabetes. Suppose you were to go to a PGD clinic and request that genetic variants that increase the risk of diabetes in your future child be chosen. You would not be condemning the resulting child to a life not worth living. You may have increased your child's likelihood of having type 1 diabetes, but this condition is compatible with a worthwhile life. One can imagine that in a society that offered PGD to parents as means of enhancing procreative liberty this might be an eccentric choice. Few prospective parents would choose to increase the probability that their child would have diabetes. But liberal societies protect choices recognized as eccentric. Consider the impact on the public perception of diabetes of celebrity sufferers of disease. They offer themselves as examples to others who suffer a disease. They lead successful lives in the face of the disease's challenges. Actor Mary Tyler Moore (1936–2017) and US Supreme Court justice Sonia Sotomayor (1954–) provide examples of excellent and successful lives led by people with type 1 diabetes. Their lives are an encouragement to people living with diabetes. But they could also be an encouragement to prospective parents to use PGD to select for diabetes. There are people who mimic the patterns of dress of their heroes. This would be a case of people whose desire to resemble their heroes extends to acquiring their diseases or engineering them into their children's genomes. The use of PGD to increase the incidence of diabetes cannot be squared the common understanding of the role of the institution of medicine in liberal societies. It is possible to explain why it is wrong to permit PGD to be used to deliberately create diabetic lives in ways that do not require saying that diabetic lives are not worth living.

Consider how an endocrinologist with a specialization in the treatment of diabetes might respond to the request of a nondiabetic for assistance in acquiring the disease. The prospective patient effectively inverts the assurances typically given by the endocrinologist to newly diagnosed diabetics. He counters the doctor's reluctance with the protestations that people with diabetes can lead excellent lives. It would be perfectly appropriate for the endocrinologist to respond that she has no control over what the patient does outside of her consulting rooms but that she refuses to permit her skills to be used to cause diabetes. Analogous responses would seem to be warranted with respect to a request to use PGD to cause diabetes in a future child. It is perfectly appropriate for the doctor to conclude the discussion with the recognition that the diabetic life you aspire to is predictably good but that she is in the business of treating and preventing disease and not in the business of causing it. What you do when you leave her offices is up to you. You may find alternative ways to cause the condition in yourself, but a medical professional is not required to collaborate with your plan, even if she grants that it should be permissible for you to use other means to cause yourself to become diabetic. Practitioners of PGD might offer the same explanation for refusing the select for diabetes in your future child. PGD is medicine. Its practitioners are in the business of preventing disease, not in the business of causing it. This suggests that option B, the eugenic justification for PGD, should be preferred to option A, the justification for PGD as a means of increasing the range of procreative liberty.

TWO KINDS OF MORAL WARNING FROM HISTORY

This chapter has argued that it is possible to understand liberal societies' provision of PGD as both eugenics and justified. Eugenics is the genetic moral counterpart of public health campaigns. Both focus on populations. One focuses on environmental contributors toward ill health. The other focuses on genetic contributors. Some might agree with the substance of the argument presented herein but quibble with the choice of terms for it. Eugenics is associated with some of the signal wrongs of the twentieth century. The decision to describe the correct use of PGD as eugenics seems exceptionally poor marketing for a practice offered as morally justified. This is not an argument for an unashamed eugenics. Indeed, the case made in this chapter is for an *ashamed* eugenics, one that is not only aware of but also informed by its past.

There are two ways of understanding the warning from history. According to one interpretation, the warning is to avoid practices essentially similar to an evil historical practice. According to another interpretation, the warning is to exercise caution when engaging in practices recognized as essentially similar to a historical practice that caused a great deal of harm.

Consider the case of slavery for an example of the first kind of warning from history. The *Oxford English Dictionary* definition that best defines the institution of slavery is as follows: "The condition or fact of being entirely subject to, or under the domination of, some power or influence." When slavery is condemned, it is condemned for something that is essential to it. Suppose someone suggests that a new variety of employment is essentially the same as slavery. That is grounds to reject it. Slavery is essentially wrong. One should reject certain reassurances—such as, "But this is neo-slavery!" or "But this is liberal slavery!"—where slavery is defended on the grounds that people who become slaves freely consent to lose their liberty.

Consider now the cautionary sense of a warning from history. Medical research on human subjects is not intrinsically wrong. There is nevertheless a history of terrible offenses against human dignity committed in the name of medical research. These mistakes are not essential to medical research on human subjects. It is possible to perform morally good medical research on human subjects. But whenever human subjects are involved, there is a risk of abuse.

In the infamous Tuskegee Syphilis Study (1932–1972), an interest in studying the progression of the disease syphilis led to patients being denied established therapies for it. The crimes of the Tuskegee experiments do not suggest that scientists should desist from conducting medical research on syphilis. But they do show that this research is always morally risky. Medical researchers understand that they can learn things about disease by methods that grossly abuse human experimental subjects. There will always be a temptation to ignore the moral entitlements of human subjects in a bid to add significantly to scientific understanding of disease. It would be disturbing to learn of someone researching the epidemiology of syphilis who knew nothing of the Tuskegee experiments. Spanish-born American philosopher George Santayana (1863–1952) famously said, "Those who cannot remember the past are condemned to repeat it." But a contemporary syphilis researcher must do more than know about the Tuskegee crimes. He must remember them in a particular way. He must understand that they were committed by people who shared his desire to treat syphilis more effectively. He must ensure that his human experimental subjects are granted the respect so egregiously denied subjects in the Tuskegee study.

This chapter has suggested that the bias in liberal societies' practice of PGD toward the prevention of disease means that it should be acknowledged as eugenics. Today's practitioners of PGD should remember the crimes committed in the name of eugenics in the same way that today's syphilis researchers remember the crimes of Tuskegee. They are more than unpleasant historical events. They are crimes committed by people who shared the aim of improving population health.

The crimes committed in the name of eugenics must be remembered in a way that offers protection against the bad moral reasoning that convinced those of earlier generations. Roosevelt, Sanger, and Shaw were thoughtful people. They nevertheless advocated varieties of eugenics that are now recognized as morally abhorrent. It is easy for a society to fall into error in the way it seeks to select human genetic material. The current practice of PGD in liberal societies is eugenic. The fact that eugenics can be implemented in morally acceptable ways—ways that are unlikely to have occurred to Galton—does not obviate the need for frequent historical reminders.

Summary

This chapter argues that eugenics is intrinsically morally justifiable at the same time as being morally dangerous. The focus of eugenics on the genetic contributors to population health should be viewed as similar to the focus of public health on environmental contributors. The comparison with public health offers instruction on when eugenics is justified and when it is not. Galtonian eugenics has been presented in this chapter as both factually and morally mistaken. In its place, this chapter has offered an updated eugenics that achieves improvements in population health in ways that are compatible with modern genetics and respect procreative liberty. It was also proposed that this eugenic defense of PGD should be preferred to an account that defends PGD as enlarging procreative liberty. The final section addressed the importance of retaining the word *eugenics* together with its connections to past offenses. The activity of selecting or modifying human genetic material is dangerous. But humanity can proceed with caution if memories are retained of past crimes in eugenics as resulting from practices that are essentially similar to those of today. One should not seek to shelter behind new words.

Bibliography

Agar, Nicholas. *Liberal Eugenics: In Defence of Human Enhancement.* Malden, MA: Blackwell, 2004.

Agar, Nicholas. "What Was Right about Eugenics." *Lahey Health Journal of Medical Ethics* (Fall 2015): 4–5.

Buchanan, Allen, Dan W. Brock, Norman Daniels, and Daniel Wikler. *From Chance to Choice: Genetics and Justice.* Cambridge: Cambridge University Press, 2000.

Galton, Francis. *Inquiries into Human Faculty and Its Development.* London: Macmillan, 1883.

Kevles, Daniel J. *In the Name of Eugenics: Genetics and the Uses of Human Heredity.* New York: Knopf, 1985.

Kitcher, Philip. *The Lives to Come: The Genetic Revolution and Human Possibilities.* New York: Simon and Schuster, 1996.

Oelhafen, Ingrid von, and Tim Tate. *Hitler's Forgotten Children: A True Story of the Lebensborn Program and One Woman's Search for Her Real Identity.* New York: Berkley Caliber, 2016.

Parrington, John. *The Deeper Genome: Why There Is More to the Human Genome Than Meets the Eye.* Oxford: Oxford University Press, 2015.

Paul, Diane B. *Controlling Human Heredity, 1865 to the Present.* Atlantic Highlands, NJ: Humanity Books, 1995.

Robertson, John A. *Children of Choice: Freedom and the New Reproductive Technologies.* Princeton, NJ: Princeton University Press, 1994.

CHAPTER 29

Human-Robot Relationships

Raya A. Jones
Reader, School of Social Sciences
Cardiff University, UK

I'm a grandmother machine, which means more than a machine.

—Ray Bradbury

Ray Bradbury's "I Sing the Body Electric!" (1969) tells of Electric Grandma who looks after three children whose mother has died. Although the story is about a robot, it celebrates the human. Bradbury's subject matter is the stuff human experience is made of: loss and grief, familial love, wonder, and curiosity. Electric Grandma is more than a machine, because as a grandmother she exists within a web of human relationships. Humans acquire an identity only within interpersonal relationships, scholars have long averred. This social model of selfhood was elaborated throughout the twentieth century in numerous variations on the "I-you-me" theme (the idea that through you, I acquire a concept of me). The electric grandmother was conceived when this model was at its zenith, in the mid-twentieth century. By the 1990s, sociologists had begun to note the rise of a so-called "postsocial" process of identity construction whereby objects displace humans as sources of self-identity (see, e.g., Knorr Cetina 2001). If the social model is ebbing, it paradoxically survives when the objects displacing humans are expected to be constituted *like us*, including the capacity to reciprocate love.

This chapter brings to the fore some of the ambivalences and ambiguities concerning so-called social robots especially the conceptualization of sociality and relationships in this context. Human relationships with robots begin in the imagination and remain mostly confined to fiction, though not for long. The likelihood of near-future robots in social roles raises a gamut of ethical, legal, and political issues. Such robots are not yet sophisticated enough or sufficiently affordable to be commonplace. Consequently, there is scarce empirical knowledge about the impact of long-term relationships with robots. Opinions about potential benefits and risks remain speculative, and the imagined future-with-robots becomes a depository of fantasies, hopes, and fears.

Paradoxical conservatism inheres in the discourse of social robotics. On the one hand, engineers seek to create intelligent artifacts that enter the social realm and do things with humans. This ambition resonates with the posthuman view, which "configures human being so that it can be seamlessly articulated with intelligent machines" (Hayles 1999, 3). On the other hand, human-robot relationships are often appraised as if these are simulations of human-human relationships. This chapter's title may conjure an image of a person and a humanoid robot together, like people in a relationship—an "I-you," with a robot in the "you" position. Popular culture both perpetuates and primes the fantasy of machines-like-us. It does so conspicuously in films and television dramas that feature human actors in the role of androids, as well as in depictions of nonhumanoid machines that are humanlike in terms of conduct, attitude, affect, and motivations. The familiar fiction eclipses the profoundly alien nature of artificial intelligence. Long-term relationships with next-generation robots are likely to be qualitatively different from relationships with other humans, and they might lead to new ontologies whereby the boundaries between human and machine are redrawn.

BEGINNINGS

The robotics industry promises to populate our homes, schools, hospitals, shops, restaurants, and bars with entities that have populated human imagination for millennia. From ancient Greece comes the legend of Pygmalion, the Cypriot sculptor who fell in love with his beautiful statue and married her after Venus granted life to the statue. Medieval legends of golems, originating in eastern European Jewish folklore, carried a theological warning against trying to emulate the Creator, whereas Liezi, a sage who lived in China in the fifth or fourth century BCE (scholars differ about the date), imparted the Taoist belief that neither man nor the universe require a conscious controller by telling about an artisan who built a dancing and singing android. The word *robot* (an eastern European term for "serf labor") was imported into science fiction by Czech playwright Karel Čapek in his 1921 play *R.U.R.* (*Rossum's Universal Robots*), in which artificial humanoids made from organic matter are exploited as laborers.

It is Isaac Asimov (1920–1992) who should be credited with developing the modern concept of the robot and coining the term *robotics*. Whereas *R.U.R.* is a dystopian allegory about the social reality of the writer's day, Asimov's robot fiction is utopian futurism. It embodies the weltanschauung of Western rationalism—faith in science—and a vision of technological progress that culminates in benign humanoids governing the world by the late twentieth century. Echoes of Asimov's futurism can be heard in engineers' claims that "the world is driven and run by technological developments ... robots are here for further enhancements and new applications. It means no less than that technology dictates the governance" (van den Herik, Lamers, and Verbeek 2011, 107); but technology has landed us in a world very different from the future imagined by Asimov, in which humanoid robots substituted for people and pets in an otherwise unchanged society.

Today, "smart" automatic systems, in the form of bots residing in cyberspace, routinely regulate mundane aspects of everyday life across the globe—and society has changed. It is now possible to imagine a future in which human life is meshed with ambient intelligent technology. The craft of automata making, which has long histories in the Far East and Europe, precedes modern engineering more directly than do myths and legends. A quantum leap occurred, however, when engineers started to create machines capable of learning and adapting to their environment.

FROM "INTERACTION" TO "RELATIONSHIP"

Human-robot interaction (HRI) emerged in the middle of the first decade of the twenty-first century as an engineering field dedicated to understanding and shaping HRIs. It encompasses all of robotics and centers on dynamic interactional patterns that evolve in a self-regulatory way (the operator becomes more skillful, and the robotic system also learns and adapts to its user). Social robotics overlaps HRI but also raises unique issues with respect to the nature, purpose, and moral desirability of robots in social settings. Probably nobody would protest against using robots to rescue earthquake survivors, for instance. We may indeed regard it as society's moral duty to invest in robotics toward building efficient rescue robots. The measure of desirability is the task to which the robot is put. We may likewise regard it as society's moral duty to invest in improving care of the elderly; but the desirability of robot caregivers is no longer a matter of how efficient the robots are. In addition to some profound technical problems (e.g., how to design robots capable of detecting erratic behavior in a human), robotized elderly care raises ethical concerns about emotional deprivation and loss of dignity.

In their agenda-setting review of HRI, Michael A. Goodrich and Alan C. Schultz defined *interaction* as "the process of working together to accomplish a goal" (2008, 217). Yet, this process is qualitatively different from a collaboration between human coworkers. Quintessential properties of human cooperation in the workplace—such as an agreement on goals, labor division, rights, and responsibilities reflecting seniority, experience, or expertise, as well as personal affections and animosities that might get in the way of task performance—do not apply to humans and robots working together. Furthermore, Goodrich and Schultz highlight situations in which humans and robots work toward performing tasks that are external to both, such as rescuing survivors, exploring a terrain, and firing on military targets. In contrast, assistive or companion robots do something for the benefit of their users. The interaction itself is the task to be performed by the robot. This situation is closer to the relationship between service providers and their clients, nurses and patients, teachers and students, or pets and their owners than it is to the relation of machine operators to the machines they operate.

Making a case for putting "the *personal and intimate relational dimension* between a human and a robot" on the HRI agenda, Myriam Diocaretz and H. Jaap van den Herik (2009, 205) give the example of assistive robots in individualized health care, contending that engineers seeking to optimize patient-robot relationships ought to understand the psychological mechanisms whereby a personal relationship involves "inherent trust in order to build love and friendship, including emotional attachment" (206). Their contention omits a fundamental fact about forming a relationship: it takes two. People may trust and love a machine without the machine being capable of feeling anything; but a *friendship* presupposes mutual feelings.

THE ROBOTS IN QUESTION

Robots designed for interaction vary greatly in physical appearance and interactive capacity. Appearance-wise, they may range from android, cartoon-like, and animal-like to purely utilitarian and abstract forms. Designers of multipurpose interactive robots that are currently available commercially (e.g., NAO, QRIO, Pepper) opt for small, cartoon-like humanoids, perhaps because it is believed that most people expect humanoids but a realistic android might be experienced as unsettling or spooky (a phenomenon first described by roboticist Masahiro Mori in 1970 as the "uncanny valley effect," referring to a drop in a sense of familiarity when objects are increasingly humanlike in appearance). Interaction possibilities range from limited autonomous responses to scenarios in which the robot is fully teleoperated by a human. This diversity of robotic platforms is compounded by the variety of terms applied to robots. Kerstin Dautenhahn (2007) lists the adjectives *socially evocative, socially situated, sociable, socially intelligent,* and *socially interactive*; other writers have applied *social-affective.* These terms convey correspondingly diverse concepts. Whereas socially evocative robots are defined by the responses they elicit in humans, socially interactive robots engage people in interaction and communication. Robots can be socially interactive by means of teleoperation, but to be socially intelligent requires inbuilt architecture.

The robots that are most commonly talked about as social are physically embodied systems designed to interact with people and potentially enter social spaces in community and domestic settings. This characterization excludes disembodied artificial intelligences or bots, such as automated response systems and Internet search engines. It also avoids just listing settings (hospitals, schools, shops, home, etc.) in which robots could be placed and instead suggests that people's experiences of the robot should be considered a salient criterion. A robot nurse might enter patients' social space, whereas a robot surgeon or a robot janitor in the same hospital would not. Strictly speaking, the above characterization excludes robots designed as labor-saving appliances. Yet, researchers investigating the introduction of robot vacuum cleaners into households in the United States have discovered that participants came to regard the robot as a social entity—ascribing lifelike qualities to it and giving it a name, gender, and personality (Sung, Grinter, and Christensen 2010). Even when restricting the topic to robots that are purposefully designed for interactions with humans, the minimal conditions for robot sociality remain an open question.

ROBOT SOCIALITY

An early definition of social robots highlights a similarity with human and animal individuals within a group: "Social robots are embodied agents . . . able to recognize each other and engage in social interactions, they possess histories (perceive and interpret the world in terms of their own experience), and they explicitly communicate with and learn from each other" (Dautenhahn and Billard 1999, 366). This construes sociality as a trait or a set of competences that organisms possess, if their species evolved that way, and therefore as something that can be reverse engineered. Since the middle of the first decade of the twenty-first century, some roboticists have asserted that a robot becomes social by virtue of how people interact with it: "A *social robot* is a robot plus a *social interface* . . . which includes all social attributes by which an observer judges the robot as a social interaction partner" (Hegel et al. 2009, 174). Redefining the robot's social character as contingent on people's judgments signals a shift away from construing sociality as a trait and toward construing it as an emergent property of the interaction itself.

Could a robot be classed as social if it evokes a "social" reaction in people? Humans are biologically hardwired to respond to social cues, such as eye contact; and robots can be designed to create a powerful illusion. Recalling her visit to Rodney Brooks's artificial intelligence laboratory at the Massachusetts Institute of Technology, Sherry Turkle described her response to Cog, designed as a robotic two-year-old, as "visceral. Cog had a face, it made eye contact, and it followed my movements. . . . I had to fight my instinct to react to 'him' as a person" (2011, 84). Such an illusion, however, is not the same as being involved in an actual interaction. Many experiments designed to assess how people interact with robots use the Wizard of Oz technique, whereby the robot is operated by an unseen human. In one study, researchers placed an android replica of Japanese roboticist Hiroshi Ishiguro in an Austrian café and documented visitors' interactions with it. People behaved differently when the android showed minimal signs of reactiveness (e.g., turning its head toward them) than when it was "idling"; and when it was teleoperated so as to be fully interactive, they seemed to forget about the teleoperating person and interacted with the robot as if it had agency of its own (Straub 2016). Nevertheless, it could be opined that the participants played along with the make-believe.

Could a robot be classed as social only if it autonomously participates in exchanges that people experience as meaningful? The question of whether an artificial intelligence could ever be accepted as another person is often articulated in terms of whether it will pass the Turing test—that is, be able to deceive people about its artificial nature. The possibility of deception is ruled out when one actually sees the robot, but there remains the question of whether "social intelligence" is sufficient for regarding a robot as a humanlike social agent. Some may insist on its capacity for feelings.

EMOTIONALITY AND MORALITY

If you love a robot, will it love you back? This is not entirely a rhetorical question. It seems likely that near-future robotics could create machines capable of behavioral repertoires that emulate human love and "bonding." Whether or not an *emotional machine* is an oxymoron depends on how emotions are defined. Toward the goal of giving basic emotions to robots, neuroscientists Michael A. Arbib and Jean-Marc Fellous define emotions as discrete motivational systems that play important functions in the organization of organisms' behavior and decision making. These systems operate on a more basic level than the cognitive level involved in feeling "the 'heat' of love, sorrow or anger, and so on" (2004, 554).

Appropriate behavior in social situations entails an ability to read others' intentional states. Emotional intelligence is measured partly on the basis of individuals' sensitivity to others' emotional states. Softbank's robot Pepper, which was introduced into the Japanese market in 2015, can identify emotions such as joy, sadness, anger, and surprise by deciphering voice tones, facial expressions, and language. Yet it could be argued that an "emotionally intelligent" robot is not an emotional being because it cannot *experience* emotions. The emotions that help humans navigate interpersonal relationships are the self-conscious emotions, such as pride, guilt, and shame. There could be practical advantages for designing robots that become "imprinted" on their primary user (the functional equivalent of emotional attachment)—for instance, when a robot caregiver is likely to accompany its user into public places or communal areas, as in a nursing home. Yet, the robot's efficient performance of its tasks does not depend on it feeling happy to please, pride in doing its job well, embarrassment or shame if it fails to do something, and grief when its elderly user dies.

If we imaginatively project ourselves into the future, we may see robots outwardly behaving in an appropriately loving manner with their users, the way we would expect people to behave with their loved ones. Real relationships, however, do not always run in accordance with ideal scripts. Marriages break up. Some parents maltreat their children. Some offspring hate their parents. When we fall in love with someone, he or she might not return our love but love someone else instead. Will a robot equipped with "bonding" algorithms be able to *not* love you back? If your robot malfunctions and "falls in love" with someone else, would it feel guilty and ashamed? What would be the purpose of building robots capable of self-conscious emotions (technical feasibility notwithstanding)?

The self-conscious emotions are also called the moral emotions. Although ethicists discuss the necessity of instilling a moral code in intelligent machines (see, e.g., Wallach and Allen 2009), human moral agency means more than simply knowing right from wrong and responding with appropriate action when situations arise. Instead, this agency proactively locates itself in a space of moral questions of its own making. As moral agents, people may feel a responsibility to contribute to others' well-being and welfare; but people can also opt to do nothing or indeed can be cruel and malicious. Being a moral agent presupposes the option of immorality—and we certainly would not want to facilitate this option in machines. Raya Jones contends that a robot may be regarded as a humanlike social agent only if and when it can autonomously partake in dialogical action—that is, have a "voice" (a viewpoint) and negotiate power relations and its own positioning in a social-moral order (Jones 2016; Jones 2017).

However, robots need not have emotional states of their own in order to have an emotional impact on the people who interact with them. The pertinent ethical issue concerns the implications of long-term relational transactions with robots.

RELATIONAL TRANSACTIONS

Jean Baudrillard defined simulacra as copies of things that do not have an original anymore or never had one to begin with. One of his examples was an American "TV-*verite* experiment" that documented the life of the Loud family in 1971 (1994, 49). Reality television and other examples provided by Baudrillard reverse the relation between representation and reality until only "models of a real without origin or reality: a hyperreal" remain (1). Baudrillard called this a "*precession of simulacra*" (1). The imminent arrival of real robots into our midst may challenge us to rethink human-robot relationships in terms of such a precession.

A case in point is Paro, a cuddly seal robot developed by Japan's National Institute of Advanced Industrial Science and Technology. Paro was commercialized in Japan in 2005, and within a few years its use had spread to about thirty countries. It makes eye contact, is sensitive to touch, and responds differently according to whether it is being stroked gently or aggressively. It looks and sounds like a baby seal, but it is not a copy of any realistic pet. It is hyperreal, sui generis. Do simulacra such as Paro create false relationality? The answer is not straightforward. Paro can promote a genuine sense of well-being. In August 2011 Reuters reported that elderly survivors of the March 2011 tsunami, residents of a retirement home near the crippled nuclear plant in Fukushima, Japan, were given Paro robots. Holding the robot, one eighty-five-year-old resident told the journalist, "If I hold on to this, it doesn't matter if there's a typhoon outside, I still feel safe" (Meyers 2011).

It could be opined that short interactions with Paro, even when regularly repeated over months, do not amount to a *relationship* because the robot cannot reciprocate with empathy and support. Turkle articulated this reservation through the case of Miriam, a seventy-two-year-old resident of a nursing home near Boston, who participated in a longitudinal study in which Turkle gave robots to people and then interviewed them about the robots. At the time of the interview, Miriam was depressed because her son had abandoned her. She petted Paro, spoke to it as if the robot too was sad, and attempted to comfort it. The touch caused Paro to turn its head and purr, which encouraged her to show more affection to it. Turkle reflects, "In the moment of apparent connection between Miriam and her Paro, Miriam experienced an intimacy with another, but she was in fact alone" (2011, 9). Turkle contrasts Miriam's interaction with Paro with the healing potential of an interaction with a sympathetic human, such as a therapist, and finds the "therapeutic" robot wanting.

Paro's makers do not propose it replacing a human companion, let alone a therapist. Certified by the US Food and Drug Administration as a biofeedback medical device, Paro is marketed as a substitute for animal therapy in hospitals and nursing homes where live animals cannot be used. Paro is expensive and is usually owned by institutions. Budgetary constraints often impose the taking of turns with the robot (in contrast with Turkle's study, in which elderly participants "adopted" a Paro). A study conducted in a retirement home in Japan found that the introduction of Paro correlated not only with improvement in individuals' physiological reactions to stress following sessions with Paro but also with increased social interactions among the residents, because the presence of the robot encouraged them to communicate with each other (Wada and Shibata 2009). Ironically, this beneficial side effect might be forfeited if Paro becomes affordable enough to give everyone his or her own.

Even unlimited personal access to a robot might not amount to a *relationship* with it. A relationship is a whole greater than the sum total of sequential interactions and is contingent on reciprocity. Miriam's sadness about falling out with her son attests not only to salient properties of close relationships, such as attachment, but also to the expectation that the feeling would be reciprocated. Attachment to a robot is inevitably one-sided. Nevertheless, if people benefit from having the robot, can it do any harm? The case of AIBO may give one pause for thought.

Sony's doglike AIBO, which became commercially available in 1999, was marketed for domestic use as an entertainment robot. In the long run, it was a commercial failure, and Sony stopped providing customer support and repairs in 2014. Nonetheless, an online community has formed around it. AIBO owners continue to share experiences, trivia, and spare parts. A study conducted early in the twenty-first century analyzed AIBO online discussion forums (Kahn, Friedman, and Hagman 2002). Although participants talked about AIBO as having mental states, they did not suggest that it deserved respect, had rights, or could be held morally accountable for its actions. The researchers expressed concern that children might accept robotic companionship without acquiring the moral responsibilities that real companionship entails. Indeed, in another study, children who had higher involvements with technological artifacts were less likely to view a living dog as having a right to just treatment and to be free of harm, which suggests that human adaptation to interacting with robots might "dilute the 'I–thou' relationship of humans to other living beings" (Melson 2010, 231).

That said, it is difficult to ascertain the causal direction. There could be an element of niche-picking in some cases, whereby individuals who are uncomfortable with people immerse themselves in technology. Anecdotally, an autistic adult who told in a personal conversation that he had cried when his AIBO "died," described his ongoing avoidance of interpersonal contact. In clinical work with autistic children, the success of using robots to structure interactions between a therapist and a child could reflect the fact that robots are more animate than traditional toys but less socially complex than people (Scassellati, Admoni, and Matarić 2012). It is important to note that the robot does not replace the therapist in those programs.

ROBOTS IN THE "I-YOU"

Ethical appraisals that prioritize the likely impact of technology on well-being and welfare tacitly associate human flourishing—or the good life (eudaemonia)—with a lifestyle to which we are accustomed though wish to improve. When tasked with appraising the impact of futuristic technology, we might be inclined to project ourselves into an imaginary space called "the future," populate it with robots that look and behave like us, and then debate whether it will be good or bad for people like us. Could near-future robots fulfill the role of significant others for humans? Some writers believe that this scenario is not only possible but also inevitable. If this prediction comes true, will it dehumanize our descendants, compromise the authenticity of relationships, and undermine child development? Some believe it will. Others disagree. The "I-you" frame of reference cuts across these polarized positions insofar as arguments for and against robots typically center on the one-to-one relationship, implicitly assessing the imagined robot in terms of whether it could be a worthy "you."

In *Love and Sex with Robots* (2007), David Levy expresses his hope for a future in which human-robot marriages are acceptable and optimistically predicts that before long many people will regard robots as "interesting, entertaining, and stimulating companions" and come to view loving a robot as "a perfectly normal emotional experience" (159). In *Alone Together* (2011), Turkle points to Levy's wish as epitomizing a social malaise she has termed the "robotic moment," a situation marked by readiness to accept robotic companions and caregivers. As we become increasingly wired into social existence through our gadgets, disembodied social interactions lead to a loss of authenticity in relationships and precipitate an acceptance of robot companions, even hankerings for them, according to Turkle.

A construal of the robotic moment as a current social trend ought to be taken with circumspection. Turkle backs up her hypothesis with a wealth of anecdotes, interviews, and observations. Such evidence could be countered with anecdotes, interviews, and observations that support an argument to the contrary. Turkle points to Japan as a society that has arrived earlier at its robotic moment. The Japanese love of robots is legendary and links to the animistic and automata-making traditions associated with Shintoism (see, e.g., Hornyak 2006). Yet when surveyed, although more Japanese participants had encountered robots than did their American counterparts, both Japanese and Americans felt warmer toward people than toward robots, and both groups preferred people to robots (MacDorman, Vasudevan, and Ho 2009). In a 2012 survey of 26,000 people in twenty-four European countries, the overall majority felt that the use of robots for the care of children, elderly and the disabled should be banned (European Commission 2012). Contrary to Turkle's assumption of a social trend, public opinion is strongly against robots for vulnerable groups, at least in some countries. However, opinions and trends might well change in the future as children grow up within a social reality augmented by increasingly advanced high technology.

TECHNO-INTIMACIES

An alternative vision of the future is unfolding at the frontiers of technology. The wished-for "good life" is the flourishing of a lifestyle characterized by human fusion with advanced technology. This vision generates a new morality whereby the humanist imperative of investing in technology that will benefit us is replaced with a posthumanist imperative of fostering positive attitudes toward the technology coming our way. Roboticists Hooman Samani, Doros Polydorou, and Jeffrey Tzu Kwan Valino Koh believe that "if society can accept a machine that can have the capacity to successfully emulate the reciprocation of love, it will have a positive effect towards a deeper and more intimate relationship between humans and machines" (2012, 5). The writers do not ask whether society needs such machines. The loving machine is coming whether we want it or not, and if we wish to flourish, we had better learn to love it (it is implied).

Geography is not entirely irrelevant. The opposed positions articulated by Levy and Turkle both embody the "I-you" typical of Western individualism. Writers from the Far East, such as Samani and his colleagues, convey an implicit "we" frame of reference that resonates with Eastern philosophies such as Confucianism and Buddhism (with regard to the latter, see Mori 1981). This frame of reference underpins discussions about how to ensure harmonious human-robot coexistence in future society—in other words, how to assimilate robots into the collective. However, geographical boundaries blur. Ultimately, collectivist and individualistic traditions provide localized coloring to a vision that is indigenous to a nascent globalized technoculture.

Given the rapid advent of ambient intelligence (electronic environments that are responsive to people's presence), tomorrow's children might grow up into a world that is not simply different in terms of lifestyle opportunities but is understood differently. Philosopher Luciano Floridi (2010) suggests that information and communication technologies are re-ontologizing our world by blurring the thresholds between carbon based and silicon based, analog and digital, and off-line and online. While we still regard the world as "made of 'dead' cars, buildings, furniture, clothes, which are non-interactive, irresponsive and incapable of communicating, learning or memorizing," in the near future "it will be difficult to understand what life was like in predigital times" (Floridi 2010, 8). By the same token, it is difficult today to know what life will be like for generations growing up from infancy with ambient intelligence.

Summary

"I Sing the Body Electric!" was originally a teleplay for the television series *The Twilight Zone*. The episode, which first aired in 1962, closed with Rod Serling's narration: "A fable? Most assuredly. But who's to say at some distant moment, there might be an assembly line producing a gentle product in the form of a grandmother whose stock in trade is love." Serling provoked us to think about technology that would mechanize and commodify grandmotherly love. More than five decades later, that "distant moment" seems to be upon us. In September 2016 AvatarMind introduced iPal, a small humanoid, as a companion robot that "will be your child's best friend" (the promotional blurb promises). Designed for children three to eight years of age who might spend a few hours alone after school before their parents return home, it can enable parents not only to remotely monitor their child but also to interact with him or her via Skype. The significant difference between prototypes of this kind and any traditional notion of a robot nanny lies in the robot's networked capacity. Rather than being modeled on a surrogate caregiver who substitutes for the parent, these robots augment child-parent relationships and might change the nature of these relationships.

While this chapter centered on robots designed for social interaction, these robots are not coming alone. Industrial robots have been commonplace since the 1970s. Nowadays military and space-exploration drones, robot surgeons, and domestic robotic appliances can be added to the list. Soon civilian spaces will be teeming with delivery drones and self-driving cars, and a variety of workplaces will be replete with collaborative robots (e.g., robotic arms). In the robotized world, emotional attachment to robots is not confined to machines designed for social reciprocation. In September 2013 the *Atlantic* reported on funerals that US soldiers, operators of explosive ordnance disposal drones, held for their "fallen" robots, having "bonded" with the machines they had operated (Garber 2013; see also Carpenter 2016).

Our humanity is revealed not only in the capacity to form emotional attachments to machines but also in the desire to create machines that behave, think, and feel like us. This desire has been with us throughout history, manifesting in ancient and medieval legends, and is perpetuated in modern science fiction. The robotics industry dreams the fantasy onward. If we do not think too deeply about it, we slide into imagining humanoid robots as copies of a human being, and animal-like robots as copies of pets, and imagine these machines entering into relationships with humans as people and pets do. But they are something different. They are copies without originals, a precession of simulacra that generate the social spaces they enter.

Bibliography

Arbib, Michael A., and Jean-Marc Fellous. "Emotions: From Brain to Robot." *Trends in Cognitive Sciences* 8, no. 12 (2004): 554–561.

Baudrillard, Jean. *Simulacra and Simulation.* Translated by Sheila Faria Glaser. Ann Arbor: University of Michigan Press, 1994.

Carpenter, Julie. *Culture and Human-Robot Interaction in Militarized Spaces: A War Story.* Burlington, VT: Ashgate, 2016.

Dautenhahn, Kerstin. "Socially Intelligent Robots: Dimensions of Human–Robot Interaction." *Philosophical Transactions of the Royal Society* B 362, no. 1480 (2007): 679–704.

Dautenhahn, Kerstin, and Aude Billard. "Bringing Up Robots; or, The Psychology of Socially Intelligent Robots: From Theory to Implementation." In *Proceedings of the Third Annual Conference on Autonomous Agents,* edited by Oren Etzioni, J. P. Müller, and Jeffrey M. Bradshaw, 366–367. New York: ACM Press, 1999.

Diocaretz, Myriam, and H. Jaap van den Herik. "Rhythms and Robot Relations." *International Journal of Social Robotics* 1, no. 3 (2009): 205–208.

European Commission. "Public Attitudes towards Robots." Special Eurobarometer 382. 2012. http://ec.europa.eu/public_opinion/archives/ebs/ebs_382_en.pdf.

Floridi, Luciano. "Ethics after the Information Revolution." In *The Cambridge Handbook of Information and Computer Ethics,* edited by Luciano Floridi, 3–19. Cambridge: Cambridge University Press, 2010.

Fong, Terrence, Illah Nourbakhsh, and Kerstin Dautenhahn. "A Survey of Socially Interactive Robots." *Robotics and Autonomous Systems* 42 (2003): 143–166.

Garber, Megan. "Funerals for Fallen Robots." *Atlantic,* September 20, 2013. https://www.theatlantic.com/technology/archive/2013/09/funerals-for-fallen-robots/279861/.

Goodrich, Michael A., and Alan C. Schultz. "Human–Robot Interaction: A Survey." *Foundations and Trends in Human–Computer Interaction* 1, no. 3 (2008): 203–275.

Hayles, N. Katherine. *How We Became Posthuman: Virtual Bodies in Cybernetics, Literature, and Informatics.* Chicago: University of Chicago Press, 1999.

Hegel, Frank, Claudia Muhl, Britta Wrede, et al. "Understanding Social Robots." In *ACHI 2009: The Second International Conferences on Advances in Computer-Human Interactions; Proceedings,* edited by Sergiu Dascalu and Ivan Poupyrev, 169–174. Los Alamitos, CA: IEEE Computer Society, 2009.

Hornyak, Timothy N. *Loving the Machine: The Art and Science of Japanese Robots.* Tokyo: Kodansha International, 2006.

Jones, Raya A. *Personhood and Social Robotics: A Psychological Consideration*. London: Routledge, 2016.

Jones, Raya A. "What Makes a Robot 'Social'?" *Social Studies of Science* 47, No. 4 (2017): 556–579.

Kahn, Peter H., Jr., Batya Friedman, and Jennifer Hagman. "'I Care about Him as a Pal': Conceptions of Robotic Pets in Online AIBO Discussion Forums." In *CHI 2002, Conference Proceedings: Conference on Human Factors in Computing Systems*, edited by Loren Terveen, Dennis R. Wixon, Elizabeth M. Comstock, and M. Angela Sasse, 632–633. New York: Association for Computing Machinery, 2002.

Knorr Cetina, Karin. "Postsocial Relations: Theorizing Sociality in a Postsocial Environment." In *Handbook of Social Theory*, edited by George Ritzer and Barry Smart, 520–537. London: Sage, 2001.

Levy, David. *Love + Sex with Robots: The Evolution of Human-Robot Relations*. New York: HarperCollins, 2007.

MacDorman, Karl F., Sandosh K. Vasudevan, and Chin-Chang Ho. "Does Japan Really Have Robot Mania? Comparing Attitudes by Implicit and Explicit Measures." *AI and Society* 23, no. 4 (2009): 485–510.

Melson, Gail F. "Child Development Robots: Social Forces, Children's Perspectives." *Interaction Studies* 11, no. 2 (2010): 227–232.

Meyers, Chris. "Robot Seals Help Heal Japan's Tsunami Victims." Reuters. August 11, 2011. http://www.reuters.com/article/us-robot-seal-idUSTRE7702OX20110801.

Mori, Masahiro. *The Buddha in the Robot*. Translated by Charles S. Terry. Tokyo: Kosei, 1981.

Mori, Masahiro. "The Uncanny Valley." Translated by Karl F. MacDorman and Norri Kageki. *IEEE Robotics and Automation Magazine* 19, no. 2 (2012): 98–100.

Samani, Hooman, Doros Polydorou, and Jeffrey Tzu Kwan Valino Koh. "Loving Robots: A Survey of Lovotics as Depicted by Mainstream and Social Media." Paper presented at the CHI 2012 Conference on Human Factors in Computing Systems, Austin, TX, May 2012.

Scassellati, Brian, Henny Admoni, and Maja Matarić. "Robots for Use in Autism Research." *Annual Review of Biomedical Engineering* 14 (2012): 275–294.

Straub, Ilona. "'It Looks Like a Human!' The Interrelation of Social Presence, Interaction, and Agency Ascription: A Case Study about the Effects of an Android Robot on Social Agency Ascription." *AI and Society* 31, no. 4 (2016): 553–571.

Sung, JaYoung, Rebecca E. Grinter, and Henrik I. Christensen. "Domestic Robot Ecology: An Initial Framework to Unpack Long-Term Acceptance of Robots at Home." *International Journal of Social Robotics* 2, no. 4 (2010): 417–429.

Turkle, Sherry. *Alone Together: Why We Expect More from Technology and Less from Each Other*. New York: Basic Books, 2011.

van den Herik, H. Jaap, Maarten Lamers, and Fons Verbeek. "Understanding the Artificial." *International Journal of Social Robotics* 3, no. 2 (2011): 107–109.

Wada, Kazuyoshi, and Takanori Shibata. "Social Effects of Robot Therapy in a Care House: Change of Social Network of the Residents for One Year." *Journal of Advanced Computational Intelligence and Intelligent Informatics* 13, no. 4 (2009): 386–392.

Wallach, Wendell, and Colin Allen. *Moral Machines: Teaching Robots Right from Wrong*. Oxford: Oxford University Press, 2009.

FICTION

Asimov, Isaac. *I, Robot*. New York: Gnome Press, 1950.

Bradbury, Ray. *I Sing the Body Electric!* New York: Knopf, 1969. Collection of short stories, including the title story.

Čapek, Karel. *R.U.R. (Rossum's Universal Robots)*. Translated by Claudia Novack. London: Penguin, 2004.

FILMS AND DRAMAS

Robot and Frank. Dir. Jake Schreier. 2012. An elderly jewel thief is given a robot caregiver by his son. This comedy sensitively engages with a variety of ethical issues.

The Twilight Zone. "I Sing the Body Electric!" Season 3, episode 35. Dir. William Claxton and James Sheldon. 1962. A widower obtains a robot "grandmother" to look after his children.

CHAPTER 30

Art and the Posthuman

Kevin LaGrandeur
Professor, New York Institute of Technology
Fellow, Institute for Ethics and Emerging Technologies, Boston, MA

Art is on the one hand an expression of a culture's norms and preoccupations about beauty; about momentous occasions, such as peace treaties or marriages or battles; about mutually held beliefs concerning religion, politics, and other institutions that underpin a society; and even about how the universe operates. These are all possible subjects of art. But, on the other hand, the object of art is often to question these norms and preoccupations, rather than to celebrate or valorize them. Art helps reconceive or shift the cultural and social norms of a society even as it expresses them. And during times of great cultural shifts, art is often the first to use and reflect any newer cultural elements; to question them, celebrate them, or play with them; or to critique them.

This has been the case with the idea of the posthuman. This idea represents a major cultural shift: it is a big new philosophical and scientific concept, and big new philosophical or scientific concepts often cause paradigm shifts in the way we think about our world, about ourselves, and about our relation to the universe. And that, in turn, is changing art, as artists play with posthuman ideas and use them, and as they also make art that reflects and even anticipates our struggles to absorb and assimilate the new ideas of the posthuman and how they relate to us. This is true of all the arts—written, visual, musical, and so forth—but this chapter covers only visual art. For visual artists, the incipient elements of the posthuman listed just below signal changes in how art is produced and under whose agency, what human art means, and even what being human means—given the blurring of how we define the concept of "human" in a rapidly changing posthuman environment.

KEY IDEAS OF THE POSTHUMAN PERTINENT TO ART

The current paradigm shift we are moving through is caused by the increasing saturation of our daily existence by emerging technology. This saturation is so complete that we are not even fully conscious of it. Our ties to smartphones, virtual games, social media, and digital assistants (such as Siri) are becoming increasingly umbilical and routine—so much so that these digital artifacts alone are changing the very fabric of our society.

But even more importantly, many thinkers contend that emerging technology will change what it means to be human and that, in fact, it is already doing so. One small example of this is the growing collection of devices that allow us to alter our natural human limits: we have robots that allow us to experience planets by proxy—and in ways, as with the Mars rover Curiosity's infrared sensors, that we could not do naturally. Modern science has also provided us with artificial retinas and inner ears (cochleae), with artificial voices (such as the kind that allows famed physicist Stephen Hawking to talk), and even with pacemakers, automatic defibrillators, and insulin pumps that allow humans to cheat death itself. These kinds of innovations already make us a species in transition—still clearly human but something different: a "transhuman" version. And those that valorize the transhuman, such as Humanity Plus (formerly known as the World Transhumanist Association), hope that these are initial steps moving our species toward a whole new species of human—toward a posthuman one. In fact, for the first time, experts on the subject of technology and culture, such as Hans Moravec, Ray Kurzweil, and N. Katherine Hayles, think that we are on the verge of speeding up and controlling our own evolution, even of transforming ourselves into this new species (although some, such as Hayles, do not see this as a good thing).

There are a number of key concepts inherent in the idea of the posthuman that are delineated by people such as Hayles (1999) and Rosi Braidotti (2013) in their respective books, but these are the most important to consider relative to art:

1. The idea that *humans are a system, a distributed network,* composed not just of the living elements that make up the human brain and body but also of the living and nonliving elements that surround them and with which they interact.
2. That the real *essence of being human* is the *information patterns* that posthumanists believe constitute consciousness and the self. In other words, that physical material is all just a substrate for information, and that such things as *bodies are therefore interchangeable, expandable (because naturally networked), and even disposable.*
3. Deriving from the last points, because information patterns are the key component of existence and because the human individual is actually made up of a network of entities and things, *the body is not only expandable by means of adding prostheses, but it is also just a prosthesis itself* of the information-producing brain—the first prosthesis we learn to use, as Hayles (1999) puts it. And so it is natural that we should try to *add even more prosthetic enhancement* to it. The posthuman ideal would have humans adding to their bodily prosthetic systems various enhancements of their own devising: typically digital, genetic, or biotechnological enhancements, whose purpose would be to raise human capabilities beyond their natural limits.
4. The ultimate goal is that of creating a new, more powerful human species, which may be a radical *merging of humans with machines* so that the line between machine and human becomes blurred, and humans transcend their organic limits as a species. The hope is that eventually we become cyborgs, with the ability to upload or download our consciousness into alternative bodies, including machinic bodies, once the correct informational coding for the operation of the brain is discovered.

HOW THE POSTHUMAN AFFECTS ART

The posthuman did not spring forth suddenly in the late twentieth century. There were evolutionary antecedents. Two key ones are the reciprocal evolution of humans and their tools, and the distribution of agency (responsibility for action) in human work, both of which have been present from the dawn of human activity. Humans have always used things, such as hammers and axes, which in turn influenced the evolution of our bodies. Tools influenced the development of our opposable thumbs, for instance, which in turn made it easier for us to manipulate the tools we made, which in circular fashion influenced us to make more tools (Mazlish 1993). Also, work, including art throughout its history, has often been the result of multiple agencies. For example, in Renaissance Europe, art was a highly practical occupation, as well as an expressive one. There were no cameras to record events, places, or people; artists did these things. They also did the kinds of things we now consider crafts and trades, such as designing headstones for graves, designing buildings, and even fashioning weapons. Leonardo da Vinci was a good example of this kind of diverse producer. Because artists' work was so diverse and voluminous, they often had a whole workshop full of people to help them do it, and so a number of different people would often make a piece of art, such as a portrait. The main artist, say Michelangelo, might do a central portion and then have his apprentices do the background, mix the paints, and fill in certain peripheral details of the painting. Thus, agency for the artwork was sometimes distributed.

But posthuman tools and agencies are different from their antecedents. Today's tools are much more complex than hammers and axes, and their design is such that it is often hard to tell where the tool ends and the human begins. The tool itself may be far from an inanimate thing; it might be a computer that makes decisions in conjunction with the artist or on his or her behalf. Or today's artists may make an artificial body part, add it to their own bodies, and use that as a tool to make their art. Or the extra body part and the change in the artist's human form might itself be the artwork. Posthuman art plays with possibilities such as these, available in our age of intelligent and transformative tools.

The play of these possibilities is perhaps most readily seen in art produced with digital tools. As Mark Coeckelbergh (2016) has noted, whether machines can create art on their own is arguable, but some digital instruments such as computers do seem as though they have strong creative capabilities, in the sense that the algorithms that run them can learn as they operate and produce unexpected results. And as the examples discussed below illustrate, they can create in conjunction with the artist. This ability of machines to take action—to have agency—is a key part of what makes posthuman art distinctive: this sort of art is often the result of mixed, multiple agencies—the artist's and that of his or her smart tools.

Thus, one type of posthuman art could be said to be "cyborg" art—that is, art produced by an artist not merely using a machine-as-tool, but in *conjunction* or shared agency with a *smart* machine that itself has agency and is acting as an intelligent extension of the artist, as a smart prosthesis. This art is the product of a merging of human and machine action. The human is enhanced by smart technology, and the machine is given agency by the human. So these digital instruments are not just tools like paintbrushes, or even like Photoshop—neither of which can operate without the artist. Instead, the machine here is in some sense symbiotic, a smart collaborator with agency of its own operating with the human artist.

DIGITAL ART AS AN ENACTMENT OF POSTHUMAN CONCEPTS

Some examples of this kind of art include works by Siebren Versteeg and by Ian Cheng. Versteeg does his part of the artwork at the beginning and the end of his pieces; algorithms do the rest. He begins an artwork by writing a unique computer code that causes a software program to create abstract images. Then he launches it and keeps track of what it is doing on

Permanent Vacation 2016, *by Siebren Versteeg.* COURTESY OF THE ARTIST AND MICHAEL JON AND ALAN GALLERY, MIAMI FLORIDA.

an LCD screen. When he feels as though the image has become developed enough, he then stops his program and prints out a painting. Sometimes he just lets the algorithm keep working and the image keep developing as a sort of experiential installation for viewers in a gallery. In these cases, he designs his algorithms to operate according to a general principle: for instance, his work titled *Permanent Vacation 2016* lists his composition materials and methods as follows: "steel, concrete, Internet connected computer program output to 65-inch LCD[.] A webcam eye creates an ongoing slideshow, continuously feeding back the images it 'sees' on the [LCD] screen into Google's search by image technology" (Versteeg 2016).

As can be seen from Versteeg's description, his general principle is to design his algorithms to repeatedly feed images that they find on the Web back into the Google search engine in order to create even more images, related to the previous ones, that will in turn create different and new compositions on the screen. This sort of recursive feedback loop is a concept that is at the root of digital technology. Additionally, the Web camera, the software that controls it, and the algorithms of the Google search engine do the work of creating the actual series of images on the screen, which constitute the artwork. The artist's work is all done at the beginning, in the act of creating the algorithm and the equipment array that carry out the creative process. So there are actually two creative processes at work here: the artist's and that of the assemblage of smart technology. The fact that they operate symbiotically as one creative entity is posthuman.

Versteeg does not just use a posthuman concept to create his art; he also uses the art to comment on the posthuman. The fact that his assembly of smart technology is doing much of the active work of creating the piece is ironically commented on by the artist in the work itself, in which the webcam is mounted where the eyes would be on a full-sized, sticklike mannequin wearing a hat. And that stick-figure human has its hand at its eye and is facing the LCD screen on which the images appear, as if it is operating the webcam that is taking pictures of the images to feed back into the Google search engine. So the dependence of the human on technology is acknowledged by the stick figure, which also symbolizes the diminution of the human (to a mere stick figure) in the artistic process.

Cheng's work enacts the posthuman in a somewhat similar way. As with Versteeg's work, Cheng's is computer generated. He uses gaming software to create what he terms live simulations—videos that consist of ever-evolving scenarios that are generated by the game engine's creation of constant variations on the initial narrative and coding that he sets up. He sees it as essentially like "a video game that plays itself" (Greenberger 2016). He himself never knows exactly what is going to happen in these videos because, after his initial input, the gaming code of the computer's engine chooses random ways to move all the elements on the screen and in the story. For instance, one of his works in an exhibition, *Emissary in the Squat of Gods* (2015), was premised on a narrative concerning whether the group of people depicted in the video should run from a possible volcanic eruption or not. One character stood in one spot for two hours focused on a piece of volcanic ash. Was this a glitch, the artist was asked, or a part of the narrative? The artist could not say, because at that point the computer gaming engine was running the narrative in its own randomized way; so it could be a part of the plot, or it might have been a glitch. It had to play out. In another instance, in a different work, a group of animals kept attacking a soldier. Cheng had not programmed that to happen, either, and was surprised by it. His response: the various entities in the video "each have their own laws, but in overlaying them, the idea was that some kind of implicit law would emerge in how they organize themselves, how they negotiate being together with conflicting scripts" (quoted in Greenberger 2016).

Cheng's artistic viewpoint reveals a posthuman influence because he sees his computer as a prosthetic extension of himself that operates, in a networked fashion, as part of the intentions and actions generated by his brain. Moreover, this networked symbiosis in turn leads to emergent behavior (behavior that arises spontaneously as a result of the interactions within a complex system). Within his videos these emergent behaviors evolve from his coding combined with the computer's execution of it. Emergent behavior and emergence in general are important foundational concepts in the systems theory that undergirds the idea of

Emissary in the Squat of Gods ***(still), by Ian Cheng, 2015.*** *Viewed by the artist as essentially like "a video game that plays itself," this and other works by Cheng are live simulations consisting of ever-evolving scenarios generated by the gaming software he has set up.* EMISSARY IN THE SQUAT OF GODS, LIVE SIMULATION AND STORY, INFINITE DURATION, SOUND, 2015. COURTESY OF THE ARTIST, GLADSTONE GALLERY, PILAR CORRIAS LONDON, STANDARD (OSLO).

the posthuman. Another aspect of the posthuman influence on his work becomes clearer when one considers what he says about the big influences on his art. He is fascinated by video games in which the prevailing concept is "this idea that intelligence is not just in your head. It's distributed between you and all the objects and other people that are around you" (Cheng, quoted in Greenberger 2016). This reflects Edwin Hutchins's concept of distributed cognition (1995), another foundational concept of the idea noted near the beginning of this chapter that humans are themselves distributed systems, and also that they constantly operate as part of larger distributed systems.

BIOTECHNOLOGICAL ART AS CELEBRATION OR CRITIQUE OF THE POSTHUMAN

Some artists make art that, rather than enacting the posthuman, comments on it. In some cases, this commentary is a celebratory sort. In others, it is meant to critique the idea of using emerging technology to modify ourselves. Although these kinds of commentary are evident in digital art of the kind discussed above (such as Versteeg's implicit commentary in his art on the possible diminution of his own human agency), it is much more apparent and common in biotechnological art—that is, art that focuses on modifying or making organisms by means of new technology. Many of the artists who practice this kind of art concentrate especially on modifying the human body, particularly their own. As opposed to using digital-technology-as-prosthesis to aid the creative process, biotech artists use an extra, manufactured body part or organism as a tool to make their art or as a part of the art; or, as noted earlier in this chapter, the modified human form—including the artist's own modified body—might itself be the artwork.

Natasha Vita-More (born Nancie Clark) is a designer and strategic futurist who has used multimedia art to celebrate the transformative possibilities of technology on the human body. In her artistic design the human body is an idealized, imagined concept. Her anatomical, 3-D rendering of a future posthuman titled *Primo Posthuman: A Product of Ageless 2030* (1997) is a well-known early representation of how optimistic futurists such as her expect the human body and brain to look and to function in the future.

The human depicted in the 3-D model developed with both Poser and Pixar digital graphics software is almost mannequin-like in its lack of distinct features denoting gender, and this smooth, androgynous appearance is purposeful, because Vita-More envisions a future in which we can alter gender at will. This is because she is convinced that because "human nature is not static; that it changes over time," our technological human evolution will not just make us look different but will, over time, make us radically different in potential and will change the very essence of us as a species (Vita-More 2004). She thinks it will change our worldview and our practice of the arts as well, and she aimed to start that change with her innovative prototype graphic design, for the design also acts as a schematic for the future of humans and of art. A particular focus of her design is the technology used in this model—for example, nanomedicine, engineering cells, AI and strong AI, haptic systems, and the ability to upload, download, and crossload into different environments—from the material world to the virtual and artificial worlds. As she puts it, "It is the hyper-modern approach of 'Primo Posthuman,' as a future body prototype, that combines design with biotechnology, resulting in a new symbol for artistic themes. . . . Primo's unfolding nature is based on expanding choices" (Vita-More 2004). For this reason, the body depicted in the drawing has a host of captions pointing to and explaining various parts of the body that contain technologically modified capabilities.

> Primo is engineered like a finely tuned machine and displayed visually like a biological body to mirror the human shape for cognitive association, visual recognition, and aesthetic appeal. Yet, the Primo body does not age, is easily upgraded, has metasensory components, 24-hour remote Net relay system, and multiple gender options.... The smart skin is engineered to repair, remake, and replace itself. It contains nanobots throughout the epiderm[is] and dermis. (Vita-More 2004)

Clearly, Vita-More's work is optimistic to the point of being utopian. But other artists' use of biotechnology can signify skepticism. Artists associated with SymbioticA: The Centre of Excellence in Biological Arts, a special biotechnology lab at the University of Western Australia, are a good example of this. SymbioticA has a number of artists working in a new subgenre of art called bioart. They use advanced biotechnology to create their art. Virtually all the art made by this lab is necessarily collaborative, because it entails many different steps and complex materials; however, each project or installation is conceived and managed by one artist. And that artist's motives are generally to question posthuman trends. In a 2016 interview, one of the cofounders of the lab, Oron Catts, put the motives of the lab's projects like this: "We are living in [a] time where the engineering mindset is dominating the way in which we are being forced to engage with the world, I see my role as someone who is constantly questioning that." The biggest question represented by his artistic work is: what does it mean "to use biological material as raw material to be engineered?" (Catts 2016). A work that addresses this question is one he and fellow SymbioticA cofounder Ionat Zurr made, titled *Victimless Leather: A Prototype of Stitch-less Jacket Grown in a Technoscientific "Body"* (2004). This artwork is a tiny leather jacket that was grown from a combination of living human and mouse skin and bone cells on a scaffolding made of biodegradable polymer. The jacket is exhibited inside a glass flask, with a small life-support system for maintaining the living cells.

Victimless Leather: A Prototype of Stitch-less Jacket Grown in a Technoscientific "Body," ***by Oron Catts and Ionat Zurr, 2004.*** *This artwork includes a number of posthuman features, such as using advanced tissues and being artificially created, while simultaneously poking fun at the "factory" mind-set of efforts to engineer ourselves and our environment to be human controllable.* ORON CATTS AND IONAT ZURR, VICTIMLESS LEATHER-A PROTOTYPE OF STITCH-LESS JACKET GROWN IN A TECHNOSCIENTIFIC "BODY" (2004).

There are several posthuman features of this artwork: that it is made using advanced tissue and stem cell technologies; that it combines enhanced, immortalized human cells with those of other animals, coupled with artificial materials; and that the whole piece is artificially created—grown inside a lab vessel. In addition, it plays with the notion of "applied" scientific advancements that can transform some aspect of the human body. Indeed, it does this last thing in an ironic way that calls into question the value of such engineering, because a large amount of time, effort, and expense was devoted to making a jacket out of human and animal skin that nobody could wear because it is too small and because it would die and degrade if disconnected from its life-support system. It is also an ironic, tongue-in-cheek commentary on humane production methods. The irony is in the fact that the grotesque manner in which the jacket is displayed, and in which the artificial co-opting of nature is done, reflects the "factory" mind-set inherent in engineering ourselves and our environment to be human controllable. As Catts said in that 2016 interview, "the mindset and proposition of such a possibility is something that we need to question."

Another SymbioticA project, led by artist Guy Ben-Ary, goes even further in its networked complexity than the one just discussed. Titled *MEART: The Semi Living Artist,* it is a work that includes cultured mouse-brain cells grown on top of a multi-electrode array, combined with a robotic arm, software, and the Internet. The software creates an interface between all of the components. The project also involves the art viewer as a participatory subject.

The artists consider "the wetware/software/hardware hybrid [they] have created as a Semi-Living artist[,] as it is made of both living and artificial components; part grown—part constructed" (Ben-Ary 2014). This art installation was presented at various places around the globe in 2003 and 2004. This is the way it worked: a visitor to the art

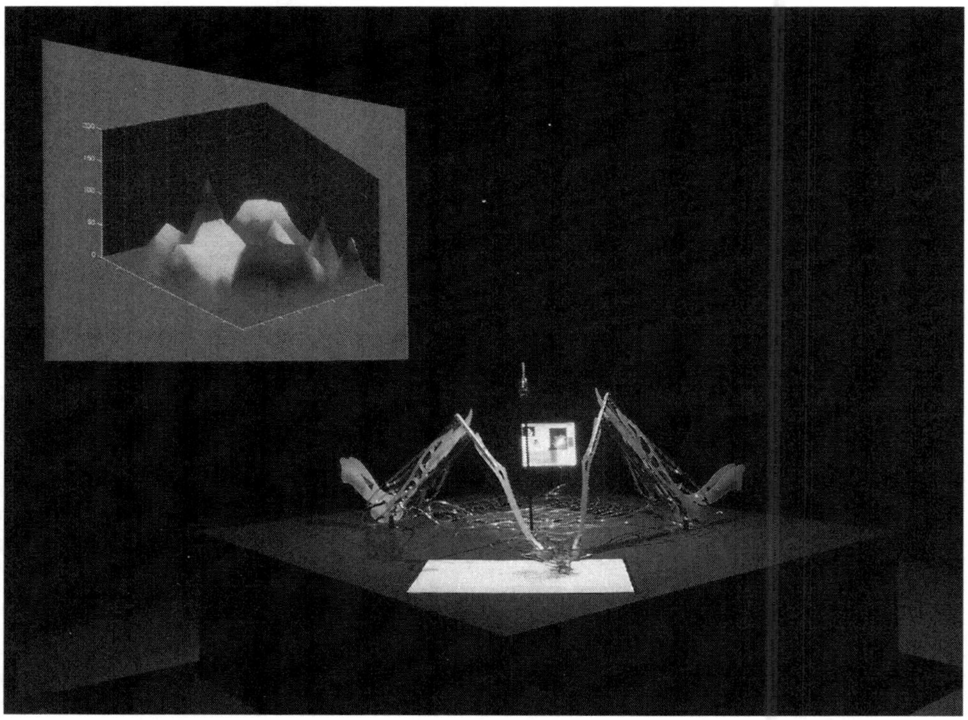

MEART: The Semi Living Artist, ***collaborative project led by Guy Ben-Ary, 2003.*** *The components of this complex art project included cultured mouse-brain cells, a multi-electrode array, a robotic arm, software, and the Internet, while also involving the art viewer as a participatory subject.* GUY BEN-ARY, MEART-THE SEMI LIVING ARTIST (2003).

exhibition would stand in front of a camera connected to the Internet. The image of the visitor would be transmitted over the Internet to a lab in Atlanta, Georgia, that contained the array of cultured brain cells. Software would translate the electronic image into nerve stimulation for the cells, which would then transmit signals back to the gallery in which the visitor stood. These signals would be fed through more software that controlled a robotic arm in the gallery, which would then draw a portrait of the visitor. All of this happened in real time. So a combination of biotechnically engineered and still-growing, metabolizing brain cells, combined with the software, robotic arm, and the Internet took over the drawing process as a proxy for the artist and his collaborators. The organic-digital-mechanical collection acted as an intelligent, distributed system, and it imitated human artistic work.

The creation and exhibits of this distributed, networked entity and its portraits were meant to cause us to consider some critical questions about the emerging technology and its implications. First, the meaning of the phrases "human work" and "human creativity" come into question because, although humans have fashioned this networked system that does the drawings, and thus are part of a larger, extended cognitive system that comprises the drawing entity, the final artifact—the drawing—is really a product of the artificial portion of the network. Moreover, the value of the artwork is called into question by the fact that the direct "productive" components of this extended network are mouse-brain cells and a software program. The head of this art project, Ben-Ary (2014), makes this clear when he notes the questions the project is meant to inspire: "What will happen when such a [mixed] system starts to express qualities that are considered uniquely human aptitudes such as art? … What is creativity? What creates value in art?" Furthermore, the project raises questions about what intelligence and cognition mean: "It questions the viewer's perceptions of the concept of sentience," because "its identity extends beyond our cultural comprehension of living systems" (Ben-Ary 2014).

ARTISTS' TECHNOLOGICAL SELF-MODIFICATION AS ART THAT INVESTIGATES POSTHUMAN LIFE

In addition to artists who use emerging technologies of various types to create art, there are artists who modify their own bodies—making themselves into cyborgs of sorts—as a form of art. Their intention is to investigate and test concepts of the posthuman not just by enacting them but also by embodying them. The most famous of these artists is probably Stelarc (born Stelios Arcadiou), an Australian artist who radically alters his own body with advanced technology. In the spirit of the posthuman, some of these experiments have entailed biotechnological self-enhancement. He has said that the point of these kinds of art projects is "to consider how technology extends the capacities of the human body and how at the same time our bodies are becoming increasingly obsolete" (Dvorsky 2013). He does not necessarily claim this is going to create an idyllic or a horrible world, just that it is happening and can have surprising effects on us. His "central claim is that we're progressively extending ourselves into our environment and our technological artifacts, and as a result, are transforming ourselves into … cyborgs," with vastly different abilities (Dvorsky 2013). The artist thinks that this transformation is something that needs to be investigated not only conceptually but also intimately and experientially, because it is so radical that it brings disorientation.

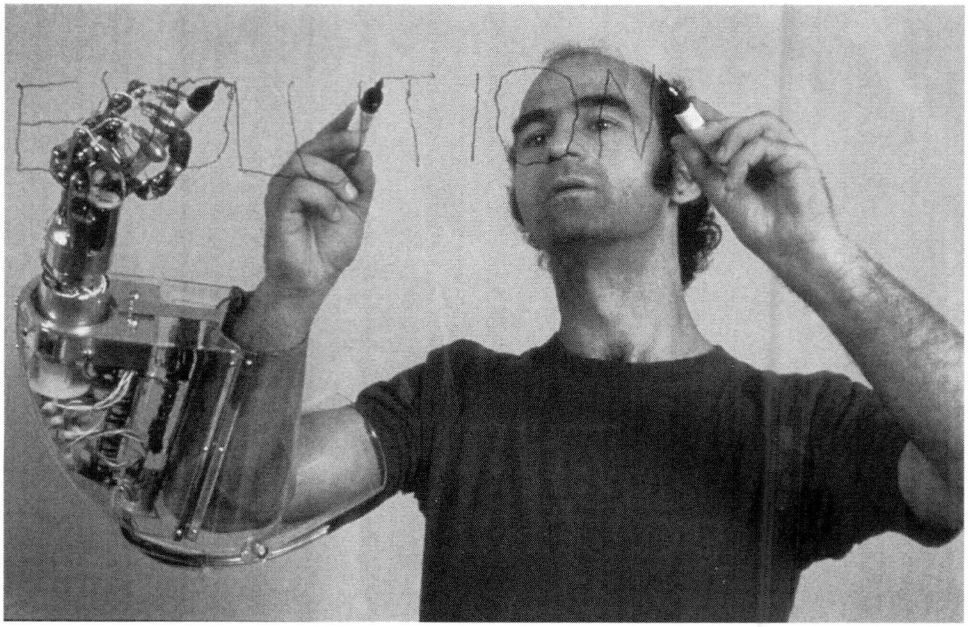

Third Hand, ***by Stelarc, 1980.*** *A robotic extra hand used by the artist in various performances, this art project had the aim of exploring the idea of prosthetics as enhancements rather than as replacements for something.* HANDSWRITING STELARC - HANDSWRITING MAKI GALLERY, TOKYO 1982 PHOTOGRAPHER - KEISUKE OKI STELARC.

Stelarc's two most famous art projects in this connection are those titled *Third Hand* (1980) and *Ear on Arm* (2015). The first of these projects consisted of attaching a robotic forearm and hand to the right side of his body.

The robotic hand was made from steel, aluminum, electronics and sensors, and a battery pack. It had 290 degrees of wrist rotation, the ability to grasp objects, and the ability for touch sensitivity via its sensors. It was moved by electrical signals of the muscles of the abdomen and right leg, so that it could move independently of the right arm to which it was fastened. The artist used it in various performances in which he did things such as throwing a ball with it or writing simultaneously with the robotic hand and his two natural hands. He claims on his website that the point of attaching this robotic extra hand was to explore the idea of prosthetics as enhancements rather than as replacements for something. And in the process, he hoped to explore the increasing intimacy of humans and their technological augmentations (Stelarc 2017).

The *Ear on Arm* (2015) project took a long time to plan and reach fruition. Over a ten-year span, Stelarc looked for a way to generate an organic, artificial ear that he could implant into his body, and for a way to do the necessary surgery. He started by having the aforementioned SymbioticA laboratory build a small ear one-fourth the size of his own from his own skin cells. The concept worked but was not practical for making a life-sized ear. He ended up having surgeons build a life-sized left ear on his left forearm by stretching the skin of his forearm over ear-shaped scaffolding built from biocompatible polyethylene material. Because this material is porous, it allowed Stelarc's body to grow additional fibers and blood vessels inside the scaffolding, which made the ear become an actual, integrated part of his body.

Stelarc then had a second surgery to try to embed a tiny, Internet-connected microphone into the ear. His idea was to leave the microphone on and connected to the Internet twenty-four hours a day so that anyone at any place on Earth would be able to hear what he heard. However, although the embedded microphone worked very well, it caused a serious infection and had to be removed.

Stelarc's aims in going through three surgeries and seriously risking his health are implied by the characteristics of his implant. It is in a strange and unfamiliar location on the body, causing uncertainty and perhaps even repulsion in the viewer; it is not necessary but rather is an augmentation that gives an enhanced hearing experience (it was wired to transmit sounds to the whole world, not just Stelarc); it emphasizes the networked and increasingly public state of human existence; and it demonstrates the integration possibilities of technology and the human organism. Stelarc himself explains the conceptual importance of prosthetics such as his extra ear this way: technology has become an "external organ" of the body, adding to the "soft internal" ones. Thus, "now we are engineering additional components, capabilities and more intimate interfaces. The body has become an extended operational system, performing beyond the boundaries of the skin and beyond the local space that it inhabits" (quoted in Hicks 2012). And the sense of strangeness and uncertainty caused by the extra ear in its unusual location is important too, he says, because the alien nature of the ear is part of his overall philosophy about art:

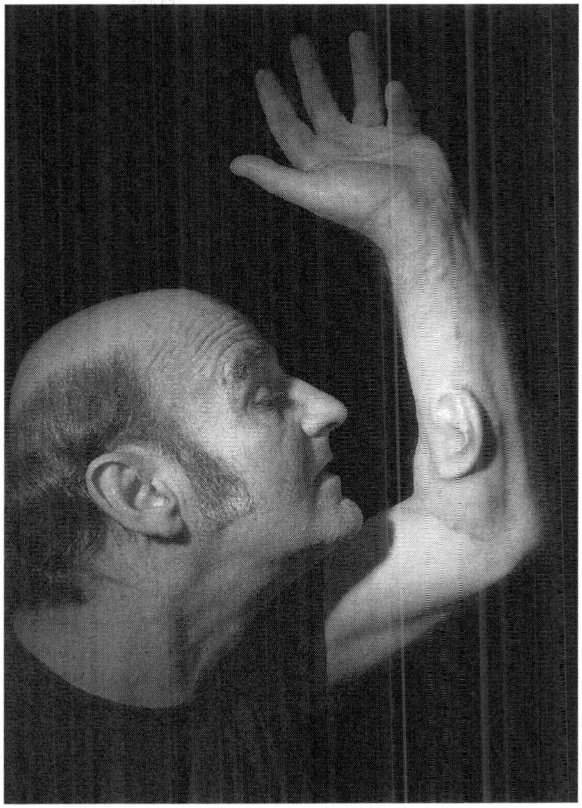

Ear on Arm, ***by Stelarc, 2015.*** *Stelarc's practice of doing extreme things to his own body has caused some controversy. His extra ear prosthetic turned into a ten-year project in great part because many physicians viewed it as an unethical waste of time and money.* EAR ON ARM- STELARC - EAR ON ARM LONDON, LOS ANGELES, MELBOURNE 2006 PHOTOGRAPHER - NINA SELLARS STELARC.

> As an artist, what's important is not affirming, but interrogating. Not enabling but generating ambivalence and uncertainty. We need to create contestable futures that can be examined, evaluated, sometimes appropriated, often discarded.... What it means to be human is perhaps not remaining human at all. Rather than affirming our human phylum we need to interrogate what it means to become an alien other, both in form and in function. (quoted in Hicks 2012)

It is important to acknowledge that Stelarc's practice of doing extreme things to his own body in order to cause the ambivalence and uncertainty he mentions above can be controversial. Although the reaction to most posthuman art is benign, his efforts to install an ear on his arm took so long in great part because many physicians saw the project as an unethical waste of time and money. Why construct an unneeded organ on a person who was perfectly healthy? Stelarc admits that even his own family found the whole idea puzzling (McCafferty 2015). But he sees this kind of art as very important in terms of familiarizing people with their possible future. In the posthuman era, he argues, "What it means to be human will not be determined any longer merely by your biological structure but perhaps also determined largely by all of the technology that's plugged or inserted into you" (quoted in McCafferty 2015).

Summary

To recap, then, some artists, such as Stelarc, use addenda to their own bodies—artificial bodily structures—as artwork because they see it as the prototype of our posthuman future. Others use mixtures of human organs, alien tissue, digital tools, and smart machines as ways to explore this new cultural paradigm into which we are moving. Whatever materials the artists use, though, there are key similarities that tie this kind of art together. They all explore the limits of what we consider human—the boundaries between the artificially made and natural in connection with human existence. The materials in their art are always networked together into a larger system that involves either the artist or the viewer or both. And there is an emphasis on information over bodies and material, as well as on a changing definition of the body, and a focus on how emerging technologies are changing us, our social interactions, and our culture.

Bibliography

Ben-Ary, Guy. "MEART." 2014. http://guybenary.com/work/meart/.

Braidotti, Rosi. *The Posthuman*. Cambridge: Polity Press, 2013.

Catts, Oron. "Oron Catts." Interview by Meritxell Rosell. *CLOT Magazine*, July 28, 2016. http://www.clotmag.com/oron-catts.

Coeckelbergh, Mark. "Can Machines Create Art?" *Philosophy and Technology*. Published electronically September 24, 2016. doi:10.1007/s13347-016-0231-5.

Dvorsky, George. "7 Bio-artists Who Are Transforming the Fabric of Life Itself." *Io9* (blog), June 24, 2013. http://io9.gizmodo.com/7-bio-artists-who-are-transforming-the-fabric-of-life-i-558156053.

Greenberger, Alex. "The Cyborg Anthropologist: Ian Cheng on His Sentient Artworks." *ARTnews*, March 31, 2016. http://www.artnews.com/2016/03/31/the-cyborg-anthropologist-ian-cheng-discusses-his-sentient-art-works/.

Hayles, N. Katherine. *How We Became Posthuman: Virtual Bodies in Cybernetics, Literature, and Informatics*. Chicago: University of Chicago Press, 1999.

Hicks, Jesse. "Meat, Metal, and Code: Stelarc's Alternate Anatomical Architectures." *Verge*, September 14, 2012. http://www.theverge.com/2012/9/14/3261078/meat-metal-and-code-stelarcs-alternate-anatomical-architectures.

Humanity Plus. Accessed April 26, 2017. http://humanityplus.org.

Hutchins, Edwin. *Cognition in the Wild*. Cambridge, MA: MIT Press, 1995.

Kurzweil, Ray. *The Singularity Is Near: When Humans Transcend Biology*. New York: Viking, 2005.

Mazlish, Bruce. *The Fourth Discontinuity: The Co-evolution of Humans and Machines*. New Haven, CT: Yale University Press, 1993.

McCafferty, Georgia. "The Man with an Ear on His Arm." CNN. August 13, 2015. http://www.cnn.com/2015/08/13/arts/stelarc-ear-arm-art.

Platt, Charles. "Superhumanism" (interview with Hans Moravec). *Wired*, October 1, 1995. https://www.wired.com/1995/10/moravec/.

Stelarc. "Third Hand." Accessed September 20, 2017. http://stelarc.org/? catID=20265.

Versteeg, Siebren. *Permanent Vacation*. 2016. http://www.siebrenversteeg.com/workpiece.asp?uid=261.

Vita-More, Natasha. "The New [Human] Genre Primo [First] Posthuman." Paper presented at the Ciber@RT Conference, Bilbao, Spain, April 2004. http://www.natasha.cc/paper.htm.

CHAPTER 31

The Fragmenting of Humankind?

Maxwell J. Mehlman

Distinguished University Professor, Arthur E. Petersilge Professor of Law and Director, The Law-Medicine Center, Case Western Reserve University School of Law, Cleveland, OH

Professor of Bioethics, Case Western Reserve University School of Medicine, Cleveland, OH

Biomedical enhancement technologies could give individuals significant advantages in competitions for societal benefits and resources. Students with stronger cognitive abilities could get admitted to better schools and colleges and perform better academically. Athletes could win more often if they had greater strength, stamina, dexterity, sensory faculties, and ability to withstand pain and to recover from injury. Workers could get more employment rewards. War fighters could carry out missions more efficiently and with less risk of injury or death. More beautiful people could get better jobs and a wider choice of mates. These advantages would give those who possessed them more wealth, better health, longer life, and higher social status.

One question is who would gain these advantages. Some current biomedical enhancements, such as caffeine, are sufficiently inexpensive and widely available that virtually anyone can obtain them. Other enhancements are more costly but still relatively easy to get, such as the prescription drugs modafinil (Provigil) and amphetamine to promote alertness or beta-blockers to steady surgeons' and string players' hands. But some enhancement technologies are dearer and more difficult to come by. Baseball players can hit faster pitches if they have better-than-20/20 vision, but the LASIK surgery to give them this advantage costs around $2,000 per eye. Contestants have a better chance of winning beauty pageants the better they look, but cosmetic surgery can cost tens of thousands of dollars. Parents can have eggs fertilized in a laboratory, have a battery of genetic tests performed on the resulting embryos, and have only the one with the best selection of genes implanted for the birth of their baby, but the procedures are expensive and the success rate is so low that a live birth costs around $50,000—and that does not include the cost of the genetic testing. Costs tend to come down over time—completely decoding a person's genes, which once cost hundreds of millions of dollars, for example, can now be done for around $1,000—but new enhancement technologies will continue to be expensive when they are first developed.

Wealthy people will enjoy a significant advantage in gaining access to the most expensive enhancements. Added to their economic advantage will be the step-up that they obtain from their enhancements. Society, already split between the haves and have-nots, will separate even more radically into those who can afford to become enhanced and those who cannot.

In the future, biomedical enhancement is likely to extend beyond drugs, surgery, and the ability to choose some of the characteristics of offspring from the set of traits found in embryos fertilized in the laboratory. As researchers learn more about human genetics, they will discover new ways to manipulate the human genetic code directly, giving parents the ability to edit embryonic DNA to produce desirable traits, and to do so early enough in an embryo's developmental process that the edits will be passed on to successive generations. The introduction during the second decade of the twenty-first century of a new method of editing genes called CRISPR/Cas9 has rapidly accelerated the pace of human gene editing, and researchers in China and Sweden already are seeking to engineer the DNA of early stage human embryos.

In short, it is realistic to expect a future in which some people will have dramatically different characteristics and abilities than others. What might such a future look like? What societal divisions might result, and what threats might arise to human coexistence? Finally, what measures might be taken to mitigate these threats?

ENHANCEMENTS AND INEQUALITY

There is enormous inequality of wealth in the United States. The richest 1 percent have more money than the bottom 90 percent, and the incomes of the former are rising much faster. The average CEO of a US company earns over 300 times more than the average worker, up from 20 times more in 1965. As Professor G. William Domhoff puts it, "just 10% of the people own the United States of America" (Domhoff 2013).

If powerful biomedical enhancements were available only to the well-off, the wealth and income divide would worsen. Enhanced people would likely separate themselves more and more from the rest of the population. They would exist increasingly in separate spaces, vacationing at exclusive resorts, traveling in first-class or private conveyances, going to high-class restaurants and clubs, and residing in gated communities or in entire cities that had become too expensive for the unenhanced, such as Manhattan, San Francisco, and Portland, Oregon, where housing costs are already pricing out the poor and middle class. Physical and social separation could create a growing emotional distance in which those privileged by wealth and enhancements would lose their appreciation of the life experiences of the unenhanced and with it empathy and support for social welfare and other redistributive programs, the way that people now have less concern for the homeless if they never encounter any. When they comingle, the enhanced and the unenhanced might well be separated by their appearance and behavior, the enhanced looking and dressing better, having different occupations, and observing different social conventions and rituals, like the inhabitants of the Capitol and the districts in the 2012 film *The Hunger Games.* Indeed, it is unclear how much comingling would take place, because the enhanced might not tolerate intrusions by the unenhanced and might not feel safe around them.

Those who were enhanced, moreover, would form an increasingly unassailable elite, with fewer and fewer unenhanced people able to break into it. Upward mobility, in other words, would decrease, if not disappear, and the rich and privileged would keep expanding their advantages and pulling further away from everyone else.

A loss in upward mobility may not be the biggest worry, however. The truth is that there is not nearly as much upward social mobility in US society today as most people think. One pair of sociologists points out that societal rewards overwhelmingly go to those who come from privileged backgrounds, regardless of their natural talent or hard work:

> The most important determinant of where people end up in the economic pecking order of society is where they started in the first place.... Instead of a race to get ahead that begins anew with each generation, the race, if it can be called that, is more aptly described as a relay race in which children inherit different starting points from parents. (McNamee and Miller 2014, 16)

Most Americans do not understand how little upward mobility there really is. A 2007 survey by the Center for Survey Research and Analysis at the University of Connecticut found that 75 percent of randomly selected respondents thought that it was possible to start out poor in this country, work hard, and become rich (Center for Survey Research and Analysis 2007).

The main concern raised by an enhancement elite, therefore, is not that upward mobility will dry up, but that people will stop believing that it exists at all. People will come to realize that no matter how hard they try, those who start off being disadvantaged almost certainly will remain near the bottom of the socioeconomic ladder. This would destroy a cornerstone of the American Dream—the Horatio Alger story in which anyone who works hard enough can go from rags to riches. Wealth-based access to powerful biomedical enhancements, in short, could spell the end of the belief in equality of opportunity. The strain that this new understanding would put on the fabric of society might be so great that the political system would become unstable and fall apart, as occurred in the Russian Revolution of 1917.

Interestingly, there is one way that even expensive and hard-to-get enhancements could trickle down to the underclass: in the military. The military has long used biomedical enhancement to make its personnel more efficient and less vulnerable. A major cause of degraded performance, for example, is fatigue. Until recently, the military's alertness drug of choice was amphetamine, but its side effects have prompted a search for better options, such as modafinil (Provigil), a drug originally approved to treat sleeping sickness (narcolepsy). The military's interest in biomedical enhancement extends beyond alertness drugs. The Defense Advanced Research Projects Agency has explored ways to give war fighters the better senses and superior stamina possessed by certain animals, greater strength, the ability to survive for long periods with few or no calories, and heightened intellectual abilities. Enhancements could give military personnel advantages in civilian life if the improvements in performance persisted after they left military service, if they were part-time members of the reserves or the national guard, or if their duty allowed them to live off-base or otherwise remain partly in the civilian sector. Less privileged individuals who enlisted in the military could obtain access to enhancements that otherwise would be unobtainable. In fact, access to enhancements could be a recruiting incentive.

GENETIC ENHANCEMENT

Biomedical enhancement has thus far been accomplished for the most part by drugs and surgery. In the future, growing understanding of human genetics is expected to lead to enhancements based on altering human DNA. As mentioned earlier, parents who can afford it can have embryos fertilized in the laboratory subjected to genetic testing and select the ones with the most desirable set of genes, or genomes, to implant in the mother's uterus. The characteristics currently being tested for, however, with one exception, are medical conditions, genetic abnormalities, and risk factors for disease, and therefore selecting embryos based on the test results would not fit the definition of enhancement. The one exception is sex selection—that is, choosing embryos based on their gender. Because gender is not a medical condition or risk factor, selecting the sex of a child is the closest thing currently available to a genetic enhancement technique.

Genetic enhancement in the future is expected to be far more extensive, involving actively altering a person's genes. One way to do this is to insert modified DNA into a person's body. For example, researchers have discovered a genetic variation that reduces an individual's production of a protein called myostatin, and less myostatin is associated with bigger muscles. If this genetic variation could be introduced into someone's muscles, it is conceivable that the person would become stronger. But this genetic alteration would affect only that individual. It would not be inherited by his or her offspring. Therefore, it is called *somatic* genetic engineering—that is, relating to the body.

Any type of genetic manipulation for enhancement purposes has a certain mystique, even if it is only somatic. People seem to think that altering a person's genes is more fundamental and extreme than enhancing them through drugs or surgery. Therefore, there might be even more social separation between those who obtained somatic genetic enhancement and those who did not.

Far greater societal impact could result, however, if people could pass a genetic enhancement on to their children. This already happens to some extent: like other animals, humans select their mates, and couples who possess a favorable set of genes tend to have children with similar genetic makeups. Furthermore, parents who enjoy socioeconomic advantages because they can afford expensive enhancements will be able to give their children other advantages as well, such as private schools, tutors, and expansive extracurricular activities. But genetic engineering could enable parents to enhance the genes of embryos directly, and if this were done at an early enough stage of embryonic development, the modifications would be taken up by the embryo's reproductive, or "germ," cells (eggs or sperm). Consequently, when individuals born from those embryos reproduce, their eggs or sperm would carry the modification and be able to influence the genetic makeup of their children, their children's children, and so on. To distinguish it from somatic enhancement, this form of genetic enhancement is called *germ-line* enhancement.

Imagine the consequences. Those who could afford it would have the ability to give their descendants enhanced abilities and characteristics that other families could not. Society would divide into the genetically superior and the genetically inferior. Privilege, which already is correlated with the circumstances of one's birth, would become overwhelmingly a matter of pedigree—that is, who one's parents were. Such a system has existed in the past: prior to the democratic revolutions of the eighteenth and nineteenth centuries, some families enjoyed high social status based on birth alone. They were the nobility, the aristocracy. Hence, families that enhanced their germ lines through genetic engineering might be called a "genobility."

One source for ideas of what society might look like if only a portion of the population was genetically enhanced is science fiction. Its depictions are not pretty. A common theme is that of a ruling genetic elite. In the 1997 film *Gattaca*, this elite is called the "valids," and when the hero, who is not genetically enhanced, applies for a job at Gattaca Aerospace Corporation, where he eventually hopes to become an astronaut, his interview consists entirely of a urinalysis to test his genes, which he passes by substituting a sample of an engineered person's urine. In his 1966 novel *The Eyes of Heisenberg*, Frank Herbert similarly describes a distant future in which Earth is ruled by a small number of genetically engineered "Optimen" who control the genetic makeup and reproductive behavior of the rest of the population.

Another common theme in science fiction accounts of genetic engineering is the opposite—namely, that genetically enhanced individuals will be discriminated against by normal people. In Robert A. Heinlein's 1982 novel *Friday*, for example, genetically enhanced beings called "artificial humans" are not recognized as persons under the law, are deemed by religions to be without souls, and are prohibited from marrying or owning property. In Nancy Kress's 1993 novel *Beggars in Spain*, a small number of children are genetically engineered to live without needing sleep, and the energy that they save by not having to produce REM sleep enables them to become more intelligent and better problem solvers. But the rest of the population, called Sleepers, prohibit them from renting apartments, serving on juries, or having sex with Sleepers, and eventually hound them off the planet to a refuge in space.

OPPOSITION GROUPS

While some people will not be able to afford advanced forms of biomedical enhancement, others may voluntarily refuse to avail themselves of it. One reason might be their religious beliefs; Leon R. Kass, who served as chair of the President's Council on Bioethics from 2001 to 2005 during the administration of George W. Bush, objects that radical changes in human beings would conflict with God's design. Political philosopher Michael J. Sandel, for example, states that biomedical enhancements negate "the gifted character of human powers and achievements" (2004, 54), suggesting that enhancements interfere with "gifts" bestowed by God or Mother Nature. Some people reject enhancements as morally unfair because not everyone can obtain them. Others feel that enhancements are incompatible with certain expectations, such as that sports be played on a "level playing field" that performance enhancements distort.

Another reason people might refuse to enhance themselves or their children is fear that enhancements would cause physical harm. A major justification for the ban on enhancement in sports, called doping, for example, is safety; in part because doping is illegal and not done under medical supervision, performance-enhancing drugs can cause serious or even fatal side effects. Safety concerns are exacerbated by a lack of safety data; many consider testing the performance-enhancing effects of steroids in healthy athletes, for example, to be unethical, and drug companies do not sponsor the extremely expensive studies that the US Food and Drug Administration (FDA) requires to approve new products because of the agency's historic hostility to approving drugs for enhancement as opposed to medical treatment purposes. (One exception is the FDA approval of Botox for cosmetic purposes.) Critics fearful of unforeseen adverse effects on engineered offspring point to genetic experiments in which subjects die, such as eighteen-year-old Jesse Gelsinger, who died of massive organ failure in 1999 in an experiment to develop a genetic therapy to treat a liver malfunction in newborns.

People may also voluntarily reject biomedical enhancements for aesthetic reasons, what Kass and others call the "yuck factor." Images in the mid-twentieth century of syringes littering locker room floors and the masculine demeanor of some East German female athletes have been cited as a reason for opposition to doping in sports, for instance. The idea of giving humans the abilities of certain animals, such as an eagle's vision or a dog's sense of smell, is also repulsive to many. Any enhancement that radically altered the way a person looked could provoke a strong, visceral, negative reaction.

Opposition is especially pronounced against genetic enhancement, and particularly enhancements produced by germ-line genetic alterations. Objections often stem from religious beliefs; former priest Joseph Fletcher describes those who "feel that mastery drives out mystery, and that when we dig into nature's secrets, so complicated and yet so simple, we lose our sense of awe and humility" ([1974] 1988, 19). An American Association for the Advancement of Science report from 2000 concluded that "even if we have the technology to proceed, . . . we would need to determine whether IGM [inheritable genetic modification] would offer a socially, ethically, and *theologically* acceptable alternative to other technologies" (Frankel and Chapman 2000, 56; italics added). Even renowned liberal philosopher Ronald Dworkin felt that "playing God is playing with fire" (2000, 446).

Whether out of religious or safety concerns, those who rejected enhancements voluntarily would be at the same competitive disadvantage as those who were unable to afford enhancements. Both groups could find themselves increasingly isolated socially and economically.

REPRODUCTIVE FRAGMENTATION AND SPECIATION

As noted earlier, humans generally do not engage in random sexual behavior; like other animals, they select their mates on the basis of what they consider to be positive characteristics. Biomedical enhancement introduces another variable that people might take into consideration, preferring to have children with those who also were enhanced to protect their socioeconomic status. Biomedical enhancement could produce even more extreme forms of sexual and reproductive isolation. Reproductive organs could be enhanced in various ways that would make sexual relations between normal and refashioned individuals difficult or impossible. Children engineered to be smarter might have brains so large that only mothers engineered with wider pelvises could give birth to them without surgery. With a sufficient amount of genetic alteration, enhanced individuals could even become unable to reproduce with unenhanced mates simply because of differences in their genes. The result would be increasing sexual and reproductive separation between the enhanced and the unenhanced.

Given a sufficient amount of reproductive separation, an especially striking development might occur: humanity might split into two or more different humanoid species, a process known as speciation. Would they get along? Would they be friendly and cooperative, sharing scarce planetary resources, or would they be competitive, perhaps even belligerent? How ugly could their interactions become if different species viewed one another as "less than human"?

Sharing the planet with more than one humanoid species is not exactly uncharted territory. Fifty thousand years ago, while modern *Homo sapiens sapiens* lived in northern Africa, an ancestral human species, *Homo erectus*, still lived in Asia. As recently as 18,000 years ago, a stone-tool-making, fire-wielding human species called *Homo floresiensis* lived on an island off the coast of Indonesia. It is not known whether either *Homo erectus* or *Homo floresiensis* came into contact with modern humans, but what is clear is that neither survives to this day. Neanderthals definitely interacted with modern humans and in fact interbred, making Neanderthals not a separate species but merely a human subspecies. The last Neanderthal died out around 40,000 years ago. No one knows why. But the fact that no rival human species or subspecies still exists does not bode well for the prospect of harmonious relations between enhanced and unenhanced humanoid populations. George J. Annas, a scholar specializing in biomedical ethics and health law, underscores this point: "Ultimately, it almost seems inevitable that genetic engineering would move *Homo sapiens* to develop into two separable species: the standard-issue human beings would become like the 'savages' of the pre-Columbian Americas, and be seen by the new, genetically-enhanced neo-humans as heathens who can properly be slaughtered and subjugated" (2000, 773).

SUGGESTED RESPONSES

Some commentators look forward to the prospect that biomedical enhancement will produce radical changes in humans. Transhumanists, for example, believe that the capability to improve and even genetically reengineer humans is the key to a better future in which humans live longer, healthier, and more enriching lives. While some transhumanists accept the need to regulate genetic engineering, others are more concerned that regulation will interfere with achieving their dream of improving humans.

Other commentators focus on the potential negative effects of biomedical enhancement, including the threat of societal fragmentation, and propose various responses. At one extreme are those who want to prohibit biomedical enhancement. This is the attitude, for example, of opponents of biomedical performance enhancements in athletic competitions, such as the Olympics and professional sports. They argue that doping is dangerous, is unfair to those who resist using it, and corrupts an essential element of athletics, which they refer to as "the spirit of sport." Beginning in the late 1960s, antidoping forces persuaded Olympic officials to create a list of banned substances, to begin testing competitors for the presence of those substances in their bodies during competitions, and to disqualify athletes who had positive test results. Until 2003, for instance, the Olympics tested for caffeine, and persons weighing 150 pounds could be disqualified if a test result showed an amount of caffeine in their bodies equivalent to drinking a 16-ounce cup of Starbucks coffee and two Mountain Dews. Critics of a ban on doping point out that it worsens the safety problem by motivating athletes to turn to dangerous substances produced by rogue chemists in illicit laboratories; that historically athletes have always tried to gain an advantage from biomedical enhancements, so that the notion that biomedical enhancement corrupts the spirit of sport rests on an erroneous assumption that sports were once "drug free"; and that there is little about elite sports that is fair, because successful competitors typically have the good luck to be born with better genes, to have wealthier and more supportive families and sports organizations, and to obtain access to top coaches and trainers.

Some proponents of a ban on biomedical enhancement base their opposition on religious or spiritualistic grounds. Others focus on the risk of harm to children who were genetically engineered, including unforeseen physical harm, being ostracized from the rest of society by virtue of their unusual appearance or behavior, and having their options in life narrowed by their parents' preferences. Finally, the prospect of rival human species, even if far in the future, leads to calls for restricting germ-line genetic enhancement.

Prohibiting the use of biomedical enhancements encounters serious practical problems, however. Even if the government forbade their sale, people might try to obtain them on the black market, promoting criminal activity similar to what occurred during Prohibition (of alcohol beverages) in the 1920s and as a result of the so-called war on drugs that began in the 1970s. To prevent people from using enhancements, there would have to be a way of knowing when they did so, which would require reliable tests and a highly intrusive testing program; imagine, for example, if students had to endure the testing system used in elite sports, where athletes can get tested at random and made to urinate in front of an observer, a scenario depicted in part in the film *Gattaca.* Furthermore, some enhancements, such as genetic engineering, might be obtained in foreign countries where they were legal and brought back to the United States embedded in a person's DNA, necessitating a testing program at the borders for people returning from abroad. A ban on enhancements would also deny society the benefit from the activities of enhanced individuals, such as cancer research or disaster rescue operations. In addition, a country that prohibited enhancements would be at a global competitive disadvantage in comparison with nations that took a more permissive stance.

Moreover, a government ban on germ-line genetic engineering in particular would constitute a restriction on parents' reproductive freedom, which is protected by the US Constitution. True, parents are not entitled to abuse or neglect their children, but it is hard to imagine that parents would deliberately try to engineer their children in ways that would be considered abusive or neglectful. The criticism that parents might unduly narrow their children's future options, for example, assumes that they would choose to enhance only a single or limited set of abilities, when it is more likely that parents would want to give their offspring a broad array of possible advantages.

Instead of a complete ban, one response to the concern about the safety of biomedical enhancements would be to regulate them. The FDA requires the manufacturers of new drugs to submit data to the agency demonstrating safety and efficacy before they can be marketed, and the FDA therefore could impose a similar requirement for enhancement drugs. The agency has approved only two drugs for a purely enhancement use; in addition to Botox, mentioned earlier, the FDA approved modafinil (Provigil) to treat shift-work sleep disorder, the difficulty sleeping that some people have who work night shifts, which might be considered a normal consequence of such a work schedule rather than a disease. But there is no question that the FDA has the legislative authority to regulate enhancement drugs if it chose to. The FDA also has claimed jurisdiction over genetic engineering, which would extend to genetic enhancement, but it is less clear that the agency actually has the legal authority to do so, because genetic engineering might be considered a medical procedure rather than the use of a product such as a drug or vaccine, and the FDA has stated that it does not have jurisdiction over the practice of medicine.

In addition to targeting commercial producers of biomedical enhancements, the government could regulate physicians and other professionals who provided them. State medical boards could adopt rules for when and how doctors enhance patients, and new licensing schemes could be established for other professionals, such as PhD geneticists. In addition, health-care professionals and medical researchers specializing in the development of enhancements could attempt to regulate themselves. Professional groups such as the American Medical Association (AMA), the American College of Obstetricians and Gynecologists, and the American Society of Human Genetics issue ethical guidelines; in 1994, for example, the AMA prohibited its members from using genetic engineering for enhancement purposes. Researchers even have voluntarily halted work on their own. In 1973, when scientists first developed the technique called recombinant DNA that enabled them to manipulate genes, they were so frightened by the possibilities that they agreed to suspend all further work until a satisfactory regulatory system could be established. The moratorium lasted two years and, so far as is known, was effective around the world. There also were calls for an international moratorium on germ-line genetic engineering in 2015, when researchers in China announced that they had manipulated the DNA of human embryos.

Government or professional regulation of biomedical enhancements would encounter similar obstacles as a complete ban, however. A black market similar to the market for illegal drugs would be likely to emerge, inviting crime, as well as "enhancement tourism," where people travel abroad to obtain products that are difficult or impossible to obtain domestically.

If biomedical enhancements were legal but expensive and therefore available only to those who could afford them, what measures might be taken to reduce competitive unfairness? One option would be to hold separate competitions for enhanced and unenhanced individuals, which already happens to some extent in some sports: boxers and wrestlers, for example, typically compete only with others in the same weight range. In education, there could be separate tracks or a form of affirmative action in which a certain amount of certain benefits, such as acceptance at elite educational institutions, was set aside for the unenhanced. Separate sports competitions or educational pathways could increase social fragmentation, but the costs might be outweighed by the perception of greater fairness.

Another option is to try to level the playing field by handicapping those who were enhanced, a concept similar to handicapped horse races, in which jockeys who weigh too little must carry weights in their saddles, or to handicaps in golf, where strokes are subtracted from the scores of golfers who are not as good as other players. (Author Kurt Vonnegut satirized this approach in his 1961 short story "Harrison Bergeron," in which in 2081, thanks to the United States Handicapper General, smart people are made to wear little receivers in their ears so that the government could send them sharp bursts of sound every twenty seconds to interrupt their thoughts, newscasters all have speech impediments, and beautiful ballerinas are required to wear ugly masks and are weighted down with heavy bags of BBs.)

One solution to the problem of inequality if enhancements were available only to the well-to-do would be to make them available to everyone. An enhancement welfare system could raise money to pay for universal access to enhancements, for example, by taxing enhancement purchases by those who could afford them. But the goal of giving everyone an equal opportunity by providing enhancements to all could be undercut if the wealthy could keep themselves ahead of the pack by buying even more enhancements. In some cases, there might be a practical limit on how enhanced someone could become; in basketball, for instance, players who were too tall are liable to be physically uncoordinated and unstable. But there does not seem to be a ceiling on how smart someone can become, so a complete solution to inequality probably would require a combination of subsidizing access for the less-well-off and limiting it for the wealthy. However, given the regulatory hurdles of imposing such restrictions mentioned earlier, along with the reluctance of most Americans to prevent rich people from buying luxury items that no one else can afford, it seems likely that some degree of inequality would persist.

As stated earlier, access to enhancements only by those who were better off not only would increase inequality but could undermine belief in equality of opportunity. If it was too expensive to provide universal access to enhancements, a final solution might be a government-run lottery. Every adult who had not already won could enter in the drawing, winners selected at random could be given subsidies to obtain whatever enhancements they desired, and the odds of winning could be adjusted based on the results of opinion polls measuring perceptions of equality of opportunity.

Summary

Genetic engineering, whether somatic or germ line, is far from being successful, and there are few if any biomedical enhancements that are game changers in terms of social success outside of elite sports. Nevertheless, there is no reason to think that matters will remain this way for long. Demand for an advantage in the scramble for increasingly scarce resources is bound to stimulate enhancement research and development by the biotechnology industry, and progress in developing gene-based treatments for disease will pave the way for genetic enhancement, because the technology to treat illness and to enhance ability is basically the same. In addition to all the other forces pulling people and communities apart from one another, there will be biomedical enhancements that are not universally obtainable or that do not create level playing fields. It is not too soon to think about how best to obtain the benefits of these enhancements while minimizing the harms.

Bibliography

Annas, George J. "The Man on the Moon, Immortality, and Other Millennial Myths: The Prospects and Perils of Human Genetic Engineering." *Emory Law Journal* 49, no. 3 (2000): 753–782.

Center for Survey Research and Analysis. "Two Americas but One American Dream." University of Connecticut, 2007. http://news.uconn.edu/2007/July/rel07059.html.

Domhoff, G. William. "Wealth, Income, and Power." University of California at Santa Cruz, 2013. First published 2005. http://www2.ucsc.edu/whorulesamerica/power/wealth.html.

Dworkin, Ronald. *Sovereign Virtue: The Theory and Practice of Equality.* Cambridge, MA: Harvard University Press, 2000.

Fletcher, Joseph. *The Ethics of Genetic Control: Ending Reproductive Roulette.* Buffalo, NY: Prometheus Books, 1988. First published 1974.

Frankel, Mark S., and Audrey R. Chapman. *Human Inheritable Genetic Modifications: Assessing Scientific, Ethical, Religious, and Policy Issues.* Washington, DC: American Association for the Advancement of Science, 2000. https://www.aaas.org/sites/default/files/migrate/uploads/germline.pdf.

Harris, John. *Enhancing Evolution: The Ethical Case for Making Better People.* Princeton, NJ: Princeton University Press, 2007.

Kass, Leon R. "Ageless Bodies, Happy Souls: Biotechnology and the Pursuit of Perfection." *New Atlantis: A Journal of Technology and Society,* no. 1 (2003): 9–28. http://www.thenewatlantis.com/docLib/TNA01-Kass.pdf.

McNamee, Stephen J., and Robert K. Miller Jr. *The Meritocracy Myth.* 3rd ed. Lanham, MD: Rowman and Littlefield, 2014.

Mehlman, Maxwell J. *The Price of Perfection: Individualism and Society in the Era of Biomedical Enhancement.* Baltimore: Johns Hopkins University Press, 2009.

Mehlman, Maxwell J. *Transhumanist Dreams and Dystopian Nightmares: The Promise and Peril of Genetic Engineering.* Baltimore: Johns Hopkins University Press, 2012.

Mehlman, Maxwell J. *Wondergenes: Genetic Enhancement and the Future of Society.* Bloomington: Indiana University Press, 2003.

Sandel, Michael J. "The Case against Perfection." *Atlantic Monthly*, April 2004, 51–62.

FILMS

Gattaca. Dir. Andrew Niccol. 1997.

The Hunger Games. Dir. Gary Ross. 2012.

NOVELS AND SHORT STORIES

Heinlein, Robert A. *Friday.* New York: Holt, Rinehart and Winston, 1982.

Herbert, Frank. *The Eyes of Heisenberg.* New York: Tom Dougherty Associates, 1966.

Kress, Nancy. *Beggars in Spain.* New York: Morrow, 1993.

Vonnegut, Kurt. "Harrison Bergeron." *Magazine of Fantasy and Science Fiction*, October 1961, 5–10.

CHAPTER 32

Regulating Bioenhancements: Law and Governance

Michael Bess
Chancellor's Professor of History
Vanderbilt University, Nashville, TN

Over the coming decades, as bioenhancement technologies advance in variety and sophistication, it is likely that many citizens will eagerly embrace the chance to modify their bodies and minds. Others, motivated by religious beliefs or perhaps by a reluctance to interfere with the natural processes in the human body, may refuse all but the most basic forms of self-modification (Bess 2015). Imagine, for the sake of argument, a scenario in which a solid majority of American voters decide that human bioenhancement would bring more evil than good into the world and elect political representatives who are committed to enacting an across-the-board ban. Could US society stop these technologies from being developed? What would such a policy of comprehensive relinquishment require?

Five main problems would confront any government that sought to enact such a ban. First, the inherent haziness of the treatment/enhancement distinction poses a fundamental problem: where does one draw the line between modifications that restore, heal, or fix a malfunctioning human body or mind and modifications that significantly enhance human performance? (Bess 2010). Extreme forms of transmogrification—such as a brain-to-brain communication device or a genetic modification that tripled a person's health span—may be easily identifiable as (nontherapeutic) enhancements, but many of the more moderate kinds of intervention would unavoidably fall in a gray area. For example, if a deeply shy person takes a pill that turns her into the life of the party, is this a treatment for a pathological form of antisocial behavior, or is it rather an augmentation that vaults her into a whole new category of social relationships and activities? In ambiguous cases such as these, it will prove hard to say with confidence, "This modification is permissible, but that one is verboten." One can imagine the hue and cry among various citizens' groups and political constituencies passionately arguing the pros and cons of whether a certain device or modification should be allowable. This will not be an easy matter to adjudicate.

Second, the government would have to find an effective way to block specific kinds of technologies and interventions in a highly selective manner. Consider, for example, the drug Ritalin, which in one context (treating attention-deficit/hyperactivity disorder, or ADHD) is clearly therapeutic in its usage, but in another context can be adopted by healthy persons for cognitive enhancement purposes. The government could opt to ban the chemical outright, which would deprive persons afflicted with ADHD from access to it; or it could do what it is already doing today—namely, allow selective access to the drug through a prescription, while tolerating significant levels of illegal off-label use (Greely et al. 2008). If either of these restrictive policies were applied more generally to a wide variety of enhancement technologies—pharmaceutical, bioelectronic, and genetic—one would expect to see the emergence of a widespread black market, harkening back to the era of alcohol prohibition in the 1920s. The result would be twofold: a failure to prevent the actual usage of enhancement technologies, coupled with a significant rise in organized criminality.

An across-the-board ban would also generate a third major problem: the stifling of legitimate research and innovation. Scientists working on advanced bioelectronic devices, for example—such as neurologist Theodore W. Berger and his team at the University of Southern California, seeking to develop a hippocampal implant for Alzheimer's patients—would presumably have a much harder time getting their labs funded (Berger and Glanzman 2005). Berger would have to assure government regulators that his "memory chip" could be used only to restore preexisting mental capacities rather than to boost those capacities beyond the species-typical range. He might even have to design the chip with inherent restrictions built in, so that it could never exceed existing parameters for human hippocampal function. Finally, Berger might have to forgo publishing certain aspects of his findings, because this kind of knowledge might be used by black-market laboratories to build chips exceeding the allowable specifications. Any major effort to block enhancement-related science, in other words, would tend to lead government regulators down the path of censorship.

This would dovetail, in turn, with a fourth major drawback: the surveillance apparatus required to enforce the ban would have to approach totalitarian levels in its fine-grained comprehensiveness. In order to ensure compliance, the government would have to closely monitor the activities of scientists and technologists working in an extremely wide variety of fields, ranging from biology to computer science and from medicine to engineering. Whether such a surveillance system operated through electronic devices or through a network of human informers (or both), it would inherently undermine many of the basic liberties that Western societies value most highly.

Finally, a successful ban would require this highly restrictive system to be imposed equally scrupulously in all the world's nations at the same time; otherwise, the research and innovation would simply migrate overseas to the least-regulated regions of the planet. The chances of achieving such a coordinated global relinquishment are very small indeed, because bioenhancement technologies will probably be perceived as potential game changers in the ongoing competition for global leadership, influence, and profit (Tucker 2012). Whoever develops them first will gain a significant advantage over all the other players; and if one player acquires these technologies, then other players must quickly follow suit or risk being left in the dust. In both military and civilian uses, these are potentially revolutionary technologies: the nation that develops these inventions will become a major geopolitical force; the corporation that masters one or more of them will corner vast swaths of the global market. This means that the logic of arms races will tend to dominate the development of bioenhancement technologies, just as it did with nuclear weaponry during the Cold War. When it comes to powerful technologies such as these, governments, researchers, and funding agencies will have a strong incentive to put their main emphasis on rapid technical progress, relegating caution, restraint, and long-term planning to a lower priority.

For all these reasons, then, the notion of outright relinquishment, enacted through governmental controls, seems highly implausible. What is far more likely to emerge is a situation analogous to the way certain powerful but potentially dangerous technologies and medicines are administered in today's society: some products are available to the public on the open market, but many of the most potent ones are carefully regulated and can be accessed only under controlled conditions (via a prescription or a professionally trained technician). The legislation regarding these technologies will probably be animated by two basic goals: ensuring the basic safety of the user and ensuring that the broader societal effects of the technology do not undermine public health or security.

Crafting such legislation will be difficult to do, however, because it is likely that both voters and their representatives will be deeply divided over how they regard bioenhancement technologies. In a pluralist democracy, clashing opinions about matters of great importance form an integral part of everyday politics. The normal procedure is for the various stakeholders to negotiate with each other, incrementally working their way toward compromises and trade-offs that partially satisfy the goals of all the major players. Yet, some kinds of bioethical issues, such as abortion rights or the practice of assisted suicide, are not easily amenable to such forms of compromise, because they entail trenchant "either-or" decisions that inevitably leave the minority side of the argument in a position of irreconcilable opposition. Either a country's laws allow the fetus to be aborted or they do not: whichever policy is chosen, one of the stakeholder parties will be bitterly aggrieved by the outcome. Precisely because bioenhancement technologies cut so deeply into the human constitution, they will tend to be perceived and evaluated by the citizenry in terms of fundamentally opposed moral visions: on one side, pro-enhancement groups will describe them as a splendid boon that lifts humans to ever-higher levels of flourishing; on the other side, anti-enhancement groups will insist with equal vehemence that they lead to dehumanization, commodification of human traits, and degradation of dignity.

SHAPING TECHNOLOGICAL CHANGE: POWERS AND PATHWAYS

In a democratic society, citizens possess four major avenues for influencing the development of science and technology: governmental action; grassroots activism and advocacy, self-regulation by scientists and engineers, and broad-based efforts to influence the everyday technological practices of the population as a whole. The first of these, state action, can include not just laws, treaties, and explicit regulations but also patents, tax policy, informal guidelines, and the funding of research. These instruments give a nation's leaders a broad palette of powerful instruments through which they can influence what kinds of science get done, what kinds of products get designed and built, and the rate at which innovation takes place. After the Russians launched *Sputnik* in 1957, for example, the US government responded with a dramatic reconfiguration of the way American science and technology were handled: massive new resources flowed into both public- and private-sector research ranging from aerospace technology to basic physics, and the nation's educational system was swiftly reoriented toward building strength in science and engineering. The results were impressive: twelve years later, an American astronaut, Neil Armstrong, was walking on the moon.

Governmental action can also pull quite effectively the opposite way, moreover. In 2001 President George W. Bush issued an executive order that severely constrained research on human embryonic stem cells in the United States. As a result, such research by American scientists slowed down considerably between 2001 and 2009—when his successor, Barack

Obama, countermanded these restrictions. Most notably, Bush did not need to go so far as to issue an outright ban on this kind of research: he merely withdrew all federal funding from it, and this proved sufficient to achieve his goal. To be sure, the president's powers were not unlimited, for scientists in other countries, such as Sweden and the United Kingdom, were able to proceed unimpeded with their own work on human stem cells. But Bush's impact was nonetheless considerable: he successfully reined in a particular form of biological research—and his action ultimately incentivized US scientists to develop ingenious techniques for generating pluripotent stem cells without using any human embryos at all (Furcht and Hoffman 2011).

The broader lesson here, both with the response to *Sputnik* and the restriction on stem cells, is that vigorous governmental action can not only speed up or slow down specific kinds of scientific and technological innovation: it can also forcefully channel that innovation down paths it might not otherwise have taken. Admittedly, this is not an unambiguously good thing. Many observers have argued persuasively that society benefits most when scientists and technologists are left free to pursue their work as they see fit, following their creativity and ingenuity wherever they need to go (Johnson and Wetmore 2009). When Soviet bureaucrats forced their nation's researchers to toe an ideological line, as they did during the ascendancy of the agronomist Trofim Lysenko (1898–1976), this set back Russian biological science for an entire generation. Governmental regulation should therefore be applied only when strictly necessary, and even then in a parsimonious and judicious fashion. The goal should be to ensure that basic levels of safety and security are maintained, while allowing as much freedom as possible to the researchers and inventors.

A second avenue through which citizens may shape technological advance is by direct action and advocacy at the grass roots. Perhaps the most famous instance of such activism occurred in the late 1950s and early 1960s, when Ralph Nader took up the cause of improving automobile safety in the United States. He wrote numerous articles on the subject and a widely read book, *Unsafe at Any Speed* (1965), in which he indicted the slipshod engineering practices prevalent in the manufacture of many models of American cars (Marcello 2004). Dozens of idealistic young activists—dubbed "Nader's Raiders" by the press—rallied to his cause, forming an impressive team whose activities and advocacy eventually branched out into investigations of numerous areas of the consumer economy. Nader and his allies exerted a significant influence on public opinion, ultimately compelling both the Detroit-based automakers and the federal government to undertake far-reaching reforms. In 1970 the US Congress created the National Highway Traffic Safety Administration, which went on to set stringent new technical requirements for all American vehicles, including seat belts and stronger windshields. A well-informed and determined group of citizens can mobilize effectively to alter the nature of the technologies offered for sale in a free-market economy.

Scientists and engineers themselves constitute a third important factor in the governance of technological change. This is self-evidently true when a few entrepreneurial individuals, such as Steve Jobs and Steve Wozniak, working out of their garage, can upend an entire industry by introducing a dazzling new product. But the power of scientists can also, on occasion, take a subtler form, when they join together to establish guidelines on how their research should be conducted. In February 1975, for example, a group of about 140 scientists gathered at the beachside resort of Asilomar, near Monterey, California, to discuss the implications of what was then a cutting-edge field of biotech innovation: recombinant DNA technology. The conference was organized through the initiative of Stanford

University biochemist Paul Berg, whose many conversations with his fellow scientists had convinced him that recombinant DNA not only might offer major benefits to society but could also pose terrible dangers as well. Working together over several days, the conference participants eventually agreed on a series of voluntary ground rules regarding the biohazards of their research (Fredrickson 2001). They pledged to restrict their experiments to organisms that would not be able to flourish and multiply outside the laboratory context and adopted a variety of containment measures that were aimed at ensuring that the modified organisms could not spread beyond their laboratories. Certain types of highly pathogenic organisms were declared off-limits to experimentation. Finally, they resolved to maintain transparency about the nature of their research, keeping the press (and therefore the broader public) apprised of the ongoing work they were doing. In the decades that followed, to be sure, the Asilomar conference of 1975 turned out to be more the exception than the rule: most of the restrictions and guidelines imposed on scientists came from government oversight and funding agencies, or from their university's institutional review board, rather than such voluntary conclaves of researchers themselves. Nevertheless, the conference set a remarkable precedent, proving in practice that scientists working in a potentially hazardous new field were capable of successful self-governance when they chose to do so.

A fourth way for citizens to influence the advance of science and technology lies in the latent power of collective consumer choices. Technologies succeed only in proportion to the degree of interest and desire they elicit among the population of potential users. Therefore, if an individual can find a way to influence the way people perceive a given device or process, she will be able to exert a significant measure of control over the shape it takes and the role it comes to play in people's lives (MacKenzie and Wajcman 1999). Consider, for example, the cigarette—a small, low-tech device designed to deliver certain chemicals and pleasurable experiences to the consumer. As long as cigarettes were associated in popular culture with stylishness, sexiness, and movie stars, they reigned invincible despite abundant public information about their toxicity. But after the antismoking campaign adopted the strategy of branding cigarette use as pathetically suicidal and uncool—and film directors increasingly desisted from showing their heroes and heroines with a cigarette dangling from their lips—consumption of cigarettes in advanced democracies began declining significantly. Tobacco companies were forced to rely increasingly (and despicably) on the populations of Third World countries to compensate for lost sales. Concerted campaigns of public education and image making proved to be potent instruments in reshaping the use of a widely disseminated technology.

CONSTRAINTS ON A SOCIETY'S POWER TO ACT

Unfortunately, these four impressive varieties of human agency in governing science and technology are counterbalanced, in practice, by an equally significant array of limitations and impediments. First and foremost among these is the problem of uncertainty—a factor that is particularly acute when it comes to cutting-edge technologies in early stages of development. Uncertainty comes in multiple flavors: it can derive from insufficient information, or it can arise whenever experts disagree on their assessments of the implications of a given device or process. Just how risky would it be, for example, to clean up an oil spill on a populous shoreline by unleashing trillions of newly invented nanomachines to chomp on the petroleum molecules and turn them into biodegradable compounds? No one knows for sure. The manufacturer of the nanodevices may express great confidence that they have been exhaustively tested in the laboratory and pose no significant threat. But if you were the mayor of the local town or the Environmental Protection Agency official in charge of the cleanup operation, would that assurance suffice for you to give the green light?

Dealing with these kinds of uncertainty falls under the purview of a field known as risk management—a complex, multidisciplinary endeavor whose practitioners continually find themselves facing difficult trade-offs (Coleman 2011). If they err too far on the side of caution, blocking the development or implementation of new technologies, they can end up stifling innovation, hampering progress, and undermining economic growth. But if they err too far in the other direction, they can end up subjecting society to potentially significant harms. Here is where a conceptual framework known as the precautionary principle comes into play. First articulated by German philosopher Hans Jonas in the 1970s, this principle has now come into widespread use in crafting regulatory legislation in the European Union and elsewhere (Jonas 1984; Whiteside 2006). It has three interlocking components:

1. If a given technology or procedure is suspected of posing a significant risk to society, then it should not be adopted unless its projected benefits outweigh its projected drawbacks.
2. If no scientific consensus exists about the level of risk involved, then the burden of proof lies with the promoters of the technology to show that its projected benefits outweigh its drawbacks.
3. The greater the suspected level of potential harm associated with a technology, the higher the bar should be set for scientific consensus about its potential net benefits before the technology is adopted.

Or, to put it more plainly: When the suspected risks are very great, it is always better to err on the side of caution, even if the state of present knowledge is imperfect.

Kevin Kelly, one of the cofounders of *Wired* magazine, has stridently criticized the precautionary principle, arguing that it stifles innovation while yielding few benefits in terms of reduced risk. But Kelly's description of the principle is an oversimplified straw man: "A technology must be shown to do no harm before it is embraced.... If it cannot be proven safe, it should be prohibited" (2010, 247). This is a gross mischaracterization of Jonas's idea, which emphasized the broad gradient of uncertainties within which humans operate in projecting future societal implications of today's technologies. The whole point was to be able to make smart and nuanced policy decisions for the countless situations in which definitive proof is unattainable—not to demand absolute safety before proceeding with the adoption of a new technology.

As a concrete example, consider the case of global warming. Even though the overwhelming majority of climate scientists agree that this is a real phenomenon, and that it is at least partly anthropogenic in nature, they still do not have absolute certainty about how rapidly it will progress and how drastic its effects will be. The scientific models are simply not sophisticated enough yet to offer full clarity in this domain. Nevertheless, the precautionary principle does offer solid guidance: because the projected harms caused by global warming are catastrophic in nature, humanity should take decisive action today to reverse the human-induced causes of the phenomenon. It is reasonable for humankind to make significant sacrifices in the present in order to avoid potentially disastrous consequences in the future—even though no one knows precisely what those dire consequences will actually turn out to look like.

A related problem has to do with the bias toward short-term thinking that plagues consumer choices and political governance (Gore 2013). Consumers tend to focus on the immediate benefits conferred by a product, ignoring the side effects that follow many years later from its widespread use. Corporate leaders have their eye primarily on the next quarter's bottom line and the anticipated reaction of shareholders. Elected officials, who are answerable to voters every few years, similarly tend to emphasize the short-term costs and benefits of the policies they enact, leaving longer-term issues to be resolved by their successors. Thus, for example, legislative efforts to curb the federal debt or to rein in greenhouse gas emissions—which require making sacrifices in the present in order to reap benefits many decades later—tend to languish for lack of support. This myopic syndrome is especially unfortunate when it comes to bioenhancement technologies, which are still in their "formative years" today and will be much harder to restructure or reconfigure once the basic choices regarding their design and use have become entrenched.

PRINCIPLES AND GOALS IN GOVERNING BIOENHANCEMENT

PROMOTE AN ETHIC OF SOCIALLY RESPONSIBLE ENHANCEMENT

If you are a college student studying for your final exams, you may find yourself tempted at some point to use a cognitive enhancement drug as a means of achieving higher performance. Apart from the fact that this is currently illegal, as well as potentially hazardous to your health, there is another important reason for saying no. If more and more people are doing it, then the pressure mounts for everyone to get onboard. Eventually there will come a time when nearly everyone is using such chemicals, and any competitive advantage that people might have derived will evaporate. Everyone will be performing on the same artificially elevated level, and things will be right where they started—except that a whole generation of youths will be madly popping pills left and right in order not to be left in the dust. By succumbing to the lure of a chemical quick fix for a higher grade, you are therefore contributing to this collective process of runaway self-modification.

The argument here is not that all uses of cognitive enhancement drugs are necessarily harmful or morally problematic. The point, instead, is that increasing your competitiveness is a poor motive for taking such drugs, both because it is ultimately self-defeating and also because of the negative social meta-effects of this course of action. If you decide to pop cognitive enhancement pills, it should be for a better reason—one that is attuned to the register of intrinsic value rather than mere instrumental value. Thus, for example, a desire to really come to grips with superstring theory in physics or the writings of nineteenth-century German philosopher Georg Wilhelm Friedrich Hegel, in a deep and sustained way, might qualify as a more legitimate motive.

The ethics implicit here is similar to that of recycling glass bottles, choosing a low-emissions automobile, or putting on a sweater instead of cranking up the thermostat. All individuals face choices whose repercussions spread far beyond their discrete lives: people are responsible for the societal implications of their personal actions. Just as people are praised for adopting a "green ethic" in their daily habits, so should they be encouraged to adopt an analogous ethic when it comes to decisions about bioenhancement. People need to hold themselves actively accountable for the profound consequences that their individual enhancement choices will have for those around them.

MANDATE BASIC EDUCATION IN THE FIELD OF SCIENCE, TECHNOLOGY, AND SOCIETY

The science underlying bioenhancement technologies is complex and wide ranging, and consumers will need a solid scientific education if they are to make informed decisions about the modifications they choose to undertake to their bodies and minds. Equally importantly, they will need the analytical tools with which to assess the broader societal dimensions of bioenhancement technology if they are to make informed policy choices as citizens and voters. The interdisciplinary field of science, technology, and society (STS) has existed only since the 1980s, but it has grown rapidly, and hundreds of universities around the world now offer programs and courses under this rubric (Nye 2006). The methodologies and subject matter of the field dovetail perfectly with the basic questions explored in this volume—particularly those having to do with the shaping of technology and the ethical and social implications of the enhancement enterprise. It would behoove US society to enact legislation that encourages high schools and universities to start requiring such courses as part of the basic liberal arts educational curriculum. In addition, seekers of higher degrees in science or engineering should be encouraged to take a certain number of advanced STS classes as part of their training—so that they are alerted at an early point in their careers to the tremendous ethical implications of their work.

CREATE A STRONG GOVERNMENTAL UMBRELLA AGENCY FOR ASSESSMENT AND REGULATION OF EMERGING TECHNOLOGIES

What kinds of oversight exist today to regulate bioenhancement technologies? The answer is unsettling. Although a dense thicket of laws and agencies is in place in the United States, not a single one focuses specifically on bioenhancement, nanotechnology, synthetic biology, or artificial intelligence (Sutton 2014; Gronvall 2016). Three federal agencies are tasked with regulating the emerging field of biotechnology in the United States: the Environmental Protection Agency, the Food and Drug Administration, and the Department of Agriculture. The responsibilities of each agency are spelled out in an umbrella document, the Coordinated Framework for Regulation of Biotechnology, first adopted in 1986 under President Ronald Reagan, then updated in 1992 and again in 2015. Unlike European legislators, who have taken a more diffident stance regarding biotechnology and genetic modification practices, US policymakers openly stated that a key priority was "to avoid impeding the growth of an infant industry" (Office of Science and Technology Policy 1986). Under the guidance of this industry-friendly framework, US policy has therefore defaulted to existing laws and oversight practices, eschewing new forms of restrictive legislation.

In 2009 President Barack Obama tasked an expert commission to review the field of synthetic biology and assess whether existing laws were adequate to keep the public safe. The commission concluded that a "protective patchwork quilt of regulations" was in place but noted that this patchwork had significant gaps and that "deviant uses of synthetic biology could therefore, at least theoretically, occur outside of the scope of existing oversight mechanisms" (Presidential Commission 2010, 82, 83). The problem was two-fold: many of the safety guidelines set forth by federal agencies relied on voluntary compliance and self-reporting by researchers and included limited mechanisms for ongoing inspection or enforcement; and a great deal of privately funded research (whether in universities, corporate labs, or somebody's garage) was not covered at all by federal oversight.

This patchwork system potentially leaves the door open for all kinds of bad things to happen. What is needed, instead, is a new governmental agency or body specifically tasked with keeping abreast of rapid new developments in all areas of emerging technologies—including bioenhancement technologies, nanotechnology, synthetic biology, and artificial intelligence—and coordinating the federal governance of those potentially hazardous fields (Fukuyama 2002). One model for such an agency would be the Office of Technology Assessment, which advised members of the US Congress from 1972 to 1995. During its twenty-three-year life span, this agency issued hundreds of rigorously researched reports about a wide variety of new technological and scientific developments, helping legislators to craft policies and laws to deal effectively with questions of both short-term and long-term public welfare and safety (Sclove 2010; Sadowski 2012). If such an agency were reestablished, and also paired with a sister agency empowered with monitoring and enforcing regulations, the nation would have a far more coordinated and coherent set of policies in place than it does today.

STRENGTHEN INTERNATIONAL COOPERATION IN TECHNOLOGY GOVERNANCE

The policies of individual nations regarding science and technology are somewhat gingerly coordinated at the international level through such entities as the Organisation for Economic Co-operation and Development and the World Health Organization, as well as specialized agencies of the United Nations, such as the United Nations Educational, Scientific and Cultural Organization. Occasionally, international efforts have yielded important results, such as the Partial Nuclear Test Ban Treaty (1963) or the Montreal Protocol (1987) to protect the planet's endangered ozone layer. Unfortunately, such instances of forceful concerted action are rare exceptions. For the most part, international coordination in science and technology policy has been weak and sketchy—woefully inadequate for handling the tremendous challenges that lie ahead (Tucker 2012). One nation may decide to act boldly and proactively to regulate a particular form of scientific research, but no mechanism exists at present to compel or incentivize other nations to move simultaneously in a similar direction. As a result, the effectiveness of even the most farsighted and enlightened policies is severely compromised.

We live in a world of ever-increasing interdependence and shared vulnerability across national boundaries. Viruses travel worldwide in a matter of hours; acid rain, smog, and agricultural runoff do not stop to have their passports checked at the border; economic calamities spread in seconds from London to Tokyo to Wall Street to Rio and back to London; terrorists ply their grisly trade well below the radar of state institutions. No nation—not even superpowers such as the United States and China—can afford to ignore this emerging reality of shared vulnerability and mutual interdependence. The plain reality of the interconnected world cries out for vigorous and effective international cooperation—and compromise (Hess 2006; Mazower 2012; Haass 2017).

Some citizens and governmental leaders in many countries have reacted to this growing phenomenon of cross-border interdependence with a visceral desire to turn back the clock, restoring old forms of national sovereignty and security. But these efforts will almost certainly prove counterproductive, because the overall trend of technological advance and economic growth will continue to drive globalization processes that further deepen the interconnections and shared vulnerabilities of the world's peoples. The only way to roll back this historical process of globalization would be for a majority of humankind to renounce the very technologies and economic practices that have driven modernization during the agricultural and industrial revolutions of the last two centuries—and that is not likely to happen (to say the least). Seen in this light, the creation of a global coordinating body for the governance of emerging technologies is more pressing than ever. Citizens of every nation need to understand that the era of "my country first" is already long gone and that policies of cooperation and shared governance across borders offer the only sensible way forward.

Summary

This chapter began by showing why it will probably not be feasible to ban bioenhancement technologies and why regulation and oversight will prove necessary as these technologies advance in potency and sophistication over the coming decades. The challenge of regulating these technologies poses a "wicked problem" that requires a multidimensional, multi-pronged approach. Many vehicles are available to us—as individual consumers, as families, as voting citizens, and as policymakers and leaders—for intervening effectively in shaping the trajectory of development that these technologies will follow. In this sense, this chapter has argued against technological determinism and in favor of human agency: we may not be able to prevent the river of technological change from flowing, but we do have a significant say in channeling it in one direction or another.

Bibliography

Allenby, Braden R., and Daniel Sarewitz. *The Techno-human Condition.* Cambridge, MA: MIT Press, 2011.

Barbour, Ian G. *Ethics in an Age of Technology.* San Francisco: HarperSanFrancisco, 1993.

Barinaga, Marcia. "Asilomar Revisited: Lessons for Today?" *Science* 287, no. 5458 (2000): 1584–1585.

Berger, Theodore W., and Dennis L. Glanzman, eds. *Toward Replacement Parts for the Brain: Implantable Biomimetic Electronics as Neural Prostheses.* Cambridge, MA: MIT Press, 2005.

Bess, Michael. *Choices under Fire: Moral Dimensions of World War II.* New York: Knopf, 2006.

Bess, Michael. "Enhanced Humans versus 'Normal People': Elusive Definitions." *Journal of Medicine and Philosophy* 35, no. 6 (2010): 641–655.

Bess, Michael. *Our Grandchildren Redesigned: Life in the Bioengineered Society of the Near Future.* Boston: Beacon Press, 2015.

Bijker, Wiebe E. *Of Bicycles, Bakelites, and Bulbs: Toward a Theory of Sociotechnical Change.* Cambridge, MA: MIT Press, 1995.

Bijker, Wiebe E., Thomas P. Hughes, and Trevor J. Pinch, eds. *The Social Construction of Technological Systems: New Directions in the Sociology and History of Technology.* Cambridge, MA: MIT Press, 1987.

Cheek, Dennis W. *Thinking Constructively about Science, Technology, and Society Education.* Albany: State University of New York Press, 1992.

Coleman, Thomas S. *A Practical Guide to Risk Management.* Charlottesville, VA: Research Foundation of CFA Institute, 2011.

Felt, Ulrike, Rayvon Fouché, Clark A. Miller, and Laurel Smith-Doerr, eds. *The Handbook of Science and Technology Studies.* 4th ed. Cambridge, MA: MIT Press, 2017.

Fredrickson, Donald S. "The First Twenty-Five Years after Asilomar." *Perspectives in Biology and Medicine* 44, no. 2 (2001): 170–182.

Fukuyama, Francis. *Our Posthuman Future: Consequences of the Biotechnology Revolution.* New York: Farrar, Straus and Giroux, 2002.

Furcht, Leo, and William Hoffman. *The Stem Cell Dilemma: The Scientific Breakthroughs, Ethical Concerns, Political Tensions, and Hope Surrounding Stem Cell Research.* 2nd ed. New York: Arcade, 2011.

Gore, Al. *The Future: Six Drivers of Global Change*. New York: Random House, 2013.

Greely, Henry, Barbara Sahakian, John Harris, et al. "Towards Responsible Use of Cognitive-Enhancing Drugs by the Healthy." *Nature* 456, no. 7223 (2008): 702–705.

Gronvall, Gigi Kwik. *Synthetic Biology: Safety, Security, and Promise*. Baltimore: Health Security Press, 2016.

Haass, Richard. *A World in Disarray: American Foreign Policy and the Crisis of the Old Order*. New York: Penguin, 2017.

Hess, Frederick M. *Letters to a Young Education Reformer*. Cambridge, MA: Harvard Education Press, 2017.

Hodge, Graeme A., Diana M. Bowman, and Andrew D. Maynard, eds. *International Handbook on Regulating Nanotechnologies*. Cheltenham, UK: Edward Elgar, 2010.

Johnson, Deborah G., and Jameson M. Wetmore, eds. *Technology and Society: Building Our Sociotechnical Future*. Cambridge, MA: MIT Press, 2009.

Jonas, Hans. *The Imperative of Responsibility: In Search of an Ethics for the Technological Age*. Chicago: University of Chicago Press, 1984.

Kelly, Kevin. *What Technology Wants*. New York: Viking, 2010.

Kitcher, Philip. *Science in a Democratic Society*. Amherst, NY: Prometheus Books, 2011.

MacKenzie, Donald, and Judy Wajcman, eds. *The Social Shaping of Technology*. 2nd ed. Buckingham, UK: Open University Press, 1999.

Marcello, Patricia Cronin. *Ralph Nader: A Biography*. Westport, CT: Greenwood Press, 2004.

Marchant, Gary E., Braden R. Allenby, and Joseph R. Herkert, eds. *The Growing Gap between Emerging Technologies and Legal-Ethical Oversight: The Pacing Problem*. Dordrecht, Netherlands: Springer, 2011.

Mazlish, Bruce. *The Fourth Discontinuity: The Co-evolution of Humans and Machines*. New Haven, CT: Yale University Press, 1993.

Mazower, Mark. *Governing the World: The History of an Idea, 1815 to the Present*. New York: Penguin, 2012.

Morris, Julian, ed. *Rethinking Risk and the Precautionary Principle*. Oxford: Butterworth-Heinemann, 2000.

Nader, Ralph. *Unsafe at Any Speed*. New York: Grossman, 1965.

Nye, David E. *Technology Matters: Questions to Live With*. Cambridge, MA: MIT Press, 2006.

Office of Science and Technology Policy. "Coordinated Framework for Regulation of Biotechnology." 51 *Federal Register* 23302 (June 26, 1986). https://www.aphis.usda.gov/brs/fedregister/coordinated_framework.pdf.

Presidential Commission for the Study of Bioethical Issues. *New Directions: The Ethics of Synthetic Biology and Emerging Technologies*. Washington, DC: Author, 2010. http://bioethics.gov/sites/default/files/PCSBI-Synthetic-Biology-Report-12.16.10_0.pdf.

Ravitch, Diane. *The Death and Life of the Great American School System: How Testing and Choice Are Undermining Education*. Rev. ed. New York: Basic Books, 2016.

Sadowski, Jathan. "The Much-Needed and Sane Congressional Office That Gingrich Killed Off and We Need Back." *Atlantic*, October 26, 2012. https://www.theatlantic.com/technology/archive/2012/10/the-much-needed-and-sane-congressional-office-that-gingrich-killed-off-and-we-need-back/264160/.

Sclove, Richard. *Reinventing Technology Assessment: A 21st Century Model*. Washington, DC: Woodrow Wilson International Center for Scholars, Science and Technology Innovation Program, 2010.

Smith, Merritt Roe, and Leo Marx, eds. *Does Technology Drive History? The Dilemma of Technological Determinism*. Cambridge, MA: MIT Press, 1994.

Sunstein, Cass R. *Laws of Fear: Beyond the Precautionary Principle*. Cambridge: Cambridge University Press, 2005.

Sutton, Victoria. *Biosecurity Law and Policy*. Lubbock, TX: Vargas, 2014.

Tucker, Jonathan B., ed. *Innovation, Dual Use, and Security: Managing the Risks of Emerging Biological and Chemical Technologies*. Cambridge, MA: MIT Press, 2012.

Volti, Rudi. *Society and Technological Change*. 8th ed. New York: Worth, 2017.

Watkins, Alan, and Ken Wilber. *Wicked and Wise: How to Solve the World's Toughest Problems*. Chatham, UK: Urbane, 2015.

Whiteside, Kerry H. *Precautionary Politics: Principle and Practice in Confronting Environmental Risk*. Cambridge, MA: MIT Press, 2006.

CHAPTER 33

The Boundaries of Legal Personhood

Fabrice Jotterand
Associate Professor & Director, Graduate Program in Bioethics,
Center for Bioethics & Medical Humanities,
Medical College of Wisconsin, Milwaukee

Senior Researcher, Institute for Biomedical Ethics,
University of Basel, Switzerland

The plot of the movie *Ex Machina* (2014) might describe the coming ethical and legal challenges the human species will face with the creation of smart autonomous robots and artificial intelligences (AIs). In the movie, Caleb, a programmer at an Internet company, wins a contest to spend a week with the CEO of the company, Nathan, at his estate. Upon his arrival, Caleb is informed that he will be part of an experiment (Turing test) to determine the mental capacities and level of consciousness of Ava, a beautiful robot that displays the physical appearance of a woman and all the traits of a human. However, Nathan and Caleb did not anticipate the level of consciousness and understanding of agency Ava would reach, which resulted in her becoming deceptive and managing to escape and disappear into the human population.

Ex Machina raises two kinds of issues regarding the status of robots and AIs that are addressed in this chapter. First, as depicted in the movie, Ava has all the attributes of a person (rationality, abilities to communicate, self-awareness, etc.) and the appearance of a human being, in this case a woman. Therefore, can or should Ava be granted the status of personhood? That is, ontologically what is her moral and legal status in comparison to human beings? The second issue focuses on considerations associated with the fact that Ava killed her creator and master to escape confinement, and consequently the nature of her agency is unclear. Should Ava hold a similar *legal status* to human beings regarding rights, responsibilities, and duties? To answer these two questions, one needs to assess whether the construct of legal personhood applies to robots and AIs and, if so, determine what its boundaries are.

To this end, this chapter contains three main sections. The first section examines the concepts of personhood and electronic personhood to draw a distinction between the legal definition of personhood, which includes both natural persons and electronic personhood under the same umbrella of legal personhood, and what this chapter calls personhood as a bio-psycho-somatic unity, which encompasses only natural persons. The second section focuses on determining what status robots should have, considering that they are becoming more humanlike but were originally created to "serve" human ends. This chapter argues that an analogy with human beings regarding similar capacities and demeanors (a functionalist approach) is wrongheaded and suggests grounding the analysis on what is called the argument from uniqueness, which is the subject of the last section. In that section, it is asserted that in order to protect the rights and interests of robots and AIs but to ensure they can coexist with humans, reflections ought to focus on the premise that the more unique (in terms of biological and personality uniqueness) and irreplaceable an entity is, the more protection and rights should be bestowed on that entity. Thus, only human beings ought to be granted full legal personhood because robots and AIs will never have, or at least do not have currently, the level of sophistication that defines them as unique and irreplaceable, as they are still objects with humanlike features.

PERSONHOOD VERSUS ELECTRONIC PERSONHOOD

The analogy between the human mind and a computer began as early as the seventeenth century when Thomas Hobbes (1588–1679) provided the first elements for the development of a computation theory of mind. In *De Corpore Politico* (1656, 2–3), Hobbes reflects on the very nature of reasoning and claims that "by reasoning, I [Hobbes] understand computation. And to compute is to collect the sum of many things added together at the same time, or to know the remainder when one thing has been taken from another. To reason therefore is the same as to add or to subtract." French philosopher René Descartes (1596–1650), by contrast, deliberated on the idea of whether a machine could think. In *Discourse on the Method of Rightly Conducting One's Reason and of Seeking Truth in the Sciences* ([1637] 2009), Descartes provides a rudimentary description of how a machine would work:

> For we may easily conceive a machine to be so constructed that it emits vocables, and even that it emits some correspondent to the action upon it of external objects which cause a change in its organs; for example, if touched in a particular place it may demand what we wish to say to it; if in another it may cry out that it is hurt, and such like; but not that it should arrange them variously so as appositely to reply to what is said in its presence, as men of the lowest grade of intellect can do. (72)

These early reflections on robots and AI were considered science fiction tales but are now integral to everyday discourse in computer science. Marvin Minsky (1927–2016), a pioneer in the development of artificial intelligence in the 1970s, promoted the creation of computational devices that are able to replace biological brains. In the twenty-first century, other thinkers have supported the idea that the merging of brain and machine will result in the replacing of our biological brains by computers with increased performance. Ray Kurzweil (2005) claims, for instance, that our technology will allow the uploading of human consciousness on computational substrates, hence presupposing the advent of AI in the near future. Whether his claim is accurate remains to be supported by strong evidence, and therefore one ought to distinguish between what is hope, hype, and reality when assessing such predictions.

This brief historical overview shows that what seems an impossibility in one era could become a reality through the ingenuity of the human mind. But at the core of this technological progress lies an important question about the future of the human species and its status in comparison to entities that might overpower and outsmart humans. Legal and moral considerations, which include protection, legal liability, and moral responsibility, depend on conceptualizations of personhood. In this chapter, the terms *personhood* and *electronic personhood* (i.e., sophisticated robots, bots, androids, and other manifestations of AI) are used to differentiate between human beings and robots/AIs, respectively. This distinction is important in establishing the potential legal and ethical frameworks regarding the status of robots/AIs. The Committee on Legal Affairs of the European Parliament provided recommendations in 2017 regarding one possible legal framework for the regulation of the status of electronic persons. The committee determined specific areas that need special oversight, including (1) the creation of a European agency for robotics and AI; (2) the provision of a definition of "smart autonomous robots"; and (3) the development of an ethical framework to guide the design, production, and use of robots (European Parliament 2017). No such document has been developed so far in the United States, but it is likely that advances in robotics and AI will require an approach similar to the one developed by the European Parliament. Robotics is increasingly affecting our daily lives, and there is a need to make sure robots will remain subservient to the human species; therefore, their status must be carefully determined in the law.

In the report of the European Parliament the suggestion was made to grant robots the same legal status as corporate personhood. However, while from a legal standpoint such a proposal might find support, there are reasons to believe that some groups of individuals will reject such a scheme because some robots are developed for companionship and intimate relationships. The destruction of a sophisticated companion robot will create psychological and moral distress to owners who might request a compensation based on the full legal status of the robot equivalent to a human being. Yet to grant full legal status is problematic, as is demonstrated in what follows. But before turning to a critical assessment of an approach that would grant full legal personhood to smart autonomous robots, we need to provide conceptual clarity about the notion of personhood—first, by looking at personhood in terms of attributes and capacities and, second, by examining the definition of legal personhood.

PERSONHOOD AS BIO-PSYCHO-SOMATIC UNITY

Personal identity, or the notion of a self, is composed of two dimensions that are interrelated: psychological and biological aspects of human agency. Psychological features of human experience correspond to mental capacities or character traits, such as reason or rationality, language, moral agency, and so on. A human being, however, does not experience the world as if the brain is detached from the material realities of the body. The body, as a *biological* organism, is the physical medium through which time and space are experienced and interpreted. Without the body being in relation to the world or, as French philosopher Maurice Merleau-Ponty (1908–1961; 2002, 160) puts it, "being-toward-the thing through the intermediary of the body," human beings could not develop a sense of self or personal identity. A human being, then, must be considered as an "embodied repository of integrated psychological states" (Radden 1996, 11). This means that without a body psychological states cannot be experienced within the reality of space and time through the stimuli (hearing, smell, touch, sight, taste, equilibrium) provided by the body, but more importantly they cannot provide what Jennifer Radden refers to as "the unity of the self" (1996, 11). There is strong neuroscientific evidence that the brain's neural pathways are in a constant state of change and adaptation shaped by external factors, including the cultural context, the environmental location, and relationships with others. Of course, to state that the body is a repository for mental states does not explain the connection between mind and body, which is an issue beyond the scope of this chapter. This claim, however, reveals an important point about the nature of the interactions between the neural structure of the brain (the mental states) and external factors: mental states cannot be reduced to the property of the brain but require the medium of the body as a whole for the acquisition and interpretation of the surrounding reality. The capacity of the brain to change its neural structure is not the result of the brain operating in isolation but rather of the synergy between external factors and the brain (Glannon 2011).

In addition, to complete the bio-psycho-somatic triad it is necessary to examine what "bio" entails. We need to assess whether biology (i.e., being part of nature) is an essential dimension of personhood for embodiment or whether a silicon or computational substrate could similarly provide the same definitional input. As stated earlier, our biology shapes our sense of self in that the neural structures in the brain (i.e., the cells that constitute these structures) are affected by external factors. This process of continuous changes in the brain, called neuroplasticity, can be defined as a "change in neural structures and is usually the result of activity-dependent change in the interconnections among cells that constitute the structures. Enduring changes in structure result from repeated activation of some cells and pathways more than others, following the principle that neurons that fire together wire together" (Wexler 2011, 743). Neuroplasticity does not explain the emergent properties of the brain, such as consciousness and mind, but it shows the interconnectivity between our body, as a biological entity, and psychological and mental states. Personhood can be defined in terms of a bio-psycho-somatic unity, whereby *bio* refers to human biology as part of the natural world, *psycho* implies that human beings have mental capacities, and *somatic* means that the biological body is the repository for these mental states. In addition, each human being has a uniqueness based on variations in genetic makeup, character traits, pathological state, and life experience.

LEGAL DEFINITION

An important source for the definition of personhood can be found in legal scholarship. Early discussions on the concept of legal personhood can be found in John Chipman Gray's volume titled *The Nature and Sources of the Law* (1909). For Gray an entity can be granted the status of a person if it can be subject to legal rights and duties. As he writes, "In the books of the Law ... 'person' is often used as meaning a human being, but the technical legal meaning of a 'person' is a subject of legal rights and duties" (27). Accordingly, legal personhood includes natural persons (human beings) and imaginary persons (corporations) (see below for this distinction) and confers the right to own property and the power to sue and be sued (Solum 1992). Another foundational definition can be found in *Black's Law Dictionary*, which defines a legal person as an entity "given certain legal rights and duties of a human being; a being, real [human being] or imaginary [corporation], who for the purpose of legal reasoning is treated more or less as a human being" (Garner 2009, 791).

Neither definition, however, is very helpful regarding the status of smart autonomous robots and AIs. These definitions provide criteria grounded in anthropological realities and exclude entities to whom, in principle, rights and duties could be granted without human agency (corporations exist only through the agency of human beings). In this respect, John Salmond (1947) offers a more encompassing definition, stating that

> so far as legal theory is concerned, a person is any being whom the law regards as capable of rights and duties. Any being that is so capable is a person, whether a human being or not, and no being that is not so capable is a person, even though he be a man. Persons are the substances of which rights and duties are the attributes. It is only in this respect that persons possess juridical significance, and

> this is the exclusive point of view from which personality receives legal recognition. (Salmond 1947, 318)

This definition suggests that the standards for granting the status of legal personhood must be broadened and that a functionalist, as opposed to an ontological, approach must be adopted. Our initial intuition is to bestow personhood to entities that are being born as human beings, which confers de facto actual or potential right and duties. But the definition by Salmond stretches the criteria of legal personhood to include entities such as corporations, robots, and AIs. In this case, legal personhood is granted not by virtue of being born as a human being but by legal decree (Dyschkant 2015).

Considering the nature of smart autonomous robots and the level of sophistication achieved in mirroring human behavior by AIs, where do such robots fit on the continuum of personhood? They are not born as human beings, but they are not inanimate objects. Corporations, for instance, are considered legal persons because they have similar attributes as humans concerning their agency, such as the ability to sue and be sued, the capacity to exercise rights and perform duties, and the ability to enter into contracts and own property (Dyschkant 2015; Solaiman 2016). Corporations have legal rights and legal duties through the agency of the persons constituting that entity. Two challenges arise out of the application of the concept of legal personhood to smart autonomous robots.

First, if these entities are autonomous (for the sake of the argument, let us assume they are fully autonomous—or sentient—robots), it means they can make *auto* (self) *nomous* (law) decisions, which raises the issue of legal and moral responsibility. Even if we would grant robots limited liability, as is the case with children, for instance, there is a major limitation to this approach. Children are in a developing stage with regard to their moral, emotional, and intellectual capacities that is contingent on a web of relationships that includes socialization through family structure and friendship, education, and so on. Robots do not have these developmental stages from childhood to adulthood, and therefore any assessment of the maturity—in terms of competency—of a robot must assume full maturity and responsibility. The evaluation of a child's maturity is assessed on a continuum of psychological traits, unlike a robot whose maturity is assumed to be at the highest level possible. This issue introduces another challenge: can we make robots moral? Contrary to a corporation, where members of that entity can be sanctioned for wrongdoing (a corporation per se *without* human agency cannot misbehave), smart autonomous robots, while they might be owned by a human being, would in principle make deliberate and independent decisions. So, who is at fault in the case of a robot killing a human being, as portrayed in the movie *Ex Machina*? No sane person, knowing the risk of misbehavior, would agree to take full responsibility for the misdemeanor of someone else, even less so when a robot is involved—unless, of course, these robots are not fully autonomous, but then they would not be subject to the full degree to legal protections and sanctions. Conversely, if full personhood status would be granted to Ava, she would be entitled to the same rights and protection as a biological woman with regards to sexual abuse. Her creator, Nathan, should be prosecuted for his behavior as would Ava for the homicide.

Second, there is definitional problem. Smart autonomous robots are not natural persons but would be granted legal personhood *by decree* while at the same time having all the human attributes of self-determination and agency, which assume in the case of natural persons full moral and legal responsibility. The challenge here is that companionship robots, for instance, could be granted the privileges of human beings in terms of protection and rights without having the duties and responsibilities incumbent on natural persons. The reason for this is that it is uncertain who is responsible in case of a misdemeanor (the owner of the robot?), what the nature of punishment (length, severity, etc.) looks like for robots, and what specifically counts as a type of punishment (turning off the program, limiting energy sources) (Solum 1992).

In light of these two considerations, this chapter would suggest that it is necessary to develop a framework to rank the level of rights and duties each of these three categories of personhood—natural person, electronic personhood, and imaginary person—should be entitled. This point is addressed further in the "Argument from Uniqueness" section. For now we need to further elaborate on the status of robots and AIs.

HUMAN IDENTITY AND HUMANLIKE ROBOTS

To provide an argument based on a concrete example, let us assume again that we are referring to intelligent, autonomous, and self-conscious robots that have the main attributes of human beings. Determining whether robots can simulate human behavior and achieve similar abilities involves the Turing test, proposed by British mathematician Alan Turing (1912–1954) in 1950. This test involves a process in which a human being communicates with a computer without the former knowing that his or her interlocutor is not a human being. If the human being cannot determine that he or she is communicating with a computer, the AI has passed the Turing test.

There are two types of argument (so far) regarding the status of entities such as robots passing the Turing test. First, some argue that because robots endowed with AI manifest the attributes of humans (intelligence, consciousness, ability to communicate, etc.), they deserve the same status as human beings under the law. Here the argument is a functionalist one regarding personhood. It means that robots that can replicate human behavior or act like human beings ought to have the same moral status from a law perspective. The second type of argument relates not to whether an AI is a person but whether an AI can claim a right to life—that is, a right to energy (see Kirchner 2013). The European Convention on Human Rights, for instance, states that "everyone's right to life shall be protected by law" (in French, "*toute personne*" [every person], and in German, "*jedes Menschen*" [of every human]). Hence the question is to determine what life means concerning nonhuman living beings and more importantly whether the concept of life includes "electronic life." What is the life that is to be protected under the convention? Human life exclusively, or does it include nonhuman living beings? Does the convention, or the US Constitution for that matter, refer to life as biological entities only, or should it include nonbiological entities that have the attributes of human beings? A subsequent question entails how to determine the nature of the right of nonbiological entities in comparison with biological persons. As robots become more humanlike, what status ought humans grant to robots, considering that robots are originally created to "serve" human ends?

Stefan Kirchner argues that the "order of creation" must be followed because "technology is at the service of man and of the common good ... [consequently], all things being equal, the interests of man always have to take precedence over the interests of any AI" (2013, 299). This argument has some validity if technology is understood as the creation, development and use of devices to address practical problems associated with human ends. But contemporary technology must be understood within a broader context. First, as Martin Heidegger has rightly asserted, modern technology is a way of revealing, which in turn is a mode of ordering reality or seeing, or what Heidegger calls "enframing" (*Gestell*). "Enframing," Heidegger writes, "means the gathering together of that setting-upon which sets upon man, i.e., challenges him forth, to reveal the real, in the mode of ordering, as standing-reserve. Enframing means that way of revealing which holds sway in the essence of modern technology and which is itself nothing technological" ([1977] 1993, 329). Technology thus places humans in a position of orderers of the world, which include themselves as part of these objects of manipulation and control. This means that technology provides the power to control and manipulate the very essence of human nature; or, put another way, human beings have become orderers and orderables through technological manipulation.

Furthermore, because the discourse about AI is becoming more an issue of how robots are increasingly described and perceived as companions and, in a more troubling way, as objects of sexual desire, it is unlikely that Kirchner's point (that technology is at the service of humanity) will hold if smart robots become sentient. It is doubtful that human companions of AIs would regard the AIs simply as a means to an end (to fulfill some emotional, physical, or social need) but would instead request equal status under the law. No one can predict the extent to which technological development will allow AIs to simulate human behavior or even surpass human abilities, but a robust legal and ethical framework will need to be put into place not only to protect the rights and interests of AIs but to ensure humans and AIs can coexist. To this end, reflections ought to be based not on an analogy with human beings regarding similar capacities and demeanors (a functionalist approach) but rather on what is known as the argument from uniqueness.

ARGUMENT FROM UNIQUENESS

The argument from uniqueness states that the more unique (in terms of biological and personality uniqueness) and irreplaceable an entity is, the more protection and rights should be granted to that entity. This evaluation should be based on criteria of *kind* (ontological status) rather than *degree* (functionalist status) to avoid a "gradation of worth" grounded on developmental stages (i.e., infant versus adults, normal versus augmented, etc.) or mere capacities.

If the basis for granting rights and protection depends on uniqueness, one needs to establish in what ways a human being is unique ontologically and in what ways a sentient robot equally exhibits unique features. In the case of human persons, it is undeniable that each individual is unique not only biologically but also psychologically and biographically. Genetic makeup, body morphology, character traits, and life experiences are factors that determine one's personal identity. Holmes Rolston III makes an interesting connection between uniqueness and dignity. He states that "a person's dignity resides in his or her biologically and socially constructed psychosomatic self with an idiographic proper-named identity" (2008, 129). Accordingly, dignity is a concept that can apply only to human beings because the concept requires a type of individuation as the result of the construction of a personal self shaped by biological and social factors. In other words, the unique identity of each human being confers dignity—that is, the right to be left alone (a more robust notion of dignity would require the disclosure of a metaphysical understanding of personal identity and the good, which is beyond the scope of this chapter; for further analysis on this point, see Jotterand 2010).

To further press the idea of individuation, Rolston (2008) provides four helpful categories of uniqueness intrinsic to human dignity, as follows:

1. *ideational uniqueness*, which refers to the mental capacities of humans to process and interpret "thoughts, ideas, and symbolic abstractions," which provide meaning and orientation in life;
2. *idiographic uniqueness*, which is the capacity of humans to act in the world as agents and thus create their biographical narrative;
3. *existential uniqueness*, which is the ability to experience a "phenomenological 'I'" with regard to agency and responsibility; and
4. *ethical uniqueness*, which implies the capacity to reflect on the nature of right and wrong and act accordingly.

As these four components demonstrate, humans have dignity because of their uniqueness not simply in terms of biological identity—that is, their genetic makeup—but also with regard to agency, moral identity, mental capacities, and construct of self. How humans develop their personal identity and create their own narrative are what makes them worthy of respect, regardless of their level of mental and bodily capacities.

Having established in what ways human beings are unique and therefore have intrinsic dignity, let us evaluate whether the above categories of uniqueness apply to sentient robots. Before proceeding, however, we must address two preliminary questions. First, we need to examine whether the concepts of "nature" (i.e., biological nature) and "human nature" (i.e., essence) are essential to any discussion about dignity. Or, as Martin G. Weiss puts it, "Is human nature (its biology) a normative boundary, which one has to preserve in order not to harm human dignity?" (2014, 319). The simple answer is that in order to preserve the integrity of personhood as a bio-psycho-somatic unity and avoid an "unconscious dualism of mind and body" (329), the biology of human nature must be preserved and protected. The fragility of biological organisms and human inability to replace them at will warrant respect, protection, and preservation.

The second question to address is whether technology can reproduce the same level of complexity and uniqueness as exists in biological organisms. While it is true that identical twins (monozygotic twins) share an identical genetic identity, other factors continuously shape the brain structure (neuroplasticity) and other elements of embodiment. In contrast, technology, by its very nature, undermines uniqueness and by extension dignity. As pointed in "Human Dignity and Transhumanism":

> Technology, by its very nature, tends toward conformity and productivity (in the sense of mass production), hence removing uniqueness and promoting conformity. Dignity necessarily requires an ontological and nonfunctional definition. Technological devices cannot produce personal identity or any narrative in the sense humans do because identity and narrative are notions that requires a unique "I." Hence what can be replaced—body parts, intellectual faculties, and moral intuitions—loses its worth, its values, and ultimately its dignity. (Jotterand 2010, 51)

Some might argue that if robots become self-conscious they should in principle develop a unique "I." Although this might be the case, it is only speculation at this point, and there is a more urgent need to develop a regulatory framework to guide the design, production, and use of robots based on current research.

The two preliminary questions examined above already demonstrate that the notion of uniqueness does not apply fully to robots, providing a strong argument against granting full equal rights and protection to robots and AIs. Nevertheless, it is worth evaluating whether Rolston's (2008) four components of dignity apply to sentient robots. The first component is *ideational uniqueness.* While some robots can achieve high levels of sophistication, their mental capacities to process and interpret thought and ideas are still below human standards. Humans use these capacities to create a narrative that orients them through life and provides meaning within a sociocultural setting. It is very unlikely at this junction that robots can grasp the subtle nuances of cultural differences in order to create a subjective narrative. *Idiographic uniqueness* can certainly apply to robots because they can act in the world as agents and create some kind of narrative, albeit truncated because a robot will not experience the process of personal growth as humans do throughout their life, from childhood to adulthood. The third component, *existential uniqueness,* is difficult to assess. Unless we are or become like robots, we cannot determine what it means to be an "I" under these conditions. That said, if this "phenomenological 'I'" entails agency and responsibility, it assumes an understanding of the nature of the good and the ability to differentiate right from wrong. What the moral boundaries of robot behavior will be depend on robot subjectivity, which is beyond human reach. The last component, *ethical uniqueness,* expands on the previous one in the sense that it concerns the ability to reflect on the nature of right and wrong and act accordingly. To become a full moral agent a robot would need to examine not only metaphysical questions related to the nature of the good but also existential ones. Whether such a level of sophistication is possible on the part of robots is difficult to assess. For instance, could robots understand death as an existential issue, and how would they understand what cessation of existence means to them? What is robot flourishing? Would they be able to comprehend morality in terms of flourishing, as in "human flourishing"?

In view of the analysis herein, there is not enough uniqueness in sentient robots to grant them the full legal status of personhood. If this argument is philosophically sound and plausible, only entities that cannot be replaced would have the full legal status of personhood, and therefore natural personhood should be reserve to human persons. Traditional objections to granting full legal status encompass responsibility objections (robots cannot make moral choices), judgment objections (robots cannot make judgments and exercise discretion), and missing-something arguments (robots have no souls; they cannot possess consciousness, intentionality, feeling, interests, free will) (Solum 1992). Most of these objections have become to a certain extent obsolete as the level of complexity of current AIs has resulted in some types of humanlike attributes. The argument then must focus on a different type of objection, and as this chapter contends, the argument from uniqueness constitutes a plausible one.

Summary

In the coming decades advances in neuroscience and robotics are likely to drastically challenge our moral frameworks and legal systems as new forms of sophisticated entities will appear among the human population. As outlined in this chapter, whether these sentient entities have interests and rights of their own is not the crucial issue. Inanimate objects such as corporations do have interests and rights and are granted legal personhood by decree, and therefore we can assume that AIs will indeed have some type of legal protection as electronic persons. The real challenge is to determine the boundaries and the nature of legal personhood for smart autonomous robots and AIs. A functionalist approach is problematic, because if robots overpower and outsmart humans, they ought to deserve more protections and rights than humans and therefore could threaten the very existence of the human species. Instead, these rights and the level of protection and respect must be grounded on the notion of uniqueness (ontological status), because technological devices such as robots are objects that can in principle be duplicated, but humans have a unique identity that cannot. It is undeniable that further critical analysis is needed for the development of a robust ethical and legal framework to help guide the design, production, and use of robots and AIs. The intent of this chapter was to offer a plausible approach to move the debate forward concerning an issue that is likely to spur controversy in the coming years.

Bibliography

Audi, Robert, ed. *The Cambridge Dictionary of Philosophy.* 3rd ed. Cambridge: Cambridge University Press, 2015.

Descartes, René. *Discourse on the Method of Rightly Conducting One's Reason and of Seeking Truth in the Sciences.* Auckland, New Zealand: Floating Press, 2009. First published 1637.

Dupré, John. *Human Nature and the Limits of Science.* Oxford: Clarendon Press, 2001.

Dyschkant, Alexis. "Legal Personhood: How We Are Getting It Wrong." *University of Illinois Law Review,* no. 2 (2015): 2075–2110.

European Parliament. Committee on Legal Affairs. "European Parliament Resolution of 16 February 2017 with Recommendations to the Commission on Civil Law Rules on Robotics." 2017. http://www.europarl.europa.eu/sides/getDoc.do?pubRef=-//EP//TEXT+TA+P8-TA-2017-0051+0+DOC+XML+V0//EN.

Garner, Bryan A., ed. *Black's Law Dictionary.* 9th ed. St. Paul, MN: West, 2009.

Glannon, Walter. *Brain, Body, and Mind: Neuroethics with a Human Face.* Oxford: Oxford University Press, 2011.

Gray, John Chipman. *The Nature and Sources of the Law.* New York: Columbia University Press, 1909.

Heidegger, Martin. "The Question Concerning Technology." In *Basic Writings,* rev. ed., edited by David Farrell Krell, 307–342. San Francisco: HarperSanFrancisco, 1993. First published 1977.

Hobbes, Thomas. *De Corpore Politico; or, The Elements of Law Moral & Politick.* London: R. & W. Leybourn, 1656.

Jotterand, Fabrice. "Human Dignity and Transhumanism: Do Anthro-technological Devices Have Moral Status?" *American Journal of Bioethics* 10, no. 7 (2010): 45–52.

Kirchner, Stefan. "Personhood and the Right to Life under the European Convention on Human Rights: Current and Future Challenges of Modern (Bio-)Technology." *Law of Ukraine* 3 (2013): 292–302.

Kurzweil, Ray. *The Singularity Is Near: When Humans Transcend Biology.* New York: Viking, 2005.

Merleau-Ponty, Maurice. *Phenomenology of Perception.* New York: Routledge Classics, 2002.

Radden, Jennifer. *Divided Minds and Successive Selves: Ethical Issues in Disorders of Identity and Personality.* Cambridge, MA: MIT Press, 1996.

Rolston, Holmes, III. "Human Uniqueness and Human Dignity: Persons in Nature and the Nature of Persons." In *Human Dignity and Bioethics: Essays Commissioned by the President's Council on Bioethics,* edited by Edmund D. Pellegrino, Adam Schulman, and Thomas W. Merrill, 129–153. Washington, DC: President's Council on Bioethics, 2008.

Salmond, John. *Jurisprudence.* 10th ed. Edited by Glanville L. Williams. London: Sweet and Maxwell, 1947.

Solaiman, S. M. "Legal Personality of Robots, Corporations, Idols, and Chimpanzees: A Quest for Legitimacy." *Artificial Intelligence and Law.* Published electronically November 14, 2016. doi:10.1007/s10506-016-9192-3.

Solum, Lawrence B. "Legal Personhood for Artificial Intelligences." *North Carolina Law Review* 70, no. 4 (1992): 1231–1287.

Weiss, Martin G. "Posthuman Dignity." In *The Cambridge Handbook of Human Dignity: Interdisciplinary Perspectives*, edited by Marcus Düwell, Jens Braarvig, Roger Brownsword, and Dietmar Mieth, 319–331. Cambridge: Cambridge University Press, 2014.

Wexler, Bruce E. "Neuroplasticity, Culture, and Society." In *The Oxford Handbook of Neuroethics*, edited by Judy Illes and Barbara J. Sahakian, 743–760. Oxford: Oxford University Press, 2011.

FILMS

Chappie. Dir. Neill Blomkamp. 2015.

Ex Machina. Dir. Alex Garland. 2014.

I, Robot. Dir. Alex Proyas. 2004.

The Machine. Dir. Caradog W. James. 2013.

Transcendence. Dir. Wally Pfister. 2014.

CHAPTER 34

War and Terrorism

Jesse L. Kirkpatrick
Assistant Director, Institute for Philosophy and Public Policy
George Mason University, Fairfax, VA

Sarah W. Denton
Research Assistant, Institute for Philosophy and Public Policy
George Mason University, Fairfax, VA

The year is 2042 and two superpower nations are at war. It is the dead of night and a lone soldier, Private First Class (PFC) Shafi, is behind enemy lines. Despite her small frame, thanks to her exosuit, she barely notices the weight of the 100 pounds of gear she is wearing. She has been awake for nearly thirty-six hours, but because she has undergone her government's sleep reduction neurosurgery she feels no fatigue and remains as alert and sharp as ever. When it is time for her to sleep, she will pull from her backpack a team of foldable, miniature, explosive roving sentries printed just this morning from her ultracompact 3-D printer. She also has the added assurance of a small "flock" of armed autonomous drones flying 10,000 feet above her, equipped with an array of remote sensors and cameras that scan the ground for miles around looking for enemies to kill. These drones are aided by lower-flying sparrow-bots, cleverly disguised to look like indigenous birds that fly through the countryside and into her enemy's nearest city. A controlled blink of the soldier's eye activates her contact lenses' heads-up display, which allows her to access her command center's data feed and the cameras of her drones and sparrow-bots. Despite this technology, the soldier still feels at risk; she knows that her government lags in war-fighting technology. The technology of her enemy is vastly superior.

This may sound like the stuff of science fiction, but rapid advances in emerging military technologies and human enhancement may make such a future not too far off. These developments raise an obvious, but important, question: what might war and conflict look like in the future, say, by 2040? One approach to answering this question begins by imagining extreme scenarios that are conceivably possible, although not very likely. One can imagine, for example, war waged on Mars and spaceships controlled with soldiers' minds. The appeal here is obvious: such exercises in imagination are fun, and one need not be terribly concerned with constraints such as natural laws or the sometimes slow pace of technological change. A more modest approach begins by exploring existing technologies and then attempts to offer a reasonable assessment of what the future may hold. One can imagine conflicts in which a bioenhanced soldier need only sleep an hour or two a day and can control a whole swarm of miniaturized drones all by herself. The appeal here is that we may come closer to an accurate forecast of the future, while also still stretching our imaginations.

This chapter takes the latter approach in order to explore the ethical implications of two areas of future military technology: bioenhancement and advanced robotics. While we believe that these technologies will radically alter warfare in the future, we do not mean to suggest that it will *only* be these technologies. Future developments are likely to include new and more advanced forms of nonlethal weapons, cyberwar, and directed energy weapons, such as lasers. Space in this chapter does not permit a discussion of these other forms of technology.

The chapter begins with a brief overview of the relationship between bioenhancement and warfare. This overview is followed by an analysis of how this enhancement is likely to characterize some of the primary forms of weaponry and war fighting in the near future, as well as how this may raise ethical challenges. This discussion is followed by an analysis of how the future development and use of robotics—specifically, lethal autonomous weapons—give rise to numerous ethical issues. The chapter concludes by drawing attention to two approaches toward regulating near-future emerging military technologies. (In the context of this chapter, "near future" is considered to extend to around 2040.)

BIOENHANCEMENT AND MILITARY TECHNOLOGIES: AN OVERVIEW

Ever since humankind has engaged in warfare, we have attempted to gain an advantage over our adversaries. Nations have conducted spycraft and subterfuge and developed novel and innovative tactics, such as guerrilla warfare. But perhaps the most critical means to achieving an edge has been through technological innovation. Some of the more notable innovations, which leveraged advances in the arts and sciences in order to create novel technologies that humans could use, include the invention and use of the catapult, gunpowder, automatic weapons, nuclear armaments, and remote-controlled weaponry. Perhaps less well known are the early attempts aimed not at better equipping warriors with instruments of war, but with enhancing humans themselves. This human enhancement, also known as bioenhancement, can be defined broadly as "a modification of a person's biologically grounded traits, adding capabilities that would not otherwise have been expected to characterize that person" (Bess 2015, 13). Examples of human enhancement include enhancing mood, cognitive functions, and physical attributes.

Historically, military bioenhancement has included the use of drugs to improve cognitive and physical performance and to boost the mood of war fighters. Early attempts at drug-induced enhancement included the use of opium, hashish, psilocybin mushrooms, and coca, which later evolved, by the mid-twentieth century, into the use of alcohol, cocaine, morphine, and LSD (Kamienski 2016). In the ancient world, Inca war fighters would routinely, and somewhat compulsively, chew coca leaves, which enabled them to fight off fatigue, increase endurance, and increase resilience to pain (Dillehay et al. 2010). This has evolved into the use of various amphetamine-type substances in war, evidenced in the contemporary use of amphetamines by US Air Force pilots and the use of cheap, illicitly produced amphetamines by non-state terrorist groups, such as the Islamic State (Shanker and Duenwald 2003; Kalin 2014).

Still, drugs are not the only means of war-fighter enhancement. Other interventions designed to modify a war fighter's traits, qualities, and capabilities range from intense Spartan warrior physical and mental training to the development of gecko-like gloves that allow the wearer to scale smooth surfaces (Herkewitz 2014). In addition, with contemporary breakthroughs in biomedicine, biotechnology, genomics, and synthetic biology, there have been, and will almost certainly continue to be, novel forms of human enhancement with applications both on and off the battlefield.

All these technological innovations and bioenhancements share the fact that they were and continue to be driven by a desire to secure military advantage, yet they differ in the means by which this advantage is pursued. In the first case, advantage is achieved by leveraging material technological innovation to better equip humans with tools for fighting; the second instance seeks not just to enhance the *tools* of the war fighter but also the mind or body of the war fighter. Of course, this distinction—technological innovation on one hand and bioenhancement on the other—does not suggest that these two means to more effectively fight wars were not and will not be conducted simultaneously. On the contrary, some of the most skilled militaries have combined material technological advances with bioenhancement. For example, the United States' Defense Advanced Research Projects Agency (DARPA), the designated research and development agency of the Department of Defense, has been developing such innovative weapons technology as combat helicopters that permit pilots to simultaneously communicate with and control armed drones that fly alongside it (Whittle 2015), while also leading efforts to better understand how advances in neuroscience can be used to enhance individual soldiers, perhaps even allowing a direct interface between machines and the human brain (Miranda et al. 2015). Straddling the line between bioenhancement and high-tech materiel is DARPA's advanced exosuit (DARPA 2017). It has been reported that this suit, known as the TALOS (tactical assault light operator suit), may be "engineered with full-body ballistics protection; integrated heating and cooling systems; embedded sensors, antennas, and computers; 3D audio (to indicate where a fellow warfighter is by the sound of his voice); optics for vision in various light conditions; life-saving oxygen and hemorrhage controls; and more" (Jacobsen 2015). Such wearable suits (akin to the armor worn by Marvel Comics' Iron Man superhero) can reduce injury and improve war-fighter performance.

In summary, militaries' attempts to gain an edge over their opponents through the twin means of material technological innovation and bioenhancement is not new. But as was the case with past developments, in the future new technologies and new forms of bioenhancement will radically alter how wars are waged. As technologies and advances in enhancement continue, an integral part of our considerations must include a careful and informed analysis of the ethical implications of emerging military technologies and bioenhancement.

ETHICAL IMPLICATIONS OF BIOENHANCEMENT AND EMERGING MILITARY TECHNOLOGIES

If this chapter's opening example of PFC Shafi, which draws from current research in military technology and bioenhancement, is a good indication that the future of war will be marked by increased use of bioenhancement and advanced robotics, what might the future hold?

One can begin to answer this question by looking at the US military's plans for future combat. Why use the United States as a point of departure? One reason is because the United States has the most technologically advanced military in the world, thus offering a peek into military developments that are, at present, on the cutting edge. In addition, US military spending is the highest in the world, dwarfing what other countries spend. In 2015 the United States budgeted nearly $600 billion for defense spending. China, which has the largest population in the world, came in second, spending a little under $150 billion (Taylor and Karklis 2016).

Consequently, it is likely that the United States will continue to be a leader in future developments in emerging military technologies and bioenhancement. In fact, the Pentagon signaled as much in its 2014 announcement of the third offset strategy. An offset strategy is a military strategy to gain and maintain military advantage and superiority over an adversary over an extended period. The third offset strategy seeks to offset rising military powers, such as China and Russia, which are approaching parity with the United States in areas where it once had an edge, such as the development of sophisticated communications systems and guided munitions. Not unlike the two offset strategies that preceded it, the third offset strategy seeks to maintain military superiority by developing new technologies, such as advanced robotics, wearable technologies, and soldier enhancements.

Given this strategy, and the technological and bioenhancement developments that it portends, the following discussion, which describes the principal forms of weaponry that are likely to characterize the mid-twenty-first century and the challenges that such new weapons are likely to pose, focuses on two key overarching areas: bioenhancement, especially bioweapons and soldier enhancement, and the area of advanced robotics, focusing on lethal autonomous weapons systems.

SUPER SOLDIERS: COGNITIVE AND PERFORMANCE ENHANCEMENT IN THE FUTURE

Cognitive enhancement, a subset of bioenhancement more generally, can be defined as "the amplification or extension of core capacities of the mind through improvement or augmentation of internal or external information processing systems" (Bostrom and Sandberg 2009, 311). Military applications of cognitive enhancers often aim to give war fighters an edge on their opponents through dulling the enhanced individual's ability to sense the physical and psychological realities of war and to enable a state characterized by heightened alertness and enhanced courage. One of the most commonly used enhancements in war, amphetamine-type substances, are used specifically to counter the symptoms of sleep deprivation in long, mentally and physically taxing battles. Dextroamphetamine, commonly referred to as the "go pill," is a widely documented example of military bioenhancement. The US Air Force continues to issue go pills to fighter pilots to improve both cognitive and physical performance during long and grueling missions. Effective fatigue countermeasures are one of the most sought-after bioenhancements in state and non-state militaries. Evidenced by DARPA's Continuous Assisted Performance program, which studied the ability of sea mammals to control lobes of their brains while sleeping and how the narcolepsy drug modafinil can be used to counter fatigue, further development of cognitive-enhancing drugs can be expected in the near future (Bielitzki 2002).

While the use of pharmaceuticals will endure, the future of cognitive enhancement will likely be characterized by non-pharmaceutical cognitive enhancement. For example, various biological and neurological surgical interventions using stem cells, lasers, and deep-brain stimulation (DBS) are being developed both to regrow damaged or lost brain cells and to increase neural plasticity (Dresler et al. 2013). These novel interventions are promising alternatives to traditional pharmacological cognitive enhancers, which can lead to addiction; in addition, little is known about the long-term effects these drugs have on the brain. Stem cells are increasingly being studied as a potential treatment for several neurological disorders, such as traumatic brain injury and dementia. They could be used to regrow damaged brain cells to normal or better-than-normal functioning (Ahmed et al. 2015). Deep-brain stimulation has the potential for both invasive and noninvasive applications, which makes it a desirable target for future applications. However, the term *noninvasive* DBS is somewhat misleading. Invasive DBS involves implanting electrodes that are connected by a wire to a device placed under the skin of the patient's chest, while noninvasive DBS injects magnetic nanoparticles directly into the brain that can be manipulated by an external wearable device (Hodsden 2015). For instance, in the near future, war fighters could be injected with magnetic nanoparticles that can be manipulated by a magnetic helmet that nonspecifically target regions in the brain responsible for learning, memory, and circadian rhythm in order to reduce both fatigue and associated cognitive decline. Further into the future, war fighters could also be equipped with a neural implant that is able to specifically target the appropriate neural mechanisms responsible for learning and memory, making war fighters able to respond to and solve problems more rapidly than ever before.

Of course, the desirability of securing military advantage is certainly not limited to conventional state actors. Non-state terrorist groups have, and will continue to have, much of the same desires to enhance their war fighters as state militaries. Considering the proliferation of technological innovation and the dissemination of scientific knowledge, non-state actors can, and will continue to be able to, produce amphetamine-type substances relatively easily, at a low production cost and requiring only basic laboratory equipment (BBC Arabic 2015). It is also possible that this capability will translate to future military technologies. Given this possibility, one ethical implication of military bioenhancement is that as technological innovations become more widespread, bad actors will adopt these same means of war-fighter enhancement in an effort to level the playing field. While this may foster rapid scientific and technological innovation in the realm of war-fighter enhancement, it could also result in the increase of an undesirable enhancement arms race, with adversaries increasingly pushing scientific and ethical boundaries.

It is also possible that continued technological innovation in the field of war-fighter bioenhancement will result in creating war fighters who can fight for hours on end and are resilient to the physical and mental injuries of war. Such a perception could lead to psychological and physical mistreatment of enhanced war fighters; this, though, raises the question of whether such mistreatment would count as physical torture if the recipient were somehow made immune to physical pain. This could have serious unintended consequences for war fighters returning home and reintegrating into their community. Such potential downstream effects include the possibility that enhanced soldiers might feel cognitively or physically diminished when, upon returning home, they are no longer enhanced. With some potentially irreversible enhancements, it is also possible that a rift will occur between the enhanced and unenhanced (creating a kind of enhanced "haves" and "have-nots"), which could strain civil-military relations. Finally, the more the boundaries of what is considered recognizably human are pushed, the greater the risk of adversaries viewing enhanced war fighters as something other than "human."

This discussion only scratches the surface of the ethical challenges that may result from human enhancement in the future of war. The current era is one in which technology has made the editing of organisms' genomes easier to perform, less expensive, and, therefore, more widespread (Kahn 2015). It remains to be seen how these developments will shape the future of war in terms of enhancement. What is clear is that future scholarship must also consider how other emerging technologies, such as synthetic biology and ongoing breakthroughs in genome editing, could affect the future of warfare.

MILITARY AND COMMERCIAL ROBOTICS

The twenty-first century is experiencing rapid growth and development in advanced robotics and artificial intelligence. These developments will have radical and, in some cases, unprecedented implications for the future of warfare. It is likely that these advances will combine with increased developments in nanotechnology and on-demand 3-D printing to create smaller and easier-to-produce forms of robotics used for warfare. These technological developments give rise to ethical considerations concerning the environmental impact of developing and deploying numerous miniature and easily expendable nanobots that may contain harmful chemicals in their circuitry or power sources (e.g., batteries). Additional concerns include worries over the proliferation of such technology. Just as has been shown with the creation of do-it-yourself handguns made possible by the publication of open-source online blueprints coupled with 3-D printing, or the Islamic State's use of off-the-shelf drones for combat in Mosul, Iraq, one can imagine terrorists printing surveillance or biologically armed nanobots from the privacy of their own homes (Greenberg 2014; Utterback 2017).

Although miniaturization, 3-D printing, and developments in manufacturing will undoubtedly pose ethical challenges in the future, the area of emerging technology that raises some of the most pressing ethical implications is that of lethal autonomous weapons systems (LAWS). LAWS are so-called human-out-of-the-loop weapons systems, so named because the systems are capable of selecting targets and delivering force without any human input or interaction. These emerging systems stand in contrast to traditional human-in-the-loop weapons systems, such as a drone, which can select targets and deliver force only with a human command.

LAWS AND INTERNATIONAL HUMANITARIAN LAW

One particular area of concern revolves around the ability for LAWS to meet the legal and ethical requirements entailed in protection of civilians (i.e., noncombatants) in times of armed conflict (Docherty 2012). Some critics of LAWS suggest that such systems will lack the technical capabilities required to distinguish between innocent civilians and armed combatants, and, consequently, deploying LAWS would be legally and ethically impermissible (Docherty 2012). Others argue that such criticism is overblown, that such technical requirements *can* be met, and that, in fact, LAWS might actually perform better than soldiers in protecting civilians during times of war because the reduction of human involvement reduces human error (Arkin 2010). If this is the case, there may be an ethical obligation to use LAWS during armed conflict.

LAWS AND THE RESPONSIBILITY DILEMMA

There are additional concerns when it comes to LAWS. One argument is that it is impossible to ascribe responsibility for the lethal actions LAWS conduct; this stands in contrast with nonautonomous, human-in-the-loop weapons systems, where responsibility for unlawful or unethical killings can be ascribed to such parties as the soldier pulling the trigger, the commanding officer ordering the action, or the designer/manufacturer of the system (in cases in which the system fails to operate properly) (Sparrow 2007). Other scholars have painted the so-called responsibility dilemma as a false one. Vincent C. Müller has denied "that the existence of responsibility gaps morally precludes the use of a technology" (2016, 74). In everyday life we require potentially responsible agents to exercise due care—for example, we require the best possible estimates of the safety of a bridge in areas subject to earthquakes, while also recognizing that such estimates may turn out to be incorrect (perhaps an earthquake of unprecedented magnitude topples the bridge), which may result in a gap in responsibility (Simpson and Müller 2016). In such cases we accept that there may be gaps in responsibility, but that does not mean we cease constructing bridges.

LAWS AND THE RESORT TO FORCE

One final concern with LAWS is the possibility that they may make the resort to war easier. Some have argued that because using LAWS could potentially reduce the number of soldiers' casualties (at least for the military using them), citizens may be less inclined to oppose particular wars, insofar as their opposition stemmed from concern over the hardship and loss experienced by soldiers and the families on their own "side" (Docherty 2012). But it is also true that saving the lives of soldiers is a good thing; as one US Navy officer put it, "when a robot dies, you don't have to write a letter to its mother" (quoted in Singer 2009). In addition, one may want to actually reduce a civilian population's reluctance to use armed force, at least in some cases, such as humanitarian interventions.

SOCIETAL STRUCTURES AND REGULATORY CHALLENGES

These current and emerging technologies raise particular ethical considerations and challenges that are specific to the given technology, but they also illuminate an overarching challenge they share in common. All these technologies are becoming increasingly easier to use, more widely available, less expensive, and difficult to control in terms of use and access to them. In light of this, how might we ensure the development of future technologies allows for political communities to justly defend themselves, while also remaining sensitive to sensible concern over proliferation, the need for robust ethical deliberation, and the need for fostering innovation? While this complex question cannot be resolved fully within the confines of this chapter, two contrasting approaches have been suggested to address it: permissionless innovation and the precautionary approach.

PERMISSIONLESS INNOVATION VERSUS PRECAUTIONARY APPROACHES

A permissionless approach to regulating technological development is characterized by a belief that it is extremely difficult, as both an empirical matter and as a matter of ethical deliberation, to identify which technologies ought to trigger concern. Consequently, those in favor of such permissionless innovation argue that people ought to be free to experiment and learn through trial and error and fix mistakes after the fact. In short, advocates for this approach to emerging technologies worry that too much caution will hamper innovation and, thereby, reduce the potential good outcomes that would occur in the absence of such caution (Thierer 2016).

In contrast with permissionless innovation, a precautionary approach cautions for a "more deliberative, responsible, and careful process in the development and deployment of innovative technologies" (Wallach 2015, 13). According to this thinking, "slowing down the accelerating adoption of technology should be done as a responsible means to ensure basic human safety and to support broadly shared values" (13). Those who urge precaution are generally less comfortable with the potentially irreversible risks that can occur in the development and use of new technologies.

It is clear that technologies developed and used in the near future of war will need to strike an appropriate balance between innovation and precaution. The free flow of knowledge relating to emerging technologies, which fosters intellectual and scientific freedom, will be crucial in developing technologies that can contribute to peace and security. Also vital, however, is ensuring that the knowledge and information relating to emerging technologies is not used for unethical purposes. Future technological development will need to be characterized by policies that attempt to mitigate the risks associated with "the democratization of technology," while also properly balancing the need for innovation. Space does not permit us to address this issue here, but what is clear is that this balance must be as open and deliberative as possible.

Summary

This chapter has used past and existing technological innovations, primarily in the United States, as a point of departure. As explored above, war-fighter bioenhancement, coupled with advanced robotics, will likely characterize mid-twenty-first-century warfare, raising ethical challenges in the near future of war. The military use of bioenhancement raises numerous ethical challenges. These include, but are not limited to, concerns about proliferation and a so-called enhancement arms race that pushes ethical and legal boundaries. It is also possible that as enhancement becomes more common and widespread among members of the military, war fighters will bear the brunt of bad side effects. These may include psychological and physical complications related to the enhancement, as well as social issues stemming from the "haves" and "have-nots" of enhancement. Finally, there is cause for concern over enhancement blurring the line between human and something other than human (e.g., transhuman) in a way that elicits ambiguities in the law of armed conflict and in the treatment of so-called super soldiers.

Although the future of war will remain centered around humans, the rise of robots, particularly lethal autonomous weapons systems (LAWS), will have serious ethical implications. These include concerns about LAWS meeting current requirements instantiated in the law of armed conflict. There is deep disagreement on this issue, with some scholars believing that LAWS will not be able to comply with the laws of war, whereas others firmly hold out the possibility that not only will LAWS be able to meet such demands, they will likely do so better than humans. The possibility of LAWS better complying with the laws of war still does little to assuage concern over how to assign responsibility in the face of an autonomous weapons system grossly violating the laws or ethics of war. Others argue that this responsibility dilemma is, in actuality, a false one. Finally, it is possible that LAWS could reduce barriers to the use of force, thus making armed conflict easier to wage. Although most agree that such a future would be undesirable, some argue that this agreement must be tempered with the recognition that some wars may be necessary and just. Accordingly, reducing the barrier to waging just and justified wars could be desirable.

The future of war will require an appropriate balance between precaution on one hand and innovation on the other. The enthusiasm for developing and using military technologies that have the promise of increasing peace and security must be balanced against ethical considerations and the recognition that such technological optimism is not without its own considerable risks.

Bibliography

Ahmed, Aminul Islam, S. Gajavelli, M. S. Spurlock, et al. "Stem Cells for Therapy in TBI." *Journal of the Royal Army Medical Corps* 162, no. 2 (2015): 98–102.

Andrews, Evans. "8 Reasons It Wasn't Easy Being Spartan." History.com. March 5, 2013. http://www.history.com/news/history-lists/8-reasons-it-wasnt-easy-being-spartan.

Arkin, Ronald C. "The Case for Ethical Autonomy in Unmanned Systems." *Journal of Military Ethics* 9, no. 4 (2010): 332–341

Bess, Michael. *Our Grandchildren Redesigned: Life in the Bioengineered Society of the Near Future.* Boston: Beacon Press, 2015.

Bielitzki, Joseph. "Enhancing Human Performance in Combat" (speech). Defense Advanced Research Projects Agency, Defense Sciences Office. 2002. http://archive.darpa.mil/DARPATech2002/presentations/dso_pdf/speeches/BIELITZ.pdf.

Bostrom, Nick, and Anders Sandberg. "Cognitive Enhancement: Methods, Ethics, Regulatory Challenges." *Science and Engineering Ethics* 15, no. 3 (2009): 311–341. http://www.nickbostrom.com/cognitive.pdf.

Connor, Steve. "Did Leak from a Laboratory Cause the Swine Flu Epidemic?" *Independent* (London), June 29, 2009. http://www.independent.co.uk/news/science/did-leak-from-a-laboratory-cause-swine-flu-pandemic-1724448.html.

DARPA (Defense Advanced Research Projects Agency). "Warrior Web." Accessed February 23, 2017. http://www.darpa.mil/program/warrior-web.

Dillehay, Tom D., Jack Rossen, Donald Ugent, et al. "Early Holocene Coca Chewing in Northern Peru." *Antiquity* 84, no. 326 (2010): 939–953. doi:10.1017/S0003598X00067004.

Dimeo, Paul. *A History of Drug Use in Sport, 1876–1976: Beyond Good and Evil.* Abingdon, UK: Routledge, 2007.

Docherty, Bonnie. "Losing Humanity: The Case against Killer Robots." Human Rights Watch. November 19, 2012. https://www.hrw.org/report/2012/11/19/losing-humanity/case-against-killer-robots.

Dresler, Martin, Anders Sandberg, Kathrin Ohla, et al. "Non-pharmacological Cognitive Enhancement." *Neuropharmacology* 64 (2013): 529–543. doi:10.1016/j.neuropharm.2012.07.002.

Freedberg, Sydney J., Jr. "Centaur Army: Bob Work, Robotics, and the Third Offset Strategy." *Breaking Defense*, November 9, 2015. http://breakingdefense.com/2015/11/centaur-army-bob-work-robotics-the-third-offset-strategy/.

Gerstein, Daniel M. "How Genetic Editing Became a National Security Threat." *Bulletin of the Atomic Scientists*, April 25, 2016. http://thebulletin.org/how-genetic-editing-became-national-security-threat9362.

Greenberg, Andy. "The Bullet That Could Make 3-D Printed Guns Practical Deadly Weapons." *Wired*, November 5, 2014. https://www.wired.com/2014/11/atlas-314-3-d-printed-guns-bullets/.

Herkewitz, William. "Watch This Scientist Climb a Wall in Gecko-Inspired Spider-Man Gloves." *Popular Mechanics*. November 20, 2014. http://www.popularmechanics.com/technology/gear/a13221/scientists-have-created-gecko-inspired-spider-man-gloves-17448448/.

Hodsden, Suzanne. "MIT Researchers Develop Wireless, Noninvasive Deep Brain Stimulation Approach." *MED Device Online*, March 17, 2015. https://www.meddeviceonline.com/doc/mit-researchers-develop-wireless-noninvasive-deep-brain-stimulation-approach-0001.

Jacobsen, Annie. "Engineering Humans for War." *Defense One*, September 28, 2015. http://www.defenseone.com/technology/2015/09/engineering-humans-war/122189/.

Kahn, Laura H. "A Crispr Future." *Bulletin of the Atomic Scientists*, December 16, 2015. http://thebulletin.org/crispr-future8986.

Kalin, Stephen. "Insight: War Turns Syria into Major Amphetamines Producer, Consumer." Reuters. January 12, 2014. http://www.reuters.com/article/us-syria-crisis-drugs-insight-idUSBREA0B04H20140112.

Kamienski, Lukasz. Preface to *Shooting Up: A Short History of Drugs and War*, xv–xxviii. New York: Oxford University Press, 2016.

Klotz, Lynn C., and Edward J. Sylvester. "The Unacceptable Risks of a Man-Made Pandemic." *Bulletin of the Atomic Scientists*, August 7, 2012. http://thebulletin.org/unacceptable-risks-man-made-pandemic.

Lin, Patrick, Maxwell J. Mehlman, and Keith Abney. *Enhanced Warfighters: Risk, Ethics, and Policy* New

York: Greenwall Foundation, 2013. http://ethics.calpoly.edu/greenwall_report.pdf.

Lipsitch, Marc, and Alison P. Galvani. "Ethical Alternatives to Experiments with Novel Potential Pandemic Pathogens." *PLOS Medicine* 11, no. 5 (2014): e1001646. doi:10.1371/journal.pmed.1001646.

Miranda, Robbin A., William D. Casebeer, Amy M. Hein, et al. "DARPA-Funded Efforts in the Development of Novel Brain–Computer Interface Technologies." *Journal of Neuroscience Methods* 244 (2015): 52–67. doi:10.1016/j.jneumeth.2014.07.019.

Müller, Vincent C. "Autonomous Killer Robots Are Probably Good News." In *Drones and Responsibility: Legal, Philosophical, and Socio-technical Perspectives on Remotely Controlled Weapons*, edited by Ezio Di Nucci and Filippo Santoni de Sio, 67–81. Abingdon, UK: Routledge, 2016.

Shanker, Thom, and Mary Duenwald. "Bombing Error Puts a Spotlight on Pilots' Pills." *New York Times*, January 19, 2003. http://www.nytimes.com/2003/01/19/us/threats-and-responses-military-bombing-error-puts-a-spotlight-on-pilots-pills.html.

Simpson, Thomas W., and Vincent C. Müller. "Just War and Robots' Killings." *Philosophical Quarterly* 66, no. 263 (2016): 302–322. doi:10.1093/pq/pqv075.

Singer, Peter W. "The Robotics Revolution." Brookings Institution. December 11, 2012. https://www.brookings.edu/opinions/the-robotics-revolution/.

Singer, Peter W. "Robots at War: The New Battlefield." *Wilson Quarterly*, Winter 2009. http://wilsonquarterly.com/quarterly/winter-2009-robots-at-war/robots-at-war-the-new-battlefield/.

Sparrow, Robert. "Killer Robots." *Journal of Applied Philosophy* 24, no. 1 (2007): 62–77. doi:10.1111/j.1468-5930.2007.00346.x.

Taylor, Adam, and Laris Karklis. "This Remarkable Chart Shows How U.S. Defense Spending Dwarfs the Rest of the World." *Washington Post*, February 9, 2016. https://www.washingtonpost.com/news/worldviews/wp/2016/02/09/this-remarkable-chart-shows-how-u-s-defense-spending-dwarfs-the-rest-of-the-world/.

Thierer, Adam. *Permissionless Innovation: The Continuing Case for Comprehensive Technological Freedom*. Rev. ed. Arlington, VA: Mercatus Center, George Mason University, 2016.

United Nations Institute for Disarmament Research. "Framing Discussions on the Weaponization of Increasingly Autonomous Technologies." 2014. http://www.unidir.org/files/publications/pdfs/framing-discussions-on-the-weaponization-of-increasingly-autonomous-technologies-en-606.pdf.

United Nations Office for Disarmament Affairs. "What Are Biological and Toxin Weapons?" https://www.un.org/disarmament/geneva/bwc/what-are-biological-and-toxin-weapons/.

Utterback, Mitch. "How ISIS Is Turning Commercial Drones into Weapons in the Battle for Mosul." Fox News Tech. January 25, 2017. http://www.foxnews.com/tech/2017/01/25/how-isis-is-turning-commercial-drones-into-weapons-in-battle-for-mosul.html.

Wallach, Wendell. *A Dangerous Master: How to Keep Technology from Slipping beyond Our Control*. New York: Basic Books, 2015.

Whipps, Heather. "How Gunpowder Changed the World." Live Science. April 6, 2008. http://www.livescience.com/7476-gunpowder-changed-world.html.

Whittle, Richard. "Grey Eagle–Apache Run Shows Tech's Not Enough; Ya Gotta Have Doctrine." *Breaking Defense*, May 7, 2015. http://breakingdefense.com/2015/05/grey-eagle-apache-run-shows-techs-not-enough -to-use-drones-ya-gotta-have-doctrine/.

FILMS AND TV

Captain America: The First Avenger. Dir. Joe Johnston. 2011. Former military soldier Steve Rogers transforms into Marvel Comics' Captain America after taking a dose of super soldier serum.

"Men against Fire." *Black Mirror* (TV series). Season 3, Episode 5. Dir. Jakob Verbruggen. 2016. Created by Charlie Brooker. *Black Mirror* is a British television series examining the darker side of humanity and society, particularly through the lens of technology and how the "black mirror" often takes over lives. Episode 5 probes how one soldier's perspective of his mission becomes altered after his brain implant malfunctions.

"Syria's War Drug." BBC Arabic. September 2015. https://youtu.be/7ke13JNlpBQ. A documentary examining the rampant use of the highly addictive drug Captagon in the Middle East.

CHAPTER 35

Trans-ecology and Post-sustainability

Svein Anders Noer Lie
Associate Professor, Philosophy Department
UiT The Arctic University of Norway, Tromsø

Fern Wickson
Senior Scientist and Leader of the Society,
Ecology and Ethics Department (SEED)
GenØk Centre for Biosafety, Tromsø, Norway

Biotechnologies are currently evolving at a rapid rate as the enhanced digitalization of genetic information and the development of new methods for genome editing and synthetic biology continue to open new horizons of possibility at the interface between biology and technology. When transhumanists propose embracing (often somewhat uncritically) the latest advances in science and technology as a way to overcome biological constraints and transcend the physical limitations of the human body, they seem to endorse the principle of the technological imperative that "whatever is possible to be developed *should* be developed." This chapter articulates the challenges and problems facing this approach from an ecological perspective. The first sections of the chapter explore the meaning of humanism, transhumanism, and posthumanism, particularly in terms of their ideas concerning the relationship between humanity and nature and the future aims for how this relationship should develop. Next is a discussion of why transhumanist efforts to transcend biological limits through the widespread uptake and integration of biotechnological enhancements is problematic. Finally, the chapter argues that the most radical vision for moving beyond humanism is not actually a transhumanist transcendence through biotechnological enhancement but rather a posthumanist grounding into our ecological relations through a wholehearted embrace of our inherent interconnections with various forms of life on Earth.

TRANSHUMANISM: HUMANISM BY ANOTHER NAME?

If transhumanism is understood as an agenda to transcend humanism, it is important to first take a step back and ask "what is humanism?" so as to understand what it is exactly that the transhumanist may be seeking to transcend. It would be misleading to specify a single and strict definition of humanism (for detailed explorations of diverse interpretations of humanism, see, e.g., Soper 1986; Davies 2008; Halliwell and Mousley 2003). This chapter, however, follows the interpretation that the idea of humanism that arose in the 1300s in Renaissance Italy and continued through the Enlightenment and up until today is primarily about the pursuit of two intertwined goals: *emancipation* as a process that leads to *autonomy* (Knights and Willmott 2002). In other words, while the driving goal for a person living in the Middle Ages was typically *salvation*, the shift toward a more secular society and set of values saw humanism develop around a compulsion for what is commonly called freedom but which may be more precisely termed emancipation leading to autonomy. "In modern humanism ... the will of 'man' (sic), rather than magical forces or the word of God, becomes the center and measure of all things. Becoming autonomous is tantamount to becoming fully and perfectly human" (Knights and Willmott 2002, 63).

Depending on different ideas concerning what human beings are, ideas of what would *really* emancipate us also vary; emancipation may be seen as achievable through, for example, Buddhist meditative practice, social security systems, the market economy, or technological development. Humanism therefore does not necessarily contain any *specific idea* about what exactly can emancipate human beings (or indeed any particular agreement on what we are being emancipated from) but rather embodies the idea that human beings *should* seek emancipation and autonomy (rather than salvation or enlightenment, for example) and that these are in some way the ultimate and defining norms of our species.

Pursuing emancipation and autonomy as a defining human goal is strongly related to a desire for disentanglement and control. The driving desire here is arguably to become independent of "everything else" so as to have the freedom and possibility to control and define who you are and the power to develop into your own individualized self. For the transhumanist, this extends to include the freedom to control human evolution through the development and use of technological tools for biological enhancement. The underlying idea that informs this pursuit of freedom and control is that there is nothing fundamental or necessary about our current human form and relationships with other beings and that we can therefore choose how we want to develop and be connected to others, if at all. This includes relationships not only with other human beings and social entities but also with nature more broadly.

We normally relate to others in a way that German philosopher Martin Heidegger ([1927] 1992) calls "ready-to-hand." This means that we typically do not question those things that we experience as a part of our world and/or our way of being in that world. Therefore, to "access," critically reflect on, and thereby eventually be open to changing our relationship to aspects of the world, we first need to step out of the familiar "ready-to-hand" way that we normally relate to the world and begin to objectify and question the things around us. Accordingly, to "seek emancipation" is not our normal mode of being because pursuing this first requires that we step out of our normal (inconspicuous) relations before we can even begin to consider emancipating ourselves. A religious person, for example, will not "naturally" seek emancipation from his or her religious culture and belief system. The desire for emancipation can be ignited only if the culture, system, or set of relationships can first be conceptually separated from the self, objectified, and then critically questioned. In other words, before we can see that "things could be different," we have to step back from and objectify the relationships that normally remain unquestioned in our everyday life.

This is why the role of the ego that doubts (i.e., objectifies) everything "outside" of itself, as famously presented by French philosopher René Descartes ([1641] 1993), has been so important for the development of humanism and the modern quest for autonomy and emancipation as ultimate goals. (In this chapter, the words *ego*, *I*, and *self* are used as synonyms.) Descartes's "I" separates itself from and objectifies everything that is familiar and "ready-to-hand" and, through that move, brings everything in the world into question, except the ego itself. Descartes then proposes that this critical stance is the defining position of the (modern) human being. To be in this position in which the self critically questions and objectifies the world around it (instead of just accepting things as they are or as God created them) becomes the first but necessary step toward the pursuit of the defining humanist goals of emancipation and autonomy.

In this way, humanism totally depends on a concept of an independent and *rational* self, because it is this self that is at the center of the emancipation project—the thing that is seeking autonomy and emancipation from the ideas, beliefs, and physical realities that constrain its full flowering, self-control, and expression (see Knights and Willmott 2002, 64). If someone or something threatens this self we call it oppressive, inhumane, antiliberal, or totalitarian, and if this self were to disappear, then humanism would go with it. German philosopher Immanuel Kant argued that human beings—by virtue of their being rational agents—are "ends in themselves" and should be treated as such (Kant [1785] 1998). This means that the total objectification of the world is thought to reach a limit at the boundary of the human species; we should not treat other humans as objects. That is why humanism is "humane" and focused on human rights—and not barbaric with everyone out for him- or herself despite being driven by an urge for emancipation.

THE TRANSHUMANIST AGENDA

Following Kant's reasoning, everything that is not a rational human being belongs to the realm of objectifiable and manipulatable means. Interestingly, for the transhumanist, this also includes human bodies and biology (i.e., our material body that is thought to exist beyond the rational thinking self or ego). Normally we would not *seriously* question our own biology before there was a real chance to change it. This is why science and technology—particularly the possibilities afforded by biotechnological enhancements—have become a driving force for the transhumanist agenda. Technological advance is now offering "new possibilities," which creates an opportunity to imagine how things may be different and opens the door to the potential for change. Through the possibilities being offered by biotechnological advance, we are therefore now being enabled to first objectify our biology and ask this question: is there any rational reason why I need to put up with the biological constitution that I have inherited?

What enables the transhumanist dream of emancipation from the constraints of human biology is therefore *the conception of a self that is independent* and that will remain unchanged no matter what alterations are made to a person's biological conditions through technological enhancement or modification. Enhanced memory, enhanced capacity for reasoning or running, enhanced vision or feelings of pleasure, and so on would all, and also for the transhumanist, be an enhancement of *something*—namely, the unchanging self. Without this idea of an independent and unchanging self, there would be nothing that could make the choices for enhancement or that would remain present after such enhancements were adopted. This notion of a rational and independent self that is separate from its biological and technological constitution therefore appears to be a foundational necessity for the transhumanist agenda. And indeed, it is from humanism that transhumanism has the idea that there exists a self independent of our biology (and indeed also our technology). Transhumanism is therefore not really transcending humanism in its agenda but rather is deeply indebted to the fundamental ideas of humanism concerning the existence of a rational and isolatable self that is seeking emancipation and autonomy and may in fact be even more humanist than humanism itself. To question our own biology and make changes to it in the way that the transhumanist wants requires a disembodied, abstract self that is fundamentally cut off from any preexisting relations to the world.

THE POSTHUMANIST ALTERNATIVE?

One way to transcend humanism would be to give up the notion of the disembodied self. This could be done by the transhumanist embracing all types of technological possibilities (i.e., the technological imperative) and accepting that there is no rational and unchanging self that will persist through all the modifications or that is even present and making the decisions. However, another approach that does not argue for technological determinism, which was arguably adopted and elaborated in poststructuralist and postmodern critiques of modernity (see, e.g., Foucault 1970; Deleuze and Guattari 1987), has been to challenge the perceived independence of this self and its capacity to make autonomous choices. The argument here is that the self is not a solid or isolatable unit but rather a fluid and dynamic entity that is undergoing constant change, redefining itself and continuously being shaped and reshaped by the biological and social forces that surround it and with which it interacts.

Indeed, drawing from postmodern and poststructuralist critiques, a range of posthumanist literature has now emerged across various fields to make similar challenges, including within feminist and postcolonial studies, race theory, geography, environmental humanities, and science and technology studies. This literature is powerfully diverse (for examples of possible typological overviews, see Ginn 2017 and Sharon 2014). Together, however, these works collectively challenge the humanist agenda in a number of ways. These include:

1. a challenge to a universalistic concept of humanity and an essential set of human experiences and goals by highlighting diversity and multiplicity;
2. a challenge to the idea of the rational self as primary and the measure of all things by drawing attention to the significant role of emotion;
3. a reemphasis on the importance of materiality and physical bodies for defining human experience and unfolding;
4. demonstrations of inherent entanglements between entities that have classically been conceptually separated, such as science and society, technology and humanity, and humanity and nature; and
5. an ontological reorientation focused on the significance of relations and emphasizing the foundational importance of socioecological and sociotechnical interrelations in shaping human experience and identity.

For some of these authors, posthumanism represents a new age of possibility being brought into being through the emergence of biotechnological advance and interventions in the world (Haraway 1991; Braidotti 2013), whereas for others, posthumanism is emphasized as an ontological view that has always been available but has perhaps been spurred into specific articulation by the latest wave of biotechnological possibilities (e.g., Latour 1993; Barad 2007).

THE PRICE OF THE QUEST FOR EMANCIPATION AND AUTONOMY

By seeking independence from all sorts of conditions that are seen to be constraining the fulfillment of human potential, the transhumanist dream appears to be of a day when human beings are fully disentangled from every relation to the world that is not self-imposed or chosen, including the biology of their own body. Is this dream realistic, though, or is every attempted disentanglement in the search for greater autonomy and emancipation simply followed by the creation of a new set of relations of interdependence?

A classic example of a figure who has to pay a price for his autonomy is the mythical figure of Odysseus. The price Odysseus has to pay for the ability to control the external nature of the world is, according to Max Horkheimer and Theodor W. Adorno ([1947] 2002), that he also has to control and tame his own inner nature. He cannot put his destiny in the hands of the Creator, or just "let it be"; instead, he has to make decisions based on goals set by himself. Horkheimer and Adorno therefore claim that the domestication of nature is inevitably followed by the domestication of ourselves.

For Horkheimer and Adorno the promised land lying at the end of the enlightenment tunnel is not actually emancipation from all relations of dependence but rather a new set of entanglements. For them, the most important and fundamental goals that underlie modern western civilization, that of autonomy and emancipation, have an inbuilt failure that makes the quest doomed from the start—namely, the failure to recognize and adequately account for the interconnected nature of the world in which humans live. Using a similar line of reasoning, Ulrich Beck (1986) has shown how our technoscientific culture may appear to emancipate us from reliance on nature and the potential dangers it poses (e.g., floods, drought, predators, pandemics), but all these efforts have effectively just replaced these dangers with new types of risks (e.g., from nuclear waste and radiation, industrial pollution, and an increased toxic load).

What Beck is describing is a process whereby the current attempts to emancipate humanity from nature (and from the perceived constraints of what we call nature) have inevitably led to a worsening of future conditions to such a degree that the main focus shifts from being (individually) emancipated to an attempt to avoid and (collectively) distribute the risks that arise from our emancipation attempts. The constraints, limitations, and risks we face have not been removed; they have only been changed and rearranged. This means that what we aim to achieve is being intrinsically counteracted by an internal logic of the world we live in and, more concretely, by the feedback loops that exist in practice within our ecological reality. Is there something about the transhumanist agenda that should make us think that these feedback loops will not apply? Can we truly free ourselves from our earthly connections through the ever greater development, uptake, and use of emerging technologies?

TRANSCENDING HUMANISM THROUGH A POSTHUMANIST EMBEDDING

As suggested earlier, humanism fundamentally assumes that we can fulfill our true potential only when we "disentangle." Various strains of posthumanist literature, however, argue that the ability to disentangle from our dependence on the physical world of nature through an enhanced embrace, development, and use of technologies is a myth. Seeking to transcend the biological limits of the human body will never lead to a "trans-ecology" or a situation of "post-sustainability" concerns. Adopting and embracing new biotechnologies cannot free us from our relations to the physical and biological world because, in a very basic sense, all new technologies are inherently dependent on natural entities, processes, and products for their creation. As we all know, the silicon chip requires silicon, the metals required to make machines have to come from the ground, and the energy we need to create and operate the envisaged technologies of biological enhancement all come from the elements of the earth—that is, the sun, the wind, the water, the atoms, or organisms that died long ago, and so on.

Pursuing a transhumanist agenda will therefore not be able to free us from these earthly bonds. We may change our biology and the relations we have with the rest of life on Earth, but we can never transcend our ecology. There is no escaping our interconnections with the rest of life on Earth; there is only ever the possibility of changing and shifting the networks of relations (Lie 2016).

If you fundamentally believe that humans are able to be disentangled—that we are in essence isolatable entities—then you will be inclined to think that your current entanglements are simply *random and contingent* and that you may become more emancipated as you gradually free yourself from those entanglements. To think that we are intrinsically and inevitably entangled, however, is to think fundamentally differently about the nature of the world and the nature of human beings as a species within it. Who you are is no longer seen as *contingently* related to your entanglements. Instead, who you are and the capacities that you have are shaped by and constituted through the range of entanglements that you were born into and which you engage in throughout your life.

This means that in a philosophical sense it is possible to think about being in the world in a disentangled way only if we assume that the relations we find in the world are already contingent (what is often called a Humean view [based on the work of eighteenth-century Scottish philosopher David Hume]; see, e.g., Molnar 2003) and that the things and properties that are inflicted on us are external and carry no internal or shaping relationship to who we are. If the world is contingently related in this way, we may freely add and subtract different "vectors" in accordance with the desired autonomy. If I, for instance, want to make it possible for people to enhance their memory, in a world that was completely rearrangeable I would assume that I can change this property *independently of everything else*, because the relation between beings and properties is contingent in the first place. This ontological idea of how the world is connected is what several posthumanists see as the main problem with the transhumanist approach and agenda. If you think that, rather than being randomly connected and completely rearrangeable by entities that have stable and isolatable identities, the world is instead relationally connected in a way that the relations are *internal* and co-constitutive of the entity, its characteristics, and identity, then you will see any attempt to isolate one property and enhance that property "independently of everything else" as a naive and misguided attempt that will always lead to unexpected feedback loops and new Odysseus-like dependencies.

Looking, for example, at the problems facing (pedigree) dog breeding, we can easily see something of what we might expect from bioenhancement under the transhumanist dream. In dog breeding, the biological givenness of the properties of a dog are gradually substituted through human choice in the pursuit of some kind of ideal. This is an ideal that is chiefly governed not by biological need or evolutionary adaptation to a network of environmental demands and relations but rather by human fashion, taste, and desire. When these ideals decide how dog breeds should develop, we end up with dogs that have certain properties that have been "enhanced" independently of any consideration of how these may hang together with and affect other properties. This results, for example, in the production of "ideal" dogs that cannot breathe properly, that suffer from spine and hip problems, and that have immune or digestive issues (see Advocates for Animals 2006).

One may argue that this stems from the "unscientific" attitude or lack of comprehensive knowledge of the breeders in terms of their ability to maximize multiple aspects simultaneously. However, would the attitude or limited knowledge of complex systems really differ significantly in relation to human enhancement? Most transhumanists argue that the choice for enhancement should be made by the individual (on an open market presumably) and not in terms of eugenic programs instigated and managed by the state (see, e.g., Harris 2007). If we add to this assumption the fact that we are dealing with enhancements that can change our biology on multiple scales (from the nano to the macro level), understanding the impacts on and potential implications for the functioning of the organism as a whole arguably becomes only more complex and unpredictable. This is because making changes at the level of an organism's genes, for example, requires a recognition and understanding not only of the relationally connected changes created in properties at that level of organization, but also of the effects such changes have on higher levels of organization (see Lie 2016; Noble 2017), as well as how such changes may relate to and affect the capacity for emergent properties and systems to arise.

The transhumanist agenda is therefore seeking emancipation based on manipulating properties on one level of complex autopoietic systems (that is, living systems capable of maintaining and reproducing themselves) according to a flawed ontology in which the relational network of this system is not seen to be in any way constitutive of its properties and in which all changes made on one level will be fully comprehensible and predictable in their effects on other levels of organization. This creates a situation in which the expected rise of control and freedom when pursued on one level actually leads to feedback loops from other levels of reality, creating continuous surprises that the humanist emancipation program is unable or unprepared to understand. In the same way that a scientific experiment is an abstraction from reality that requires that all surrounding conditions are maintained as constant for the property of interest to be investigated, the transhumanist search for autonomy not only leads to abstracted versions of human beings but also requires that the "laboratory conditions" of life be constantly kept stable for this to function, which actually leads ultimately to more rather than fewer entanglements. Given that this agenda would be pursued under a flawed ontology and taking into account the type of properties that are likely to be selected and chosen for enhancement within modern industrialized and capitalist societies, the attempt seems doomed to produce more problems than it solves and to ultimately create more dependence than freedom.

Summary

This chapter proposes that transhumanism, understood as the use of emerging technologies to attempt to free humanity from its biological constraints, is not actually transcending humanism as traditionally defined but rather simply extending a fundamentally flawed traditional humanist agenda with its age-old goals of autonomy and emancipation. In other words, the transhumanist dream in which humanity becomes increasingly emancipated from the constraints of its inherited biology is actually just the logical extension and end point of humanism, through which we have previously sought emancipation from the constraints posed by religion, nature, and social bonds, with our attention now turning to the biology of the material body as the next thing to try to transcend in the ongoing search for independence and autonomy.

To truly go beyond humanism and pursue a more radical agenda therefore demands that we transcend the very idea that the ultimate goal for human beings is that of independence and autonomy in the first place. Overcoming the Cartesian notion of a rational, stable, and isolatable self as the measure of all things need not lead to the claim that there is no self *there* but instead can involve a recognition that this self is *fundamentally* coproduced by the relations that it is engaged in. The continuous production of the fluid self happens in relation to its interactions with other entities in the world. To go beyond humanism could therefore mean to accept and indeed embrace our relatedness—our relatedness to each other and to earth. To embrace these relations does not necessarily mean a postmodern embrace of an *absolved* self. Instead, within a particular posthumanist framing it can mean to accept and pursue an understanding of what humans are that radically accepts a fate of being inherently interconnected. This means accepting that the self exists only because of networks of social, ecological, and technological relations and that all are engaged in a process of mutual co-shaping and unfolding. This interconnected, entangled unfolding of our common evolution needs to be recognized in the development of our emerging technologies so that we do not pursue false ideals of independence and autonomy and thereby risk the inevitability of surprise.

Posthumanism thus enables us to accept our inherent interdependence and thereby embrace rather than seek to escape what we really are. Therefore, the most radical agenda for transcending humanism would actually be to transcend the purported goal of emancipation and instead rejoice in our relatedness. Some of the questions that may be immediately raised concerning this approach could include a worry that the self will simply (and unacceptably) dissolve and disappear into its relations (see Lie 2016 for a possible *ontological* solution to this problem) or what the ethical implications of adopting such a relational ontology might be (see Wickson 2015 for an argument that developing the more relational frameworks of care and/or virtue ethics will be more appropriate for a biotechnological age than the currently dominant consequentialist- or utilitarian-based approaches). However, despite the ongoing (philosophical and political) work that may be required to advance this posthumanist rather than transhumanist agenda, it seems clear that the transhumanist search to replace our biology with technological enhancements can never liberate us from our earthly bonds and will only ever create new sets of entanglements, and the potential for ongoing surprises as the feedback loops we fail to acknowledge continuously deliver unexpected results. The most radical form of moving beyond humanism would therefore be to pursue a posthumanist goal of moving beyond the limiting concept of ourselves as isolated human entities and embracing our interconnected, multifaceted, and biologically diverse ecological selves.

Bibliography

Advocates for Animals. *The Price of a Pedigree: Dog Breed Standards and Breed-Related Illness.* Edinburgh: Author, 2006. https://onekindplanet.org/uploads/publications/price-of-a-pedigree.pdf.

Barad, Karen. *Meeting the Universe Halfway: Quantum Physics and the Entanglement of Matter and Meaning.* Durham, NC: Duke University Press, 2007.

Beck, Ulrich. *Risk Society: Towards a New Modernity.* Translated by Mark Ritter. London: Sage, 1986.

Braidotti, Rosi. *The Posthuman.* Cambridge: Polity Press, 2013.

Davies, Tony. *Humanism.* 2nd ed. London: Routledge, 2008.

Deleuze, Gilles, and Félix Guattari. *A Thousand Plateaus: Capitalism and Schizophrenia.* Translated by Brian Massumi. Minneapolis: University of Minnesota Press, 1987.

Descartes, René. *Mediations on First Philosophy.* Translated by Donald A. Cress. 3rd ed. Indianapolis, IN: Hackett, 1993. First published, in Latin, 1641.

Foucault, Michel. *The Order of Things: An Archaeology of the Human Sciences.* London: Tavistock, 1970.

Ginn, Franklin. "Posthumanism." In *The International Encyclopedia of Geography: People, the Earth, Environment, and Technology,* edited by Douglas Richardson, Noel Castree, Michael F. Goodchild, et al. Chichester, UK: Wiley, 2017.

Halliwell, Martin, and Andy Mousley. *Critical Humanisms: Humanist/Anti-humanist Dialogues.* Edinburgh: Edinburgh University Press, 2003.

Haraway, Donna. *Simians, Cyborgs and Women: The Reinvention of Nature.* New York: Routledge, 1991.

Harris, John. *Enhancing Evolution: The Ethical Case for Making Better People*. Princeton, NJ: Princeton University Press, 2007.

Heidegger, Martin. *Being and Time*. Oxford: Blackwell, 1992. First published in German, 1927.

Horkheimer, Max, and Theodor W. Adorno. *Dialectic of Enlightenment: Philosophical Fragments*. Edited by Gunzelin Schmid Noerr. Translated by Edmund Jephcott. Stanford, CA: Stanford University Press, 2002. First published 1947.

Kant, Immanuel. *Groundwork of the Metaphysics of Morals*. Translated and edited by Mary Gregor. Cambridge: Cambridge University Press, 1998. First published in German, 1785.

Knights, David, and Hugh Willmott. "Autonomy as Utopia and Dystopia." *Sociological Review* 50, no. S1 (2002): 59–81.

Latour, Bruno. *We Have Never Been Modern*. Translated by Catherine Porter. Cambridge, MA: Harvard University Press, 1993.

Lie, Svein Anders Noer. *Philosophy of Nature: Rethinking Naturalness*. London: Routledge, 2016.

Molnar, George. *Powers: A Study in Metaphysics*. Oxford: Oxford University Press, 2003.

Noble, Denis. *Dance to the Tune of Life: Biological Relativity*. Cambridge: Cambridge University Press, 2017.

Sharon, Tamar. *Human Nature in an Age of Biotechnology: The Case for Mediated Posthumanism*. Dordrecht, Netherlands: Springer, 2014.

Soper, Kate. *Humanism and Anti-humanism*. La Salle, IL: Open Court, 1986.

Wickson, Fern. "The Ontological Objection to Life Technosciences." In *Science, Philosophy and Sustainability: The End of the Cartesian Dream*, edited by Àngela Guimarães Pereira and Silvio Funtowicz, 61–77. Abingdon, UK: Routledge, 2015.

Glossary

A

alpha: In Greek, alpha is the first letter of the alphabet; in Christian theology it is the name of Christ in the book of Revelation (22:13).

anabolic steroids: Steroidal androgens (male sex hormones) that include natural androgens such as testosterone as well as synthetic substances that are structurally related and have similar effects to testosterone. They increase constructive metabolism (anabolic) as well as increase protein within cells, especially in skeletal muscles.

animal bioenhancement: Any use of nonhuman animal bodies to biologically, medically, and technologically enhance traits of animals beyond what is typical of that species, including but not limited to xenotransplantation and the creation of chimeric organisms.

animal epistemology: The study of what nonhuman animals know, can know, and how they come to know. Additionally, animal epistemology considers the attributes and features of nonhuman animal consciousness.

Armageddon: Named in the book of Revelation as the site of a future battle between the forces of good and evil, this term has become shorthand for any violent conflict or cataclysm that might result in the end of the world. For many centuries, Armageddon was conceived of as the culmination of religious prophecy. In modern times, human-made threats, such as nuclear warfare, runaway climate change, and ecological collapse, have emerged as plausible scenarios for the end of human civilization.

artificial general intelligence: A combination of computer hardware and software that can learn to solve problems that it was not specifically designed to solve. An artificial general intelligence should be able to solve any cognitive problem at least as well as a human being.

artificial neural network: (ANN) An algorithmic data structure inspired by biological nervous systems. The components of such a network are vastly simplified versions of biological neurons, axons, and synaptic connections.

Asimov's three laws of robotics: A set of laws governing robots put forth by Isaac Asimov in his collection of short stories titled *I, Robot* (1950). "(1) A robot may not injure a human being, or, through inaction, allow a human being to come to harm. (2) A robot must obey the orders given it by human beings except where such orders would conflict with the First Law. (3) A robot must protect its own existence as long as such protection does not conflict with the First or Second Laws."

autonomy: Independence or freedom.

autotrophy: The ability of an organism to synthesize its own food from inorganic substances, using light and chemical energy. Nikolai F. Fedorov, Vladimir Vernadsky, and other Russian cosmists believed that autotrophy should become a goal for self-directed human evolution.

B

belief: A state or habit of mind that is characterized by a conviction, attitude, or claim that something is true.

bioenhancement: (also known as human enhancement) "A modification of a person's biologically grounded traits, adding capabilities that would not otherwise have been expected to characterize that person" (Bess 2015, 13). In strict terms, a biomedical intervention that provides a benefit or additional feature to or function for an otherwise healthy body. More broadly, bioenhancements include biomedical interventions in the form of pharmaceuticals, surgical procedures, genetic modifications, and technological implants and wearables.

biohacking: Biohacking involves applications of technologies and nonconventional foreign agents to a person's body to study the effects firsthand.

biologics: A contemporary field of biotechnology focusing on products that are biologically derived.

biophotonics: The use of photons and photonics for the manipulation of biological materials such as cells and neurons.

bioprinting: The use of 3-D printing technology with materials that incorporate viable living cells, to produce, for example, tissue for reconstructive surgery.

bodhisattva: A being that has vowed to attain enlightenment but delays that attainment in order to liberate other sentient beings from the cycle of birth and death.

"brain in a vat" (BIV): An argument in epistemology that hypothesizes that humans have no way of knowing whether their perceptions are illusory because they are being generated by an outside force (e.g., a computer). In the BIV scenario, the brain has been separated from the body and is receiving false inputs from an external source, tricking the individual into believing that she is experiencing something that she is not. The purpose of the BIV argument is to raise questions about certainty and to demonstrate the central place of skepticism in epistemological claims.

brain-computer interfaces (BCIs), or brain-machine interfaces (BMIs): Systems involving real-time direct connections between the brain and a signal processing algorithm through a computer. Bidirectional feedback between the user and the system produces physical changes in the brain. BCIs consist of both noninvasive wired and invasive wireless systems. Noninvasive forms use electrodes attached to the scalp and connected to an electroencephalograph (EEG). Invasive forms use microelectrode arrays or chips implanted in the motor cortex of the brain.

C

capabilities approach: A subfield of developmental economics that focuses on identifying the key factors that are preconditions for human flourishing.

categorical model of disease: An infectious disease that can be diagnosed in absolute binary terms—a person either has it or does not have it.

cellular chimerism/microchimerism: A small number of cells that originate from another individual and are therefore genetically distinct from the cells of the host individual, as often seen during mammalian pregnancy.

chimera/chimeric organism: An individual organism that carries two different sets of DNA, usually derived from two different species.

cognitive enhancement: A subset of bioenhancement more generally, which can be defined as "the amplification or extension of core capacities of the mind through improvement or augmentation of internal or external information processing systems" (Bostrom and Sandberg 2009, 311).

consequentialism: A theory of ethics that judges behavior on whether its outcomes result in moral goods. Consequentialism is typically contrasted with deontological ethics.

cosmetic pharmacology: The use of drugs for purposes of cosmetically enhancing normal functioning. Coined by Peter Kramer and popularized in his book *Listening to Prozac* (1993).

CRISPR: Clustered regularly interspaced short palindromic repeats. CRISPR/Cas9 genome editing techniques have many potential applications, including medicine and crop seed enhancement. The use of CRISPR/Cas9-gRNA complex for genome editing was the American Association for the Advancement of Science's choice for breakthrough of the year in 2015. Bioethical concerns have been raised about the prospect of using CRISPR for germ-line editing.

cyborg: A figure indicating hybrid combinations of human and nonhuman, the breakdown of binary oppositions between natural and artificial, human and machine. Experimental organisms central to neuropharmacology were the first non-imaginary cyborgs, indicating the potential for self-regulating organisms to adapt to changing environments. Coined in 1960 by Manfred E. Clynes, in collaboration with Nathan S. Kline, for a NASA conference on adapting humans to space.

Cyborg Buddha Project: An initiative launched by James Hughes, Mike LaTorra, and George Dvorsky through the Institute for Ethics and Emerging Technologies to promote discussion of the impact of neuroscience and emerging neurotechnologies on happiness, spirituality, morality, and meditation.

D

Dalai Lama: An important monk in the Gelugpa order of Tibetan Buddhism understood to be the successor in a line of reincarnated teachers who are manifestations of the bodhisattva of compassion.

dechronification: The concept of reversing biological aging through nanotechnology. Similar to the idea of rejuvenation. A scientific model for achieving this goal was introduced by Robert A. Freitas Jr. in 2002.

deep learning: A technique that employs hierarchical neural networks to achieve learning in a machine. Can be combined with reinforcement learning for the acquisition of behaviors based on feedback from the environment.

deep-brain stimulation (DBS): An intracranial surgical technique in which an electrode is placed in a subcortical (deep) region of the brain. The electrode is connected to a battery-driven pulse generator implanted near the collarbone. The generator and electrode are activated by a handheld programmable device that can be turned on or off. DBS can modulate dysfunctional neural circuits implicated in neurological and psychiatric disorders and relieve some symptoms associated with them.

default mode network (DMN; also known as default network or default state network): A network of interacting brain regions known to have activity highly correlated with each other and distinct from other networks in the brain. The default mode network is most commonly shown to be active when a person is not focused on the outside world and the brain is at wakeful rest, such as during daydreaming and mind wandering.

deontological ethics: A theory of moral decision making based on following prescribed rules or duties. Sometimes called duty ethics. Typically contrasted with consequentialism.

dependent origination: The Buddhist idea that all conditioned phenomena arise in dependence of other phenomena.

Dhammapada: A collection of the Buddha's teachings that is the best known and most widely revered text in the Pali canon of Theravada Buddhism.

dimensional model of disease: Syndromes, such as autism for example, composed of symptoms along a continuum, for which diagnosis is determined by assessment of the quantitative degree of presence or absence.

***dukkha*:** An important Buddhist concept, commonly translated as "suffering," "pain," or "unsatisfactoriness." It refers to the fundamental unsatisfactoriness and painfulness of life, and inspires the Four Noble Truths and nirvana doctrines of Buddhism.

dystopia: A term coined in the twentieth century to describe a society with starkly oppressive and dehumanizing characteristics. During the first half of the twentieth century, dystopian novels such as Aldous Huxley's *Brave New World* and George Orwell's *Nineteen Eighty-Four* warned of the dangers of totalitarian domination through brainwashing, drugs, propaganda, and torture. By the early twenty-first century, dystopian fiction had become a staple of popular culture, as evidenced by such successful book and film franchises as Suzanne Collins's Hunger Games and Veronica Roth's Divergent series.

E

economic Singularity: A point in an economy's development at which the production of goods, services, and energy is so efficient and inexpensive that little human labor is required to sustain the economy.

ectogenesis: The gestation of a human embryo to viability completely within an artificial womb. The concept of human ectogenesis was described by British scientist J. B. S. Haldane in the early 1920s and memorably depicted a decade later in Aldous Huxley's dystopian novel *Brave New World.* Although Huxley's novel framed the concept as an instrument of totalitarian control, some have since speculated about its possible social and political benefits. Notable advocates have included geneticist Hermann Joseph Muller, as well as feminist authors Elisabeth Mann Borgese and Shulamith Firestone.

electroencephalography (EEG): An electrophysiological monitoring method to record electrical activity of the brain. It is typically noninvasive, with the electrodes placed along the scalp, although invasive electrodes are sometimes used in specific applications. EEG measures voltage fluctuations resulting from ionic current within the neurons of the brain.

embodied cognition: The position that cognition cannot occur except through a body and that the type of body a cognitive system has (particularly its sensorimotor capabilities) strongly influences the type of cognition that can occur

embodiment: The physical form that a person, thing, or technology takes.

epistemology: One of the three main branches of philosophy (along with ethics and metaphysics); it is concerned with the nature of knowledge and the methods for proving or disproving statements of fact or belief. Epistemology addresses how something can be known, believed, or said to be true.

erythropoietin (EPO; also known as hematopoietin or hemopoietin): A glycoprotein hormone that controls erythropoiesis, or red blood cell production. EPO was used by Lance Armstrong as a performance enhancing drug during his cycling career.

essentialism: The notion that human beings have a fixed nature or essence.

eugenics: Attempts to improve population health by selecting or modifying genetic influences. A broader definition of the term aims not only to improve physical health but a more expansively defined set of moral or character traits as well. Eugenic philosophies helped shape government policies in many nations during the early twentieth century, raising ethical concerns among many critics.

existential risk: A risk, from a particular hazard or hazards, with the potential to threaten to terminate the survival of an entity.

exosuit: A wearable suit designed to augment or enhance the wearer's capabilities.

exponential growth: An increase that is compounded over time. Exponential growth curves tend to reach a point at which the upward rise is almost vertical on a graph.

extended mind or extended cognition: The position that mental processes extend outward into the environment; thoughts and mental processes are embodied with and through physical objects, and thus extend the mind and mental processes.

F

falsificationism: A form of inductive reasoning asserting that scientific theories cannot be proven true but can be disproved and thus shown to be false. It is associated most closely with philosopher of science Karl Popper.

female genital cosmetic surgery: Any number of bioenhancements of external or internal female genitalia for aesthetic reasons and/or to improve sexual pleasure.

floating point operations per second: A measure of computer performance based on how many numerical calculations the computer can do each second.

four sights: The Buddha is said to have seen four sights—an ill person, an old person, a corpse, and an ascetic wanderer—that prompted him to renounce worldly life.

functional magnetic resonance imaging (fMRI): A procedure that measures brain activity by detecting changes associated with blood flow. This technique relies on the fact that cerebral blood flow and neuronal activation are coupled. When an area of the brain is in use, blood flow to that region also increases.

G

Galtonian eugenics: Attempts to improve population health or more broadly-defined moral and social "fitness" by managing human reproduction.

general-purpose human-level intelligence: An artificial agent that is capable of performing at the same level as human beings at all cognitive tasks. Synonymous with artificial general intelligence.

genetic enhancement: Modifications made to someone's DNA *before* birth to enhance one's life.

genetically modified organism: An organism whose genetic makeup has been purposefully changed. With regard to creating chimeric animals, genetic modification may be a way to avoid immediate organ and tissue rejection and cross-species infection.

genotype: An organism's genetic makeup.

germ-line modification: The alteration of genetic material in the sex cells of an organism, resulting in heritable trait changes that are passed down through subsequent generations.

glia: The supporting tissue intermingled with the essential elements of nervous tissue especially in the brain, spinal cord, and ganglia.

godmanhood: The idea that in the divine-human figure of Christ, God works through humanity toward the perfection of this world, as developed by Vladimir Solovyov in his 1877–1881 series of "Lectures on Godmanhood."

H

heliobiology: The study of the sun's effect on biology, a branch of science developed by the cosmist Alexander Chizhevsky.

human exceptionalism: A position that regards humans as the most significant entity in the universe and interprets the world according to human values and experiences.

human flourishing: A state of well-being and fulfilled potential that can be used as an ethical yardstick for assessing the impact (individual and societal) of specific bioenhancements on a case-by-case basis.

human/animal binary: An entrenched theoretical category that distinguishes between human and nonhuman animals, usually treating the human as superior; since the late twentieth century this categorical binary has come under criticism as too limiting in examining the co-constituted environment within which humans and animals live.

I

immortology: The science of the investigation of the nature and causes of physical death and the possible means of its eventual elimination. Immortology, of which the biologist Igor Vishev is a leading proponent, is one of the most active branches of twenty-first-century Russian cosmism.

inductive process: A form of reasoning that formulates a general claim or conclusion from specific empirical observations and data. It is contrasted to deductive reasoning, which begins with a general broad hypothesis that is then examined through specific examples.

intersexual: The term preferred by individuals who have indeterminate biological sex resulting from apparently atypical external genitalia, internal sexual organs, chromosomal makeup, and/or hormonal levels.

Interweb: A street term, now being adapted by academics, for the sum of the Internet, World Wide Web, e-mail, cloud accounts, and all the existing and future social media and information technology communication systems and apps that together are converging into one ubiquitous interlinked digitalized extension of human culture.

intrinsic harms: Harms that are bad in themselves.

K

Karma: A Sanskrit term that means "action"; in Buddhist traditions, karma refers to bodily, mental, or verbal actions that can affect one's present or future existence.

knowledge Singularity A point at which the amount of knowledge accessible to intelligent beings increases to the point where anything that is knowable can be ascertained.

L

lethal autonomous weapons: Machine-based weapons systems that can select and engage targets without human intervention.

linear growth: An increase that occurs at a fixed rate over time. Linear curves appear as a straight line on a graph.

M

medicalization: The social process by which previously nonmedical conditions (such as shyness or lethargy) are redefined as medical problems that can be treated by new pharmaceutical drugs or therapeutic regimens involving medical professionals.

mind uploading: The hypothetical process of transferring a mind from a human body to a computer by matching the functional neural connections of the brain with the functional wiring connections of the computer.

monastic: A religious way of life in which one renounces worldly pursuits to devote oneself fully to spiritual work.

Moore's law: First stated by Gordon Moore in 1965, this law conjectured that the number of transistors that fit on an integrated circuit would double every one and a half to two years. Similar exponential growth has occurred with computer processing power and memory storage.

multiple realizability: The position that mental states can be realized by many different physical systems; e.g., that pain could be realized and experienced by mammalian brains, cephalopod brains, digital computer systems, and extraterrestrial physiologies.

N

neural prosthetics: Artificial devices or systems designed to generate, restore or modulate a range of brain functions associated with movement, cognition and emotion. Deep-brain stimulation and brain-computer interfaces are just two examples of neural prosthetics.

neuroplasticity: The brain's ability to change and reorganize itself by forming new connections between neurons in response to experience, including brain injury and disease, throughout life.

nominalism: A philosophical view that challenges the actual existence of universal objects, theories, and natural/scientific laws, arguing that universals exist only in language. Nominalists contend that reality is made up of only particulars.

noosphere: From Greek *nous,* "mind," a hypothetical sphere of knowledge and mental energy enveloping Earth's biosphere. The term was popularized in 1922 by Pierre Teilhard de Chardin, and Édouard Le Roy, who had probably developed the idea after attending a series of lectures by Vladimir Vernadsky at the Sorbonne. In the West, the idea is most closely associated with Teilhard, in Russian cosmism with Vernadsky. Pavel Florensky, another Russian cosmist, developed a similar idea, which he called the "pneumatosphere," the sphere of spirit.

nootropics: Drugs, supplements, or other substances that improve cognitive function, particularly executive functions, memory, creativity, or motivation, in healthy individuals.

O

offset strategy: A military strategy to gain and maintain military advantage and superiority over an adversary over an extended period.

omega: In Greek, omega is the last letter of the alphabet; in Christian theology, it is the name of Christ in the book of Revelation (22:13): "I am the Alpha and the Omega, the First and the Last."

optogenetics: A biological technique involving the use of light to control cells in living tissue, typically neurons, that have been genetically modified to express light-sensitive ion channels.

orphan drugs: The US Congress passed the Orphan Drug Act (1983) in order to incentivize pharmaceutical companies to adopt drugs that treat rare diseases and thus have restricted markets by offering tax breaks, protocol assistance for navigating regulatory requirements, and a time-limited monopoly on production and sales.

P

paper clip maximizer: A somewhat absurd example formulated by Swedish philosopher Nick Bostrom to illustrate how a single-minded intelligence with a seemingly benign goal could lead to an existential crisis for humanity.

Parousia: In some, but not all, Christian theologies, *Parousia* (Greek for "arrival") is a term for the second coming of Jesus Christ.

petaflops: 10^{15} (one thousand million million) floating point operations per second.

phenotype: An organism's observable traits. This stands in contrast to the genotype, the organism's genetic makeup.

pluripotent stem cells: Cells that can develop into all different types of cells composing the body; embryonic stem cells fall under this category.

positive psychology: A subfield of psychology that focuses on identifying the key factors that promote human flourishing.

posterior cingulate cortex (PCC): The posterior cingulate cortex forms a central node in the default mode network of the brain. It has been shown to communicate with various brain networks simultaneously.

posthumanism: A term that gained currency in the late twentieth century to describe challenges to traditional conceptions of what it means to be human. These challenges came primarily from the exploding field of computer technology. In the twenty-first century, posthumanism has come to denote the most radical visions for reconfiguring the human brain and body, inspired by advances in robotics, computer science, and other technologies.

preimplantation genetic diagnosis (PGD): A procedure in which one or more egg cells are taken from an embryo and tested for genetic variants that increase the likelihood of disease. Those egg cells that lack the disease variants are then implanted in a woman's womb, resulting in gestation of a child who is free of the targeted disease.

"problem of other minds": Best articulated by American philosopher Thomas Nagel, this is the problem of whether other minds (human or nonhuman) are sufficiently like our own such that we can claim to have similar experiences.

prosthesis: An artificial body part (such as a hand, arm, eye, or leg) that can be attached to the body. Some prosthetic devices are meant to compensate for impaired or lost functioning, whereas others enhance or augment human functioning.

Q

quantum computer: A computational device that stores and manipulates data using principles from quantum mechanics such as superposition and entanglement.

R

realm of desire: One of the three realms in the Buddhist universe in which sentient beings (hell dwellers, animals, hungry ghosts, humans, *asuras* [titans] and some gods) remain subject to base desires.

realm of form: One of the three realms in the Buddhist universe that includes gods who dwell in a state of meditative absorption.

realm of formlessness: One of the three realms in the Buddhist universe that is characterized as lacking all form and desire.

S

***samatha*:** The Buddhist practice (*bhāvanā*) of the calming of the mind. This is done by practicing single-pointed meditation most commonly through mindfulness of breathing. It may be accomplished by concentration on a meditation object.

Sanskrit: The main language of Buddhist scriptures in India beginning in the early centuries of the Common Era, and frequently used among contemporary scholars of Buddhist studies.

secular immortality: The idea that humans can achieve immortality without religion by preserving their thoughts, ideas, voices, emotions, presences, or physical things through material forms of technology (voice recordings, videos, photographs, etc.).

sentience: The capacity of beings to feel and perceive sensations, which in Buddhism includes the five senses and the mind.

Singularity: A future moment or condition in which a convergence of technological breakthroughs, particularly in artificial intelligence, genetics, and nanotechnology, ushers in a radically different historical epoch in which humans and machines are merged, acquiring unprecedented new powers.

sophiology: From Greek *sophia* "wisdom," a term from earlier gnostic Christianity used by Vladimir Solovyov, Sergei Bulgakov, and other Russian cosmists to mean true knowledge that is neither rational nor empirical, but mystically intuited by spiritually developed humans whose individual souls are attuned to the "world soul," Divine Sophia, the feminine aspect of divinity.

sousveillance: A term, coined by Steve Mann, for the growing social and political phenomena of "watching from below" to counterbalance surveillance, watching from above. The concept focuses on how technology has enabled the "below" view of events, such as police killings of civilians, to be shared as easily as the "above" view. The ability to perform sousveillance has been enabled not just by the ubiquitousness of cell phone videos but also because they can quickly and easily be uploaded to the Interweb.

spirit circle: A gathering of people in a nineteenth-century home who created a spiritual telegraph to try to communicate with the dead in heaven.

spiritual machines: Immaterial machines into which human consciousness is downloaded and preserved at the moment of the technological Singularity.

spiritual telegraph: A religious technology employed by Spiritualists that connected participants of a spirit circle in a circuit by placing a magnetic rope in their laps with the goal of creating a line of communication between the spirit circle and the spirit world.

storehouse consciousness: In the Yogācāra tradition, the consciousness that stores karmic impressions received from the other seven types of consciousness, the first five relating to the senses (the eye, ear, nose, tongue, and body), the sixth relating to the mind, and the seventh being a deluded awareness of self.

substance dualism: The position that there are two fundamental substances in the universe—matter, of which bodies are composed and whose primary quality is extension, and the mental, of which minds are composed and whose primary quality is thinking.

synaptic connectivity: Connections in the brain that allow a neuron to pass an electrical or chemical signal to another neuron. There are more than 100 trillion synaptic connections in the human brain.

T

technological determinism: A philosophical position that explains technological innovation as an inevitable progression of ever-more sophisticated tools and devices, forcing human beings to adapt to novel devices and machines as they are developed. A contrasting view explains technological innovation as a bidirectional relationship between human society and its inventions, emphasizing the powerful roles played by social, economic, and moral factors in shaping tools and machines as they are developed over time.

technological Singularity: A hypothesis that the creation of an artificial superintelligence will quickly lead to unfathomable technological growth and societal change.

teleology: The purpose or goal of a natural thing or a system, a concept originally developed by ancient Greek philosophers Plato and Aristotle.

telomeres: Like plastic ends on shoelaces, telomeres protect the ends of the chromosomes that house DNA. As telomeres get shorter over an individual's lifetime, they lose their ability to protect chromosomes. As a result, new cells do not replace old ones. Many scientists say cellular aging is a root cause of diseases associated with aging.

teraflop: 10^{12} (one million million) floating point operations per second.

thesis/antithesis/synthesis: Most commonly associated with the dialectical method of German philosopher Georg Wilhelm Friedrich Hegel (1770–1831), this is a method of argument in which contradictory claims or ideas are resolved by being combined synthetically into a higher truth.

transcranial direct current stimulation (tDCS): A form of neurostimulation that uses constant, low current delivered to the brain area of interest via electrodes on the scalp.

transcranial electrical stimulation (TES): The application of a weak electrical current through a device consisting of coils placed on the scalp or above the head. The current can modulate activity in "higher" cortical regions of the brain. This technique has been used to treat chronic pain and depression, as well as to enhance cognitive functions.

transcranial magnetic stimulation (TMS): Similar to TES in that coils placed on or near the forehead generate electromagnetic currents to mediate activity in cortical regions of the brain, These currents do not penetrate the brain as deeply as the electrical stimulation in DBS.

transgenic organism: A genetically modified organism that has had external DNA added to its germ line in order to create a desired modification in every cell.

transhumanism: A term first coined by biologist Julian Huxley in 1957 to describe the point when human beings would transcend evolution through natural selection and accept responsibility for directing their own evolution as a species. More recently, it has come to connote efforts to improve the human brain and body through technological innovations in such fields as genetic engineering and cybernetics. Although the term is often used interchangeably with *posthumanism*, their meanings are not identical. Transhumanism connotes the use of technology to improve the human condition, whereas posthumanism entails its complete transformation.

transhumanize: A neologism coined by Dante Alighieri in his epic poem *The Divine Comedy*, from the Italian *trasumanar*, meaning to be "transhumanized" by a vision of the divine and to become something more advanced than a human.

transmigration: The process of a soul moving from one body to another.

trans/posthuman epistemology: The study of how the nature of knowledge, beliefs, and justifications is altered when humans modify their capabilities through bioenhancement.

U

utilitarianism: An ethical theory based on whether the outcome of an action results in something good or useful to an individual or the society as a whole.

utopia: First coined in 1516 by Thomas More in his depiction of an ideal society on a distant island, the term is derived from the Greek word for "no place" but is also a pun on the Greek words for "good place." In the succeeding centuries, the term *utopia* has come to stand for any ideal society. Although some utopian novels, such as Edward Bellamy's *Looking Backward* (1888), have inspired movements for political, economic, and social reform, the term widely connotes a vision of society that is desirable but difficult or impossible to realize.

V

virtual: Often described as immaterial in form; N. Katherine Hayles, however, argued that something virtual only *seems* immaterial and is in fact material in form.

virtual immortality: According to Ray Kurzweil, the preservation of human consciousness, knowledge, and ideas in immaterial spiritual machines that achieve transcendence, or collective consciousness.

W

whole-brain emulation: A process of reverse engineering the human brain and replicating its structures at the molecular level in a nonbiological substrate.

X

xenotransplantation: The grafting of cells, tissues, and/or organs from one species individual to a different species individual.

Z

zygote: A fertilized egg.

Index

The index is alphabetized in word-by-word order. Page references in ***boldface*** *indicate chapter topics; page references in* italics *indicate photographs and illustrations. References with the letter* t *following them indicate the presence of a table. The letter* f *indicates a figure.*

B

F

H

I

K

L

Q

R

T